QUANTUM NANOCHEMISTRY

(A Five-Volume Set)

Volume V: Quantum Structure–Activity Relationships (Qu-SAR)

QUANTUM NANOCHEMISTRY

(A Five-Volume Set)

Volume V: Quantum Structure–Activity Relationships (Qu-SAR)

Mihai V. Putz

Assoc. Prof. Dr. Dr.-Habil. Acad. Math. Chem.
West University of Timişoara,
Laboratory of Structural and Computational Physical-Chemistry
for Nanosciences and QSAR, Department of Biology-Chemistry,
Faculty of Chemistry, Biology, Geography,
Str. Pestalozzi, No. 16, RO-300115, Timişoara, ROMANIA
Tel: +40-256-592638; Fax: +40-256-592620

&

Principal Investigator of First Rank, PI1/CS1
Institute of Research-Development for Electrochemistry
and Condensed Matter (INCEMC) Timisoara,
Str. Aurel Paunescu Podeanu No. 144 ,
RO-300569 Timişoara, ROMANIA
Tel: +40-256-222-119; Fax: +40-256-201-382

E-mail: mv_putz@yahoo.com
URL: www.mvputz.iqstorm.ro

AAP APPLE ACADEMIC PRESS

Apple Academic Press Inc.
3333 Mistwell Crescent
Oakville, ON L6L 0A2 Canada

Apple Academic Press Inc.
9 Spinnaker Way
Waretown, NJ 08758 USA

Exclusive worldwide distribution by CRC Press, a member of Taylor & Francis Group
No claim to original U.S. Government works
Printed in the United States of America on acid-free paper
International Standard Book Number-13: 978-1-77188-137-1 (Hardcover)
International Standard Book Number-13: 978-1-4987-2957-4 (eBook)

Library and Archives Canada Cataloguing in Publication

Putz, Mihai V., author
Quantum nanochemistry / Mihai V. Putz (Assoc. Prof. Dr. Dr. Habil. Acad. Math. Chem.) West University of Timişoara, Laboratory of Structural and Computational Physical-Chemistry for Nanosciences and QSAR, Department of Biology-Chemistry, Faculty of Chemistry, Biology, Geography, Str. Pestalozzi, No. 16, RO-300115, Timişoara, ROMANIA, Tel: +40-256-592638; Fax: +40-256-592620, & Institute of Research-Development for Electrochemistry and Condensed Matter (INCEMC) Timişoara, Str. Aurel Paunescu Podeanu No. 144, RO-300569 Timişoara, ROMANIA Tel: +40-256-222-119; Fax: +40-256-201-382, E-mail: mv_putz@yahoo.com, URL: www.mvputz.iqstorm.ro.

Includes bibliographical references and index.
Contents: Volume I: Quantum theory and observability -- Volume II: Quantum atoms and periodicity -- Volume III: Quantum molecules and reactivity -- Volume IV: Quantum solids and orderability -- Volume V: Quantum structure–activity relationships (Qu-SAR).
Issued in print and electronic formats.
ISBN 978-1-77188-133-3 (volume 1 : hardcover).--ISBN 978-1-77188-134-0 (volume 2: hardcover).--ISBN 978-1-77188-135-7 (volume 3 : hardcover).-- ISBN 978-1-77188-136-4 (volume 4 : hardcover).--ISBN 978-1-77188-137-1 (volume 5 : hardcover).--ISBN 978-1-4987-2953-6 (volume 1 : pdf).--ISBN 978-1-4987-2954-3 (volume 2 : pdf).--ISBN 978-1-4987-2955-0 (volume 3 : pdf).--ISBN 978-1-4987-2956-7 (volume 4 : pdf).--ISBN 978-1-4987-2957-4 (volume 5 : pdf) 1. Quantum chemistry. 2. Nanochemistry. I. Title.
QD462.P88 2016 541'.28 C2015-908030-4 C2015-908031-2

Library of Congress Cataloging-in-Publication Data

Names: Putz, Mihai V., author.
Title: Quantum nanochemistry / Mihai V. Putz.
Description: Oakville, ON, Canada ; Waretown, NJ, USA : Apple Academic Press, [2015-2016] | "2015 | Includes bibliographical references and indexes.
Identifiers: LCCN 2015047099| ISBN 9781771881388 (set) | ISBN 1771881380 (set) | ISBN 9781498729536 (set ; eBook) | ISBN 1498729533 (set ; eBook) | ISBN 9781771881333 (v. 1 ; hardcover) | ISBN 177188133X (v. 1 ; hardcover) | ISBN 9781498729536 (v. 1 ; eBook) | ISBN 1498729533 (v. 1 ; eBook) | ISBN 9781771881340 (v. 2 ; hardcover) | ISBN 1771881348 (v. 2 ; hardcover) | ISBN 9781498729543 (v. 2 ; eBook) | ISBN 1498729541 (v. 2 ; eBook) | ISBN 9781771881357 (v. 3 ; hardcover) | ISBN 1771881356 (v. 3 ; hardcover) | ISBN 9781498729550 (v. 3 ; eBook) | ISBN 149872955X (v. 3 ; eBook) | ISBN 9781771881364 (v. 4 ; hardcover) | ISBN 1771881364 (v. 4 ; hardcover) | ISBN 9781498729567 (v. 4 ; eBook) | ISBN 1498729568 (v. 4 ; eBook) | ISBN 9781771881371 (v. 5 ; hardcover) | ISBN 1771881372 (v. 5 ; hardcover) | ISBN 9781498729574 (v. 5 ; eBook) | ISBN 1498729576 (v. 5 ; eBook) Subjects: LCSH: Quantum chemistry. | Chemistry, Physical and theoretical. | Nanochemistry. | Quantum theory. | QSAR (Biochemistry)
Classification: LCC QD462 .P89 2016 | DDC 541/.28--dc23
LC record available at http://lccn.loc.gov/2015047099

[The] natural tendency of the mind [is] to give to the shape of a graph some intrinsic value…to its ultimate consequences…because qualitative and empirical deduction already gives them sufficient framework for experiment and prediction.

(Thom, 1975)

To XXI Scholars

CONTENTS

LIST OF ABBREVIATIONS

AIDS	acquired immunodeficiency deficiency syndrome
AIP	AhR inhibitory protein
ANOVA	analysis of variance
ARNT	aryl hydrocarbon nuclear translocator
ASA	activities of the structural alerts
B	butterfly
BaI	Balaban index
BCRP	breast cancer resistance protein
BraS	branching SMILES
C	Cusp
CA	cation-anion
CCD	chemical category database
CFD	compact finite differences
COSs	chemical orthogonal spaces
CRIs	co-receptor inhibitors
CS	Cauchy-Schwarz
CYP	cytochrome P450
CYP1A	cytochrome P450 1A gene/protein
DDT	dichlorodiphenyltrichloroethane
DFT	density functional theory
DMVC	dimethylvinyl chloride
DREs	dioxin responsive elements
E	enzyme
EC	effective concentration
EE	electronegativity equalization
ELISAs	enzyme-linked immunoassays
EP	enzyme-product
ES	enzyme-substrate
ETOT	total molecular energy
ETU	ethylene thiourea

EU	elliptic umbilic
F	fold
FIs	cell entry (or fusion) inhibitors
FITDAGA	fill-in-the-data-gaps
G	normal/Gaussian
GA	genetic algorithms
HAART	highly active antiretroviral therapy
HaI	Harary index
HF	Hartree-Fock
HIV	human immunodeficiency virus
HOMO	highest occupied molecular orbitals
HPV	high production volume
HSAB	hard and soft acids and bases
HU	hyperbolic umbilic
IARC	International Agency for Research on Cancer
IEMAD	inter-endpoint molecular activity difference
IEND	inter-endpoint norm difference
IL	ionic liquids
INIs	integrase inhibitors
IP	ionization potential
LoSMoC	longest SMILES molecular chain
LUMO	lowest unoccupied molecular orbitals
MDR	multidrug resistance
MOA	mechanism of action
MTD	minimal topological difference
MW	molecular weight
MX	mitoxantrone
NG	non-normal/non-Gaussian
NN	neuronal-network
NNRTIs	non-nucleoside reverse transcriptase inhibitors
NtRTIs	nucleotide reverse transcriptase inhibitors
OECD	Organization for Economic Cooperation and Development
P	product (of enzyme kinetics)
PAHs	polycyclic aromatic hydrocarbons
PaI	Platt index

PCA	principal component analysis
PIs	protease inhibitors
PLS	partial least squares
POL	polarizability
PU	parabolic umbilic
QRAR	quantitative reactivity-activity relationships
QSAR	quantitative structure-activity relationship
QSSA	quasi-steady-state approximation
RaI	Randić index
RT	reverse transcriptase
S	substrate (of enzyme kinetics)
SA	structural alerts
SAR	structure-activity relationship
SEE	standard error of estimation
SLR	spectral-like resolution
SMILES	simplified molecular-input line-entry system
SP	Standard Padé
SPECTRAL-SAR	SPECial TRace of ALgebraic Structure-Activity Relationship
ST	swallow tail
SzI	Szeged index
TM	topological modeling
TST	transition state theory
VEGST	vibrationally enhanced ground-state tunneling theory
VOCs	volatile organic compounds
WEp	reactive-parabolic colored Wiener index
WiI	Wiener index
XRE	xenobiotic response element

PREFACE TO FIVE-VOLUME SET

Dear Scholars (Student, Researcher, Colleague),

I am honored to introduce *Quantum Nanochemistry*, a handbook comprised of the following five volumes:

Volume I: Quantum Theory and Observability
Volume II: Quantum Atoms and Periodicity
Volume III: Quantum Molecules and Reactivity
Volume IV: Quantum Solids and Orderability
Volume V: Quantum Structure–Activity Relationships (Qu-SAR)

This treatise, a compilation of my lecture notes for graduates, postgraduates and doctoral students in physical and chemical sciences as well as my own post-doctoral research, will serve the scientific community seeking information in basic quantum chemistry environments: from the fundamental quantum theories to atoms, molecules, solids and cells (chemical–biological/ligand–substrate/ligand–receptor interactions); and will also creatively explain the quantum level concepts such as observability, periodicity, reactivity, orderability, and activity explicitly.

The book adopts a three-way approach to explain the main principles governing the electronic world:

- firstly, *the introductory principles* of quantumchemistry are stated;
- then, they are analyzed as *primary concepts* employed to understand the microscopic nature of objects;
- finally, they are explained through *basic analytical equations* controlling the observed or measured electronic object.

It explains the first principles of quantum chemistry, which includes quantum mechanics, quantum atom and periodicity, quantum molecule and reactivity, through two levels:

- *fundamental* (or *universal*) character of matter in isolated and interacting states; and
- the primary concepts elaborated for a beginner as well as an advanced researcher in quantum chemistry.

Each volume tells the "story of quantum chemical structures" from different viewpoints offering new insight to some current quantum paradoxes.

- The **first volume** covers the concepts of nuclear, atomic, molecular and solids on the basis of quantum principles—from Planck, Bohr, Einstein, Schrödinger, Hartree–Fock, up to Feynman Path Integral approaches;
- The **second volume** details an atom's quantum structure, its diverse analytical predictions through reviews and an in-depth analysis of atomic periodicities, atomic radii, ionization potential, electron affinity, electronegativity and chemical hardness. Additionally, it also discusses the assessment of electrophilicity and chemical action as the prime global reactivity indices while judging chemical reactivity through associated principles;
- The **third volume** highlights chemical reactivity through molecular structure, chemical bonding (introducing bondons as the quantum bosonic particles of the chemical field), localization from Hückel to Density Functional expositions, especially how chemical principles of electronegativity and chemical hardness decide the global chemical reactivity and interaction;
- The **fourth volume** addresses the electronic order problems in the solid state viewed as a huge molecule in special quantum states; and
- The **fifth volume** reveals the quantum implication to bio-organic and bio-inorganic systems, enzyme kinetics and to pharmacophore binding sites of chemical–biological interaction of molecules through cell membranes in targeting specific bindings modeled by celebrated QSARs (Quantitative Structure–Activity Relationships) renamed here as Qu–SAR (Quantum Structure–Activity Relationships).

Thus, the five-volume set attempts, for the first time ever, to unify the introductory principles, the primary concepts and the basic analytical equations against a background of quantum chemical bonds and interactions (short, medium and long), structures of matter and their properties: periodicity of atoms, reactivity of molecules, orderability of solids, and activity of cells (through an advanced multi-layered quantum structure–activity unifying concepts and algorithms), and observability measured throughout all the introduced and computed quantities (Figure 0.0).

It provides a fresh perspective to the "quantum story" of electronic matter, collecting and collating both research and theoretical exposition the "gold" knowledge of the quantum chemistry principles.

The book serves as an excellent reference to undergraduate, graduate (Masters and PhDs) and post-doctoral students of physical and chemical sciences; for it not only provides basics and essentials of applied quantum theory, but also leads to unexplored areas of quantum science for future research and development. Yet another novelty of this five-volume set is the intelligent unification of the quantum principles of atoms, molecules, solids and cells through the qualitative–quantitative principles underlying the observed quantum phenomena. This is achieved through unitary analytical exposition of the quantum principles ranging from quanta's nature

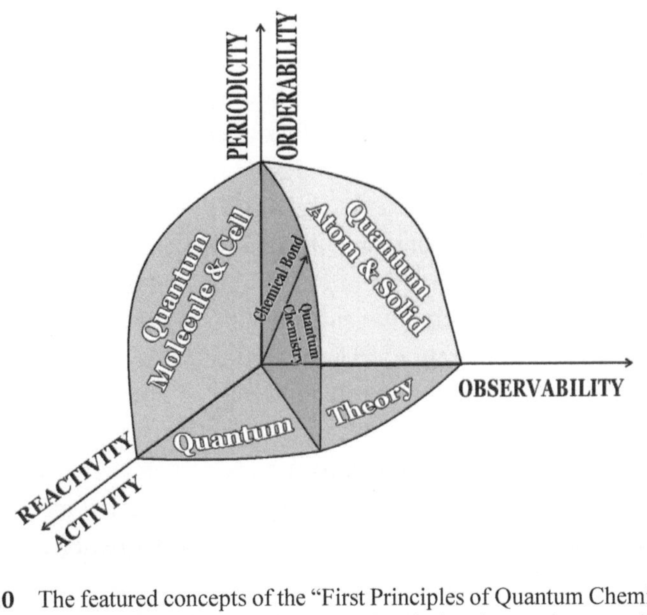

FIGURE 0.0 The featured concepts of the "First Principles of Quantum Chemistry" five-volume handbook as placed in the paradigmatic chemical orthogonal space of atoms and molecules.

(either as ondulatory and corpuscular manifestations) to wave function, path integral and electron density tools.

The modern quantum theories are reviewed mindful of their implications to quantum chemistry. Atomic, molecular, solid-state structures along cell/biological activity are analytically characterized. Major quantum aspects of the atomic, molecular, solid and cellular structure, properties/activity features, conceptual and quantitative correlations are unitarily reviewed at basic and advanced physical-chemistry levels of comprehension.

Unlike other available textbooks that are written as monographs displaying the chapters as themes of interests, this book narrates the "story of quantum chemistry" as *an extended review paper*, where theoretical and instructional concepts are appropriately combined with the relevant schemes of quantization of electronic structures, through path integrals, Bohmian, or chemical reactivity indices. The writing style is direct, concise and appealing; wherever appropriate physical, chemical and even philosophical insights are provided to explain quantum chemistry at large.

The author uses his rich university teaching experience of 15 years in physical chemistry at West University of Timisoara, Romania, along with his research expertise in treating chemical bond and bonding through conceptual and analytical quantum mechanical methods to explain the concepts. He has been a regular contributor to many physical-chemical international journals (*Phys Rev, J Phys Chem, Theor Acc Chem, Int J Quantum Chem, J Comp Chem, J Theor Comp Chem, Int J Mol Sci, Molecules, Struct Bond, Struct Chem, J Math Chem, MATCH*, etc.).

In a nutshell, the book amalgamates an analysis of the earlier works of great professors such as Sommerfeld, Slater, Landau and Feynman in a methodological, informative and epistemological way with practical and computational applications. The volumes are layered such that each can be used either individually or in combination with the other volumes. For instance, each volume reviews quantum chemistry from its level: as quantum formalisms in Volume I, as atomic structure and properties in Volume II, as detailed molecular bonding in Volume III, as crystal/solid state (electronic) in Volume IV, and as pharmacophore activity targeting specific bindings in Volume V.

To the best of my knowledge, such a collection does not exist currently in curricula and may not appear soon as many authors prefer to publish well-specialized monographs in their particular field of expertise. This multiple volumes' work, thus, assists academic and research community as a complete basic reference of conceptual and illustrative value.

I wish to acknowledge, with sincerity, the quantum flaws that myself and many researchers and professors make due to stressed delivery of papers using computational programs and software to report and interpret results based on inter-correlation. I feel, therefore, the need of a new comprehensive quantum chemistry reference approach and the present five-volume set fills the gap:

- *Undergraduate students* may use this work as an *introductory and training textbook* in the quantum structure of matter, for basic course(s) in physics and chemistry at college and university;
- *Graduate (Master and Doctoral) students* may use this work as the *recipe book* for analytical research on quantum assessments of electronic properties of matter in the view of chemical reactivity characterization and prediction;
- *University professors and tutors* may use this work as a *reference textbook* to plan their lectures and seminars in quantum chemistry at undergraduate or graduate level;
- *Research (Academic and Institutes) media* may use this work as a *reference monograph* for their results as it contains many tables and original results, published for the first time, on the atomic-molecular quantum energies, atomic radii and reactivity indices (e.g., electronegativity, chemical hardness, ionization and electron affinity results). It also has a collection of original, special and generally recommended literature, integrated results about quantum structure and properties.
- *Industry media* may use this work as a *working tool book* while assessing envisaged theoretical chemical structures or reactions (atoms-in-molecule, atoms-in-nanosystems), including molecular modeling for pharmaceutical purposes, following the presented examples, or simulating the physical–chemical properties before live production;
- *General media* may use this work as an *information book* to get acquainted with the main and actual quantum paradigms of matter's electronic structures and in understanding and predicting the

chemical combinations (involving electrons, atoms and molecules) of Nature, because of its educative presentation.

I hope the academia shares the same enthusiasm for my work as the author while writing it and the professionalism and exquisite cooperation of the Apple Academic Press in publishing it.

Yours Sincerely,

Mihai V. Putz,
Assoc. Prof. Dr. Dr.-Habil. Acad. Math. Chem.
West University of Timişoara
& R&D National Institute for Electrochemistry and Condensed Matter Timişoara
(Romania)

ABOUT THE AUTHOR

Mihai V. PUTZ is a laureate in physics (1997), with an MS degree in spectroscopy (1999), and PhD degree in chemistry (2002), with many postdoctorate stages: in chemistry (2002-2003) and in physics (2004, 2010, 2011) at the University of Calabria, Italy, and Free University of Berlin, Germany, respectively. He is currently Associate Professor of theoretical and computational physical chemistry at West University of Timisoara, Romania. He has made valuable contributions in computational, quantum, and physical chemistry through seminal works that appeared in many international journals. He is Editor-in-Chief of the *International Journal of Chemical Modeling* (at NOVA Science Inc.) and the *New Frontiers in Chemistry* (at West University of Timisoara). He is member of many professional societies and has received several national and international awards from the Romanian National Authority of Scientific Research (2008), the German Academic Exchange Service DAAD (2000, 2004, 2011), and the Center of International Cooperation of Free University Berlin (2010). He is the leader of the Laboratory of Computational and Structural Physical Chemistry for Nanosciences and QSAR at Biology-Chemistry Department of West University of Timisoara, Romania, where he conducts research in the fundamental and applicative fields of quantum physical-chemistry and QSAR. In 2010 Mihai V. Putz was declared through a national competition the Best Researcher of Romania, while in 2013 he was recognized among the first Dr.-Habil. in Chemistry in Romania. In 2013 he was appointed Scientific Director of the newly founded Laboratory of Structural and Computational Physical Chemistry for Nanosciences and QSAR in his alma mater of West University of Timisoara, while from 2014, he was recognized by the Romanian Ministry of Research as Principal Investigator of first rank/degree (PI1/CS1) at National Institute for Electrochemistry and Condensed Matter (INCEMC) Timisoara. He is also a full member of International Academy of Mathematical Chemistry.

FOREWORD TO *VOLUME V: QUANTUM STRUCTURE–ACTIVITY RELATIONSHIPS (QU-SAR)*

The last few years has seen an increased interest in modeling the quantitative structure–activity/property relationship (QSAR/QSPR). It was due, on a hand, to the need for quick estimation of properties of a large number of real or hypothetical chemical structures, for example, in drug discovery and hazard assessment of chemicals and, on the other hand, to the availability of a large number of topological indices (TIs), easily calculable on molecular graphs and properly weighted to capture chemically relevant information, to be used as parameters for QSAR/QSPR.

Various TIs have been defined to characterize different aspects of molecular structure and successfully used in the ordering of isomeric and closely related structures, to quantify the degree of molecular branching, the molecular shape, etc.; attempts have been made to define parameters with increased discriminatory power. Other molecular descriptors have been derived from the Quantum-mechanical theory to evaluate the aromaticity, electronegativity, or the molecular hardness, among others. Such molecular descriptors have been used in the QSAR/QSPR of congeneric sets of chemicals when the ordering derived from them paralleled the ordering of the molecules with respect to certain physicochemical properties or biological activity of interest.

Methods for measuring similarity/dissimilarity of chemicals, selecting analogs of chemicals from large and diverse databases, clustering real and virtual chemical libraries, enabled particularly good prediction of molecular properties/activities and molecular classification of high importance in drug design. It is worthy to be noted, the basic assumption in QSAR is that similar molecules have similar activities. This principle, also called Structure–Activity Relationship (SAR) is not necessarily applied to all molecules: there are similar molecules having not similar activities. This is known as the *SAR paradox*.

The biological activity of molecules is usually measured in assays to establish the level of inhibition of particular metabolic pathways. Drug discovery often involves the use of QSAR to identify chemical structures that could have good biological activity and low toxicity. "Discovery consists of seeing what everybody else has seen and thinking what nobody else has thought" (Albert Szent-Györgi, 1893–1986).

Recently, a variety of 3D-QSAR developments of the original QSAR, by C. Hansch and T. Fujita: JACS 86, 1616 (1964), appeared. 3D-QSAR refers to the application of force field calculations requiring three-dimensional structures, eventually based on protein crystallography or molecule superimposition. It uses computed potentials, for example, the Lennard–Jones potential, or other computed quantum-mechanical parameters. It examines steric fields (shape of the molecule), hydrophobic properties (water-soluble surfaces), electrostatic fields or reactivity properties. Among these, Comparative Molecular Field Analysis (CoMFA) and its versions include the use of GRID and partial least-squares (PLS) tools, as was recently joined in the 3-D QSAutogrid/R engine promoted by the Rome Center for Molecular Design (RCMD) team (www.rcmd.it). While many QSAR analyses involve the interactions of a family of molecules with an enzyme or receptor-binding site, it can also be used to study the interactions between the structural domains of proteins. Protein–protein interactions can be quantitatively searched for structural variations, for example, for site-directed mutagenesis.

The global *Organization of Economical and Cooperation Development* (OECD) and *European Chemical Agency* (ECMA) already credited QSAR methodology as the only and certain source of computational design for the tested compounds with biological, pharmacological or environmental impact.

Is this the context in which the book of Mihai V. Putz, *Quantum Structure–Activity Relationship* (*Qu-SAR*) add a plus-value and insight in QSAR methodology, especially in establishing the chemical–biological mechanism of interaction, so fulfilling the celebrated fifth amendment of the OECD–QSAR directives, otherwise very difficult to assess in general as based on a multivariate equation. To this aim, new theoretical approaches, like Spectral-SAR, propose purely algebraic alternatives to the traditional statistic QSAR, allowing, through the new concepts introduced (e.g., the orthogonal space of variables, the vectorial length of the biological activity, or the algebraic correlation factor for the chemical–biological interaction) the building of an optimized chart of the molecular action pathways grounded on the *minimum spectral path principle*, within a generalized space of the action norms and correlation factors.

The actual volume envisages QSAR as a tool offering a hierarchy in predicting the strength of quantum chemical bonding at the level of molecules, biomolecules (enzymes, receptors) and organisms (single- and multi-cellular) by means of ligand–substrate interaction scheme in the reactivity mechanisms and leading to the understanding, prediction and control of the bio-, eco-, and pharmacological effects. QSAR examples are given to illustrate the theoretical concepts promoted in the progressive chapters of this volume: (1) *Logistic Enzyme Kinetics* provides the theoretical background useful in the description of competitive multi- and bi-substrate enzyme catalyzes, appearing, for example, in the regulation of carbohydrate metabolism; (2) *Statistical Space for Multivariate Correlations* introduces to fundamental statistical advanced frameworks, necessary for a better understanding of the classical multilinear regression analysis; (3) *Chemical Orthogonal Spaces for Structure–Activity Relationship, COS–SAR,* presents a palette of QSAR versions (Spectral-, Diagonal-, Projective-, Catastrophe-, Residual-, SMILES, Topo-Reactive, Logistic- SARs or Quantum-SAR (Qua-SAR)), aimed to assign a sort of wave function to the congeneric active molecular series (rather than for a single molecule) in modeling the structure–metabolic interaction, rather at quantum quantitative correlation picture, allowing in establishing the specific "quantum paths" (or mechanisms) on the map of ligand–receptor bonding, for a given chemical–biological interaction.

This book may reveal some yet unexplored connections between Statistics and Algebra, as work on QSAR by the original paradigm of relation molecular structure–property/activity and the basic Physics at sub-atomic level, namely the involvement of bondons (also introduced by him in previous volumes of this *Quantum Nanochemistry* set) in chemical observables, such as reactivity/stability or biological activity, with spatial complementarity being an important premise in the effector–

receptor interaction. However, the causality is not envisaged by the regression equations working in QSAR; this is the contribution of Physics, as an advanced tool of investigation of the nature, to bring the light on consecutive events that could manifest similarly within a set of molecules, as a generalization of the intimate phenomena appearing at the level of biological receptor, as the main application of QSAR in modeling the biological activity of molecules.

By all these, this *Qu–SAR* book is of vanguardist value by offering analytical insight and control to QSAR modeling by algebraic (inherently quantum) approach as alternative to already classical statistical methods, not being disjoint but rather complementary to it, therefore highly contributing to the QSAR field toward offering a true theory of chemical–biological interactions and mechanisms.

Thus, I heartily recommend Putz's *Qu–SAR* book for its use and study equally by freshman (including graduate and Master students) in quantum eco-pharmacotoxicity and QSAR as well by advanced researchers (as PhD students and post-docs) for modeling, ordering and interpreting the chemical–biological data by the *nano*-structural (i.e., quantum nanochemical) causes of the biological activities as recorded (observable) effects.

Mircea V. Diudea
Department of Organic Chemistry
Faculty of Chemistry and Chemical Engineering
Babes-Bolyai University of Cluj-Napoca
Romania
November 2015

PREFACE TO *VOLUME V: QUANTUM STRUCTURE–ACTIVITY RELATIONSHIPS (QU-SAR)*

THE SCIENTIFIC PREMISES

In the last few years, the world scientific research focused on the so called *green chemistry*, comprising efforts to reduce or eliminate the use or production of the dangerous substances (with toxic potential) in synthesis, main stream and application of the chemical compounds through pre-industrial or computational design.[1] As such on all meridians new specific organizations and laws of validation of the entered compounds in environment or everyday and medical life have raised: the first taxonomical groups emerged in 1991 in United States by the *Environmental Protection Agency,*[2] followed by the European agency of Environment (*Umweltbundesamt*) in 1997, and by the *Environment Canada* in 1999. However, at the level of European Union, the Strategy on Management of Substances (SOMS),[3] the first step by the European Commission (EC) on 23 October 2003, triggered the program toward establishing the *Registration, Evaluation, Authorization and Restriction of Chemicals* (REACH) norms,[4] with Regulation no. 1907 directive stating that starting

[1] Benfenati, E. (2007). Predicting toxicity through computers: a changing world. *Chem. Central, J.* 1, 32.

[2] EPA (2002). US EPA AQUIRE (AQUatic toxicity Information REtrieval). U.S. Environmental Protection Agency.2002. Version 3.0 [http://www.epa.gov/ecotox/]. ECOTOX User Guide: ECOTOXicology Database System, 2002.

[3] Strategy on Management of Substances) (2001). Ministry of Housing Spatial Planning and Environment; The Hague. http://www2.minvrom.nl/Docs/internationaal/soms_engels.pdf

[4] EC (2006). European Commission. Regulation No. 1907/2006 of the European Parliament and of the Council of 18 Dec. 2006 concerning the registration, evaluation, authorization and restriction of chemicals (REACH), establishing a European Chemicals Agency, amending directive 1999/45/EC and repealing Council Regulation (EC) No. 1488/94 as well as Council Directive 76/769/EEC and commission directives 91/155/EEC, 93/67/EEC, 93/105/EC and 2000/21/EC. Off. J. Eur. Union, L 396/1 of 30.12.2006; Office for Official Publication of the European Communities (OPOCE): Luxembourg.

from 2009 any substance with carcinomic or mutagenic potential entering the life-cycle through open-market is permissible only with authorization of the *European Chemical Agency* (ECMA) at Helsinki. Such international law-scientific documents complement the *Rotterdam Convention* (10 September 2003) being part of the so called *Prior Informed Content* (PIC) procedure relating the prior consent about the risk or toxicity degree of specific chemical that will be circulating or imported across the EU countries, for instance. Moreover, the REACH normative at European level armonize themselves with the global *Organization of Economical and Cooperation Development* (OECD) that already credits the quantitative structure–activity relationship (QSAR) methodology as the only and certain source of computational design for the tested compounds with bio-, eco-, and pharmacological impact.[5]

Therefore, a certain conceptual-computational analysis of a compound of a series of compounds in the view of assigning its toxicity degree naturally works at two levels: one addresses the atomic–molecular structure together with related quantum properties while the other envisages the correlations of these properties, for example, hydrophobicity, polarizability, steric effects, etc., with the bio, eco- or pharmacological observed activities. Finally, it gets out the molecular mechanistic of the reactions involved in the studied chemical–biological interaction or in other words, of the quantum chemical strength established between the ligand (the effector or the chemical) and receptor (in the target site or organism). Still, either the structure or the quantum chemical binding aspects require the advanced studies upon them, firstly in a separate manner, and then combined both at the intrinsic structural level and for correlating the interaction, based on the versatility of the atomic and molecular world to generate surprising structures and interactions just because the quantum character involved (i.e., ondulatory, thus allowing the tunneling even for the energetic inaccessible potential barriers) when forming new apparently not explicated or controllable compounds by means of macroscopic procedures.

For instance, regarding the chemical structure of the compounds and their ondulatory character, recent discoveries give credits to the necessity of generalization of the old-fashioned valence paradigm of the chemical

[5] OECD (2004). Annexes to the Report on Principles for Establishing the Status of Development and Validation of (Quantitative) Structure-Activity Relationships [(Q)SARs], Paris, France (www.oecd.org).

bond. Reviewing only some of the pre-eminent examples worth noting the sigmatropic reactions through designing the *sigmatropic shiftamers* (a kind of polymers in which the σ and π bonds are migrating back and forth along the hydrocarbon framework);[6] there are evidence for the formal existence of the zero-valent carbon in the compounds where the electrons remain as two lone pairs not engaged in bonding, for example, $C(PPh)_2$ with Ph-phenyl group.[7] A possibility of higher orders existence is further suggested when additional shells of atomic orbitals are involved, such as *f* orbitals reaching this way through the *charge-shift bonding* concept – a new binding class between the electron pairs different from the ionic and covalent bonds in the sense that it is seen as a kind of their resonance appearing in the molecular systems like F_2, O_2, N_2 (with impact on the environmental chemistry) or in polar compounds like C–F (specific to ecotoxicology) or in reactions implying a competition between the exchange in the hydrogen or halogen (e.g., HF).[8] Typical quantum results have been recorded in the *biosynthesis of antibiotics*, for example, antitumoral agent enediyne in which the CH of the parent enediyne is replaced with $Os(PH)_3H$ or C with $(OsPH_3)_3$ resulting in a decrease of 13 kcal/mol in barrier rearrangement that may suggest a new family of organometallic enediynes that could provide useful biochemical and pharmacological properties if synthesized.[9] All these *exotic* cases of structure and chemical bonding can be controlled and designed through proper quantum paradigm for further uses in biological actions and toxicity.

Turning to the bio-, eco-, or pharmacological structure–activity relationships, the success of the QSAR methods[10] was further certified when it was accepted as the official algorithm by the EU for validating new

[6] Tantillo, D., Hoffmann, R. (2006). Snakes and ladders. the sigmatropic shiftamer concept. *Acc. Chem. Res.* 39, 477–486.

[7] Tonner, R., Oexler, F., Neumueller, B., Petz, W., Frenking, G. (2006). Carbodiphosphoranes: the chemistry of divalent carbon(0). *Angew. Chem. Int. Ed.* 45, 8038–8042.

[8] Hiberty, P. C., Megret, C., Song, L., Wu, W., Shaik, S., (2006). Barriers of hydrogen abstraction vs. halogen exchange: an experimental manifestation of charge-shift bonding. *J. Am. Chem. Soc.* 128, 2836–2843.

[9] Brzostowska, E. M., Hoffmann, R., Parish, C. A., (2007). Tuning the Bergman cyclization by introduction of metal fragments at various positions of the enediyne. Metalla-bergman cyclizations. *J. Am. Chem. Soc.* 129, 4401–4409.

[10] Cherkasov, A. (2005). *Inductive* descriptors: 10 successful years in QSAR. *Curr. Comp. Aided Drug Des.* 1, 21–42.

chemical substances.[11] Still, whatever the computational procedure applied, Hansch type, 3D, decisional, or orthogonal ones, the use of QSAR analysis for delivering the molecular interaction mechanism was furnished only recently,[12] and applied in ecotoxicology for the first time – the anionic–cationic interaction study of some ionic liquid upon *Vibrio fischeri*[13] and *Daphnia magna*.[14] This algorithm, called Spectral-SAR, proposes a purely algebraic rethinking of the traditional statistic QSAR and allows, through new concepts, e.g., the orthogonal space of variables, the vectorial length of the biological activity, or the algebraic correlation factor as an intensity measure of the chemical–biological interaction, the building of an optimized chart of the molecular action pathways grounded on the *minimum spectral path principle*, $\delta[A, B] = 0$ with A and B the endpoints within a generalized space of the action norms and correlation factors. The results have been spectacular and have opened new opportunities in the ligand–substrate correlation.[15]

The present volume, thus, adopts a dual approach to establishing the chemical–biological mechanism of interaction: the molecular structure–activities correlation offering a hierarchy in predicting and controlling of the quantum chemical bonding strength at molecular, biomolecular (enzymes with primary metabolic role), and organism (single- and multicellular) levels through implementation, and generalization of the ligand–substrate interaction scheme in the reactivity mechanisms leading to the understanding, prediction, and ultimately control of the bio-, eco-, and pharmacological effects.

---

[11] OECD (2004). Annexes to the Report on Principles for Establishing the Status of Development and Validation of (Quantitative) Structure-Activity Relationships [(Q)SARs], Paris, France (www.oecd.org).

[12] Putz, M. V., Lacrămă, A. M., Ostafe, V. (2007). Spectral–SAR ecotoxicology of ionic liquids. The *Daphnia magna* case. *Int. J. Ecol. (former Res. Lett. Ecol.)* Article ID12813/5 pages (DOI: 10.1155/2007/12813).

[13] Lacrămă, A.-M., Putz, M. V., Ostafe, V. (2007). A Spectral—SAR model for the anionic—cationic interaction in ionic liquids: application to *Vibrio fischeri* ecotoxicity. *Int. J. Mol. Sci.* 8(8, 842–863 (DOI: 10.3390/i8080842).

[14] Putz M. V., Lacrămă, A. M. (2007). Introducing Spectral Structure Activity Relationship (S-SAR) analysis. Application to ecotoxicology, *Int. J. Mol. Sci.* 8(5, 363–391 (DOI: 10.3390/i8050363)

[15] Jastorff, B., Molter, K., Behrend, P., Bottin-Weber, U., Filser, J., Heimers, A., Ondurschka, B., Ranke, J., Scaefer, M., Schroder, H., Stark, A., Stepnowski, P., Stock, F., Stormann, R., Stolte, S., Welz-Biermann, U., Ziegert, S., Thoming, J. (2005). Progress in evaluation of risk potential of ionic liquids-basis for an eco-design of sustainable products. *Green Chem.* 7, 362–372.

VOLUME LAYOUT

The present volume is the *fifth* in the five-volume set *Quantum Nanochemistry* comprising:

Volume I: Quantum Theory and Observability
Volume II: Quantum Atoms and Periodicity
Volume III: Quantum Molecules and Reactivity
Volume IV: Quantum Solids and Orderability
Volume V: Quantum Structure–Activity Relationships (Qu–SAR)

The book is organized chapterwise as follows:

Chapter 1 (Logistic Enzyme Kinetics): The logistic temporal solution of the generalized Michaelis–Menten kinetics is employed to provide a quantum basis for the tunneling time and energy evaluations of Brownian enzymic reactions. The mono-substrate and mixed inhibition cases are treated and the associated quantum diagrams of the reaction mechanisms are depicted in terms of intermediate enzyme complexes. The methodology is suited for practically controlling of the enzymic activity throughout absorption spectroscopy. For treatment of *in vitro* enzyme kinetics, the Michaelis–Menten equation is generalized to a logistic form. From the new probabilistic viewpoint, the classical Michaelis–Menten kinetics resemble the first-order expansion of the logistic one with respect to the bound substrate concentration. The probabilistic approach has three advantages. First, it better describes the quasi steady state approximation of catalysis. Second, it substitutes a logistic analytical solution for the closed-form *W*-Lambert solution for the progress curves of the substrate decay or product formation. This formalism provides an alternative time-dependent fitting curve for estimating kinetic parameters that replaces the earlier linear plot representation with a first-order time expansion. With such treatment, the Michaelis–Menten mechanism of the reversible enzymatic kinetics is also studied to provide the close form solution under the *W*-Lambert function as well as the equivalent new proposed logistic ansatz. The logistic modeling produces useful tool with which the characteristic parameter estimation can be made in an accurate straightforward manner thereby decreasing the number of associated experimental assays. Moreover, the conceptual and practical issues regarding the reduction of the Haldane–Radić enzymic mechanism, specific for cholinesterase kinetics, to the consecrated or logistically modified Michaelis–Menten kinetics, specific for some mutant enzymes, are also clarified as due to the limited initial substrate concentration, through detailed initial rate and progress curve analysis, even when other classical conditions for such equivalence are not entirely fulfilled. Besides its generality, the reliability of the present approach is also proved through applications on the competitive multi- and bi-substrate enzyme catalysts. Such approach serves as a proper analytical framework for describing many important biochemical pathways, for example, the regulation of carbohydrate metabolism as any regenerative, cyclical – and necessarily reversible – natural process to be further characterized by quantitative structure–activity relationship (QSAR) involvement (see Chapter 3).

Chapter 2 (Statistical Space for Multivariate Correlations): Aims to prepare the conceptual-computational ground for correlating chemical structure with biological activity by the celebrated quantitative structure–activity relationships (QSARs). Additionally, the fundamental statistical advanced frameworks are detailed to best understand the classical multilinear regression analysis generalized by an algebraic (in quantum Hilbert space) reformulation in terms of data vectors and orthogonal conditions (explained in see Chapter 3).

Chapter 3 (Chemical Orthogonal Spaces for Structure–Activity Relationship, COS–SAR): With ever increasing interest in correlating chemical structures with biological activities, the quantitative structure–activity relationships (QSARs) is presented with a plethora of novel and fruitful picture of regression analysis aiming to closely approach the quantum interpretation of data, and of ligand–receptor interaction by means of systematic orthogonal and scalar (dot) product of either molecular (chemicals or toxicants) descriptors between them and with the observed (recorded, measured) activities. The resulted spectral-, diagonal-spectral-, projective-, catastrophe-, residual-, SMILES (simplified molecular-input line-entry system)-, topo-reactive and logistic- SARs may be conceptually and computationally considered as realization forms of the general Quantum-SAR (Qu-SAR) that widely employs the present data as whole vectors associated in principle with the eigen-states in quantum Hilbert space. It opens the way for assigning a sort of wave function or wave packet for the congeneric active molecular series rather than for a single molecule as used to be, thus allowing specific interactions to be eventually modeled by structure (intrinsic)–metabolic (extrinsic) quantum rather than quantitative correlation picture. This further helps in establishing specific "quantum paths" (customarily known as the mechanistic map of ligand–receptor bonding) for a given or designed chemical–biological interaction.

The special features of the present volume are:

- An analytical presentation of the chemical complex interaction with biological, ecotoxicological and pharmacological environment within the ligand–receptor paradigm explained at two levels: the enzyme-substrate interaction, and the effector–receptor level of chemical drug–biological cell/organism couplings;
- Provides quantum approach of chemical phenomena and chemical reactivity principles (of electronegativity, chemical hardness, chemical action, and electrophilicity—see Volume III of the present book) at the cell's level by means of the superposition principle, applied as the multilinear superposition of chemical descriptors' influence/causes in recorded bio-eco-pharmacological action;
- Introduces two frameworks in modeling the chemical–biological interactions: the logistic approach for enzyme–substrate interaction and the algebraic (vectorial) spectral/path models of ligand–receptor interaction for the molecules (under testing or designed for prospection) transducted chemical reactivity in the cellular structure biological activity;
- Characterization of chemical–biological interaction through probability completeness space of substrate–products in enzyme kinetics and by orthogonal condition of chemical structure descriptors driving/modeling the recorded or predicted biological action/effect;
- Review of the main principles of statistics and correlation analysis such that one can better understand the probability orthogonal space of chemical complex interaction, along with the normal distribution concepts, validation requirements for a statistical analysis, confidence interval significance, all in agreement with the main QSAR–OECD principles;
- Formulation of complementary pictures of modeling biological activity by various QSAR realizations and combinations: from

spectral-SAR to spectral-diagonal, spectral-SMILES, topo-reactivity QSAR, residual- and projective-QSAR for a chemical fragments' interaction analysis, to even non-linear QSAR with the aid of the vanguardist catastrophe theory polynomials, while making useful prediction in anti-inflammatory studies;

- Formalizing the chemical–biological interaction by variational principles of the least path from the minimal to complex chemical structure–reactivity descriptors involved in a spectra of QSAR, by means of minimization of Eulerial distance between the most relevant (i.e., non-redundant, yet ergodic) models of chemical orthogonal space of bonding with biological site. This being the ultimate QSAR–OECD desiderate in assessing a mechanistic picture of the chemical complex interaction in bio–eco–pharmacological bonding, with the final prediction and control of the observed–envisaged effects in cells and organisms;
- Prediction of the biological activity by appropriate combination between the logistic enzyme kinetics and vectorial/algebraic/spectral-QSAR in the so-called logistic-QSAR approach of chemical–biological interaction, by employing the time asymptotic cutoff in natural evolution of the coupled chemical–biological systems.

I would like thank individuals, my international research group, universities, institutions, and publishers that inspired and supported the topics included in the present volume. I am just including a few of them here for want of space:

- *Supporting individuals*: Prof. Mircea Diudea (Babes-Bolyai University of Cluj-Napoca); Prof. Eduardo A. Castro (University La Plata, Buenos Aires); Prof. Adrian Chiriac (Department of Biology-Chemistry of West University of Timişoara); Prof. Ante Graovac (Rudjer-Boskovic Institute, Zagreb); Prof. Alexandru Balaban (Department of Marine Sciences, Texas A&M University at Galveston); Prof. Lionello Pogliani (University of Calabria and University of Girona); John Liebman (Department of Chemistry and Biochemistry, University of Maryland); Prof. Ivan Gutman (Faculty of Mathematics and Natural Sciences, University of Kragujevac); Dr. Adrian Chicu (Koln, Germany); Dr. Florian Nachon (Départment de Toxicologie CRSSA, Grenoble, France).

- *Members of my research group*:[16] Dr. Ana-Maria Putz (b. Lacrămă) (Timişoara Institute of Chemistry, Romanian Academy) for challenging me for developing alternative enzymes' kinetics and QSAR models for what became the actual logistic enzyme kinetics (Chapter 1) and Spectral-SAR (Chapter 3) approaches; Dr. Corina Duda-Seiman (Biology-Chemistry Department, West University of Timişoara) for keeping our research with latest topics in medical–pharmaceutical developments; Dott. Ottorino Ori (Actinium Chemical Research, Rome) for qualitatively appreciating the role of developed Spectral-SMILES-SAR (of Chapter 3) in chemistry of complex interactions and for the insight in topological modeling of chemical systems; my first officially conducted PhD student – Dr. Nicoleta Duda (Biology-Chemistry Department, West University of Timişoara and Theoretical High School "Gheorghe Lazar" Pecica – Arad County) for being with me while developing the effective SMILES-QSAR (see Chapter 3); my current PhD student Marina Alexandra Tudoran (Biology-Chemistry Department, West University of Timişoara) for enthusiasm on enriching Spectral-SMILES-SAR with topo-reactivity chemical information;
- *Supporting university*: West University of Timişoara (Faculty of Chemistry, Biology, Geography/Biology-Chemistry Department/ Laboratory of Computational and Structural Physical Chemistry for Nanosciences and QSAR);
- *Supporting institutions & grants*: CNCSIS (Romanian National Council for Scientific Research in Higher Education) by Grant: AT54/2006–2007; CNCS-UEFISCDI (Romanian National Council for Scientific Research) by Grant: TE16/2010–2013;
- *Supporting publishers*: Multidisciplinary Digital Publishing Institute – MDPI (Basel, Switzerland); University of Kragujevac (Serbia); World Scientific (Singapore); Chemistry Central (London-UK); Nova Science Inc. (New York); Bentham Science Publishers (Sharjah, U.A.E.; Oak Park-Illinois, USA).

Special thanks to my family and especially my lovely little daughters *Katy and Ela*, for always providing me with necessary energy and for

[16] http://www.mvputz.iqstorm.ro/pagina.php?id=31

creating the work-and-play atmosphere, which I am hopeful of transmitting to the readers and students too – in their quest for the scientific knowledge and method of thinking/approaching quantum *specific and complex interactions' modeling*, and of their *chemical–biological structure–activity* manifestation.

Last but not the least, the author especially thank the Publisher, Apple Academic Press (AAP), and in particular to Ashish (Ash) Kumar, the AAP President and Publisher, and Sandra (Sandy) Jones Sickels, Vice President, Editorial and Marketing, for professional handling of the volume and the book through production.

Modeling and applications of enzyme kinetics and quantitative structure–activity relationship methods is a vast field and its importance in the years to come in bioinfo and chemoinformatics sciences and biotechnology is only going to increase. The author welcomes and thanks the readers in advance for any constructive observations, corrections and suggestions, which can be incorporated and updated in its further editions.

Keep close and think high!

Yours sincerely,

Mihai V. Putz,
Assoc. Prof. Dr. Dr.-Habil. Acad. Math. Chem.
West University of Timişoara
& R&D National Institute for Electrochemistry and Condensed Matter Timişoara
(Romania)

CHAPTER 1

LOGISTIC ENZYME KINETICS

CONTENTS

ABSTRACT

The logistic temporal solution of the generalized Michaelis-Menten kinetics is employed to provide a quantum basis for the tunneling time and energy evaluations of Brownian enzymic reactions. The mono-substrate and mixed inhibition cases are treated and the associated quantum diagrams of the reaction mechanisms are depicted in terms of intermediate enzyme complexes. The methodology is suited for practically controlling of the enzymic activity throughout absorption spectroscopy. For treatment of in vitro enzyme kinetics the Michaelis-Menten equation is generalized to a logistic form. From the new probabilistic viewpoint, the classical Michaelis-Menten kinetic resembles the first order expansion of the logistic one with respect to the bound substrate concentration. The probabilistic approach has three advantages. First, it better describes the quasi steady state approximation of catalysis. Second, it substitutes a logistic analytical solution for the closed-form *W*-Lambert solution for the progress curves of the substrate decay or product formation. This formalism provides an alternative time-dependent fitting curve for estimating kinetic parameters that replaces the earlier linear plot representation with a first order of time expansion. With such treatment the Michaelis-Menten mechanism of the reversible enzymatic kinetics is also studied to provide the close form solution under the *W*-Lambert function as well as the equivalent new proposed logistic ansatz. The logistic modeling produces useful tool with which the characteristic parameter estimation can be made in an accurate straightforward manner so decreasing the number of associated experimental assays. Moreover, the conceptual and practical issues regarding the reduction of the Haldane-Radić enzymic mechanism, specific for cholinesterase kinetics, to the consecrated or logistically modified Michaelis-Menten kinetics, specific for some mutant enzymes, are also clarified as due to the limited initial substrate concentration, through detailed initial rate and progress curve analysis, even when other classical conditions for such equivalence are not entirely fulfilled. Beside its generality, the reliability of the present approach is also proved through applications on the competitive multi- and bi-substrate enzyme catalyzes. Such approach serves as a proper analytical framework for describing of many important biochemical pathways, for example, the regulation of carbohydrate metabolism, as any regenerative, cyclical – and necessarily reversible – natural process to be further characterized also by quantitative structure-activity relationship (QSAR) involvement (see Chapter 3).

1.1 INTRODUCTION

According to Charles Darwin's famous paradigm of evolution, the principle of natural selection prescribes "the survival of the fittest" (Mayr, 2001).

With the advent of the general theory of models in biology (Thom, 1975), as genomic, proteomic, and metabolomics scales are approached, the fitting concept resembles the equation of the net production of the species "*i*" (Crampin et al., 2004):

$$\frac{d}{dt}[X]_i = f_i([X]_i, a_i) \tag{1.1}$$

whose solution, i.e., the time-dependent concentrations $[X]_i=[X]_i(t)$, depends on the particular parameters a_i specific for the particular processes considered.

However, even displaying the temporal character, the master equation (1.1) differs at the micro-scale from the consecrated Hamilton-Jacobi

$$\hat{H}S_H = -\frac{\partial S_H}{\partial t} \tag{1.2}$$

or Schrödinger

$$\hat{H}\psi_H = i\hbar\frac{\partial \psi_H}{\partial t} \tag{1.3}$$

ones that drive the atomic and molecular quantum evolutions. While the equations of quantum mechanics pose the feature of being linear to superposition (Mandl, 1992), the function f_i in Eq. (1.1) has to be non-linear in variable $[X]_i$ due to the complexity of the structure of the bio-systems and of the biochemical kinetics ; the present discussion follows (Putz et al., 2006a).

Still, a mechanistic study of a biochemical network can be performed by a two-folded analysis. First, a “wiring diagram” of intermediates is proposed and then, by considering the individual interactions, a certain kinetic model is proposed (Crampin et al., 2004).

With these principles, the most elementary biochemical model can be understood in the world of the almost mystic field of enzymatic reactions – notoriously complex in mechanism and kinetics. It is well known that the rate of an enzyme-catalyzed reaction in which a substrate S is converted into product P is found to depend on the concentration of enzyme E even though the enzyme undergoes no net change (Schnell & Maini, 2003). As a mechanism, it is assumed that the substrate enzyme forms an intermediate ES, with the rates k_1 and k_{-1}, which then irreversibly breaks down into the product and the enzyme (Brown, 1892, 1902; Henri, 1901; Michaelis & Menten, 1913):

$$E + S \underset{k_{-1}}{\overset{k_1}{\leftrightarrow}} ES \overset{k_2}{\rightarrow} E + P \tag{1.4}$$

So far, kinetic studies for the reaction in Eq. (1.4) have been conducted in the context (or with the help) of the Michaelis-Menten model, due to this model’s flexibility in characterizing complex mechanisms derived from this type of reaction. For instance, when an analogue substrate blocks the action of a specific enzyme the so-called inhibited reaction takes place, with a major function in chemotherapeutic trainings (Voet & Voet, 1995; Schnell & Mendoza, 2000a). On the other side, when an enzyme catalyzes the transfer of a specific functional group from one substrate to another in a many-substrate environment, the multiple alternative substrate type of reactions occur, highlighting the economical industrial synthesis of the enantiomerically pure compounds as well as the environmental issues (Schnell & Mendoza, 2000b). Therefore, having a complete analytical picture of the elementary Michaelis-Menten reaction (1.4) becomes crucial in treating the more complex enzymic reactions derived from it.

The mechanism (1.4) is solved when the involved concentrations, i.e., $[E](t)$, $[S](t)$, $[ES](t)$, and $[P](t)$, are analytically known from the nonlinear differential equations of type (1.1) (Cornish-Bowden, 1979; Segel, 1975). However, beyond approaching the progress curves of species in Eq. (1.4) through graphical methods (Ritchie & Prvan, 1996) or by powerful computers (Zimmerle & Frieden, 1989), the analytical solutions have to be shaped in such a manner as to be further compatible with the temporal non-linear fitting when assaying experimental data (Szedlacsek et al., 1990; Goudar et al., 1999; Câteau & Tanaka, 2002).

The present chapter exposes the way in which the basic Michaelis-Menten kinetics is modified under logistic form when in vitro conditions are assumed, i.e., when the reaction parameters (temperature, solvent, pH, etc.) are held constant, as it can often be assumed in the laboratory (Goudar et al., 1999). The proposed logistic ansatz is then applied to real enzymic systems governed by competitive alternative substrates (Schnell & Mendoza, 2000b), with a particular emphasis on inhibitive bi-substrate enzyme-catalyzed reactions (Schnell & Mendoza, 2000a). This way, it follows that the present approach is a two-fold one viz. both through its logistic (mathematical) analysis and due to its applications to real systems (Putz et al., 2006a).

1.2 MICHAELIS-MENTEN ENZYME KINETICS AND LIMITED *W*-LAMBERT SOLUTION

When the law of mass action is considered for the reaction (1.4), the time evolution scheme can be draw as the system of the coupled nonlinear differential equations (Gray & Scott, 1990); the present discussion follows (Putz et al., 2006a):

$$\frac{d}{dt}[S] = -k_1[E][S] + k_{-1}[ES] \tag{1.5}$$

$$\frac{d}{dt}[E] = -k_1[E][S] + (k_{-1} + k_2)[ES] \tag{1.6}$$

$$\frac{d}{dt}[ES] = k_1[E][S] - (k_{-1} + k_2)[ES] \tag{1.7}$$

$$\frac{d}{dt}[P] = k_2[ES] \tag{1.8}$$

with initial conditions $([S],[E],[ES],[P])=([S_0],[E_0],0,0)$ at the time $t = 0$.

The set of equations (1.5)–(1.8) can be simplified in three steps.

First, it can be seen that when the Eqs. (1.6) and (1.7) are added, the conservation law for enzyme is obtained:

$$[E](t)+[ES](t)=[E_0] \tag{1.9}$$

while the combination of Eqs. (1.5), (1.7) and (1.8) leads to the conservation law for the substrate:

$$[S](t)+[ES](t)+[P](t)=[S_0] \tag{1.10}$$

With the help of identities (1.9) and (1.10), the system of differential equations (1.5)–(1.8) takes the reduced form:

$$\frac{d}{dt}[S]=-k_1[S]([E_0]-[ES])+k_{-1}[ES] \tag{1.11}$$

$$\frac{d}{dt}[ES]=k_1[S]([E_0]-[ES])-(k_{-1}+k_2)[ES] \tag{1.12}$$

in terms of substrate and substrate enzyme concentrations only, $[S]$ and $[ES]$, respectively.

Then, employing the in vitro conditions, the enzyme can always be saturated with the substrate, so that the *quasi-steady-state* (or equilibrium) *approximation* (QSSA) may apply to the intermediate formed complex in Eq. (1.4). It implies imposing on Eq. (1.12) the mathematical constrain (Segel, 1975, 1988, 1989):

$$\frac{d}{dt}[ES]\cong 0 \tag{1.13}$$

yielding with its equivalent form:

$$[ES]=\frac{[E_0][S]}{[S]+K_M} \tag{1.14}$$

where the reaction parameter

$$K_M=\frac{k_{-1}+k_2}{k_1} \tag{1.15}$$

is known as the *Michaelis-Menten constant* (Michaelis & Menten, 1913).

Now, plugging relation (1.14) into the Eq. (1.11), we get the decoupled differential equation for the substrate consumption rate:

$$\frac{d}{dt}[S] = -\frac{V_{max}[S]}{[S]+K_M} \tag{1.16}$$

where

$$V_{max} = k_2[E_0] \tag{1.17}$$

has been set as the *maximum velocity of reaction.*

At this point, the system (1.5)–(1.8) achieves its minimum dimension consisting in one equation for the substrate concentration. However, by combining the Eqs. (1.8) and (1.14), the velocity of the product formation also comes out

$$v = \frac{d}{dt}[P] = \frac{V_{max}[S]}{[S]+K_M} \tag{1.18}$$

as the famous *Michaelis-Menten equation* (Henri, 1901; Michaelis & Menten, 1913).

However, Eq. (1.18) reveals the first shortcoming of the Michaelis-Menten kinetic: when used without explicit temporal dependency of concentrations, it accounts only for the velocity of the initial instants of the reaction. In other words, the information outside the first moments of the progress curve $[S](t)$ is virtually lost or neglected as long as its analytical form is not known for any moments of time (Duggleby & Morrison, 1977; Duggleby, 2001).

Therefore, the necessity of a fully temporal analysis for the enzymatic processes stands as a natural imperative when further fitting with experiment is envisaged.

The temporal problem is to formulate a viable analytical solution $[S](t)$ for the differential equation (1.16). Once that has been done, the progress curves of the rest of species in Eq. (1.4) can be accordingly formulated employing the conservation laws (1.9) and (1.10) together with the relation (1.14) for the substrate enzyme complex.

However, it is worth noting that, for the expression (1.14), a more general temporal formulation can be cast as (Schnell & Mendoza, 1997):

$$[ES](t) = \frac{[E_0][S]}{[S]+K_M}\{1-\exp[-k_1 t([S_0]+K_M)]\} \ ,0 \le t < \infty \tag{1.19}$$

becoming identically zero at initial time, $t \to 0$, and recovering the former expression (1.14) in the long-range regime, $t \to \infty$, respectively.

Going to analytically solve equation (1.16) it is firstly rearranged as

$$\left(\frac{K_M}{[S]}+1\right)d[S] = -V_{\max}dt \tag{1.20}$$

and then integrated to give (Rubinow, 1975):

$$[S]+K_M \ln[S] = [S_0] - V_{\max}t + K_M \ln[S_0] \tag{1.21}$$

Unfortunately, the Eq. (1.21) shows another limitation of the Michaelis-Menten enzymic description. Having a transcendental form, Eq. (1.21) does not allow for explicitly writing the dependency $[S](t)$. In these conditions, many biochemists prefer to rearrange Eq. (1.21) under a sort of double plot equation (Haldane & Stern, 1932; Lineweaver & Burk, 1934; Cornish-Bowden, 1975), for instance:

$$\frac{t}{[S_0]-[S](t)} = \frac{1}{V_{\max}} + \frac{K_M}{V_{\max}}\left(\frac{1}{[S_0]-[S](t)}\ln\frac{[S_0]}{[S](t)}\right) \tag{1.22}$$

from where an intercept of $1/V_{max}$ and a slope of K_M/V_{max} provide the kinetic parameters V_{max} and K_M, respectively. Still, this approach has been criticized (Goudar et al., 1999; Schnell & Maini, 2003), and it is worthwhile investigating whether the exact solution of Eq. (1.21) can be obtained for fitting a non-linear progress curve.

In this respect, once the substitution (Putz et al., 2006a)

$$\varphi([S]) = \frac{[S]}{K_M} \tag{1.23}$$

is performed in Eq. (1.21), it leads to the equivalent equation:

$$\varphi([S]) + \ln\varphi([S]) = \frac{[S_0]}{K_M} - \frac{V_{\max}t}{K_M} + \ln\left(\frac{[S_0]}{K_M}\right) \tag{1.24}$$

The closed-form solution of Eq. (1.24) was recognized by Schnell and Mendoza through the analogy with the famous Lambert type equation (Barry et al., 2000):

$$W(x)+\ln W(x)=\ln x,\ x\geq -1/e \tag{1.25}$$

By comparing Eqs. (1.24) and (1.25), the formal temporal solution for the substrate concentration can be achieved as (Schnell & Maini, 2003; Schnell & Mendoza, 1997):

$$[S]_W(t)=K_M W\left(\frac{[S_0]}{K_M}e^{\frac{[S_0]-V_{\max}t}{K_M}}\right) \tag{1.26}$$

With the W-Lambert dependence (1.26) of the kinetic solution of the reaction (1.4), we arrive at the mathematical disadvantages of the traditional Michaelis-Menten analysis. For example, it can return multiple values for the same argument or result in an infinitely iterated exponential function (Hayes, 2005).

Here we place some consideration about W-Lambert function, in general, and on the solution (1.26), in particular. Direct time integration of the Michaelis-Menten equation (1.20) leads with the intermediate form (1.21) reaching the substrate W-Lambert solution (1.24); although the W-Lambert closed form is solved with ingenious numerical schemes, see Refs. [Huang & Niemann, 1951; Jennings & Niemann, 1953, 1955; Duggleby & Morrison, 1978, 1979; Duggleby, 1986; Duggleby & Wood, 1989; Boeker, 1984, 1985, 1987], the Eq. (1.24) bears the transcendent nature through the Lambert function $W(x)$ (Corless et al., 1996), eventually rearranged as

$$W(x)=\ln\left(x/W(x)\right),\ x\geq -1/e \tag{1.27}$$

while through successive repeating of the procedure it arrives at the infinite continuous fraction:

$$W(x)=\ln\frac{x}{\ln\left(W(x)\right)}=\ln\frac{x}{\ln\frac{x}{\ln\left(W(x)\right)}}=\ldots=\ln\frac{x}{\ln\frac{x}{\ln\frac{x}{\ln\frac{x}{\ddots}}}} \tag{1.28}$$

In other words, W-Lambert solution (1.26) has no finite closed form solution. This, along the lack of published tables of $W(x)$ numerical evaluations highly motivates for a reformulation of the enzyme kinetics description in a more workable way, eventually with an analytical solution, as suggested in following (Putz et al., 2006a).

1.3 LOGISTIC FORMULATION AND EXPLICIT ENZYME KINETICS' SOLUTION

In the post genomic era the development of kinetic models that allow simulation of complicated metabolic pathways and protein interactions is becoming increasingly important (Noble, 2002; Crampin et al., 2004). Unfortunately, the difference between an *in vivo* biological system and homogeneous *in vitro* conditions is large, as shown by Schnell and Turner (2004). Mathematical treatments of biochemical kinetics have been developed from the law of mass action *in vitro* but the modifications required to bring them in line with the stochastic *in vivo* situation are still under development (Savageau, 1969, 1976; Turner et al., 2004).

We use a probabilistic approach, based on the law of mass action, to characterize *in vitro* enzymatic reactions of type (1.4); the present discussion follows (Putz et al., 2007):

$$1 = \rho_{\text{REACT}}([S]_{bind}) + \rho_{\text{UNREACT}}([S]_{bind}) \tag{1.29}$$

In Eq. (1.29), $\rho_{\text{REACT}}([S]_{bind})$ is the probability that the reactions (1.4) proceed at a certain concentration of substrate binding to the enzyme $[S]_{bind}$. The limits are:

$$\rho_{\text{REACT}}([S]_{bind}) = \begin{cases} 0 \ , & [S]_{bind} \to 0 \\ 1 \ , & [S]_{bind} >> 0 \end{cases} \tag{1.30}$$

$\rho_{\text{REACT}}([S]_{bind})$=0 when the enzymatic reaction does not proceed or when it stops because the substrate fails to bind or is entirely consumed. Conversely, $\rho_{\text{REACT}}([S]_{bind})$=1 when the enzymatic reaction proceeds, and it is related to the standard QSSA. The probability of the occurrence of products in reaction (1.4) lies between these limits. Similarly, in the case where enzymatic catalysis does not take place, $\rho_{\text{UNREACT}}([S]_{bind})$, the limits are:

$$\rho_{\text{UNREACT}}([S]_{bind}) = \begin{cases} 1 \ , & [S]_{bind} \to 0 \\ 0 \ , & [S]_{bind} >> 0 \end{cases} \tag{1.31}$$

This probabilistic treatment of enzymatic kinetics is based on the chemical bonding behavior of enzymes that act upon substrate molecules through diverse mechanisms and it may offer the key to the quantitative treatment of different types of enzyme catalysis (Voet & Voet, 1995).

To expand the terms of Eq. (1.29) to analyze reactions in the Eq. (1.4), we first recognize that the binding substrate concentration can be treated as the instantaneous substrate concentration: $[S]_{bind}=[S](t)$.

Maintaining the quasi-steady-state conditions for in vitro systems, we may assume constant association-dissociation rates so that probability of reaction is written as the rate of consumption of the substrate,

$$v(t) = -\frac{d}{dt}[S](t) \tag{1.32}$$

to saturation (Putz et al., 2007):

$$\rho_{\text{REACT}}([S](t)) = \frac{v(t)}{V_{\max}} = -\frac{1}{V_{\max}}\frac{d}{dt}[S](t) \tag{1.33}$$

after the initial transient of the enzyme-substrate reaction in Eq. (1.4).

We know only that expression (1.33) behaves like a probability function, with values in the realm [0–1]. Given expressions (1.18), (1.29) and (1.33) we derive an expression for the unreacted probability term, $\rho_{\text{UNREACT}}([S](t))$.

The expression (Putz et al., 2007):

$$\rho_{\text{UNREACT}}([S](t))^{\text{MM}} = \frac{K_M}{[S](t) + K_M} \tag{1.34}$$

satisfies all of the probability requirements, including the limits in Eq. (1.31), and, when combined with Eqs. (1.33) and (1.29), gives the instantaneous version of the classical Michaelis-Menten equation (1.18). Remarkably, expression (1.34) can be seeing as generalization of the efficiency of the Michaelis-Menten reaction under steady-state conditions

(Laidler, 1955). Originally, the efficiency depends on two parameters: K_M that embodies the thermodynamic conditions of the enzymic reaction and the initial substrate concentration $[S_0]$; it determines the ratio of the free to total enzyme concentration in the reactions (1.4); that is, when the efficiency is equal to one, we cannot expect to find substrate free in the reaction, i.e., the reactions in (1.4) are all consumed so that first branch of the limits (1.31) is fulfilled as no further binding will occur (Figure 1.1).

It is clear that the Michaelis-Menten term (1.34) is just a particular choice for a probabilistic enzymatic kinetic model of the conservation law (1.29). A more generalized version of Eq. (1.34) that preserves all of the above probabilistic features is

$$P_{\text{UNREACT}}([S](t))^* = e^{-\frac{[S](t)}{K_M}} \tag{1.35}$$

from which the Michaelis-Menten term (1.34) is returned by performing the $[S](t)$ first order expansion for the case where the bound substrate approaches zero (Putz et al., 2007):

$$P_{\text{UNREACT}}([S](t))^* = \frac{1}{e^{\frac{[S](t)}{K_M}}} \overset{[S](t)\to 0}{\cong} \frac{1}{1+\frac{[S](t)}{K_M}} = P_{\text{UNREACT}}([S](t))^{\text{MM}}$$

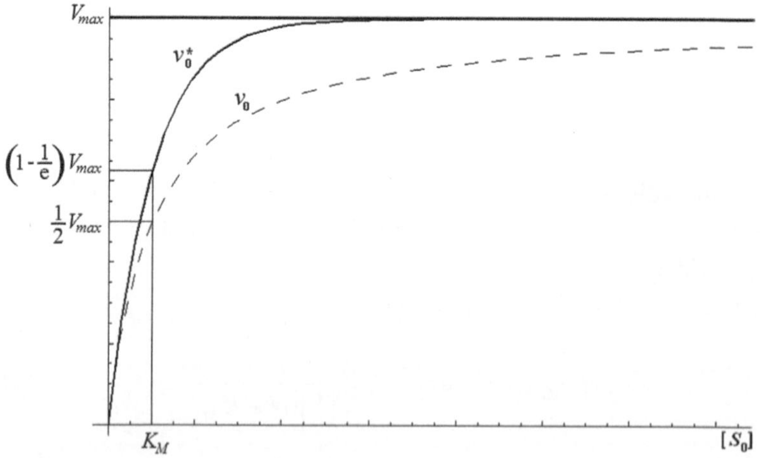

FIGURE 1.1 Initial Michaelis-Menten and logistic velocities plotted against initial substrate concentration for the reaction (1.4). The dashed curve corresponds to the Michaelis-Menten equation (1.18) while the continuous thick curve represents its logistic generalization: $v_0^* = V_{max}\left[1-\exp\left(-[S_0]/K_M\right)\right]$, see (Putz et al., 2007).

(1.36)

Worth noting that there is no monotonically form between 0 and 1 other that that of equation (1.35) to reproduce basic Michaelis-Menten term (1.34) when approximated for small x = $[S](t)/K_M$. For instance, if one decides to use $\exp(-x^2)$ then the unreactive probability will give $1/(1+x^2)$ as the approximation for small x, definitely different of what expected in basic Michaelis-Menten treatment (1.34). This way, the physico-chemical meaning of Eq. (1.36) is that the Michaelis-Menten term (1.34) and its associated kinetics apply to fast enzymatic reactions, i.e., for fast consumption of $[S](t)$, which also explains the earlier relative success in applying linearization and graphical analysis to the initial velocity equation (1.18).

Use of Eq. (1.35) instead of Eq. (1.34) expands the range of reaction rates and provides a new kinetic equation, in the form of a logistic expression (Putz et al., 2007)

$$-\frac{1}{V_{\max}}\frac{d}{dt}[S](t)=1-e^{-\frac{[S](t)}{K_M}} \tag{1.37}$$

based on probability and from the derived equations (1.29), (1.33), and (1.35).

At initial conditions, logistic equation (1.37) gives an initial velocity of reaction (v_0^*) that is uniformly higher than that calculated by Michaelis-Menten (1.18) at all initial concentrations of the substrate, except for the case where $[S_0]\to 0$, when both are zero, see Figure 1.1.

Reliability tests of the logistic form of Michaelis-Menten kinetics (1.37) are reported in Section 1.4.

1.4 RELIABILITY OF THE LOGISTIC ENZYME KINETIC

1.4.1 QUASI STEADY-STATE APPROXIMATION ANALYSIS

One of the fundamental assumptions made in deriving basic Michaelis-Menten kinetics, except in the initial so-called transient phase of the reaction, is the quasi steady state approximation of the $[ES]$ concentration, i.e., the rate of synthesis of the ES complex must equal its rate of consumption until the substrate is nearly exhausted. It has been demonstrated that the QSSA is equivalent with the physiologically common condition that the substrate is in great excess over the enzyme, as firstly shown by Laidler (1955):

$$[S_0] >> [E_0] \tag{1.38}$$

Let us investigate whether condition (1.38) may arise within the proposed probabilistic enzymatic kinetics and what consequences that has for applicability of the logistic treatment.

For reaction (1.4) to proceed with a high probability it is necessary that (Putz et al., 2007)

$$P_{\text{REACT}}([S]_{bind}) \to 1 \Leftrightarrow P_{\text{UNREACT}}([S]_{bind}) \to 0 \tag{1.39}$$

or, the probability of the enzymatic reaction proceeding increases to one as the probability that the substrate will not bind with the enzyme approaches zero. Analytically, we use the limiting case (1.39) where reaction (1.4) proceeds:

$$P_{\text{REACT}}([S](t)) = 1 \tag{1.40}$$

Then, by combining equation (1.40) with the general in vitro form (1.33), we derive the time dependent equation

$$-\frac{1}{V_{\max}} \frac{d}{dt}[S](t) = 1 \tag{1.41}$$

Substituting $V_{\max} = k_2[E_0]$ and integrating produces the linear portion of the substrate depletion curve:

$$[S](t) = [S_0] - k_2[E_0]t \tag{1.42}$$

The substrate condition $[S](t) >> 0$ corresponds to the binding case for which Eq. (1.40) is valid under the conditions given in expression (1.30). Applying this substrate condition to Eq. (1.42) during the rate limiting step when

$$t \cong \frac{1}{k_2} \tag{1.43}$$

ensures that almost all of the substrate is being transformed into product via reactions (1.4), resulting in the QSSA condition (1.38).

We have proved that the left side of the probabilistic equivalence (1.39) is valid for QSSA and we must do the same for the right side. The more closely $\rho_{\text{UNREACT}}([S](t))$ approaches zero as $\rho_{\text{REACT}}([S](t))$ approaches one, the better QSSA is obtained.

Recalling the two forms presented for the non-binding reactivity, the Michaelis-Menten in Eq. (1.16) and the logistic in Eq. (1.37), we can clearly see that the following hierarchy exists; the present discussion follows (Putz et al., 2007)

$$\rho_{\text{UNREACT}}([S](t))^{\text{MM}} = \frac{1}{1+\dfrac{[S](t)}{K_M}} > \frac{1}{e^{\frac{[S](t)}{K_M}}} = \rho_{\text{UNREACT}}([S](t))^{*} \quad (1.44)$$

regardless of the time at which they are compared. Therefore, because the logistic probability ρ_{UNREACT} is lower than the Michaelis-Menten at all times, QSSA is better satisfied using the logistic approach.

1.4.2 FULL TIME COURSE ANALYSIS

Many biochemists use the velocity equations for kinetic parameter estimates despite the fact that the rates are difficult to determine experimentally. In practice either the substrate depletion or the product formation is measured as a function of time and the rates are calculated by differentiating the data, leading to an inexact analysis (Schnell & Mendoza, 1997, 2000a). Alternatively, the differential equations governing the biochemical reactions may be solved or approximated to obtain reactant concentration as function of time. This approach decreases the number of experimental assays by at least a factor of five, as proved by Schnell and Mendoza (2001), because multiple experimental points may be collected for each single reaction.

Unfortunately, until now, the most general analytical time-dependent solution for reaction (1.4) used the closed form (1.26) that has many mathematical disadvantages. For example it can return multiple values for the same argument (Hayes, 2005) or result in an infinitely iterated exponential function (Schnell & Mendoza, 2000b).

To test whether the logistic kinetic equation (1.37), which is a natural generalization of the Michaelis-Menten equation, may provide a workable analytical solution in an elementary form we first integrate the equation

$$\int_{[S_0]}^{[S](t)} \frac{d[S](t)}{\exp(-[S](t)/K_M)-1} = \int_0^t V_{max} dt \tag{1.45}$$

generating the new equation to be solved:

$$[S_0]-[S](t)+K_M \ln\left(e^{-\frac{[S_0]}{K_M}}-1\right)-K_M \ln\left(e^{-\frac{[S](t)}{K_M}}-1\right)=V_{max}t \tag{1.46}$$

Although apparently more complex than the previous version (1.21), Eq. (1.46) can be solved exactly. This can be demonstrated by substituting

$$\varphi([S](t))=\frac{[S](t)}{K_M} \tag{1.47}$$

into Eq. (1.46) to get the simple equation:

$$-\varphi([S](t))-\ln\left(e^{-\varphi([S](t))}-1\right)=\psi(t) \tag{1.48}$$

where we have also introduced the functional notation:

$$\psi(t)=\frac{1}{K_M}(V_{max}t-[S_0])-\ln\left(e^{-\frac{[S_0]}{K_M}}-1\right) \tag{1.49}$$

Now the exact solution of Eq. (1.48) is a logistic expression:

$$\varphi([S](t))=\ln\left(1-e^{-\psi(t)}\right) \tag{1.50}$$

Finally, substituting function (1.49) into expression (1.50) gives the logistic progress curve for substrate consumption in an analytically elementary form; the present discussion follows (Putz et al., 2007):

$$[S]_L(t)=K_M \ln\left(1+e^{-\frac{V_{max}t}{K_M}}\left(e^{\frac{[S_0]}{K_M}}-1\right)\right) \tag{1.51}$$

This time-dependent solution (1.51) substitutes an elementary logarithmic dependency for the *W*-Lambert function. It is remarkable that the solution of a generalized logistic kinetic version of the Michaelis-Menten instantaneous equation provides an analytically exact solution.

The cutting test is in the comparison of the progress curves generated by the *W*-Lambert (1.26) and logistic solutions (1.51), respectively. To do this, the following working formulas for the instantaneous complex $[ES](t)$, product $[P](t)$ and enzyme $[E](t)$ concentrations are employed in both the *W*-Lambert (1.26) and logistic (1.51) versions of the binding substrate concentration, $[S]_{W,L}$, according with Schnell and Mendoza (1997):

$$[ES]_{W,L}(t) = \frac{[E_0][S]_{W,L}(t)}{[S]_{W,L}(t) + K_M}\{1 - \exp[-k_1 t([S_0] + K_M)]\} \tag{1.52}$$

$$[P]_{W,L}(t) = [S_0] - [S]_{W,L}(t) - [ES]_{W,L}(t) \tag{1.53}$$

$$[E]_{W,L}(t) = [E_0] - [ES]_{W,L}(t) \tag{1.54}$$

$$s_{W,L}(t) = \frac{[S]_{W,L}(t)}{[S_0]}, e_{W,L}(t) = \frac{[E]_{W,L}(t)}{[E_0]},$$

$$es_{W,L}(t) = \frac{[ES]_{W,L}(t)}{[E_0]}, p_{W,L}(t) = \frac{[P]_{W,L}(t)}{[S_0]} \tag{1.55}$$

The transformation:

$$\tau = 1 - \frac{1}{\ln(t + e)} \tag{1.56}$$

allows us to use scaled time for the abscissa so that an infinite time range can be mapped onto the interval [0–1].

Figure 1.2 shows the plots of the *W*-Lambert and logistic progress curves (1.55) for an enzyme-catalyzed reaction in vitro where $k_{-1}=k_2=10^2 s^{-1}$, $k_1=10^6 M^{-1}s^{-1}$, $[S_0]=10^{-4}M$, and $[E_0]=10^{-6}M$. The quantitative behavior of the reactant concentrations in both the *W*-Lambert and logistic cases are strikingly similar. In addition, time-dependent product curves may be used instead of the initial velocity curves in Figure 1.1.

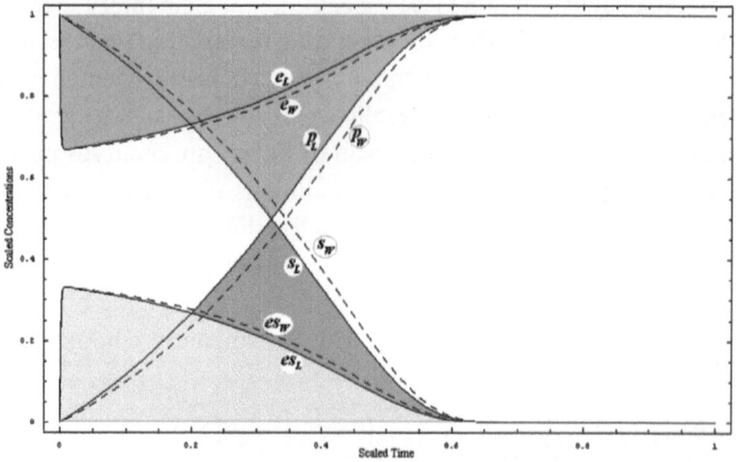

FIGURE 1.2 Time dependent behavior of the reactant scaled concentrations (1.55) for the paradigmatic enzyme-substrate reaction (1.4) when the basic (dashed lines) and generalized logistic (solid lines) versions of the Michaelis-Menten kinetic are employed, with the parametric values $k_{-1}=k_2=10^2 s^{-1}$, $k_1=10^6 M^{-1}s^{-1}$, $[S_0]=10^{-4}M$, and $[E_0]=10^{-6}M$, against the scaled time (1.56) (Putz et al., 2007).

However, the logistic product curves are smoother and at higher concentrations than those obtained from the *W*-Lambert approach due to the higher probability of reaction (see the discussion from the Section 1.3).

Having proved the reliability of the logistic time-dependent form of the substrate depletion expression (1.51) compared to the *W*-Lambert-based expression (1.26) we propose the general transformation (Putz et al., 2006a):

$$f_1 W\left(f_2 e^{f_2} e^{-f_3 t}\right) \rightarrow f_1 \ln\left(1+\left(e^{f_2}-1\right)e^{-f_3 t}\right) \tag{1.57}$$

where f_1, f_2, f_3 are factors that depend on K_M and V_{max}, which is used to transform the closed form solutions of enzymatic kinetics into elementary analytical expressions. The particular relevance of the replacement (1.57) may be visualized from the Figure 1.3.

As shown in Figure 1.3, the difference in the shape of the curves generated by the general *W*-Lambert and natural logarithm functions (curve a) is almost completely removed when the *W*-Lambert time-dependent solution is replaced with the logistic one transformed as in Eq. (1.57) (curve b). This result suggests that using this logistic transformation (1.57) we get

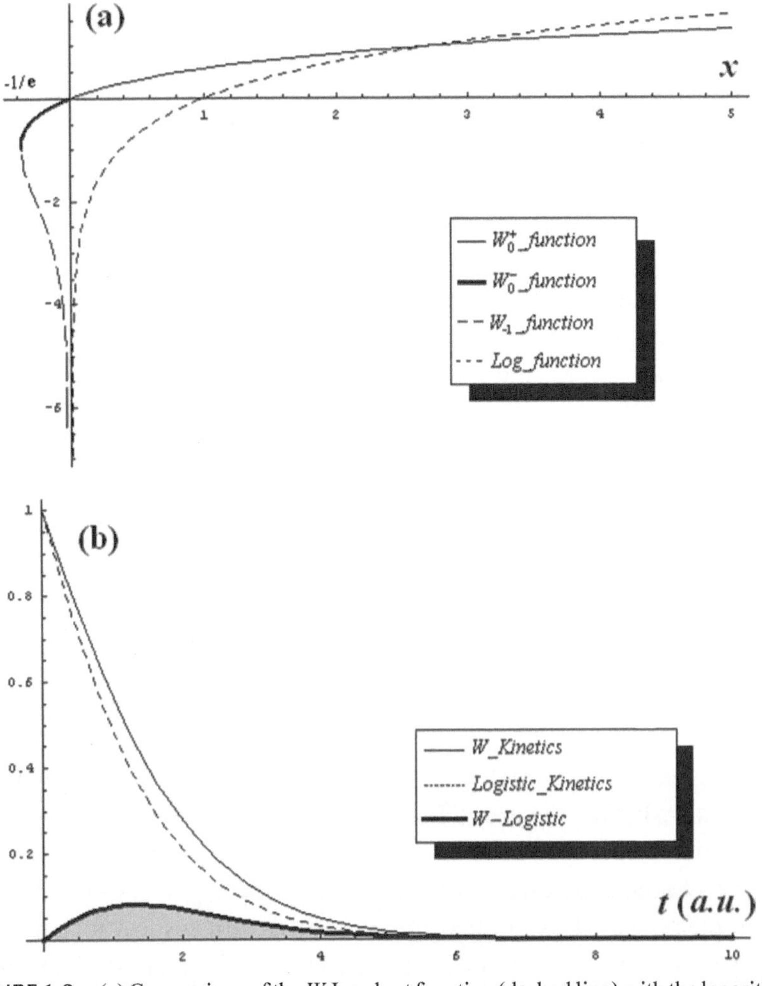

FIGURE 1.3 (a) Comparison of the *W*-Lambert function (dashed line) with the logarithmic function (solid line) against positive ranged simple arguments; (b) Comparison of the *W*-Lambert function (dashed line) with logistic function (solid line) when the arguments include the temporal dependencies as in Eq. (1.57), respectively, being all involved factors fixed to unity. The time abscise scale in (b) is taken in arbitrary units (Putz & Putz, 2011).

a good time-dependent representation over a broad range for enzymatic kinetics in vitro.

This procedure can be directly applied to the existing *W*-Lambert type solutions for many enzymatic reactions in vitro, for example, for enzyme inhibitors, for fully competitive enzyme reactions, for the enzyme kinetics of multiple alternative substrates, see Section 1.5 and Putz et al. (2006a), or for reversible enzyme kinetics, see Section 1.7 and Putz et al. (2006b), making them more useful for fitting laboratory data (Goudar et al., 1999; Schnell & Maini, 2003; Tzafriri & Edelman, 2004).

1.4.3 ANALYSIS OF FITTING CURVES

Although they are able to use the progress curves for analysis of the data obtained from experimental assays, many biochemists prefer to use linear representations of enzyme kinetics. Instead of using the time-dependent solution (1.26), they rearrange the time-dependent equation (1.21) to a sort of time-dependent regression expression, for example, the reciprocal double plot equation:

$$\frac{t}{[S_0]-[S](t)} = \frac{1}{V_{\max}} + \frac{K_M}{V_{\max}}\left(\frac{1}{[S_0]-[S](t)}\ln\frac{[S_0]}{[S](t)}\right) \tag{1.58}$$

A plot of Eq. (1.58) will yield a straight line with an intercept of $1/V_{max}$ and a slope of K_M/V_{max} from which the kinetic parameters K_M and V_{max} can be obtained.

However, this approach has been criticized and it is worthwhile to investigate whether the exact logistic solution (1.51) may be better for fitting a linear curve.

First, we take advantage of the fact that the logistic solution (1.51) has an elementary form to take its derivative with respect to time. This provides an expression for the instantaneous velocity (1.16), which can be transformed to the finite difference $([S_0]\text{-}[S](t))/t$.

Inversion of the result yields the expression; the present discussion follows (Putz et al., 2007)

$$\frac{t}{[S_0]-[S](t)} = \frac{1}{V_{\max}} + \frac{1}{V_{\max}\left(e^{\frac{[S_0]}{K_M}}-1\right)}e^{\frac{V_{\max}}{K_M}t} \tag{1.59}$$

Equation (1.59) is not a linear function, although it may be used for fitting the experimental time series data to determine the kinetic parameters K_M

and V_{max}. To obtain a linear equation from expression (1.59), recall that, from the probabilistic perspective of enzymatic kinetics, the Michaelis-Menten equation is valid for fast reactions. Performing a first order expansion with respect to time on Eq. (1.59) gives the linear equation (Putz et al., 2007):

$$\frac{t}{[S_0]-[S](t)} = \frac{e^{\frac{[S_0]}{K_M}}}{V_{\max}\left(e^{\frac{[S_0]}{K_M}}-1\right)} + \frac{1}{K_M\left(e^{\frac{[S_0]}{K_M}}-1\right)}t \tag{1.60}$$

Figure 1.4 shows the comparison between linear fitting equation (1.58) and the new logistic based expression (1.60) along with the nonlinear form (1.59), for the same parameters used in Figure 1.2 above. Generation of the curve for expression (1.58) required that the *W*-Lambert time-dependence of the substrate depletion be substituted for the time dependent substrate concentration.

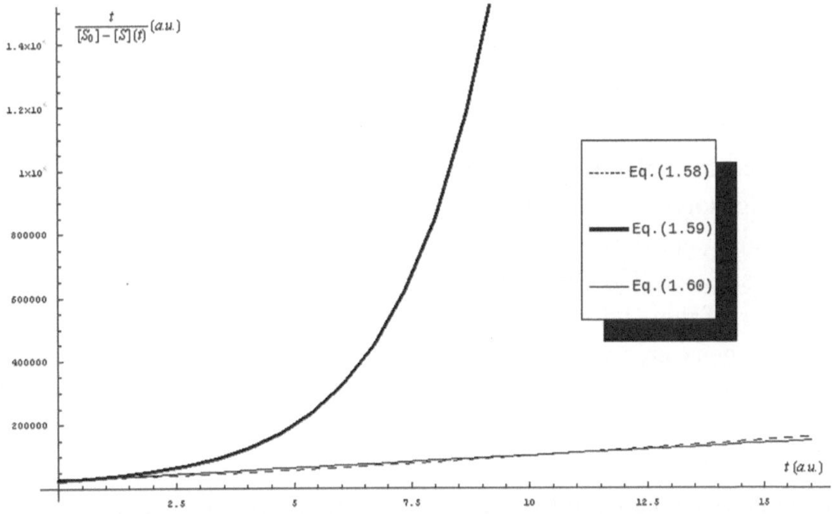

FIGURE 1.4 Time dependent representation of the fitting curves (1.58)–(1.60) for the parametric values $k_{-1}=k_2=10^2 s^{-1}$, $k_1=10^6 M^{-1}s^{-1}$, $[S_0]=10^{-4}M$, and $[E_0]=10^{-6}M$. The dashed line corresponds to equation (1.58) and involves the *W*-Lambert closed form solution (1.26). The thin continuous line is the representation of linear equation (1.60) while the thick continuous line is the plot of non-linear equation (1.59) being both based on the logistic solution (1.51). Abscise and ordinate scales are given in arbitrary units (Putz et al., 2007).

It is clear that the linear logistic curve (1.60) is nearly coincident with the Michaelis-Menten curve (1.58), both providing linear approximations of the general non-linear logistic curve (1.59). This is the first instance of treating substrate-enzyme binding probabilistically and it has the advantage of avoiding use of the *W*-Lambert function, which is impossible to evaluate exactly. The resulting linear fitting curves are essentially the same with either approach.

With the logistic ansatz (1.37), we now have a consistent recipe for temporal modeling of, in principle, any scheme of enzymatic reactions in vitro. This algorithm consists of two steps: first, the associated kinetics is solved within basic Michaelis-Menten picture until the *W*-Lambert solution is achieved; then, the logistic transformation (1.57) is performed leading with an elementary analytical form that can be further used for theoretical predictions and numerical fitting of the experimental assays. However, to emphasize the reliability of the logistic transformation (1.57) for complex enzymatic reactions as well, the case of the enzyme kinetic of the multiple alternative substrates will be presented next and then particularized to the fully competitive enzyme catalysis (Putz et al., 2006a, 2007).

1.5 MULTIPLE ALTERNATIVE ENZYME-SUBSTRATE REACTIONS

After a century of supremacy, the central dogma of biology, i.e., the fact that the genotype can not be in any way affected through protein supply and interaction (Mattick, 2004; Silverman, 2004), is currently being taken under discussion (Goodman et al., 2005). It started with the landmark contributions of the, 1950s and, 1960s scientists Koshland, Monod, Wyman, and Changeux proposing the "induced fit" model with the help of which they rationalized the competing needs of substrate binding affinity. It was concluded that the metabolic protein, in general, and enzymatic, in particular, activities can be regulated by small molecules other than the substrates, the inhibitors or activators (Cantor & Schimmel, 1980; Copeland, 2000).

As a consequence, the developed theory of allosteric regulation (from Greek: *allos*, other + *stereos*, space) prescribes that, within a cooperative interaction, the binding of one ligand (substrate) at a specific site is influenced by the binding of another ligand (inhibitor) at a different or allosteric site on the protein (or enzyme). However, actually, such behavior is generalized at the level of organismal and cellular regulation in which the cell converts the comparison of the proteins with organisms needs into metabolic process. It follows that the proteins and gene expression, far from being the endpoint, are rather a bridge from where begin the process of editing RNA transcripts, altering and maintaining the genome, over and over again by signaling other cells or bio-inspired nano-implants (Curran et al., 2005). In this process of cell differentiation, proliferation and programming, the receptors (substrates and inhibitors) and enzymes perform the task of molecular messengers. Therefore, studying the cooperative effects of the inhibitors on the enzymatic reactions, here at the theoretical level, should be most valuable for the forefront of biomedical researches; the present discussion follows (Putz et al., 2006a).

Basically, an alternative n-substrate system consists of n-reactions of the Michaelis-Menten type (1.4)

$$S_i + E \underset{k_{-i}}{\overset{k_i}{\leftrightarrow}} ES_i \overset{k_{2i}}{\rightarrow} E + P_i \quad , i = \overline{1,n} \tag{1.61}$$

which, nevertheless, generate a system with $3n$+1 differential equations, viewed as the direct expansion of the single-substrate one (1.5)–(1.8), with the initial temporal constraints $([S]_i,[E],[ES]_i,[P]_i)_{t=0}=([S_0]_i,[E_0],0,0)$.

The associated conservation laws now look as generalizations of the basic ones given in Eqs. (1.9) and (1.10):

$$[E](t) = [E_0] - \sum_{i=1}^{n} [ES]_i(t) \tag{1.62}$$

$$[P]_i(t) = [S_0]_i - [S]_i(t) - [ES]_i(t) \tag{1.63}$$

Following the deduction line of the foreground enzymes kinetic, the specific Michaelis-Menten i-constants

$$K_M^i = \frac{k_{-i} + k_{2i}}{k_i} \tag{1.64}$$

and the maximum velocity for the i-reaction

$$V_{\max}^i = k_{2i}[E_0] \tag{1.65}$$

are firstly introduced.

Additionally, a few new notations are considered here (Schnell & Mendoza, 2000b), namely the *first order rate i-constants*,

$$\kappa_i = \frac{V^i_{max}}{K^i_M} \tag{1.66}$$

and the *reduced i-concentrations*,

$$[X']_i = \frac{[X]_i}{K^i_M} \tag{1.67}$$

in order to shortcut the script of further emerging equations.

With these amendments, the above temporal equations are accompanied by the actual form of the enzyme-substrate i-complex concentration (Schnell & Mendoza, 2000b):

$$[ES]_i = \frac{[E_0][S']_i}{1+\sum_{j=1}^{n}[S']_j}\left[1-\exp\left(-k_i t K^i_M\left(1+\sum_{j=1}^{n}[S'_0]_j\right)\right)\right] \tag{1.68}$$

from which its simple form (1.19) can be recovered, since only one substrate reaction is retained from the scheme (1.61).

Certainly, as before, the kinetics is not solved until temporal analytical solution for the i-substrate concentration is derived. To achieve this goal, in this particular case, we first need to solve the generalized system of coupled equations for the alternative substrates in reaction (1.61) (Rubinow & Lebowitz, 1970; Schnell & Mendoza, 2000b):

$$\frac{d}{dt}[S']_i = -\frac{-\kappa_i[S']_i}{1+\sum_{j=1}^{n}[S']_j} \tag{1.69}$$

$$[S']_j = [S'_0]_j\left(\frac{[S']_i}{[S'_0]_i}\right)^{\delta_{ij}} \tag{1.70}$$

when the participating substrates are interrelated through the parameter (Putz et al., 2006a)

$$\delta_{ij} = \frac{\kappa_j}{\kappa_i} = \frac{V_{\max}^j K_M^i}{V_{\max}^i K_M^j} \tag{1.71}$$

also referred to as the *competition matrix*, due to its ability to measure the degree of competition among the substrates involved in the reaction with the enzyme.

As a note, one can easily check that relations (1.69)–(1.71) become the basic Michaelis-Menten equation (1.16) when dealing with single-substrate reaction. Unfortunately, the general system (1.69)–(1.71) has no explicit solution unless the competition matrix is specified in some particular cases.

As such, a first case assumes the so-called *even competition* when $\delta_{ij} \cong 1$. In this frame, the system (1.69)–(1.71) can be integrated and the result rearranged so that the proper comparison with the *W*-Lambert equation (1.25) to be employed. This causes the *W*-Lambert transcendent solutions for the system (1.69)–(1.71) to take the closed forms (Schnell & Mendoza, 2000b):

$$[S']_i^W(t) = \frac{[S'_0]_i}{\sum_{j=1}^{n}[S'_0]_j} W\left(\sum_{j=1}^{n}[S'_0]_j \exp\left(\sum_{j=1}^{n}[S'_0]_j - \frac{V_{\max}^i}{K_M^i}t\right)\right) \tag{1.72}$$

as a direct generalization of the mono-substrate Michaelis-Menten temporal solution (1.26). Finally, the logistic transformation (1.57) can be directly applied on Eq. (1.72) leading to the elementary analytic expressions (Putz et al., 2006a):

$$[S']_i^L(t) = \frac{[S'_0]_i}{\sum_{j=1}^{n}[S'_0]_j} \ln\left(1 + \left[\exp\left(\sum_{j=1}^{n}[S'_0]_j\right) - 1\right]\exp\left(-\frac{V_{\max}^i}{K_M^i}t\right)\right) \tag{1.73}$$

Looking at the mathematical form of even competition solutions (1.72) and (1.73) observing the benchmark single-substrate ones, (1.26) and (1.51), it appears that at any time the reduced substrate concentrations keep the proportion determined from their initial reduced concentrations.

Consequently, the time evolutions of the set of alternative reactants are very similar to those considered in the mono-substrate reaction.

A more interesting case regards the so-called *weak competition* when the reactants are not catalyzed with the same efficiency from the enzyme. In this situation, the competition matrix (1.71) ranges as $0 < \delta_{ij} << 1$. However, in this case the first order of the Taylor expansion of Eq.(1.70) in Eq. (1.69) can be retained and, by repeating the previous integration and rearrangement procedure the *W*-Lambert closed form solution can be cast as (Schnell & Mendoza, 2000b):

$$[S']_i^W(t) = \left(1+\sum_{j\neq i}[S'_0]_j\right) W\left(\frac{[S'_0]_i}{1+\sum_{j\neq i}[S'_0]_j}\exp\left(\frac{[S'_0]_i}{1+\sum_{j\neq i}[S'_0]_j} - \frac{V_{max}^i t}{K_M^i\left(1+\sum_{j\neq i}[S'_0]_j\right)}\right)\right) \tag{1.74}$$

which, in turn, allows its transcription under an elementary analytical form through performing the logistic transformation (1.57) (Putz et al., 2006a):

$$[S']_i^L(t) = \left(1+\sum_{j\neq i}[S'_0]_j\right) \ln\left(1+\left[\exp\left(\frac{[S'_0]_i}{1+\sum_{j\neq i}[S'_0]_j}\right)-1\right]\exp\left(-\frac{V_{max}^i t}{K_M^i\left(1+\sum_{j\neq i}[S'_0]_j\right)}\right)\right) \tag{1.75}$$

Certainly, similar mathematical analyses and logistic transformations can be considered for various types of enzymatic reactions, no matter how complex the biochemical network may be. However, in order to prove that the logistic ansatz closely follows the *W*-Lambert implicit solutions for all species when a complex kinetics is under study, the special bi-substrate case of weak competition, i.e., the case of competitive inhibition, will be presented in detail next.

1.6 APPLICATION ON COMPETITIVE INHIBITION

When an inhibitor acts to reduce the concentration of the available fee enzyme for the substrate binding, it is said that competitive inhibition takes place. An eminent example is that of succinate dehydrogenase, which is competitively inhibited by malonate to convert succinate to fumarate within the citric acid cycle (Walsh, 1979; Voet & Voet, 1995).

With fully competitive interaction, the associate network model is particularized from the scheme (1.61) by retaining two channels of alternative enzyme-substrate reactions only; the present discussion follows (Putz et al., 2006a):

$$\begin{array}{ccccccccc} S & + & E & \underset{k_{-1}}{\overset{k_1}{\leftrightarrow}} & ES & \overset{k_2}{\rightarrow} & E & + & P_S \\ & & + & & & & & & \\ & & I & \underset{k_{-3}}{\overset{k_3}{\leftrightarrow}} & EI & \overset{k_4}{\rightarrow} & E & + & P_I \end{array} \tag{1.76}$$

It is worth noting that the present assumed model for competitive inhibition represents an improved version of the commonly accepted one, in which the inhibitor-enzyme complex *EI* of Eq. (1.76) undergoes no further reaction or specific product formation (Voet & Voet, 1995).

To set the competitive inhibition's characteristics, the general alternative substrate kinetic parameters (1.64) and (1.65) now become:

- the respective Michaelis-Menten constants for the substrate and inhibitor branches of Eq. (1.76):

$$K_M^S = \frac{k_{-1} + k_2}{k_1} \tag{1.77}$$

$$K_M^I = \frac{k_{-3} + k_4}{k_3} \tag{1.78}$$

- the respective maximum velocities for the substrate and inhibitor branches of Eq. (1.76):

$$V_{max}^S = k_2[E_0] \tag{1.79}$$

$$V_{max}^I = k_4[E_0] \tag{1.80}$$

Nevertheless, the kinetic information comprised in the parameters (1.77)–(1.80) can be combined in a single quantity through the competition matrix (1.71), which now takes the specialized form

$$\delta = \frac{V_{max}^I K_M^S}{V_{max}^S K_M^I} \tag{1.81}$$

Focusing in what follows on the case of weak competition exclusively, in which the competition index fulfills the kinetic condition $\delta << 1$, the respective reduced initial and instantaneous concentrations of the substrate and inhibitor, particularizing the general definition (1.67) for the reaction channels of (1.76),

$$[S'_0] = \frac{[S_0]}{K_M^S},\ [S'](t) = \frac{[S](t)}{K_M^S} \tag{1.82}$$

$$[I'_0] = \frac{[I_0]}{K_M^I},\ [I'](t) = \frac{[I](t)}{K_M^I} \tag{1.83}$$

provide the keys with which the overall bi-substrate kinetic is solved.

This way, the W-Lambert time dependent closed solutions for the substrate and inhibition progress curves unfold with the respective forms (Schnell & Mendoza, 2000a):

$$[S']_W(t) = (1+[I'_0])W\left(\frac{[S'_0]}{1+[I'_0]}\exp\left(\frac{[S'_0]}{1+[I'_0]}\right)\exp\left(-\frac{V_{\max}^S t}{K_M^S(1+[I'_0])}\right)\right) \tag{1.84}$$

$$[I']_W(t) = [I'_0]\left(\frac{[S']_W(t)}{[S'_0]}\right)^{\delta} \tag{1.85}$$

by specializing the general multi-substrate formulas (1.74) and (1.70) to the present analysis.

In order to get the analytical counterparts of Eqs. (1.84) and (1.85), actually, two-folded methods can be considered. One is to particularize the already obtained generalized logistic form (1.75) to the actual bi-substrate alternative scheme; equally, one can directly apply the logistic transformation (1.57) to the specific W-Lambert solution of the weakly competitive inhibition of the substrate progress curve (1.84).

Using either of these two methodologies, the logistic expression that shapes the decrease (or consumption) in substrate concentration in reaction (1.76) can be obtained with the elementary form (Putz et al., 2006a):

$$[S']_L(t) = (1+[I'_0])\ln\left(1+\left[\exp\left(\frac{[S'_0]}{1+[I'_0]}\right)-1\right]\exp\left(-\frac{V_{\max}^S t}{K_M^S(1+[I'_0])}\right)\right) \tag{1.86}$$

being as well accompanied by the logistic version of the inhibitor progress curve of (1.85):

$$[I']_L(t) = [I'_0]\left(\frac{[S']_L(t)}{[S'_0]}\right)^{\delta} \tag{1.87}$$

Having formulated the *W*-Lambert and logistic functions of substrate and inhibitor progress curve for the enzymic processes of Eq. (1.76), the complete kinetic picture can be revealed for all the species.

For instance, the progress curves for the substrate-enzyme and inhibitor-enzyme complexes of Eq. (1.76) can be obtained by means of adapting the general formula (1.68), respectively as:

$$[ES]_{W,L}(t) = \frac{[E_0][S]_{W,L}(t)}{[S]_{W,L}(t) + K_M^S\left(1 + [I]_{W,L}(t)/K_M^I\right)}$$
$$\times \left\{1 - \exp\left[-k_1 t\left([S_0] + K_M^S\left(1 + [I_0]/K_M^I\right)\right)\right]\right\} \tag{1.88}$$

$$[EI]_{W,L}(t) = \frac{[E_0][I]_{W,L}(t)}{[I]_{W,L}(t) + K_M^I\left(1 + [S]_{W,L}(t)/K_M^S\right)}$$
$$\times \left\{1 - \exp\left[-k_3 t\left([I_0] + K_M^I\left(1 + [S_0]/K_M^S\right)\right)\right]\right\} \tag{1.89}$$

written compactly for both the *W*-Lambert and logistic temporal solutions.

With expressions (1.88) and (1.89), the conservation laws (1.62) and (1.63) can be further employed with their actual particular progress curves:

$$[P_S]_{W,L}(t) = [S_0] - [S]_{W,L}(t) - [ES]_{W,L}(t) \tag{1.90}$$

$$[P_I]_{W,L}(t) = [I_0] - [I]_{W,L}(t) - [EI]_{W,L}(t) \tag{1.91}$$

$$[E]_{W,L}(t) = [E_0] - [ES]_{W,L}(t) - [EI]_{W,L}(t) \tag{1.92}$$

for the product from substrate, product from inhibitor and for the enzyme, respectively.

However, in order to underline the equivalence of the *W*-Lambert and logistic at all levels of an enzymic kinetic in vitro, within the present weak competition conditions, Figure 1.5 shows the scaled shapes of the progress curves (1.84)–(1.87) of all species of the biochemical network (1.76) (Putz et al., 2006a):

$$s_{W,L}(\tau) = \frac{[S]_{W,L}(\tau)}{[S_0]}, \quad i_{W,L}(\tau) = \frac{[I]_{W,L}(\tau)}{[I_0]} \tag{1.93}$$

$$es_{W,L}(\tau) = \frac{[ES]_{W,L}(\tau)}{[E_0]}, \quad ei_{W,L}(\tau) = \frac{[EI]_{W,L}(\tau)}{[E_0]} \tag{1.94}$$

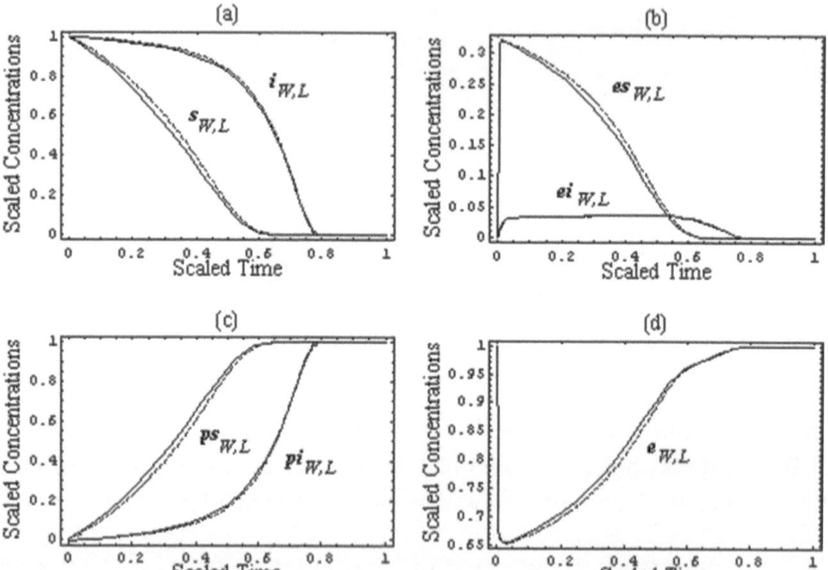

FIGURE 1.5 The scaled progress curves (1.93)–(1.96) of the species concentrations involved in competitive reaction (1.76) for the pilot test with $k_{-1} = k_2 = 10^2\,s^{-1}$, $k_1 = 10^6\,M^{-1}s^{-1}$, $k_{-3} = k_4 = 10s^{-1}$, $k_3 = 10^5\,M^{-1}s^{-1}$, $[S_0] = 10^{-4}M$, $[I_0] = 10^{-5}M$, and $[E_0] = 10^{-6}M$, arranged as follows: (a) for the leading and inhibitory substrates concentrations, according to Eq. (1.93); (b) for the substrate-enzyme and inhibitor-enzyme complexes concentrations, according to Eq. (1.94); (c) for the products of the leading and inhibitory substrates concentrations, according to Eq. (1.95); and (d) for the enzyme concentration, according to Eq. (1.96), within the *W*-Lambert (dashed lines) and logistic (solid lines) Michaelis-Menten kinetics against the scaled time (1.56), respectively (Putz et al., 2006a).

$$ps_{W,L}(\tau) = \frac{[P_S]_{W,L}(\tau)}{[S_0]}, pi_{W,L}(\tau) = \frac{[P_I]_{W,L}(\tau)}{[I_0]} \tag{1.95}$$

$$e_{W,L}(\tau) = \frac{[E]_{W,L}(\tau)}{[E_0]} \tag{1.96}$$

against the scaled time (1.56) for a pilot computational test in which the parametric values were chosen as $k_{-1} = k_2 = 10^2\ s^{-1}$, $k_1 = 10^6\ M^{-1}s^{-1}$, $k_{-3} = k_4 = 10s^{-1}$, $k_3 = 10^5\ M^{-1}s^{-1}$, while the initial condition have been set to $[S_0] = 10^{-4}M$, $[I_0] = 10^{-5}M$, and $[E_0] = 10^{-6}M$, respectively.

From the plots in Figure 1.5, it is clear that for each envisaged species of Eq. (1.76), the *W*-Lambert and logistic progress curves display outstanding similar, or almost coincident, shapes.

We have thus formulated and illustrated all the analytic steps for achieving the complete temporal picture of enzymatic catalyzed reactions in vitro, leading to the framework in which the experimental assay can be fitted to obtain the characteristic parameters, the algorithm presented being applicable, in principle, at any level of biochemical network complexity.

Moreover, the present approach may be found useful in predicting or validating the assumed kinetic schemes by employing the logistic progress curves to the non-linear fitting of the experimentally recorded data.

1.7 APPLICATION ON REVERSIBLE KINETICS

One of the most yet powerful theories of enzyme kinetics regards the Michaelis-Menten mechanism introduced since beginnings of twentieth century by the cornerstone works of Brown (1902); Henri (1901); Michaelis & Menten (1913); Slyke & Cullen (1914). It assumes that the enzyme *E* reacts with a substrate *S* to reversibly form the intermediate complex *ES* by the constant rates k_1, k_1 that undergoes to final product *P* and recovering the enzyme *E* through the irreversible dissociation with the constant rate k_2. Despite much in vivo enzymatic catalysis may be modeled by such a mechanism there are still many enzymatic reactions high reversible (Duggleby, 1994), with a small free energy of reaction, thus having product that back reacts to form substrate with the rate k_2 (Voet & Voet, 1995; Putz et al., 2006b):

$$E + S \underset{k_{-1}}{\overset{k_1}{\leftrightarrow}} ES = EP \underset{k_{-2}}{\overset{k_2}{\leftrightarrow}} E + P \tag{1.97}$$

Solving the complete kinetic for the reaction type (1.97) was always a task and the analytic solutions were formulated only in some special cases, for instance when $k_1 = k_{-2}$ (van Slyke & Cullen, 1914). For general case of Eq. (1.97), although the condition for QSSA for the *ES* complex synthesis

$$\frac{d}{dt}[ES] \cong 0 \tag{1.98}$$

in terms of its concentration [*ES*], have become a common tool to investigate the analytical solution, the complete temporal full course of the reactants has remain untracked.

This work is devoted to furnish, for the first time, the complete elementary analytical solution for the progress reactant concentrations of the reversible enzymatic reaction (1.97). Such an approach may be found suited for treating the metabolic regulating mechanisms where regenerative, cyclical and thus necessarily reversible processes occur (Tzafriri & Edelman, 2004).

As earlier stated, though the employment of the QSSA condition (1.98) for the law of mass action on the equation (1.97),

$$0 = \frac{d}{dt}[ES] = k_1[E][S] + k_{-2}[P][E] - (k_{-1} + k_2)[ES] \tag{1.99}$$

one firstly gets the reciprocal relation between the enzyme and intermediate enzyme complex as:

$$[E] = \frac{k_{-1} + k_2}{k_1[S] + k_{-2}[P]}[ES] \tag{1.100}$$

Then, relation (1.100) can be further combined with the enzymatic conservation,

$$[E_0] = [E] + [ES] \tag{1.101}$$

to obtain the complex concentration,

$$[ES] = \frac{k_1[S] + k_{-2}[P]}{k_{-1} + k_2 + k_1[S] + k_{-2}[P]}[E_0] \tag{1.102}$$

in variables of substrate and product concentrations only, being $[E_0]$ the initial concentration of the enzyme in the system.

Next, with relations (1.100) and (1.102) the instantaneous velocity of product formation can be calculated by means of the law of mass action of reaction (1.97), consecutively as:

$$v = \frac{d}{dt}[P] = -\left(\frac{d}{dt}[S]\right) = -(k_{-1}[ES] - k_1[S][E]) \tag{1.103}$$

until the result; the present discussion follows (Putz et al., 2006b):

$$v = \frac{k_1 k_2[S] - k_{-1}k_{-2}[P]}{k_{-1} + k_2 + k_1[S] + k_{-2}[P]}[E_0] \tag{1.104}$$

Nevertheless, the velocity of product formation (1.104) can be with more meaning rearranged via introducing the appropriate combination of the involved rate constants, as

$$K_M^S = \frac{k_{-1} + k_2}{k_1} \tag{1.105}$$

and

$$K_M^P = \frac{k_{-1} + k_2}{k_{-2}} \tag{1.106}$$

named the substrate and product Michaelis-Menten constants, respectively, together with the forwarded

$$V_{\max}^f = k_2[E_0] \tag{1.107}$$

and rewarded

$$V_{\max}^r = k_{-1}[E_0] \tag{1.108}$$

maximum velocities of reaction (1.97), respectively.

With the Eqs. (1.105)–(1.108) notations the compact form of Eq. (1.104) looks like (Putz et al., 2006b):

$$v = \frac{V_{\max}^{f}\dfrac{[S]}{K_M^S} - V_{\max}^{r}\dfrac{[P]}{K_M^P}}{1 + \dfrac{[S]}{K_M^S} + \dfrac{[P]}{K_M^P}} \tag{1.109}$$

The simple analysis of the relation (1.109), by contrast with the ordinary Michaelis-Menten version of (1.97) in which $k_{-2} = 0$, the typical velocity of the reversible reactions contains two terms in the numerator, at least one positive and at least one negative, leading with no possibility of linearization, as is often performed in enzyme kinetic, being so less convenient to handle. There is therefore an imperative of trying to solve out the equation (1.97) for substrate or product in order to deliver useful predictive solution for experimental assay analysis. Such goal is in next sections addressed.

Going to solve the Eq. (1.109) in terms of substrate concentration, firstly the conservation law of substrate-product concentrations is substituted,

$$[P](t) = [P_0] + [S_0] - [S](t) \tag{1.110}$$

leaving with the one variable equation for the temporal evolution of the substrate concentration (Putz et al., 2006b):

$$-\frac{d}{dt}[S](t) \equiv \frac{V[S](t) - R}{[S](t) + K} \tag{1.111}$$

where the new introduced shortcuts looks like:

$$V = \frac{\dfrac{V_{\max}^{f}}{K_M^S} + \dfrac{V_{\max}^{r}}{K_M^P}}{\dfrac{1}{K_M^S} - \dfrac{1}{K_M^P}} \tag{1.112}$$

$$R = \frac{[P_0] + [S_0]}{\dfrac{1}{K_M^S} - \dfrac{1}{K_M^P}} \frac{V_{\max}^{r}}{K_M^P} \tag{1.113}$$

$$K = \frac{1 + \dfrac{[S_0] + [P_0]}{K_M^P}}{\dfrac{1}{K_M^S} - \dfrac{1}{K_M^P}} \tag{1.114}$$

In order to integrate the Eq. (1.111), it is firstly rearranged as the time integration to can be performed directly,

$$\frac{dt}{d[S]} = -\frac{[S] + K}{V[S] - R} \tag{1.115}$$

to provide the intermediate form:

$$-t = \frac{1}{V} \int_{[S_0]}^{[S]} \frac{[S] + K}{[S] - R/V} d[S] \tag{1.116}$$

and with the final result:

$$-tV = [S] - [S_0] + \left(K + \frac{R}{V}\right)\left[\ln\left([S] - \frac{R}{V}\right) - \ln\left([S_0] - \frac{R}{V}\right)\right] \tag{1.117}$$

since one applies the elementary rule of integration:

$$\int \frac{x+a}{x+b} dx = x + (a-b)\ln(x+b) + ct. \tag{1.118}$$

However, the problem with the Eq. (1.117) is that it has to deliver the $[S](t)$ as the function of time. To do that to relation (1.117) the term R/V is added and subtracted first,

$$\left([S] - \frac{R}{V}\right) + \left(K + \frac{R}{V}\right)\ln\left([S] - \frac{R}{V}\right) = \left([S_0] - \frac{R}{V}\right) + \left(K + \frac{R}{V}\right)\ln\left[[S_0] - \frac{R}{V}\right] - Vt \tag{1.119}$$

the new identity being simplified by $(K+R/V)$, to get the form (Putz et al., 2006b):

$$\frac{[S] - \dfrac{R}{V}}{K + \dfrac{R}{V}} + \ln\left(\frac{[S] - \dfrac{R}{V}}{K + \dfrac{R}{V}}\right) = \frac{[S_0] - \dfrac{R}{V}}{K + \dfrac{R}{V}} + \ln\left(\frac{[S_0] - \dfrac{R}{V}}{K + \dfrac{R}{V}}\right) - \frac{Vt}{K + \dfrac{R}{V}} \tag{1.120}$$

Now, the left side form of Eq. (1.120) is recognized to match with of the Euler-Lambert type equation (1.27); Then, by combining the Eq. (1.120) with Eq. (1.27) the next system is formed

$$\begin{cases} W(x) = \dfrac{[S]-R/V}{K+R/V} \\ x = \dfrac{[S_0]-R/V}{K+R/V}\exp\left(\dfrac{[S_0]-R/V-Vt}{K+R/V}\right) \end{cases} \tag{1.121}$$

from where follows the closed form temporal solution of the substrate concentration (Putz et al., 2006b):

$$[S]_W(t) = \frac{R}{V} + \left(K+\frac{R}{V}\right)W\left(\frac{[S_0]-R/V}{K+R/V}\exp\left(\frac{[S_0]-R/V-Vt}{K+R/V}\right)\right) \tag{1.122}$$

Indeed, once the substrate concentration time dependency is known the product temporal evolution is then derived from the relation (1.110) and, together, provide the enzyme-substrate complex and the enzyme concentrations through the relations (1.102) and (1.100), respectively.

However, despite the consistency in the above determination of the substrate concentration under the closed-form solution (1.122) its use in further temporal fitting for experimental estimation of the kinetic parameters (1.105)–(1.108) is limited since no availability of analytical elementary expression for the *W*-Lambert function (Barry et al., 2000). This important drawback is to be removed in what next.

Aiming to find an elementary equivalency of the closed form solution (1.122) recently has been proposed that a suitable analytical transformation is represented by the general logistical substitution (1.57); in these conditions the actual logistical form solution for the substrate concentration of the reversible enzymatic reaction (1.97) shaped as (Putz et al., 2006b):

$$[S]_L(t) = \frac{R}{V} + \left(K+\frac{R}{V}\right)\ln\left\{1+\left[\exp\left(\frac{[S_0]-R/V}{K+R/V}\right)-1\right]\exp\left(-\frac{Vt}{K+R/V}t\right)\right\} \tag{1.123}$$

Having both the closed and the analytical form solutions (1.122) and (1.123), respectively worth comparing them through the progress curves of the reactants of (1.97). Such representations are most convenient achieved by appealing to the respective scaled concentrations (Putz et al., 2006b):

$$s_{W,L}(t)=\frac{[S]_{W,L}(t)}{[S_0]},\quad p_{W,L}(t)=\frac{[P]_{W,L}(t)}{[S_0]} \tag{1.124}$$

$$es_{W,L}(t)=\frac{[ES]_{W,L}(t)}{[E_0]},\quad e_{W,L}(t)=\frac{[E]_{W,L}(t)}{[E_0]} \tag{1.125}$$

mapped from the (0, ∞) interval to the 0,1 range through the temporal relation (1.56).

The scaled substrate, product, enzyme-substrate and enzyme concentrations are achieved since the *W*-Lambert and logistical (1.122) and (1.123) substrate solutions are combined with the relations (1.110), (1.100) and (1.102), respectively for a working set of kinetic constants.

The typical founded kinetic parameters to be used for full time course analyses of the reactants (1.124) and (1.125) are given by the set of data:

$$k_{-1}=100s^{-1},\ k_2=1000s^{-1},\ k_1=0.1M^{-1}s^{-1},\ k_{-2}=0.2M^{-1}s^{-1},$$

$$E_0]=10^{-4}M,\ [S_0]=10^{-2}M,\ [P_0]=0M \tag{1.126}$$

with the help of which the plots of Figure 1.6 are obtained.

First, the pictures of Figure 1.6 clearly reveal that the logistical based progress curves, the right ones, almost coincide with those drawn using the *W*-Lambert approach, the left ones. Such equivalent behavior gives credit to relations (1.123) for further use in kinetically parameter estimations for the reversible enzymatic reactions of type (1.97) grounded on the considerably reduced number of experimental assays (Putz et al., 2006b).

Another interesting feature of the Figure 1.6 plots reside in the special character of the reversible reactants evolution curves. In such there is observed that the substrate appears not entirely exhausted *ad infinitum* while the product do not form up to the initial substrate concentration. The explanation for this phenomenon reside in the reversible nature of the undergo reaction and is also reflected from the enzyme-substrate complex and the enzyme concentration curves in Figure 1.6 which, instead, do not present complete consumption and conservation, respectively, *ad infinitum*.

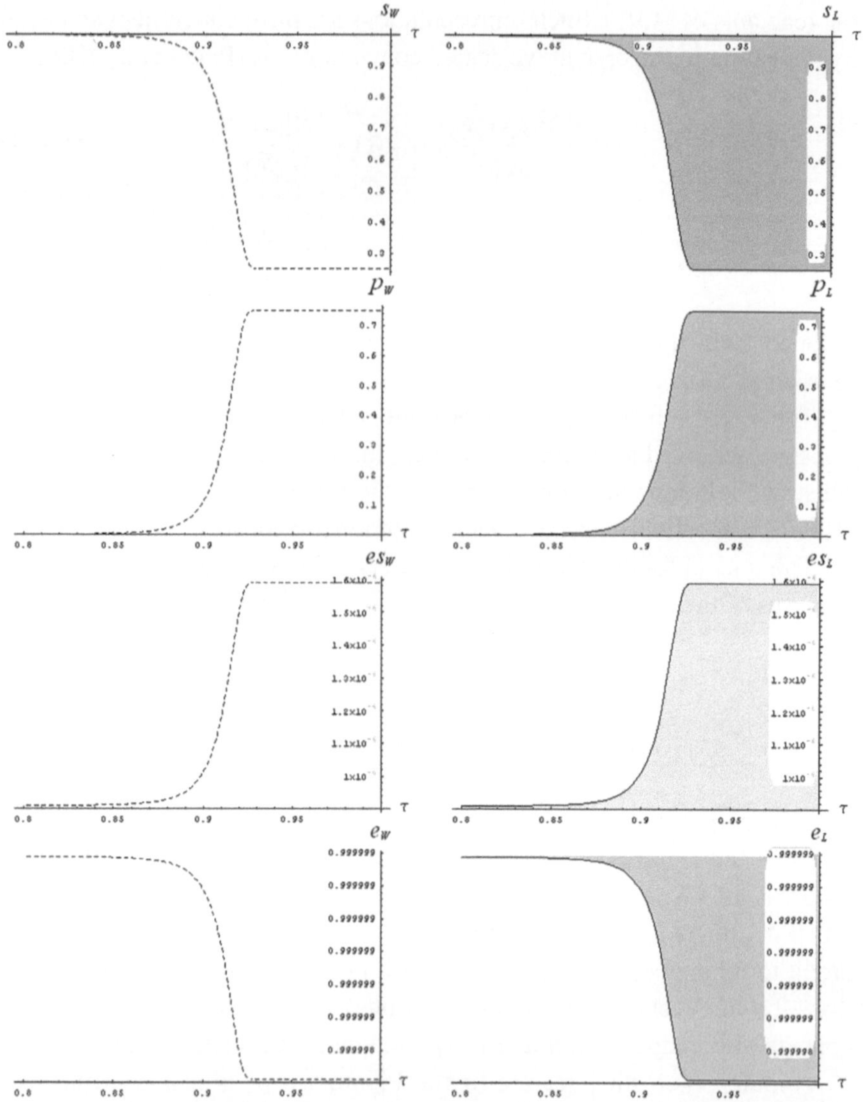

FIGURE 1.6 Comparative trend of scaled concentrations (1.124) and (1.125) of the substrate (upper first two plots), product (second line two plots), enzyme-substrate (third line two plots) and enzyme (bottom two plots) against time, according to Eq. (1.179) mapping from $(0, \infty)$ to (0,1) realm, for the *W*-Lambert (left side plots) and logistic (right side plots) approaches by combining the substrate solutions (1.122) and (1.123) with the relations (1.110), (1.102) and (1.100), respectively, within the conditions fixed by Eq. (1.126) (Putz et al., 2006b).

There is so concluded that in a reversible reaction, fraction of the enzyme remains permanently bounded with the substrate in the enzyme-substrate complex preventing thus the complete substrate consumption, product formation, and entirely enzyme separation *ad infinitum* (Putz et al., 2006b).

Moreover, from all plots of Figure 1.6 there can be concluded that the reversible reactions do not definitely flows to the product formation, until considerable high temporal intervals are leaved behind. Also important, the reaction (1.97) delivers the late but suddenly almost entirely dissociation of the enzyme-substrate complex since all the curves of Figure 1.6 behaves closely with step kind functions.

With all these features emerges out that the analysis of the reversible reaction can in no way to be treated otherwise that under temporal reactant courses making the present approach most useful when dealing with regenerative metabolic pathways.

1.8 LOGISTIC APPROACH OF HALDANE-RADIĆ ENZYME KINETICS

Nevertheless, Michaelis-Menten equation (1.4) is not a universal paradigm for enzyme modeling, especially when considering kinetics of cholinesterases from various sources. However, even for that, it was revealed, for instance, that the $Glu^{199} \rightarrow Asp^{199}$ mutation in the sequence Phe-Glu-Ser-Ala-Gly at the active center of the three-dimensional structure of *Torpedo californica* acetyl cholinesterase (AChE; EC 3.1.1.7) surprisingly appears to affect similarly the binding of the peripheral and active center site ligand; as a consequence the allosteric coupling between the sites is diminished, and the substrate inhibition is no longer observed indicating the substrate inhibition constant (K_{SS}) becomes infinitely large; the present discussion follows (Putz, 2011):

$$K_{SS} \rightarrow \infty \tag{1.127}$$

while the catalytic activity exceeds that for the wild-type enzyme for the high substrate concentration towards Michaelis-Menten kinetics (Radić et al., 1992). Equally effects and elements of substrate activations were recorded for butyrylcholinesterases (BuChE) specific mutants such as the $F_{297}I$, $F_{338}G$,

$F_{297}Y$, or $D_{74}N$ of mouse AChE-BuChE enzyme chimeras (Radić et al., 1993). Such allosteric studies introduced the conceptual need the substrate (S) may combine at two discrete sites of an enzyme forming two binary complexes ES and SE that both end up within the ternary complex SES whose hydrolysis efficiency relative to the Michaelis-Menten binary complex ES is now quantified by the catalytic parameter (b), see Figure 1.7.

Note that in Figure 1.7 it was assumed, paradigmatically, that the substrate combines equally well with enzyme and the complex ES. Therefore, Figure 1.7 principally reduces to that of Michaelis-Menten (1.4) in cases where there is little substrate inhibition, as provided by Eq. (1.127) above, or when the value b approaches unity (Putz, 2011):

$$K_{SS} \to \infty \Leftrightarrow b \to 1 \tag{1.128}$$

Further variants of Figure 1.7 were also considered in regarding the modeling of inhibition of *Drosophila melanogaster* acetylcholinesterase active site gorge trying to furnish a putative model for the essentially not-Michaelis-Menten kinetics of cholinesterases in general, and those of insects in special, such that to combine activation and inhibition for a large range of substrate concentrations (Stojan et al. 1998, 2004; Nachon, 2008). Nevertheless, all these inhibition models originate into the classical Haldane equation (Haldane, 1930):

$$-\frac{d[S]}{dt} = \frac{V_{max}[S]}{K_S + [S] + [S]^2 / K_{SS}} \tag{1.129}$$

having also been used, besides enzyme kinetics, to describe biodegradation and respirometric studies involving inhibitory substrates (Ellis et al., 1996; Grady et al., 1999); yet, due to the transcendental equation furnished upon

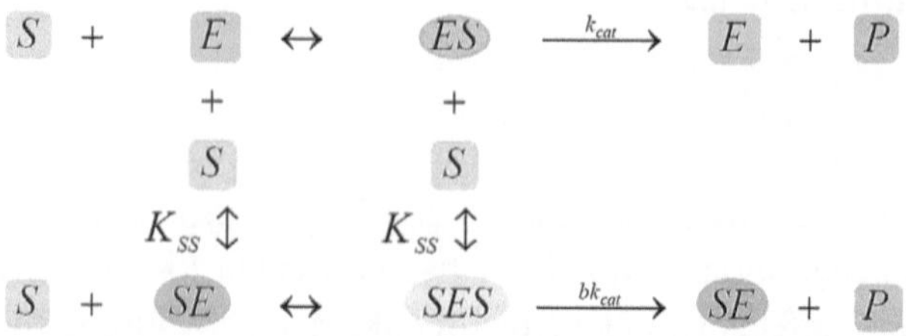

FIGURE 1.7 The Haldane- Radić enzyme kinetics flowing mechanism.

direct integration, Eq. (1.129) has no analytical solution, unless approximate serial decomposition method is used (Sonad & Goudar, 2004), with a reliability strongly depending by time-intervals considered, while the kinetic parameters are determined based on an initial estimate followed by recursive improvements.

In this context, the present Section explores the temporal solution for the substrate traffic in the Haldane-Radić enzyme kinetics presented in Figure 1.7 as it will be formulated either by closing W-Lambert analogously form of Eq. (1.26) o even as analytical progress curves for identifying the cases its reduction to the Michaelis-Menten enzyme kinetics of Eq. (1.4) may be validated (Putz, 2011).

1.8.1 HALDANE-RADIĆ EQUATION

Here we explore the working kinetic equation for the enzyme model as of Figure 1.7. One starts with considering the specific kinetic parameters such as the maximum velocity:

$$V_{\max} = k_{cat}[E_0] \tag{1.130}$$

and those of equilibrium constants:

$$K_S = \frac{[E]\cdot[S]}{[ES]} = \frac{[SE]\cdot[S]}{[SES]}, K_{SS} = \frac{[SE]\cdot[S]}{[SES]} = \frac{[E]\cdot[S]}{[SE]} \tag{1.131}$$

Next, by employing the global velocity expression:

$$v = k_{cat}[ES] + bk_{cat}[SES] = k_{cat}[ES] + bk_{cat}\frac{[S]\cdot[ES]}{K_{SS}} = k_{cat}[ES]\left(1 + \frac{b[S]}{K_{SS}}\right) \tag{1.132}$$

there appears the need for ES concentration knowledge; it can be nevertheless determined through the enzymic conservation equation:

$$[E_0] = [E] + [ES] + [SE] + [SES]$$

$$= \frac{K_S[ES]}{[S]} + [ES] + \frac{K_S[SES]}{[S]} + [SES]$$

$$= ([ES]+[SES])\left(1+\frac{K_S}{[S]}\right)$$

$$= \left([ES]+\frac{[S]\cdot[ES]}{K_{SS}}\right)\left(1+\frac{K_S}{[S]}\right)$$

$$= [ES]\left(1+\frac{[S]}{K_{SS}}\right)\left(1+\frac{K_S}{[S]}\right) \tag{1.133}$$

with the form:

$$[ES] = \frac{[E_0]}{\left(1+\frac{K_S}{[S]}\right)\left(1+\frac{[S]}{K_{SS}}\right)} \tag{1.134}$$

Finally by substituting Eq. (1.134) into Eq. (1.132) one gets the Haldane-Radić equation for substrate excess inhibition ($b < 1$) and activation ($b > 1$) (Putz, 2011):

$$v = k_{cat}[ES]\left(1+\frac{b[S]}{K_{SS}}\right) = k_{cat}\frac{[E_0]\left(1+\frac{b[S]}{K_{SS}}\right)}{\left(1+\frac{K_S}{[S]}\right)\left(1+\frac{[S]}{K_{SS}}\right)} = \frac{V_{\max}[S]}{K_S+[S]}\frac{(K_{SS}+b[S])}{(K_{SS}+[S])} \tag{1.135}$$

as a natural generalization for the Haldane equation (1.129). There is immediate the equivalence of the Haldane-Radić equation (1.135) with the Michaelis-Menten counterpart of Eq. (1.4) when the equivalent conditions of Eq. (1.128) apply. Once learned how Haldane-Radić enzyme equation and mechanism reduces to that of Michaelis-Menten, one further likes to have the solution of the Eq. (1.135) for its substrate temporal evolution. To this aim the preliminary benchmark Michaelis-Menten progress curve analysis will be next exposed within the introduced probabilistic method in enzyme kinetics, see Sections 1.3 and 1.4, and Refs. (Putz et al., 2006a, 2007, 2008; Putz & Putz, 2011).

1.8.2 PROBABILISTIC FORM OF THE HALDANE-RADIĆ EQUATION

The probability form (1.29) may be immediately inferred by considering the reactive term as in Eq. (1.34) and rewriting the Haldane-Radić equation (1.135) as:

$$\frac{v(t)}{V_{max}} = 1 - \frac{K_S(K_{SS}+[S]) + [S]^2(1-b)}{(K_S+[S])(K_{SS}+[S])} \tag{1.136}$$

From Eq. (1.136) the unreactive term may be recognized as being composed by two parts: the Michaelis-Menten contribution (1.34) superimposed on the specific Haldane-Radić term, namely (Putz, 2011):

$$\wp_{UNREACT}([S])^{HR} = \frac{K_S}{[S]+K_S} + \frac{[S]^2(1-b)}{(K_S+[S])(K_{SS}+[S])} \tag{1.137}$$

Interestingly, when performing the limits prescribed by general conditions (1.31) one gets:

$$\wp_{UNREACT}([S])^{HR} = \begin{cases} 1 & , \ [S] \to 0 \\ 1-b & , \ [S] >> 0 \end{cases} \tag{1.138}$$

while noting the persistent non-zero non-reactive behavior for higher substrate concentration – a feature that accounts for the inhibition character calling the Haldane specificity. However, the general conditions (1.31) are fully recovered by sending the b parameter to 1, which corresponds from Eq. (1.137) with resembling the Michaelis-Menten unreactive term (1.34) (Putz, 2011):

$$\wp_{UNREACT}([S])^{HR} \xrightarrow{b=1} \wp_{UNREACT}([S])^{MM} \tag{1.139}$$

Up to now, the presented probabilistic analysis shows it is qualitatively compatible with the general Haldane-Radić to Michaelis-Menten conditions (1.128); the quantitative issue will be in the sequel addressed.

1.8.3 W-LAMBERT SOLUTION OF HALDANE-RADIĆ EQUATION

The starting kinetic equation under the form (1.135) is firstly rewritten as:

$$-\frac{d[S]}{dt}=\left(\frac{1+\alpha[S]}{1+\beta[S]}\right)\left(\frac{V_{max}[S]}{[S]+\gamma}\right) \quad (1.140)$$

with the introduced notations:

$$\alpha=\frac{b}{K_{SS}},\beta=\frac{1}{K_{SS}},\gamma=K_{S} \quad (1.141)$$

It may be further rearranged as the ordinary differential equation (Putz, 2011):

$$-V_{max}dt=\frac{(1+\beta[S])([S]+\gamma)}{(1+\alpha[S])[S]}d[S]=\frac{\beta[S]^2+(1+\gamma\beta)[S]+\gamma}{\alpha[S]^2+[S]}d[S] \quad (1.142)$$

Next, Eq. (1.142) may be transformed through performing the polynomial ratio with the result:

$$-V_{max}dt=\frac{\beta}{\alpha}d[S]+\frac{\left(1+\gamma\beta-\frac{\beta}{\alpha}\right)[S]+\gamma}{(1+\alpha[S])[S]}d[S] \quad (1.143)$$

which is formally ready for integration.

However, another notation will simplify the analytical discourse, namely:

$$\rho=1+\gamma\beta-\frac{\beta}{\alpha} \quad (1.144)$$

with the help of which the integration of Eq. (1.143), between the initial conditions $0,[S_0]$ an the current one $t,[S](t)$, it firstly yields:

$$-\int_0^t V_{max}dt=\frac{\beta}{\alpha}\{[S](t)-[S_0]\}+\int_{[S_0]}^{[S](t)}\frac{\rho[S]+\gamma}{(1+\alpha[S])[S]}d[S] \quad (1.145)$$

while, by means of the right hand last term decomposition:

$$\frac{\rho[S]+\gamma}{(1+\alpha[S])[S]}=\frac{\gamma}{[S]}-\frac{\alpha\gamma-\rho}{1+\alpha[S]} \tag{1.146}$$

it leaves with the result:

$$-V_{\max}t=\frac{\beta}{\alpha}\{[S](t)-[S_0]\}+\gamma\{\ln[S]-\ln[S_0]\}$$
$$-\frac{\alpha\gamma-\rho}{\alpha}\{\ln(1+\alpha[S])-\ln(1+\alpha[S_0])\} \tag{1.147}$$

Now, Eq. (1.147) may be seen under the form:

$$a_1[S]+a_2\ln[S]+a_3\ln(1+a_4[S])=a_5 \tag{1.148}$$

that may be simplified to the compact expression:

$$\exp(a_1[S])[S]^{a_2}(1+a_4[S])^{a_3}=\exp a_5 \tag{1.149}$$

with the shortcuts:

$$a_1=\frac{\beta}{\alpha};\ a_2=\gamma;\ a_3=\frac{\rho-\alpha\gamma}{\alpha};\ a_4=\alpha;$$
$$a_5=-V_{\max}t+a_1[S_0]+a_2\ln[S_0]+a_3\ln(1+a_4[S_0]) \tag{1.150}$$

Worth noting that Eq. (1.149) looks as a modified form of the Euler-Lambert equation (1.25). Yet, the polynomial appearance in (1.149), viz.:

$$P([S])=(1+a_4[S])^{a_3} \tag{1.151}$$

produces a modification in the Euler-Lambert equation (1.25) that now becomes:

$$P(W(x))\exp(W(x))=x \tag{1.152}$$

with essentially non-algebraic analytic solutions so far. Instead, if we consider a generalization form of the polynomial (1.151) such that:

$$P([S])=\exp(a_3a_4[S]) \tag{1.153}$$

one recovers the W-type equation:

$$[S]^{a_2} \exp((a_1 + a_3 a_4)[S]) = \exp a_5 \tag{1.154}$$

that may be solved to give (Putz, 2011):

$$[S] = \frac{a_2}{a_1 + a_3 a_4} W\left(\frac{a_1 + a_3 a_4}{a_2} \exp\left(\frac{a_5}{a_2}\right)\right) \tag{1.155}$$

Note that the equivalence between Eqs. (1.151) and (1.153) relays on the limiting case (Putz, 2011):

$$a_4[S] \cong 0 \Leftrightarrow \alpha[S] = \frac{b[S]}{K_{SS}} \cong 0 \tag{1.156}$$

being such condition adding also the low substrate environment to those characterizing the Haldane-Radić to Michaelis-Menten reduction scheme, see Eq. (1.128). However, the question whether condition (1.156) suggests the low substrate constraint as sufficient or alternatively to the classical conditions (1.128) for that the Haldane-Radić to Michaelis-Menten reduction may be achieved is to be further explored.

Going back to the notations in Eq. (1.150) the W-Lambert form of the substrate depletion (1.155) looks like (Putz, 2011):

$$[S]_W^{HR}(t) = \frac{\gamma}{1 + \beta\gamma - \alpha\gamma} W\left(\left(\frac{1}{\gamma} + \beta - \alpha\right)[S_0](1 + \alpha[S_0])^{\frac{\rho - \alpha\gamma}{\alpha\gamma}} e^{\frac{\beta[S_0]}{\alpha\gamma}} e^{\frac{V_{max} t}{\gamma}}\right) \tag{1.157}$$

or in original kinetic parameters, throughout the shortcuts (1.141) and (1.144), casts as (Putz, 2011):

$$[S]_W^{HR}(t) = \frac{1}{\frac{1}{K_S} + \frac{1-b}{K_{SS}}} W\left(\left(\frac{1}{K_S} + \frac{1-b}{K_{SS}}\right)[S_0]\left(1 + b\frac{[S_0]}{K_{SS}}\right)^{\frac{(1-b)(bK_S - K_{SS})}{b^2 K_S}} e^{\frac{[S_0] - bV_{max} t}{bK_S}}\right) \tag{1.158}$$

Further use of this progress curve is in next discussed.

1.8.4 LOGISTIC SOLUTION OF HALDANE-RADIĆ EQUATION

Despite the formal solution (1.158) was achieved, it still suffers from a lack in analytical shape since the *W*-Lambert poses an implicit functional character. In order to improve such implicit solution one can consider in Eq. (1.157) the same binomial to exponential transformation for initial substrate $[S_0]$ as previously performed for the instantaneous free substrate $[S](t)$, see Eqs. (1.151) and (1.153), viz.:

$$(1+\alpha[S_0])^{\frac{\rho-\alpha\gamma}{\alpha\gamma}} \to e^{[S_0]\frac{\rho-\alpha\gamma}{\gamma}} \tag{1.159}$$

Under these conditions, Eq. (1.157) becomes (Putz, 2011):

$$[S]_W^{HR}(t) = \frac{\gamma}{1+\beta\gamma-\alpha\gamma} W\left([S_0]\left(\frac{1}{\gamma}+\beta-\alpha\right)e^{[S_0]\left(\frac{1}{\gamma}+\beta-\alpha\right)} e^{-\frac{V_{max}t}{\gamma}}\right) \tag{1.160}$$

Equation (1.160) may be turned into an analytical expression once the *logistic transformation* is employed according with the general recipe of Eq. (1.57) to provide the logistic substrate temporal form:

$$[S]_L^{HR}(t) = \frac{\gamma}{1+\beta\gamma-\alpha\gamma} \ln\left(1+\left(e^{[S_0]\left(\frac{1}{\gamma}+\beta-\alpha\right)}-1\right)e^{-\frac{V_{max}t}{\gamma}}\right) \tag{1.161}$$

or, with replacement of the kinetic parameters from Eqs. (1.141) and (1.144), the progress curve (Putz, 2011):

$$[S]_L^{HR}(t) = \frac{K_{SS}}{1-b+K_{SS}/K_S} \ln\left(1+\left(e^{\left(\frac{1}{K_S}+\frac{1-b}{K_{SS}}\right)[S_0]}-1\right)e^{-\frac{V_{max}}{K_S}t}\right) \tag{1.162}$$

Worth noting that the logistic solution (1.162) finely tunes the extreme conditions for the substrate kinetics, namely:

$$[S]_L^{HR}(t) = \begin{cases} [S_0], & t \to 0 \\ 0, & t \to \infty \end{cases} \tag{1.163}$$

a matter not easy to verify using the *W*-Lambert expressions (1.26) or (1.160).

From now on, with the help of expression (1.162) the temporal course of the kinetics (1.135) may be formulated in an analytical manner by employing the required temporal derivative of the substrate (Putz, 2011):

$$\frac{d[S]_L^{HR}}{dt} = -\frac{V_{max}}{K_S} \cdot \frac{\left(e^{\lambda \cdot S_0} - 1\right)e^{-\frac{V_{max}}{K_S}t}}{\lambda\left(1 + \left(e^{\lambda \cdot S_0} - 1\right)e^{-\frac{V_{max}}{K_S}t}\right)} \tag{1.164}$$

with the working parameter:

$$\lambda = \frac{1}{K_S} + \frac{(1-b)}{K_{SS}} \tag{1.165}$$

Thus, the initial velocity for product formation, i.e., at $t = 0$, is (Putz, 2011):

$$\frac{d[P]_L^{HR}}{dt} = -\frac{d[S]_L^{HR}}{dt} = v_L^{HR} = \frac{V_{max}}{K_S}\frac{\left(e^{\lambda \cdot S_0} - 1\right)}{\lambda \cdot e^{\lambda \cdot S_0}} = \frac{V_{max}}{K_S}\frac{\left(1 - e^{-\lambda \cdot S_0}\right)}{\lambda} \tag{1.166}$$

This curve may be used to perform the non-temporal fit for kinetic parameters, with the caution however that it gives best results in the low initial substrate range, according with the equivalence (1.156). For instance, for human AChE, using acetylthiocholine as the substrate, one gets the full bell-shaped curve fit of Figure 1.8(a) by using Haldane-Radić equation (1.135) for the working parameters $K_S = 190 \pm 30$ μM; $K_{SS} = 8{,}700 \pm 2{,}200$ μM; $V_{max} = 2.45 \pm 0.15$ ΔOD/min; $b = 0.12 \pm 0.03$; instead, with the same parameters in logistic related derived velocity of Eq. (1.166) the departure is recorded for initial substrate concentrations higher than 100 μM, see Figure 1.8(b), targeting the Michaelis-Menten kinetics of Figure 1.8(c) (Putz, 2011).

This is not surprisingly, since the actual Haldane-Radić *W*-Lambert and logistic progress curves were obtained through modification of the analytic conditions of the Figure 1.7 such that being "reduced" or "absorbed" to the Michaelis-Menten scheme (1.4) for the lower concentration of the substrate. To check the consistency of this hypothesis also for the progress curves the Figures 1.9–1.12 display the fitting of the above *W*-Lambert and logistic equations (1.160) and (1.162) for various experimental enzymic

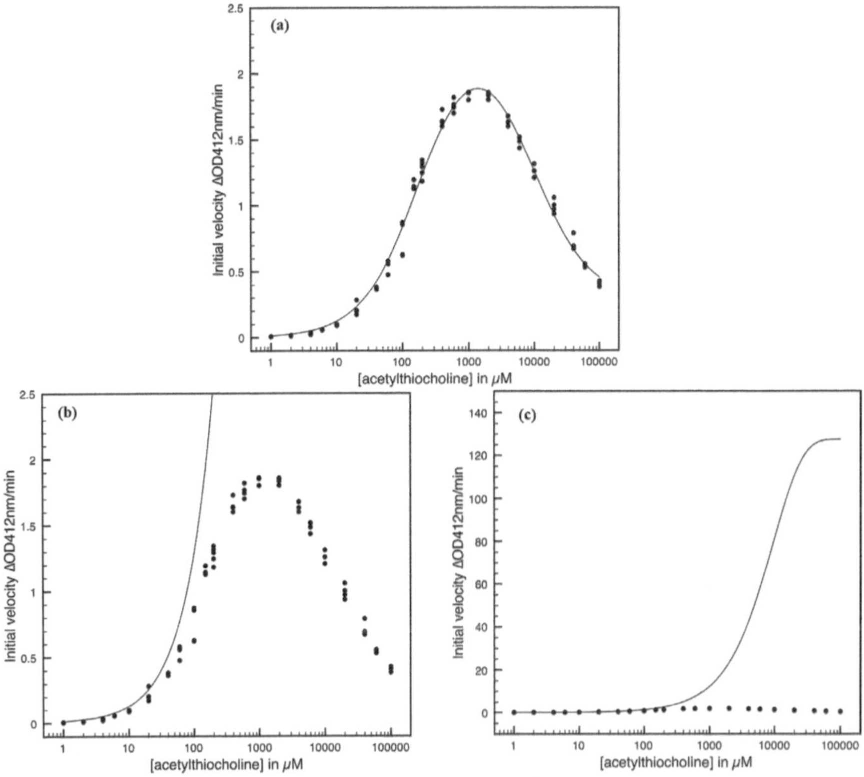

FIGURE 1.8 (a) The fitting curves for original Haldane-Radić velocity equation (1.135) corresponding with the Figure 1.7 for large human acetylthiocholine substrates' concentration intervals; (b) and (c) the same fits with logistic based velocity equation (1.166) corresponding to the "reduction" of the Figure 1.7 to the consecrated Michaelis-Menten mechanism of Eq. (1.4); the kinetic fitting parameters are $K_S = 190 \pm 30$ μM; $K_{SS} = 8{,}700 \pm 2{,}200$ μM; $V_{max} = 2.45 \pm 0.15$ ΔOD (optical density)/min; $b = 0.12 \pm 0.03$ (Putz, 2011).

kinetics with the fitting parameters of Eqs. (1.141) and (1.144) determined for the lower substrate concentration and then tested for higher and higher values of it (Putz, 2011).

The analysis of the plots of Figures 1.9–1.12 illustrates the interesting recorded behavior (Putz, 2011):

- The hAChE-ATC kinetics (Figure 1.9) differs from hBChE-ATC kinetics (Figure 1.10) essentially only in the lowering the V_{max} and increasing b parameters for the last case, in accordance with the prescription associated with activation mechanism; moreover, the *W*-Lambert and logistic curves depart clock-wise from experimental record and more quickly for logistic case;

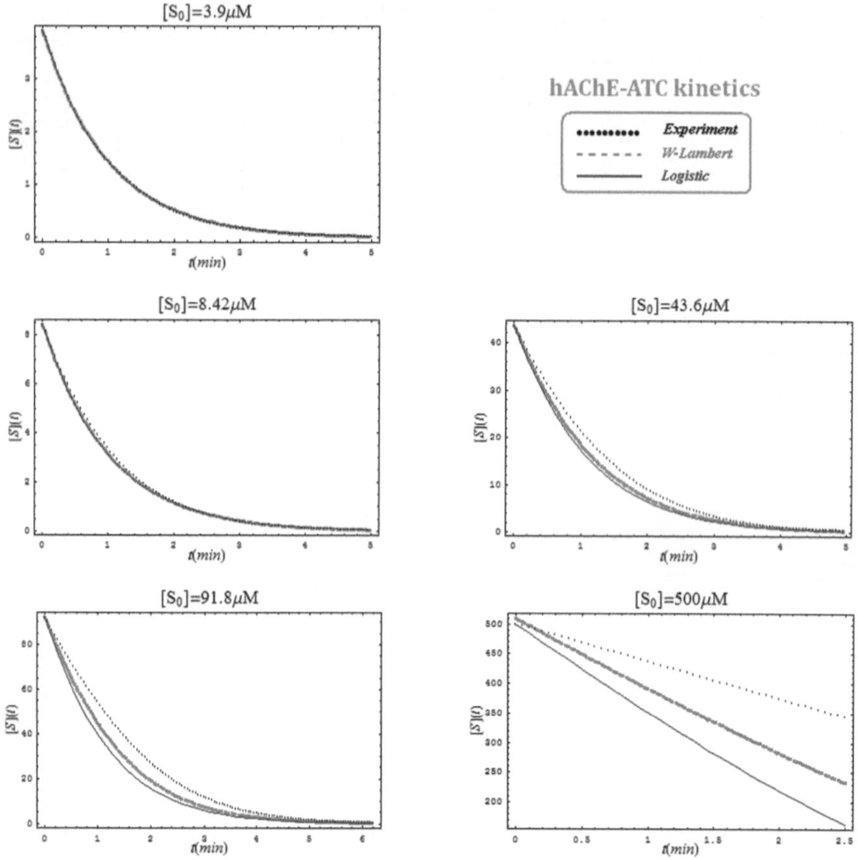

FIGURE 1.9 The W-Lambert and logistic progress curves as they fit with experimental data for the hAChE-ATC kinetics, according to the Eqs. (1.160) and (1.162), through considering the kinetic parameters from Eqs. (1.141) and (1.144) with the actual values K_S= 160 μM; K_{SS} = 8,700 μM; V_{max} = 162.45 μM/min; b = 0.12, for various initial substrate concentrations (Putz, 2011).

- The hBChE-ATC kinetics (Figure 1.10) differs from hBChE-BTC kinetics (Figure 1.11) essentially by further lowering the V_{max} accompanied by decrease of K_S parameter for the BTC kinetics, while the W-Lambert and logistic computationally fitting curves show in Figure 1.11 a departure tendency in anti-clock-wise respecting the experimental evidence; here is also recorded the clear failure of the numerical W-Lambert progress curve to reach the initial substrate

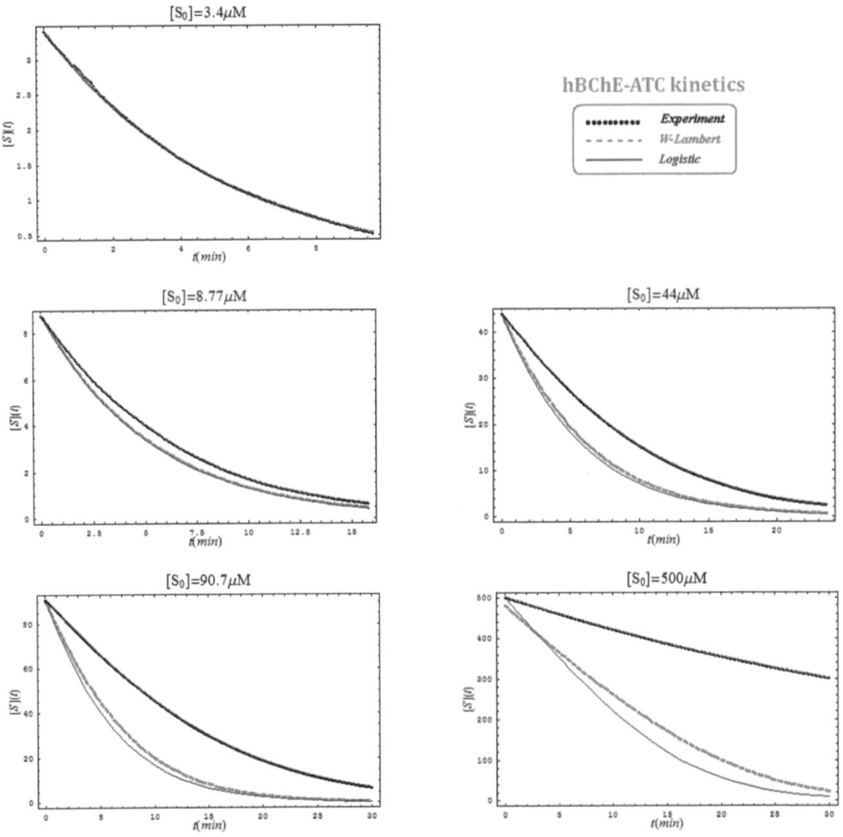

FIGURE 1.10 The same type of plots as in Figure 1.9, here for hBChE-ATC kinetics and parameters K_S = 160 μM; K_{SS} = 8,700 μM; V_{max} = 31.0 μM/min; b = 3 (Putz, 2011).

concentration, a matter fully satisfied by the logistic counterpart instead;

- Comparison between hBChE-BTC kinetics (Figure 1.11) and BSCh-BTC hydrolysis (Figure 1.12) reveals that by maintaining the same kinetic parameters between these two cases, in the latter, the computational fitting with respect the experimental data oscillate from clock-wise to anti-clock-wise departure of the logistic model as the initial substrate concentration goes from lower (<100 μM) to higher (>100 μM) values, respectively; here, again, the initial time discrepancy between *W*-Lambert and logistic kinetics is obviously in the favor of the latter approach.

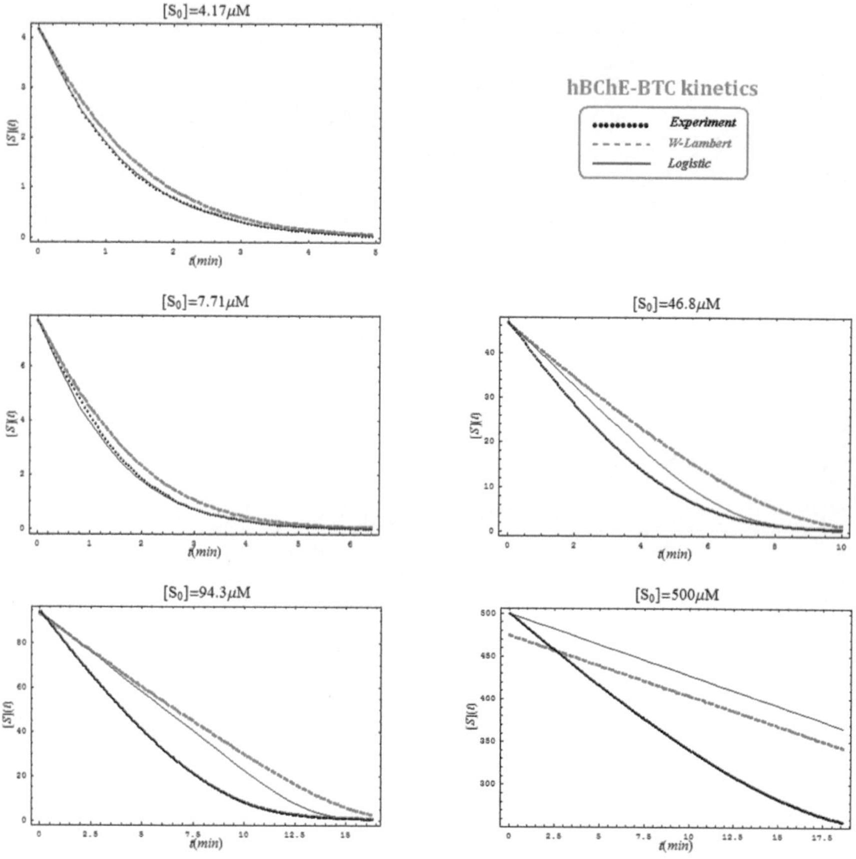

FIGURE 1.11 The same type of plots as in Figure 1.9, here for hBChE-BTC kinetics and parameters K_S = 7.5 μM; K_{SS} = 8,700 μM; V_{max} = 7.2 μM/min; b = 3 (Putz, 2011).

However, the present analysis allows the general rules to be formulated as (Putz, 2011):

- The Haldane-Radić kinetics may be quite well modeled by its Michaelis-Menten counterpart progress curves for substrate kinetics below 100 μM in all studied cases, being this condition susceptible to be a general fact that is independent of ideal approach for the Haldane-Radić kinetic parameters K_{SS} and b as prescribed in Eq. (1.128);
- Haldane-Radić kinetics display full specificity in looping S-E mechanisms of inhibition/activation for higher concentration of the substrate, i.e., within the milli-molar range;
- The *W*-Lambert logistic fails to behave correctly at initial time of kinetics in the case of higher initial substrate concentrations

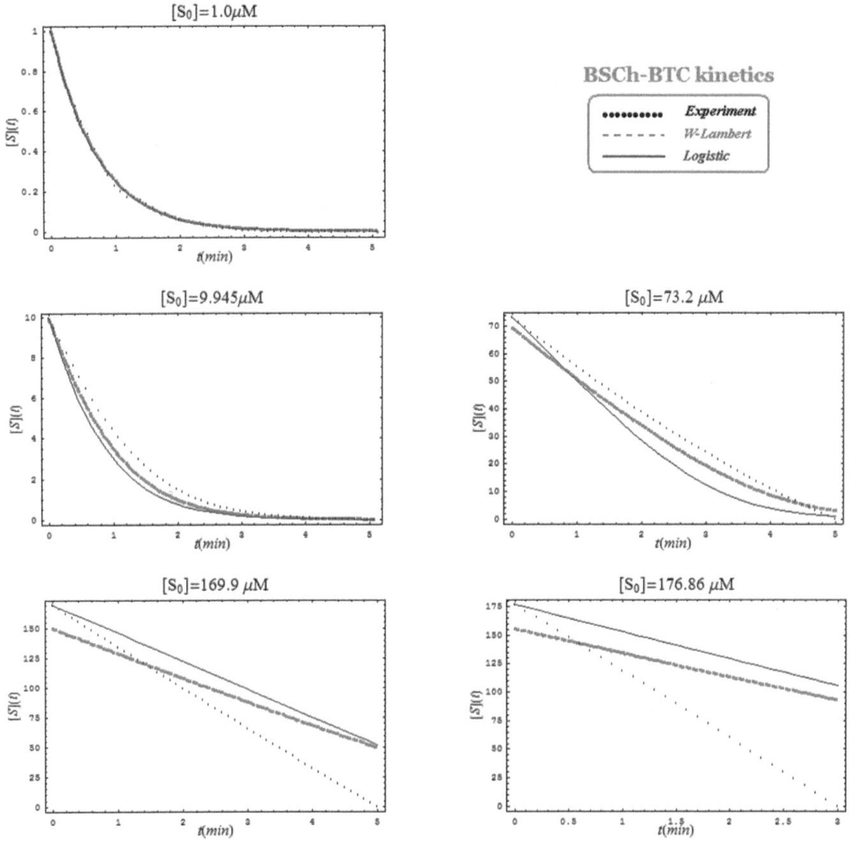

FIGURE 1.12 The *W*-Lambert and logistic progress curves as their fit with the experimental data for hydrolysis of various concentrations of butyrylthiocholine by a fixed concentration of butyrylcholinesterase according to the Eqs. (1.160) and (1.162), through considering the kinetic parameters from Eqs. (1.141) and (1.144) with the actual values K_S= 7.5 μM; K_{SS} = 8,700 μM; V_{max} = 7.2 μM/min; b = 3, for various initial substrate concentrations (Putz, 2011).

(see Figures 1.9–1.12); however, the logistic counterparts always correct this flaw due the analytical limit (1.163).

1.9 QUANTUM ENZYME KINETICS

Although in the first century from their discover the enzymes were mainly studied for elucidation of their kinetics (Schnell & Maini, 2003), emphasizing on how their structure is changed with the chemical modifications of functional groups (Hirs, 1967)), or for experiencing the "forced evolution" (Rigby et al., 1974), in current years the focus was on controlling them towards biotechnological roles through the knowledge based methods such as the site-directed mutagenesis (Graham et al., 1994; Tyagi et al., 2005) or gene-shuffling techniques (Stemmer, 1994). However, aiming to create a better enzyme, with improved specificity near the "catalytic perfection" (Albery & Knowles, 1976), raises the intrinsic difficulty to rationalize a general model for its activity since the relatively poor level of comprehension about the enzyme machinery (Nixon et al., 1998). As such, deviations from classical behavior were reported for enzymes yeast alcohol dehydrogenase (Cha et al., 1989), bovine serum amine oxidase (Grant & Klinman, 1989), monoamine oxidase (Jonsson et al., 1994), glucose oxidation (Kohen et al., 1997), or for enzyme lipoxygenase (Jonsson et al., 1996), in which it was shown that H-transfer is catalyzed by quantum tunneling process. These, and other experimental (Bahnson & Klinman, 1995) and computational (Bala et al., 1996; Hwang & Warshel, 1996; Alper et al., 2001; Astumian et al., 1989; Ross et al., 2003) indications of conformational fluctuations during protein dynamics, suggested the attractive hypothesis that quantum tunneling and the enzyme catalysis are inter-correlated (Ringe & Petsko, 1999; Sutcliffe & Scrutton, 2000).

The solvent dynamics, i.e., the in vitro and in vivo conditions, and "natural breathing", i.e., the quantum fluctuations in the active site, of the enzyme molecule need to be counted in a more complete picture of enzymic catalysis. However, the quantum (fluctuating) nature of the enzymic reactions can be visualized by combining the relationship between the catalytic rate (k_{cat}) and temperature (T) (DeVault & Chance, 1966) with that between the reaction rate and the turnover number or the effective time of reaction (Δt) via Heisenberg relation

$$\frac{1}{k_{cat}} \propto \Delta t \cong \frac{\hbar}{\Delta E_{tunnelling}} = \frac{\hbar}{k_B T} \tag{1.167}$$

were $\hbar$ and k_B stand for the reduced Planck and Boltzmann constants, respectively. Of course, in relation (1.167) the equivalence between quantum statistics and quantum mechanics was physically assumed when equating the thermal and quantum (tunneling) energies, $k_B T$ and ΔE, respectively, see the Volume II of the present five-volume book (Putz, 2016).

Nevertheless, relation (1.167) is the basis of rethinking upon the static character of the energetic barrier, recalling the so-called steady state approximation, usually assumed in describing enzymic catalysis (Laidler, 1955), within the transition state theory (TST) (Glasstone et al., 1941).

Basically, when applied to enzymic reactions the recent developments suggest that the "textbook" TST is, at least in some situations, necessarily flawed. This because TST primarily treat the enzymes as being only particle-like entities, completely ignoring their electronic and protonic constitution when mediate chemical information-transfer when act on substrate. On contrary, as electrical insulators the proteins can transfer their electrons only by means of wave-like properties or tunneling processes.

However, while electron transfer occurs at large distances, up to ca. 25Å, the same tunneling probability may be achieved by the protium (C–H group) at the distance of 0.58Å, the specific range for enzyme-substrate binding site. Such picture is sustained also by the electrostatic complementary of the catalytic site hypothesis, first suggested by Pauling (Pauling, 1946), and then refined by Marcus theory of electron transfer in chemical reactions (Marcus, 1993), stating that the dynamic fluctuations of the environment develop the driving force for that chemical reactions proceed.

Actually, the wave-particle duality of matter allows designing new pathway from reactants (enzyme E and substrate S) to products (enzyme and product P) in a Brownian enzymic reaction (Brown, 1902)

$$E + S \leftrightarrow ES \xrightarrow{delay} EP \rightarrow E + P \tag{1.168}$$

by means of passing through the barrier between the ground states of enzyme-substrate (ES) and enzyme-product (EP) complexes, employing the wave-like manifestation, instead of passing over it, as the TST predict for the particle-like manifestation of enzymes, see Figure 1.13. In this context, the thermal activation is realized on the basis of vibrational enhancement, at its turn sustained by quantum fluctuations of the enzyme-binding substrate active site, followed by the tunneling effect. As a result, an increase in the catalytic efficiency is produced.

This quantum analysis was systematized within the vibrationally enhanced ground-state tunneling theory (VEGST) (Bruno & Bialek, 1992), experimentally verified on reactions catalyzed by the bacterial enzyme methylamine dehydrogenase (Basran et al., 1999), despite some limitations for those enzyme in which the tunneling process acts so close

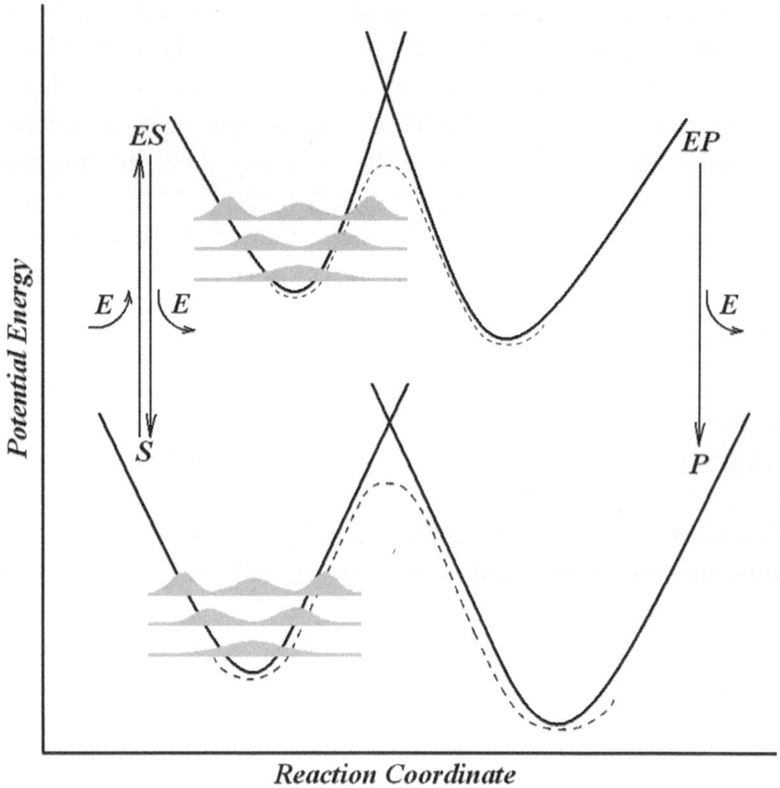

FIGURE 1.13 Reaction coordinates representation for a Brownian enzyme-catalyzed reaction (1.168) while molecular vibrations of the reactive substrate and enzyme-substrate complex are superimposed on the energy profile for the reaction. Tunneling of the vibrational states of active site of substrate becomes most favorable within the enzyme-substrate complex (ES) leading with the penetration of the enzyme-product complex (EP) curve below the classical transition state at the saddle-point of the potential energy profile (Putz & Putz, 2011).

to the saddle point of energy surface so that reactants passes over barrier before to penetrate it (Truhlar & Gordon, 1990).

Within this framework, the quest of analytical recovering of the tunneling effect is in the current work envisaged through generalizing the classical enzyme kinetics. The present strategy relies on employing the progress curves of the substrate to provide the fluctuation time, and consequently the needed tunneling energy, that assures the intrinsic motor of the enzyme catalyzed reactions. The degree with which the present venture produces a viable conceptual and computational basis to address the quantum control of the enzyme activity is discussed in various catalytic circumstances.

1.9.1 QUANTUM ENZYME CATALYSIS

Searching for suitable markers to screen the new biomaterials for adverse effects in cell and tissues (e.g., overt toxicity causing cell death around an implant, chronic inflammatory reactions due to the neoplastic changes, or undergoing degradation of the material in vivo) the enzymes' activity has been found to be the most reliable because their catalytic properties are sensitively affected by exposure to a biomaterial. Accordingly, the biophysical chemistry analysis of cell fractions from tissue culture experiments by the enzyme-linked immunoassays (ELISAs) or spectrophotometry techniques provides exact quantization of the released enzymes as an accurate measure of cell membrane integrity and cell viability (Williams, 1986; Allen & Rushton, 1994).

In this respect, we can make use of the unique feature of absorption spectroscopy as an experimental controlling tool since the substrate time dependent concentration, which entirely follow the course of an enzymatic reaction, can be related with the absorbance of the substrate molecules (at a particular wavelength) through the adapted Beer-Lambert law (Cantor & Schimmel, 1980)

$$A_S(t) = a_M l[S](t) \tag{1.169}$$

where l is the path length of the sample traversed by the light beam while a_M stands for the molar absorptivity as an intrinsic constant of the substrate molecule. However, considering the free substrate absorption

$$A_0 = a_M l[S]_0 \tag{1.170}$$

we can deal, for convenience, with the normal absorptivity of the substrate defined

$$a_S(t) = \frac{A_S(t)}{A_0} = \frac{[S](t)}{[S]_0} \tag{1.171}$$

At this point we can employ the associate absorbance progress curves for substrate (or for the product) when the *W*-Lambert or logistic solutions (1.26) or (1.51) are implemented, respectively. The starting point consists in writing of the normal absorbance for the product in the same way as for the substrate (1.171)

$$a_P(t) = \frac{A_P(t)}{A_0} = \frac{[P](t)}{[S]_0} \tag{1.172}$$

Then, the link with the substrate concentration is made through combining the free substrate and enzyme conservation equations

$$[S]_0 = [S](t) + [ES](t) + [P](t) \tag{1.173}$$

$$[E]_0 = [E](t) + [ES](t) \tag{1.174}$$

with the Michaelis-Menten constant viewed as the dynamic dissociation constant

$$K_M = \frac{[E](t)[S](t)}{[ES](t)} \tag{1.175}$$

of the first part of the reaction (1.4) (Cantor & Schimmel, 1980; Voet & Voet, 1995).

With appropriate substitutions of relations (1.173)–(1.175) in Eq. (1.172) the normal absorption curve for product takes the expression

$$a_P(t) = 1 - a_S(t) - \frac{\varepsilon[S](t)}{[S](t) + K_M}, \; 10^{-7} < \varepsilon \equiv \frac{[E]_0}{[S]_0} < 10^{-2} \tag{1.176}$$

While the parameter ε fixes the in vivo-to-in vitro regimes as it decreases to zero the product's normal absorbance can be approximated as

$$a_P(t) \cong 1 - a_S(t) \tag{1.177}$$

this way providing the quasi-equivalence of the absorption differences

$$\Delta a(t) = a_S^W(t) - a_S^{Log}(t) \cong a_P^{Log}(t) - a_P^W(t) \tag{1.178}$$

when the *W*-Lambert and logistic expressions in substrate calculation is employed.

Next, for making an idea on how the enzyme-substrate and enzyme-product complexes information are encoded within the *W*-Lambert and logistic time dependent substrate solutions, and displayed in Figure 1.14 as their associated absorbance curves together with the absorption differences of (1.178) for a trial enzymatic reaction are simulated. At first sight, as expected, there are no significant differences between the *W*-Lambert and logistical plots of normal absorption of substrate depletion and product formation in various enzymic environment, from the in vitro, when $\varepsilon \in (10^{-6}, 10^{-4})$, to the in vivo conditions, when $\varepsilon \geq 10^{-2}$. As a note, there is observed that as the in vivo conditions are approached, substrate catalysis begins earlier, with an increase in the actual time of the enzymatic reaction by which the substrate

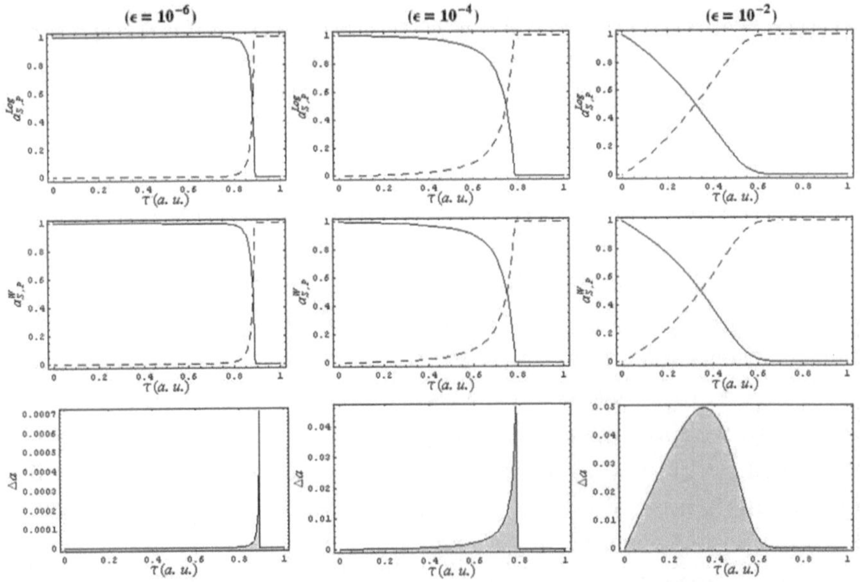

FIGURE 1.14 The first two rows show the normal absorptions for substrate (continuous curves) and product (dashed curves) concentrations of the enzyme-catalyzed reaction without inhibition, based on the logistical temporal approximation, and on the *W*-Lambert temporal solution, respectively; the third row depicts the difference between *W*-Lambert and logistical counterpart for substrate or product normal absorption progress curves; on the columns, the plots are presented for the enzyme/substrate ratio ε taking the in vitro and almost the in vivo values, from 10^{-6} to 10^{-4} and equal or greater than 10^{-2}, respectively; the employed kinetic parameters are the maximum velocity of enzyme reaction $V_{max}=10^{-4}M\times s^{-1}$ and the Michaelis constant $K_M=2\times10^{-4}M$, while the total enzyme concentration is set at $[E]_0=10^{-6}M$; the time infinite range has been mapped onto the interval (0,1) with the aid of the exponential time scale $\tau = 1-1/\ln(t+e)$, being expressed in arbitrary units (a.u.) (Putz & Lacămă, 2007).

becomes the product. Nevertheless, when about the absorbance differences a clear signal over the time interval of the enzymatic reaction in the third row of Figure 1.14 is detected. This stands as the computational proof that considering the logistic kinetics and its substrate solution the difference respecting the classical Michaelis-Menten one is transposed at the spectroscopic level in small (however, important for tunneling) difference of substrate or product concentration that is engaged or emerged, respectively, throughout of the energetic barrier of the enzyme-transition state complex, see Figure 1.15(a), leading with consequent activation of whole reaction.

Worth noting that the product's difference in (1.178), namely $\Delta a(t) \cong a_P^{Log}(t) - a_P^{W}(t)$, gives the best visualization of how the logistic solution comprise the difference information over W-Lambert counterpart solution, that information that characterizes the EP complex against the ES in the course of transition-state stabilization through tunneling.

Putted different, given that we can define the effective time of reaction as the width at the half height of the recorded signal $\Delta a(t)$, the difference absorption (1.178) provides the predicted effective time of reaction Δt and the turnover number – the catalytic rate, as estimated from the first part of the chain relation (1.167). As well, the estimate of the tunneling energy ΔE^{00}, which makes the enzymic reaction proceeding according with the Brownian mechanism of Eq. (1.168) and Figure 1.13, also follows from the Heisenberg relation in Eq. (1.167).

This way, we faced with two new perspectives of enzymic kinetics.

Firstly, we showed that the quantum fluctuation and tunneling may be simulated when among the Michaelis-Menten also the logistic kinetics and its solution is considered, a feature confirmed by applying the Beer-Lambert law of absorption spectroscopy. Such picture is in accordance with the observed enhanced rate of the vibrationally states of ES by means of quantum tunneling when considered within the Brownian mechanism (1.168) and Figure 1.13.

Secondly, we have the opportunity to design new energy diagrammatic representation of the enzyme catalysis at the quantum level. For that worth noting that, as a catalyst, enzyme molecule is found at the end of reaction unaffected respecting its initial free status. This characteristic permits assimilating enzyme with the "photon role" in quantum absorption and emission processes. Then, we can consider the molecular ground states of all involved species and representing them with

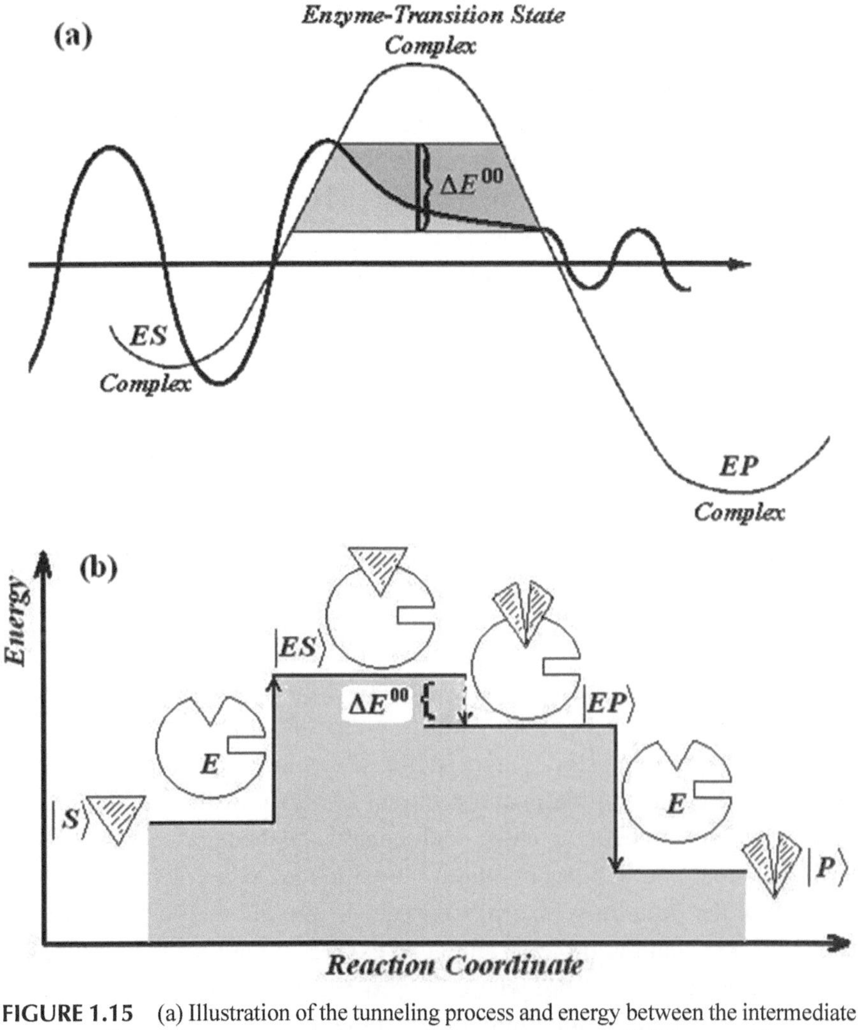

FIGURE 1.15 (a) Illustration of the tunneling process and energy between the intermediate complexes ES and EP across the enzyme-transition state complex. (b) Diagram of the energy levels of the Brownian enzyme-catalyzed reaction E+S↔ES→EP→E+P by means of the intermediate enzyme-complexes, $|ES\rangle$ and $|EP\rangle$, separated by the tunneling energy of reaction ΔE^{00}. All the reactants, enzyme (E), substrate (S), and product (P), and the intermediate molecular complexes (ES, EP) are depicted by their symbolic representations. The arrows in the diagrams indicate the final direction of the reactions they refer to (Putz & Putz, 2011).

the associate eigen-energies while the eigen-states will be appropriately symbolized, within Dirac notation, as $|S\rangle$, $|ES\rangle$, $|EP\rangle$, and $|P\rangle$ for the free substrate, enzyme-substrate complex, enzyme-product complex, and free product, respectively.

With these, and taking account of quantum tunneling process and energy, in Figure 1.15(b) the resulted quantum enzymic diagram of the Brownian mechanism (1.168) is depicted. Such diagrams are qualitatively different by the ordinary thermodynamic ones (Copeland, 2000) having the advantage of considering all TS information in tunneling energies, thus providing a clear overlook of the course of enzyme reaction and of associate mechanism.

From the experimental point of view, the present algorithm may prescribe the quantum control of the enzymatic activity through absorption, while the computational simulation may indicate the time range of the type spectroscopy to be used, from pressure and temperature jumps to the electron paramagnetic resonance and electric field jumps, depending on the output of the tunneling times fixing the fast-reaction ranges (Cantor & Schimmel, 1980).

The actual model can also be extended to the cases when different types and degrees of enzymatic inhibition are considered leading with important effects of biological regulations as will be in next exposed.

1.9.2 APPLICATION TO MIXED INHIBITION

The developed theory of allosteric regulation (from Greek: *allos*, other + *stereos*, space) prescribes that, within a cooperative interaction, the binding of one ligand (substrate) at a specific site is influenced by the binding of another ligand (inhibitor) at a different or allosteric site on the protein (or enzyme). However, actually, such behavior is generalized at the level of organismal and cellular regulation in which the cell converts the comparison of the proteins with organisms needs into metabolic process.

It follows that the proteins and gene expression are far from being the endpoint but a bridge from where the process of editing RNA transcripts begins, altering and maintaining the genome, over and over again by signaling other cells or bio-inspired nano-implants (Goodman et al., 2005; Hartwell, 2005). In this process of cell differentiation, proliferation, and programming, the receptors (substrates and inhibitors) and enzymes play the role of molecular messengers. Therefore, studying the effects of the inhibitors on the enzymatic reactions, here at the theoretical level, should be most valuable for the forefront of biomedical researches.

The Figure 1.16 displays the main types of allosteric interactions involving enzymes. Main point of the presented scheme is introduction into the reaction a further ligand called inhibitor, *I*, which interacts with

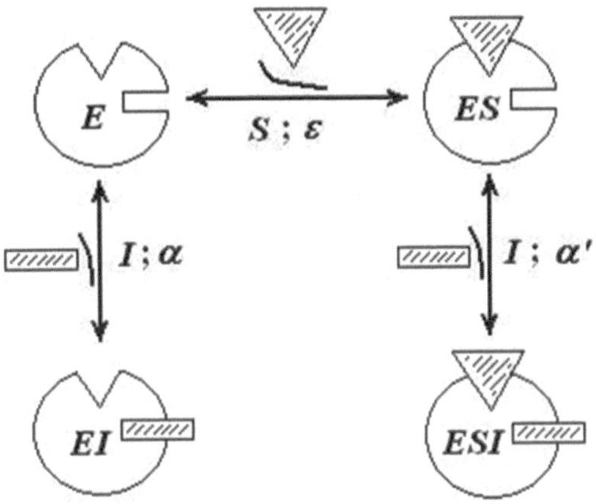

FIGURE 1.16 Symbolic representation of the competitive and uncompetitive inhibitions acting through the inhibitor (I) on the enzyme (E) or on enzyme-substrate complex (ES), being quantified by parameters α and α' respectively, adjacent to the baseline Michaelis-Menten reaction E+S↔ES→E+P (Putz & Lacrămă, 2007).

enzyme *E* leading to form *EI* while the *ES* complex also interacts with *I* giving the intermediate complex *ESI*. The discrimination on the degree with which the favorite formation is that of the complex *EI* rather than that of *ESI* one fixes the type of inhibition; the present discussion follows (Putz & Lacrămă, 2007).

The *competitive inhibition*, quantified by the parameter

$$\alpha = \frac{[E]+[EI]}{[E]} \tag{1.179}$$

appears when the inhibitor has a molecular structure analogue to that of the substrate and binds with the enzyme on this complementarily ground. Typical examples are the anti-metabolites, for example, the sulphanilamide as the competitive antagonist of the para-amino-benzoic acid (known as the H-vitamin), which display bacterio-static effects due to the blocking of microorganisms growth and multiplication. On the other way, when the inhibitor is acting upon the enzyme-substrate complex the *uncompetitive inhibition* type, quantified by the parameter

$$\alpha' = \frac{[ES]+[ESI]}{[ES]} \tag{1.180}$$

regulates the enzymatic catalysis. This situation is specific to the in vivo occasions when, for instance, the adenosine three phosphatase inhibitor blocks the enzymes of glycol so controlling the energetic release in cell. However, mixed inhibitions can also appear since both competitive and uncompetitive inhibitions take place in the course of complex biosynthesis. With these, the basic Michaelis-Menten equation (1.16) is reconsidered for the kinetic parameters of the mono-substrate reactions of Figure 1.16 and the kinetic parameters for the maximum velocity of reaction and the Michaelis-Menten constant, respectively become (Voet & Voet, 1995)

$$V_{\max}^{mixed} = \frac{1}{\alpha'} V_{\max}^{mono} \tag{1.181}$$

$$K_M^{mixed} = (\alpha / \alpha') / K_M^{mono} \tag{1.182}$$

In these conditions, the actual working forms for the *W*-Lambert and logistic substrate solutions can be immediately generalized from the expressions (1.26) and (1.51) to be:

$$[S](t)^W = \frac{\alpha}{\alpha'} K_M W\left(\frac{\alpha'[E]_0}{\alpha \varepsilon K_M} \exp\left(\frac{\alpha'[E]_0}{\alpha \varepsilon K_M} \right) \exp\left(-\frac{tV_{\max}}{\alpha K_M} \right) \right) \tag{1.183}$$

$$[S]^{Log}(t) = \frac{\alpha}{\alpha'} K_M \ln\left(1 + \left(\exp\left(\frac{\alpha'[E]_0}{\alpha \varepsilon K_M} \right) - 1 \right) \exp\left(-\frac{V_{\max}}{\alpha K_M} t \right) \right) \tag{1.184}$$

Actually, the quantum enzyme mechanisms of the mixed reactions from Figure 1.16 are driven by the tunneling times and energies prescribed by the difference between the absorptions induced by the solutions (1.183) and (1.184). This picture represents the natural generalization of the Brownian enzyme reaction (1.168) to include the allosteric effects.

Going to simulate the quantum fluctuations appeared in mixed enzyme catalysis of Figure 1.16 the absorption difference curves (1.178) are computed, in various in vitro-to-in vivo mixed inhibition combinations, with the results in Figure 1.17 displayed.

By comparing the plots of the Figure 1.17 with those corresponding to the no inhibition case (00) of Figure 1.14, the hierarchy of the tunneling times for competitive (α0), uncompetitive (0α'), and mixed (αα') Brownian reactions is worked out:

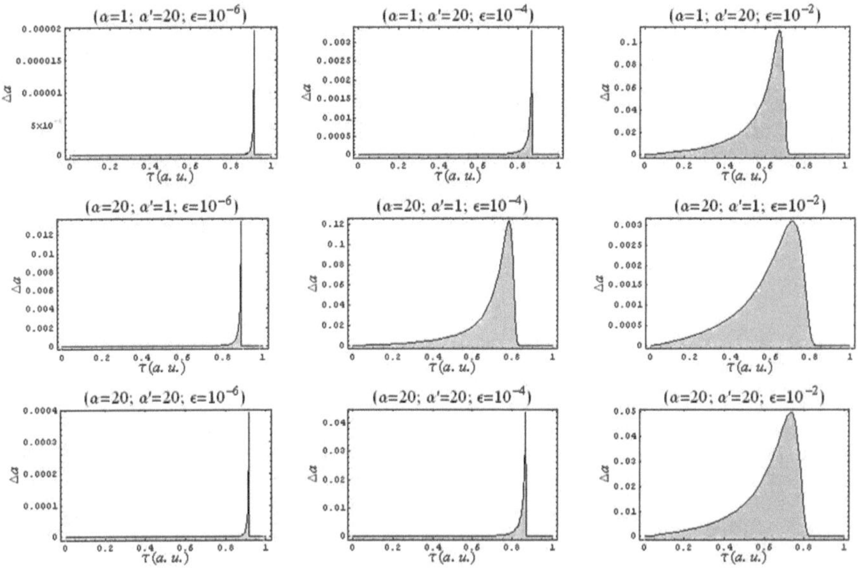

FIGURE 1.17 Differences between *W*-Lambert and logistical counterparts for substrate or product normal absorption progress curves; on the columns, the plots are presented for the enzyme/substrate ratio ε taking the in vitro and almost the in vivo values, from 10^{-6} to 10^{-4} and equal or greater than 10^{-2}, respectively; on rows different competitive and uncompetitive inhibition combinations quantified by the α and α' parameters, taking the values equal with 1 and 20, are respectively considered; the employed kinetic parameters and temporal scales are the same as used in Figure 1.3 (Putz & Lacrămă, 2007).

$$\Delta\tau^{0\alpha'} < \Delta\tau^{\alpha\alpha'} < \Delta\tau^{00} < \Delta\tau^{\alpha 0},\ \varepsilon \in \left(10^{-6}, 10^{-4}\right) \tag{1.185}$$

$$\Delta\tau^{\alpha 0} < \Delta\tau^{\alpha\alpha'} < \Delta\tau^{00} < \Delta\tau^{0\alpha'},\ \varepsilon \geq 10^{-2} \tag{1.186}$$

for the in vitro and in vivo environments, respectively, while for the tunneling energies for the associated enzyme-transition state complexes the pecking orders of (1.185) and (1.186) are reversed:

$$\Delta E^{0\alpha'} > \Delta E^{\alpha\alpha'} > \Delta E^{00} > \Delta E^{\alpha 0},\ \varepsilon \in \left(10^{-6}, 10^{-4}\right) \tag{1.187}$$

$$\Delta E^{\alpha 0} > \Delta E^{\alpha\alpha'} > \Delta E^{00} > \Delta E^{0\alpha'},\ \varepsilon \geq 10^{-2} \tag{1.188}$$

due to the spectroscopic Heisenberg relation in (1.167).

With these results, we may build the quantum diagrams for the considered types of enzymic reactions by employing the following conceptual-computational algorithm. Firstly, as already considered in treating the mono-substrate case without inhibition, two general rules are considered (Putz & Lacrămă, 2007; Putz & Putz, 2011):

(i) the enzyme assumes the "photon role", as before, since as a catalysis enters and outs unaffected, i.e., caring the same amount of energy in and out of the reaction;
(ii) all other involved molecular complexes, including intermediates, are considered in their ground state and represented by their Dirac-ket vector $|\bullet\rangle$;

while for the particular in vivo-in vitro competitive-uncompetitive cases the next particular rules apply:

(iii) the noninhibited transition state $|ES\rangle$ is always placed between inhibited intermediary complexes, $|EI\rangle$ and $|ESI\rangle$, as conceptually revealed from the Figure 1.16;
(iv) the energy of inhibitor state $|I\rangle$ is presumably higher in vivo than in vitro, as compared with the energy level of substrate $|S\rangle$, due to the crowding of the in vivo environment;
(v) competitive inhibition is firstly considered, due to the direct attack on the enzyme;
(vi) consequently, the states $|EI\rangle$ and $|ES\rangle$ are coexisting and undergo quantum combination in a mixed new intermediary state $|EI\rangle \otimes |ES\rangle$;
(vii) the uncompetitive inhibited state $|ESI\rangle$ is obtained from the state $|ES\rangle$ so that the (iii) rule above is complied with;
(viii) when both competitive $|EI\rangle \otimes |ES\rangle$ and uncompetitive $|ESI\rangle$ states are present, they further combine and the new mixed intermediary quantum $|EI\rangle \otimes |ES\rangle \otimes |ESI\rangle$ state arises, which always has to lay above the $|ES\rangle$ one;
(ix) all new inhibited intermediary states decay on the same $|EP\rangle$ state as in the case of no-inhibition (00);

It follows that the energetic differences as compared with the no-inhibition case are due to the tunneling induced by the inhibited transition states which regulate the delayed times of mixed catalysis.

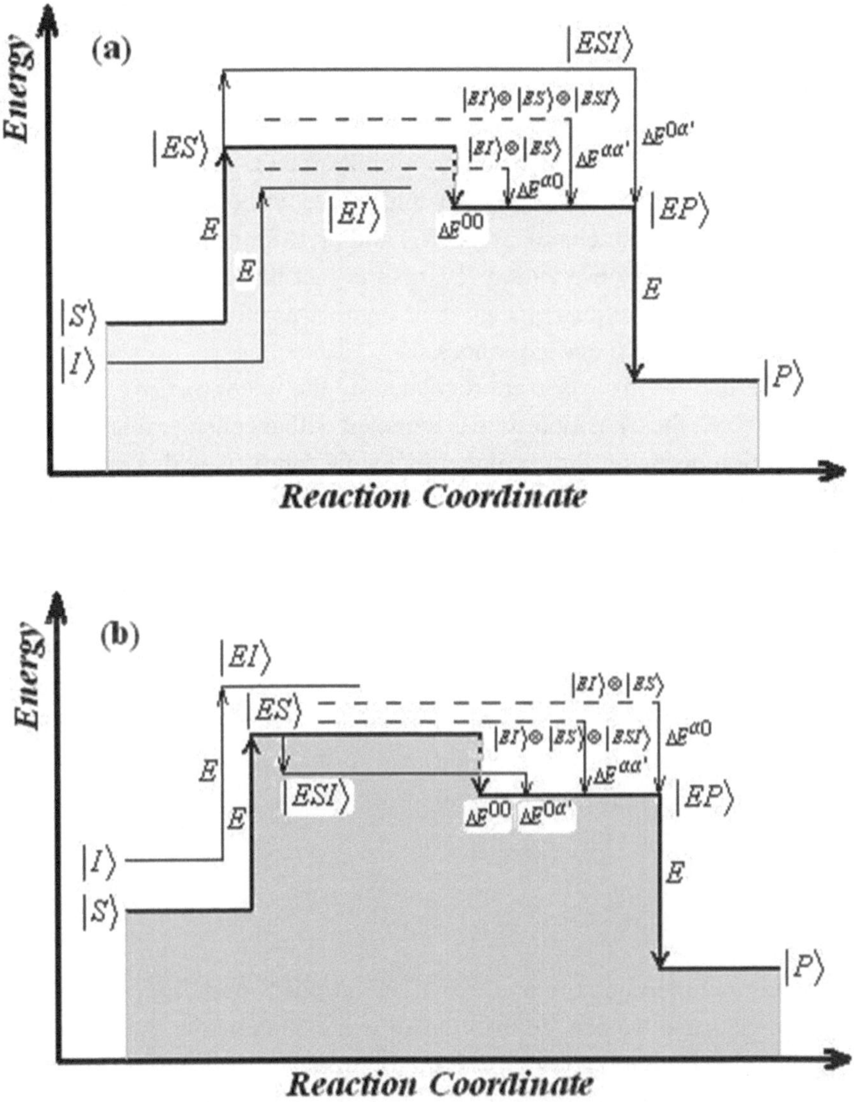

FIGURE 1.18 Quantum in vitro (a) and in vivo (b) energetic diagrams of the generalized Brownian enzyme-catalyzed reaction E+S↔ES→EP→E+P in the presence of the inhibitor (I) acting competitively on the enzyme (E), uncompetitively on enzyme-substrate complex (ES), or by mixed competition on both the enzyme and enzyme-substrate (the thin lines of mechanisms), in relation with the no inhibition case (the thick lined mechanism), providing the tunneling time of specific reaction and the associated tunneling energies of formed enzyme-transition state complexes, $\Delta E^{\alpha 0}$, $\Delta E^{0\alpha}$, $\Delta E^{\alpha\alpha'}$, and ΔE^{00}, respectively. The arrows in the diagrams indicate the final direction of the reactions they refer to (Putz & Lacrămă, 2007).

These rules can be phenomenological visualized in diagrams of Figure 1.18, in full agreement with the computational tunneling energies of Eqs. (1.187) and (1.188) for the in vitro and in vivo Brownian enzymic reactions, respectively.

An interesting fact is that the no-inhibition case always lies between mixed and some particular case of inhibition, without being at the end sides of the energetic chains of (1.187) and (1.188), as should roughly be presumed. Another observation regards the fact that the competitive and uncompetitive inhibitions change their dominant role in passing from the in vitro to the in vivo circumstances.

While the in vitro cases are regulated by the uncompetitive fast reactions in which the transition state is tunneled with a considerable energetic stabilization decay until the competitive slow catalysis with a small energetic width of tunneling of the enzyme-transition states – see relations (1.185) and (1.187), within the in vivo cases the order of times and energetic widths tunneling is vice versa for inhibition regulation situations – see the relations (1.186) and (1.188) (Putz & Lacrămă, 2007; Putz & Putz, 2011).

The present general discussion may serve for practical estimates of temporal scales of tunneling when controlling the enzymatic activity (with inhibition) at the spectroscopic level. It can also be extended in the line of quantum computation of the conditional probability of the altered heritable gene expression from the product formation by delays in proteomic and enzymic interactions.

1.10 CONCLUSION

Enzymes and their activity are known, in various forms, from the ancient history. Shortly, from a description of wine making in the Codex of Hammurabi (Babylon, cca. 2100 B.C.) to the early civilizations of Rome, Greece, Egypt, China, and India the use of microorganisms as enzyme supply for fermentation was a common practice among antique people. However, early enzymology had started with the studies of Réaumur (1683–1757) and Spallanzani (1729–1799) who showed that the digestion is a chemical rather physical process by performing the first experimental demonstration of enzyme specificity.

Then, due to do Emil Fischer, the modern enzymology was developed by the cornerstone 1890 discovery of “lock and key” mechanistic model for the stereochemical relationship between enzymes and their substrate. Soon after, Brown, in 1902, made the insightful observation that enzyme-catalysis is not simple diffusion-limited reaction but governed by the formation of an enzyme-substrate complex. His remark is still most valuable for the present study as was revealed by the present logistic vs. *W*-Lambert progress curves: "*it is quite conceivable... that the time elapsing during molecular union (ES: enzyme-substrate binding complex) and transformation (EP: enzyme-product binding complex) may be sufficiently prolonged to influence the general course of the action*" (Brown, 1902); this is the conceptual background that motivated the in depth analysis of the enzyme kinetics, al the level and beyond the mono-substrate interaction, which by its transformation delay hidden also structural conformations' fluctuations (here interchanging between ES and EP complexes, respectively between *W*-Lambert and logistic progress curves for S- and P- absorption spectra respectively) with much fundamental consequences, towards the quantum inclusion phenomena, see Section 1.9.

In this context, the present enzymic logistic approach proposes a way of simulating the quantum fluctuation by supplementing the basic Michaelis-Menten kinetic and of its temporal *W*-Lambert solution with its logistical extension and solution. Beside the fact that the proposed logistic kinetics furnishes an analytical exact temporal solution for the substrate depletion it provides also the way of computing the tunneling time and energy for a given enzymic reactions. It leaves also with the possibility of drawing of a quantum diagram for the concerned enzymic mechanism.

That is, once the classical Michaelis-Menten is solved until its *W*-Lambert closed form the associate logistical temporal solution can be immediately draw by a simple correspondence rule, see (1.57). The conceptual richness of the actual approach resides on the fact that the Michaelis-Menten kinetics was proved be recovered in the first order expansion of the logistical one respecting the substrate depletion. Such behavior supports the further assumption that the Brownian mechanism may be modeled by logistical temporal solution while the Michaelis-Menten by the *W*-Lambert temporal closed form. Nevertheless, considering the molecular picture of the involved reactants, the difference between the two-substrate temporal solutions provides also the computation basis from extracting the temporal range of tunneling throughout their absorbance progress curves. Then, the involving of the Heisenberg spectroscopic relation gives out also the tunneling energies. With all these the quantum diagram for the studied enzyme mechanism is revealed, in which the enzyme plays the "photonic" role that activates the whole reaction and drives the combined intermediate complexes.

The present algorithm was firstly tested on the mono-substrate enzymic catalysis and extended for the inhibition mixed cases, both for the in vitro and in vivo environments: it can be further applied on more complex enzyme and proteomic combinations and suitable linked with real experimental data following the same line of analysis since the recursive feature of the presented algorithm and the close connection with the enzymatic activity through the spectroscopic absorption method.

Then, although with great impact in modeling the synergism between the active gorge and peripheral sites of wild and mutant enzymes, the complex Haldane-Radić kinetics mechanism was found to be reduced to the more tractable Michaelis-Menten mechanism under special conditions of substrate-enzyme interactions. Actually, through performing a detailed analytical analysis of the initial velocity and progress curves, at both conceptual-analytical and computational-numerical levels for a paradigmatic cholinesterase system, there follows that such reduction is possible in either of the kinetic conditions (Putz, 2011, 2013):

- Higher dissociation constant for inhibition/activation substrate site interaction to enzyme;
- Equal catalytic efficiency of inhibition/activation substrate-enzyme loop as provided by E-S hydrolysis;
- Lower substrate concentration, typically in the range up to the 100 μM.

There was proved that all these three conditions may be regarded as equivalent in reducing Haldane-Radić to Michaelis-Menten enzymic kinetics, yet, being the last one a new one added, specific to some mutant *in vivo* enzyme kinetics of cholinesterases, though the present detailed theoretical and fitting analysis show its sufficiency the envisaged reduction taking place even when the first two conditions do not apply.

Overall, one can remark that from the discovery of the quantum principles and the elucidation subsequent limitation of knowledge through measurements no molecular science that attempts be analytical is complete until acquires a certain degree of quantum comprehension (Putz & Mingos, 2013). While standard enzymology was based on studying how the enzymes facilitates the passage of a reaction over a static potential-energy barrier, with the new millennium mankind exploitation of the biological catalyst the quantum tunneling and the quantum fluctuation effects of the protein dynamics provides the alternative hypothesis that the enzyme activation is produced through rather than over the enzyme transition state, see (Putz, 2012a, b) and the Chapter 3 of the present volume. However, this field of investigation is just opening now and requires much synergetic work to assess the future quantum theory of enzymic reactions.

KEYWORDS

- **competitive inhibition**
- **enzyme kinetics**
- **enzyme-substrate interaction**
- **fitting curves**
- **full time course analysis**
- **Haldane-Radić enzyme reaction**
- **logistic solution**
- **Michaelis-Menten reaction**
- **mixed inhibition**
- **multiple alternative enzyme-substrate reactions**
- **quantum enzyme kinetics**
- **quasi-steady state approximation**
- **reversible kinetics**
- ***W*-Lambert solution**

REFERENCES

AUTHOR'S MAIN REFERENCES

Putz, M. V. (2016). Quantum Nanochemistry. A Fully Integrated Approach: Vol. I. Quantum Theory and Observability. Apple Academic Press & CRC Press, Toronto-New Jersey, Canada-USA.

Putz, M. V. (2013). Bonding in orthogonal space of chemical structure: from in cerebro to in silico. *Int. J. Chem. Model.* 5(4), 369–395.

Putz, M. V., Mingos, D. M. P., Eds. (2013). *Applications of Density Functional Theory to Biological and Bioinorganic Chemistry*, Structure and Bonding Series Vol. 150, Springer Verlag, Heidelberg-Berlin.

Putz, M. V. (2012a). *Chemical Orthogonal Spaces*, Mathematical Chemistry Monographs Vol. 14, University of Kragujevac.

Putz, M. V., Ed. (2012b). *QSAR & SPECTRAL-SAR in Computational Ecotoxicology*, Apple Academics, Toronto.

Putz, M. V. (2011). On reducible character of Haldane-Radić enzyme kinetics to conventional and logistic Michaelis-Menten models. *Molecules* 16(4), 3128–3145 (DOI:10.3390/molecules16043128).

Putz, M. V., Putz, A. M. (2011). Logistic vs. *W*-Lambert information in quantum modeling of enzyme kinetics. *Int. J. Chemoinform. Chem. Eng.* 1(1), 42–60 (DOI: 10.4018/ijcce.2011010104).

Putz, M. V., Putz, A. M., Chiriac, A., Vlad-Oros, B. (2008). *Elements of Homogeneous Chemical, Classical and Logistic Enzyme Kinetics* (original in Romanian: Elemente de Cinetică Chimică Omogenă, Enzimatică Clasică si Logistică), Mirton Publishing House, Timişoara.

Putz, M. V., Lacrămă, A.-M. (2007). Enzymatic control of the bio-inspired nanomaterials at the spectroscopic level. *J. Optoelectron. Adv. Mater.* 9(8), 2529–2534.

Putz, M. V., Lacrămă, A.-M., Ostafe, V. (2007). Introducing logistic enzyme kinetics. *J. Optoelectron. Adv. Mater.* 9(9), 2910–2916.

Putz, M. V., Lacrămă, A.-M., Ostafe, V. (2006a). Full analytic progress curves of the enzymic reactions in vitro. *Int. J. Mol. Sci.* 7(11), 469–484 (DOI: 10.3390/i7110469).

Putz, M. V., Lacrămă, A.-M., Ostafe, V. (2006b). Full time course analysis for reversible enzyme kinetics. *Proceedings of the VIII-th International Symposium Young People and Multidisciplinary Research* (11–12 Mai 2006 Timisoara, Romania), Welding Publishing House, Association of Multidisciplinary Research of the West Zone of Romania, Timişoara (ISBN-10 973–8359–39–2, ISBN-13 978–8359–39–0), pp. 642–649.

SPECIFIC REFERENCES

Albery, W. J., Knowles, J. R. (1976). Evolution of enzyme function and the development of the catalytic efficiency. *Biochem.* 15, 631–5640.

Allen, M. J., Rushton, N. (1994). Use of the CytoTOX 96(TM) assay in routine biocompatibility testing in vitro. *Promega Notes Magazine* 45, 7–10.

Alper, K. O., Singla, M., Stone, J. L., Bagdassarian, C. K. (2001). Correlated conformational fluctuations during enzymatic catalysis: Implications for catalytic rate enhancement. *Protein Sci.* 10, 1319–1330.

Astumian, R. D., Chock, P. B., Tsong, T. Y., Westerhoff, H. V. (1989). Effects of oscillations and energy-driven fluctuations on the dynamics of enzyme catalysis and free-energy transduction. *Phys. Rev. A* 39, 6416–6435.

Bahnson, B. J., Klinman, J. P. (1995). Hydrogen tunneling in enzyme catalysis. *Methods Enzymol.* 249, 373–397.

Bala, P., Grochowski, P., Lesyng, B., McCammon, J. A. (1996). Quantum-classical molecular dynamics simulation of proton transfer processes in molecular complexes and in enzymes. *J. Phys. Chem.* 100, 2535–2545.

Barry, D. A., Parlange, J.-Y; Li, L., Prommer, H., Cunningham, C. J., Stagnitti, F. (2000). Analytical approximations for real values of Lambert *W*-function. *Math. Comp. Simulation* 53, 95–103.

Basran, J., Sutcliffe, M. J., Scrutton, N. S. (1999). Enzymatic H-transfer requires vibration-driven extreme tunneling. *Bichem.* 38, 3218–3222.

Boeker, E. A. (1984). Integrated rate equations for enzyme-catalyzed first-order and second-order reactions. *Biochem. J.* 223, 15–22.

Boeker, E. A. (1985). Integrated rate equations for irreversible enzyme-catalyzed first-order and second-order reactions. *Biochem. J.* 226, 29–35.

Boeker, E. A. (1987). Analytical methods for fitting integrated rate equations. A discontinuous assay. *Biochem. J.* 245, 67–74.

Brown, A. J. (1892). Influence of oxygen and concentration on alcohol fermentation. *J. Chem. Soc. Trans.* 61, 369–385.

Brown, A. J. (1902). Enzyme Action. *J. Chem. Soc. Trans.* 81, 373–388.

Bruno, W. J., Bialek, W. (1992). Vibrationally enhanced tunneling as a mechanism for enzymatic hydrogen transfer. *Biophys. J.* 63, 689–699.

Cantor, C. R., Schimmel, P. R. (1980). *Biophysical Chemistry. Part III. The Behavior of Biological Macromolecules*, W. H. Freeman and Company, San Francisco.

Cha, Y., Murray, C. J., Klinman, J. P. (1989). Hydrogen tunneling in enzyme reaction. *Science* 143, 1325–1330.

Copeland, R. A. (2000). *Enzymes-A Practical Introduction to Structure, Mechanism, and Data Analysis*, Wiley-VCH, New York.

Corless, R. M., Gonnet, G. H., Hare, D. E. G., Jeffrey, D. J., Knuth, D. E. (1996). On the Lambert W function. *Adv. Comput. Math.* 5, 329–359.

Cornish-Bowden, A. (1975). The use of the direct linear plot for determining initial velocities. *Biochem. J.* 149, 305–312.

Cornish-Bowden, A. (1979). *Fundamentals of Enzyme Kinetics*, Butterworths.

Crampin, E. J., Schnell, S., McSharry, P. E. (2004). Mathematical and computational techniques to deduce complex biochemical reaction mechanisms. *Prog. Biophys. Mol. Biol.* 86(1), 77–112.

Curran, J. M., Gallagher, J. A., Hunt, J. A. (2005). The inflammatory potential of biphasic calcium phosphate granules in osteoblasts/macrophage co-culture. *Biomaterials* 26, 5313–5320.

Câteau, H., Tanaka, S. (2002). Kinetic Analysis of Multisite Phosphorylation Using Analytic Solutions to Michaelis-Menten Equation. *J. Theor. Biol.* 217, 1–14.

DeVault, D., Chance, B. (1966). Studies of photosynthesis using a pulsed laser. I. Temperature dependence of cytochrome oxidation rate in chromatium. Evidence for tunneling. *Biophys. J.* 6(6), 825–847.

Duggleby, R. G. (1986). Progress-curve analysis in enzyme kinetics. Numerical solution of integrated rate equations. *Biochem. J.* 235, 613–615.

Duggleby, R. G. (1994). Product inhibition of reversible enzyme-catalyzed reactions. *Biochim. Biophys. Acta.* 1209, 238–240.

Duggleby, R. G. (2001). Quantitative analysis of the time courses of enzyme-catalyzed reactions. *Methods* 24, 168–174.

Duggleby, R. G., Morrison, J. F. (1977). The analysis of progress curves for enzyme-catalyzed reactions by non-linear regression. *Biochim. Biophys. Acta* 481, 297–312.

Duggleby, R. G., Morrison, J. F. (1978). Progress curve analysis in enzyme kinetics: model discrimination and parameter estimation. *Biochim. Biophys. Acta* 526, 398–409.

Duggleby, R. G., Morrison, J. F. (1979). The use of steady-state rate equations to analyze progress curve data. *Biochim. Biophys. Acta* 568, 357–362.

Duggleby, R. G., Wood, C. (1989). Analysis of progress curves for enzyme-catalyzed reactions. Automatic construction of computer programs for fitting integrated equations. *Biochem. J.* 258, 397–402.

Ellis, T. G., Barbeau, D. S., Smets, B. F., Grady, C. P. L. Jr. (1996). Respirometric technique for determination of extant kinetic parameters describing biodegradation. *Wat. Env. Res.* 68, 917–926.

Glasstone, S., Laidler, K. J., Eyring, H. (1941). *The theory of the rate processes*, McGraw-Hill, New York.

Goodman, A. F., Bellato, C. M., Khidr, L. (2005). The uncertain future for central dogma. *The Scientist* 19(12), 20–21.

Goudar, C. T., Sonnad, J. R., Duggleby, R. G. (1999). Parameter estimation using a direct solution of the integrated michaelis-menten equation. *Biochim. Biophys. Acta* 1429, 377–383.

Grady, C. P. L. Jr., Daigger, G. T., Lim, H. C. (1999). *Biological Wastewater Treatment*, 2nd ed., Marcel Dekker, New York.

Graham, L. D., Haggett, K. D., Hayes, P. J., Schober, P. A., Jennings, P. A., Whittaker, R. G. (1994). A new library of alpha-lytic protease S1 mutants generated by combinatorial random substitution. *Biochem. Mol. Biol. Int.* 32, 831–839.

Grant, K. L., Klinman, J. P. (1989). Evidence that protium and deuterium undergo significant tunneling in the reaction catalyzed by bovine serum amine oxidase. *Biochem.* 28, 6597–6605.

Gray P; Scott, S. K. (1990). *Chemical Oscillations and Instabilities. Non-linear Chemical Kinetics*, Clarendon Press, Oxford.

Haldane, J. B. S. (1930). *The Enzymes*, Longmans-Green: London.

Haldane, J. B. S., Stern, K. G. (1932). *Allgemeine Chemie der Enzyme*, Dresden, Verlag von Steinkopff.

Hartwell, L. (2005). How to build a cancer sensor system. *The Scientist* 19(18), 18–19.

Hayes, B. (2005). Why W? *American Scientist* 93, 104–108.

Henri, V. (1901). Über das gesetz der wirkung des invertins. *Z. Phys. Chem.* 39, 194–216.

Hirs, C. H. W. (Ed.) (1967). Automatic computation of amino acid analyzer data. Methods Enzymol. (Enzyme Structure) 11, 27–31.

Huang, H. Y., Niemann, C. (1951). The kinetics of the α-chymotrypsin catalyzed hydrolysis of acetyl- and nicotinyl-l-tryptophanamide in aqueous solutions at 25° and *p*H 7.9. *J. Am. Chem. Soc.* 73, 1541–1548.

Hwang, J.-K., Warshel, A. (1996). How important are quantum mechanical nuclear motion in enzyme catalysis? *J. Am. Chem. Soc.* 118, 11745–11751.

Jennings, R. R., Niemann, C. (1953). The kinetics of the α-chymotrypsin catalyzed hydrolysis of acetyl-l-hexahydrophenylalaninamide in aqueous solutions at 25° and pH 7.9. *J. Am. Chem. Soc.* 75, 4687–4692.

Jennings, R. R., Niemann, C. (1955). The evaluation of the kinetic constants of enzyme-catalyzed reactions by procedures based upon integrated rate equations. *J. Am. Chem. Soc.* 77, 5432–5433.

Jonsson, T., Edmondson, D. E., Klinman, J. P. (1994). Hydrogen tunneling in the flavoenzyme monoamine oxidase, B. *Biochem.* 33, 14871–14878.

Jonsson, T., Glickman, M. H., Sun, S., Klinman, J. P. (1996). Experimental evidence for extensive tunneling of hydrogen in the lipoxygenase reaction: implication for enzyme catalysis. *J. Am. Chem. Soc.* 118, 10319–10320.

Kohen, A., Jonsson, T., Klinman, J. P. (1997). Effects of protein glycosylation on catalysis: changes in hydrogen tunneling and enthalpy of activation in the glucose oxidase reaction. *Biochem.* 36, 2603–2611.

Laidler, K. J. (1955). Theory of transient phase in kinetics, with special reference to enzyme systems. *Can. J. Chem.* 33, 1614–1624.

Lineweaver, H., Burk, D. (1934). The Determination of the enzyme dissociation constants. *J. Am. Chem. Soc.* 56, 658–666.

Mandl, F. (1992). *Quantum mechanics*, John Wiley & Sons, Chichester.

Mattick, J. S. (2004). The hidden genetic program of complex organisms. *Sci. Am.* 291, 60–67.

Mayr, E. (2001). *What evolution is?* The Orion Publishing Group Ltd., "Science Masters" Brockmann Inc.

Michaelis, L., Menten, M. L. (1913). Die kinetik der invertinwirkung. *Biochem. Z.* 49, 333–369.

Nachon, F., Stojan, J., Fournier, D. (2008). Insights into substrate and product traffic in the *Drosophila melanogaster* acetylcholinesterase active site gorge by enlarging a back channel. *FEBS J.* 275, 2659–2664.

Nixon, A. E., Ostermeier, M., Benkovic, S. J. (1998). Hybrid enzymes: manipulating enzyme design. *TIBTECH* 16, 258–264.

Noble, D. (2002). The rise of computational biology. *Nat. Rev. Mol. Cell. Biol.* 3, 459–463.

Radić, Z., Gibney, G., Kawamoto, S., MacPhee-Quigley, K., Bongiorno, C., Taylor, P. (1992). Expression of recombinant acetylcholinesterase in a baculovirus system: Kinetic properties of glutamate 199 mutants. *Biochemistry* 31, 9760–9767.

Radić, Z., Pickering, N. A., Vellom, D. C., Camp, S. (1993). Palmer Taylor three distinct domains in the cholinesterase molecule confer selectivity for acetyl- and butyrylcholinesterase inhibitors. *Biochemistry* 32, 12074–12084.

Rigby, P. W. J., Burleigh, B. D., Hartley, B. S. (1974). Gene duplication in experimental enzyme evolution. *Nature* 251, 200–204.

Ringe, D., Petsko, G. A. (1999). Quantum enzymology: Tunnel vision. *Nature* 399, 417–418.

Ritchie, R. J., Prvan, T. (1996). A simulation study on designing experiments to measure the km of michaelis-menten kinetics curves. *J. Theor. Biol.* 178, 239–254.

Ross, G., Loverix, S., De Proft, F., Wyns, L., Geerlings, P. (2003). A computational and conceptual DFT study of the reactivity of anionic compounds: implications for enzymatic catalysis. *J. Phys. Chem. A* 107, 6828–6836.

Rubinow, S. I. (1975). *Introduction to Mathematical Biology*, Wiley, New York.

Rubinow, S.I; Lebowitz, J. L. (1970). Time-dependent michaelis-menten kinetics for an enzyme-substrate-inhibitor system. *J. Am. Chem. Soc.* 92, 3888–3893.

Savageau, M. A. (1969). Biochemical systems analysis: I. Some mathematical properties of the rate law for the component enzymatic reactions. *J. Theor. Biol.* 25, 365–369.

Savageau, M. A. (1976). *Biochemical System Analysis: A Study of Function and Design in Molecular Biology*, Addison-Wesley, Reading, MA.

Schnell, S., Maini, P. K.A (2003). Century of enzyme kinetics: reliability of the k_m and v_{max} estimates. *Comm. Theor. Biol.* 8, 169–187.

Schnell, S., Mendoza, C. (1997). Closed form solution for time-dependent enzyme kinetics. *J. Theor. Biol.* 187, 207–212.

Schnell, S., Mendoza, C. (2000a). Time-dependent closed form solution for fully competitive enzyme reactions. *Bull. Math. Biol.* 62, 321–336.

Schnell, S., Mendoza, C. (2000b). Enzyme kinetics of multiple alternative substrates. *J. Math. Chem.* 27, 155–170.

Schnell, S., Mendoza, C. (2001). A fast method to estimate kinetic constants for enzyme inhibitors. *Acta Biotheoretica* 49, 109–113.

Schnell, S., Turner, T. E. (2004). Reaction kinetics in intracellular environments with macromolecular crowding: simulations and rate laws. *Prog. Biophys. Mol. Biol.* 85, 235–260.

Segel, I. H. (1975). *Enzyme Kinetics: Behavior and Analysis of Rapid Equilibrium and Steady-State Systems*, Wiley, New York.

Segel, L. A. (1988). On the validity of the steady state assumption of enzyme kinetics. *Bull. Math. Biol.* 50, 579–593.

Segel, L. A., Slemrod, M. (1989). The quasi-steady-state assumption: a case study in perturbation. *SIAM Rev.* 31, 446–477.

Silverman, P. H. (2004). Rethinking Genetic Determinism. *Scientist 18(10),* 32–33.

Sonad, J. R., Goudar, C. T. (2004). Solution of the Haldane equation for substrate inhibition enzyme kinetics using the decomposition method. *Math. Comput. Model.* 40, 573–582.

Stemmer, W. P. C. (1994). Rapid evolution of a protein in vitro by DNA shuffling. *Nature* 370, 389–391.

Stojan, J., Brochier, L., Aliès, C., Colletier, J. P., Fournier, D. (2004). Inhibition of *Drosophila melanogaster* acetylcholinesterase by high concentrations of substrate. *Eur. J. Biochem.* 271, 1364–1371.

Stojan, J., Marcel, V., Estrada-Mondaca, S., Klaebe, A., Masson, P., Fournier, D. (1998). A putative kinetic model for substrate metabolisation by *Drosophila* acetylcholinesterase. *FEBS Lett.* 440, 85–88.

Sutcliffe, M., Scrutton, N. (2000). Enzymology takes a quantum leap forward. *Phil. Trans. R. Soc. Lond. A* 358, 367–386.

Szedlacsek, S. E., Ostafe, V., Duggleby, R. G., Serban, M., Vlad, M. O. (1990). Progress-Curve Equations for Reversible Enzyme-Catalyzed Reactions Inhibited by Tight-Binding Inhibitors. *Biochem. J.* 265, 647–653.

Thom, R. (1975). *Structural Stability and Morphogenesis – An Outline of a General Theory of Models*, W. I. Benjamin Inc., Reading, Massachusetts.

Truhlar, D. G., Gordon, M. S. (1990). From force-fields to dynamics – classical and quantal paths. *Science* 249, 491–498.

Turner, T. E., Schnell, S., Burrage, K. (2004). Stochastic approaches for modeling in vivo reactions. *Comput. Biol. Chem.* 28, 165–198.

Tyagi, R., Lee, Y.-T., Guddat, L. W., Duggleby, R. G. (2005). Probing the mechanism of the bifunctional enzyme ketol-acid reductoisomerase by site-directed mutagenesis of the active site. *FEBS J.* 272, 593–602.

Tzafriri, A. R., Edelman, E. R. (2004). The total quasy-steady-state approximation is valid for reversible enzyme kinetics. *J. Theor. Biol.* 226, 303–313.

van Slyke, D. D., Cullen, G. E. (1914). The mode of action of urease and of enzymes in general. *J. Biol. Chem.* 19, 141–180.

Voet, D., Voet, J. G. (1995). *Biochemistry* (second edition), John Wiley & Sons Inc., New York, Chapter 13.

Walsh, C. (1979). *Enzymatic reaction mechanisms*, Freeman.

Williams, D. F. (Ed.) (1986). *Techniques of Biocompatibility Testing*, vol. II, CRC Press, Boca Raton.

Zimmerle, C. T., Frieden, C. (1989). Analysis of Progress Curves by Simulations Generated by Numerical Integration. *Biochem. J.* 258, 381–387.

CHAPTER 2

STATISTICAL SPACE FOR MULTIVARIATE CORRELATIONS

CONTENTS

ABSTRACT

Aiming to prepare the conceptual-computational ground for correlating chemical structure with biological activity by the celebrated quantitative structure-activity relationships (QSARs), the fundamental statistical advanced frameworks are here detailed to best understand of the classical multilinear regression analysis to be next generalized by an algebraic (in quantum Hilbert space) reformulation of in terms of data vectors and orthogonal conditions (to be exposed in Chapter 3).

2.1 INTRODUCTION

The EU White Paper policy on chemical (from 2001 ahead) wants to identify hazards of all chemicals on the market greater than 1 tone (EU-COM, 2008). In this respect, (Q)SARs methods have been found most useful for identifying these hazards to limit animal testing, since the current policy for the risk evaluation of chemicals is to involve more (Q)SARs rather than animals testing (Hulzebos et al., 2005). Nevertheless, in 2003 the European Commission (EC) adopted a legislative proposal for a new chemical management system called REACH (Registration, Evaluation and Authorization of Chemicals) according which, for instance, all chemicals marketed in Europe in volumes greater than 1 tone per year should be granted within 11 years of entry into force, while physico-chemical and toxicity data for High Production Volume (HPV) chemicals should be already available, i.e., within 7 years from the regulations (Pavan & Netzeva, 2006; EC, 2003).

The main idea behind these global regulations is to use rapid, low cost computational models prior going to the experimental testing. However, most of the published models until now lack details of their statistical validation and definitions of the model domains of applicability. So, since such models do not fully meet the OECD (Organization for Economic Cooperation and Development) principles for QSAR validations (OECD, 200468), the global organization is likely to interrupt their use for regulatory purposes (Pavan & Netzeva, 2006). In this respect, Huzelbos and Posthumus concluded that the current risk assessment procedures for chemicals in EU are time consuming and complex processes on the reason that the minimum data requirements for performing risk evaluation for HPV chemicals appeared to be insufficient. The entire risk assessment to human and environment is carried out only for some well-selected substances (Lacrămă et al., 2008; Putz & Putz, 2011; Putz, 2012a-b). The resonance between industry and several EU member states involve very time consuming processes as well, although (Q)SARs and in vitro testing have already been advanced as alternatives for animal testing (Hulzebos & Posthumus, 2003).

With these there is clear that there is an increasing interest in the using of toxicological-based quantitative structure activity relationships as non-animal methods to provide data for priority setting, risk assessment and chemical classification and labeling (Schultz et al., 2003; Worth & Balls, 2002). The interspecies QAA(activity-activity)Rs are useful because they can be exploited to predict unknown toxicity and to verify the validity of other toxicity tests, quoting that strong trend of increasing toxicity with increasing hydrophobicity underpins such data sets (Cronin et al., 2004). However, QSARs strategies are increasingly being used as a tool to assist regulatory agencies in toxicological assessment of chemical and require appropriate validation at all stages of development (Cronin et al., 2002). When assessing to model ecotoxicological batteries reliable toxicity data are required for all trophic levels of the environment to ensure appropriate risk assessment of chemicals. With all these, there are currently insufficient published toxicity data and models to meet the needs of regulation and QSAR mechanistically interpretation (Cronin et al., 2004; Cronin & Schultz, 2003). Usually, in the case of certain endpoints, the biological variability may be too large to enable reasonable quantitative predictions to be made, so the modeler may decide to convert the data into one or more categories of toxic effect (Worth & Cronin, 2003).

On the other side, being used in Chemistry during the second half of 20th century as an extended statistical analysis (Anderson, 1958; Shorter, 1973; Box et al., 1973; Green & Margerison, 1978; Topliss, 1983; Seyfel, 1985; Kubinyi, 1996; Chatterjee, 2000), the QSAR method had attained in recent years a special status, officially certified by European Union as the main computational tool (within the so-called "*in silico*" approach) for the regulatory assessments of chemicals by means of non-testing methods (Worth et al., 2005; Benigni et al., 2007).

However, while QSAR primarily uses the multiple regression analysis (Putz & Putz, 2011), alternative approaches as such neuronal-network (NN) or genetic algorithms (GA) have been advanced to somehow generalize the QSAR performance in delivering a classification of variables used, in the sense of principal component analysis (PCA) and partial least squares (PLS) methodologies; still, the claimed advantage of the NN over QSAR techniques is limited by the fact the grounding physical-mathematical philosophies are different since highly non-linear with basic multilinear pictures are compared, respectively (So & Karpuls, 1996; Kubinyi, 1994a, 1996; Teko et al., 1996; Haegawa et al., 1999; Zheng & Tropsha, 2000; Lucic & Trinajstic, 1999; Duchowicz & Castro, 2008).

Actually, the chemical-physical advantage of QSAR stands in its multi-linearity correlation that resembles with superposition principle of quantum mechanics, which allow meaningful interpretation of the structural (inherently quantum) causes associated with the latent or unobserved variables (sometimes called as *common factors*) into the observed effects (activity) usually measured in terms of 50%-effect concentration

(EC_{50}), associated with various types of bioaccumulation and toxicity (Zhao et al., 1998).

Nevertheless, many efforts have been focused on applying QSAR methods to non-linearity features from where the "expert systems" emerged as formalized computer-based environments, involving knowledge-based, rule-based or hybrid automata able to provide rational predictions about properties of biological activity of chemicals or of their fragments; it results in various QSAR based databases: the model database (QMDB) – inventorying the robust summaries of QSARs that can be appealed by envisaged endpoint or chemical, the prediction database (QPDB) – when data from QMDB are used for further prediction to be stored, or together towering the chemical category database (CCD) documentation (Pavan et al., 2008; Tsakovska et al., 2008; Gallegos Saliner et al., 2008; Patlewicz et al., 2008; Netzeva et al., 2008; Cronin & Worth, 2008).

Therefore, a certain conceptual-computational analysis of a compound of a series of compounds in the view of assigning its toxicity degree naturally two levels: one addresses the atomic-molecular structure together with related quantum properties while the other envisages the correlations of these properties, for example, hydrophobicity, polarizability, steric effects, etc., with the bio, eco- or pharmacological observed activities. Finally, it gets out the molecular mechanistic "picture" of the reactions involved in the studied chemical-biological interaction or, with other words, of the quantum chemical strength established between the ligand (the effector or the chemical) and receptor (in the target site or organism). Still, either the structure or the quantum chemical binding aspects require the advanced studies upon them, firstly in a separate manner, and then combined both at the intrinsic structural level and for correlating the interaction, based on the versatility of the atomic and molecular world to generate surprisingly structures and interactions just because the quantum character involved (i.e., undulatory, thus allowing the tunneling even for the energetic inaccessible potential barriers) when forming new apparently not explicated or controllable compounds by means of macroscopic procedures.

Still, whatever the computational procedure approached, either of that of Hansch type, 3D, decisional, or orthogonal ones, see the Section 2.5 of this Conclusion for detailed presentation, the problem of delivering the molecular interaction mechanism as a QSAR analysis result was only recently furnished by the so-called Spectral-SAR that proposes a purely algebraic rethinking of the traditional statistic QSAR, which allows, through the new concepts introduced (e.g., the orthogonal space of variables, the vectorial length of the biological activity, or the algebraic correlation factor as an intensity measure of the chemical-biological interaction) the building of an optimized chart of the molecular action pathways grounded on the *minimum spectral path principle*, $\delta[A,B]=0$ with A and B the endpoints, within a generalized space of the action norms and correlation factors (Putz & Lacrămă, 2007; Putz & Putz, 2011; Putz, 2012a-b).

In this context, the present chapter is devoted to paving the conceptual and computational forms of QSAR implementation by fundamental analytics by exposing the foreground mathematical basis of sample statistics (characteristic for the chemical space of molecules) and of correlation (characteristic to the orthogonal space of chemical descriptors).

2.2 STATISTICAL CORRELATION ANALYSIS

2.2.1 CORRELATION COEFFICIENT

One likes to interpret the measurement data by employment of the recorded measures over a given sample; accordingly, the Table 2.1 models an experiment for two random variables (x, y).

TABLE 2.1 The Custom Table of Variables (x's) and the Records (y's) in a Given Experiment

$x \rightarrow$ / $y \downarrow$	x_1	x_2	…	x_n
y_1	p_{11}	p_{12}	…	p_{1m}
y_2	p_{12}	p_{22}	…	p_{2m}
⋮	⋮	⋮	⋮	⋮
y_m	p_{n1}	p_{n2}	…	p_{nm}

In Table 2.1 one has as the working variables:

- *m*: the stations for assessing an observable value
- *n*-quantities to be measured on each station above
- p_{ij} the recorded probability of the event (*a quantity measured to a station*)

$$\begin{cases} x = x_i \\ y = y_j \end{cases} \tag{2.1a}$$

so that over all possibilities the unity covers all possibilities of certain measurement (covering all possible range of an observation); that is equivalent to normalizing the events' probabilities as

$$\sum_{i=1}^{m} \sum_{j=1}^{n} p_{ij} = 1 \tag{2.1b}$$

Thus we can define for (continuous) random variable (x, y) the probability density function $f(x, y)$ as:

$$P[(x;y) \subset D] = \iint_{D(omain)} f(x,y)dxdy \tag{2.2}$$

so that

$$f(x,y) \geq 0 \ \& \int_{-\infty}^{+\infty}\int_{-\infty}^{+\infty} f(x,y)dxdy = 1 \tag{2.3}$$

Recall that from quantum theory one has: the average observations

$$\langle \hat{A} \rangle = \int \Psi^* \hat{A} \Psi d\tau = \int \hat{A} \wp d\tau, \quad \wp = \Psi^* \Psi = |\Psi|^2 \tag{2.4}$$

so that for the continuous random variables (x, y) one has the average observations:

$$\begin{cases} \langle x \rangle = \iint_{-\infty}^{+\infty} x \cdot f(x,y)dxdy \\ \langle y \rangle = \iint_{-\infty}^{+\infty} y \cdot f(x,y)dxdy \end{cases} \tag{2.5}$$

while for the discrete random variables (x_i, y_i) they respectively unfold as

$$\begin{cases} \bar{x} = \sum_i \sum_j x_i p_{ij} \\ \bar{y} = \sum_i \sum_j y_i p_{ij} \end{cases} \tag{2.6}$$

The obtained averages $(\bar{x}, \bar{y})$ or $(\langle x \rangle, \langle y \rangle)$ provide the center of dispersion of a system of random variables (x, y) so that assuring the discrete-continuous statistical linkage.

Next, one may introduce the *variance* (or *dispersion*) as the expectation of the mean shift from the center of dispersion (from where the dispersion meaning) defined as

$$\begin{cases} D(\langle x \rangle) = \int\int (x - \langle x \rangle)^2 \cdot f(x,y)dxdy \\ D(\langle y \rangle) = \int\int (y - \langle y \rangle)^2 \cdot f(x,y)dxdy \end{cases} \tag{2.7}$$

or

$$\begin{cases} D(\bar{x}) = \sum_{i,j} (x_i - \bar{x}) p_{ij} \\ D(\bar{y}) = \sum_{i,j} (y_i - \bar{y}) p_{ij} \end{cases} \tag{2.8}$$

Note that if x and y are *independent* variables then

$$\mathrm{p_{ij}} = \begin{cases} \mathrm{p_i} & \text{for x} \\ \mathrm{p_i} & \text{for y} \end{cases} \tag{2.9}$$

and even further as

$$\mathrm{p_{ij}} = \begin{cases} \dfrac{1}{\mathrm{n}} & \text{\& for discrete uniform distribution} \\ \dfrac{1}{\mathrm{b - a}} & \text{\& for continuous distribution on the domain } \mathrm{D = [a - b]} \end{cases} \tag{2.10a}$$

The same applies for the other probability

$$p_j = \begin{cases} \dfrac{1}{m} \\ \dfrac{1}{b - a} \end{cases} \tag{2.10b}$$

Then, we can write for the respective dispersions

$$\begin{cases} D(\bar{x}) = \sum_i \frac{1}{n}(x_i - \bar{x}) = \sigma_x^2 \\ D(\bar{y}) = \sum_j \frac{1}{m}(y_j - \bar{y}) = \sigma_y^2 \end{cases} \tag{2.11}$$

so we can introduce the correspondent *standard deviations*

$$\sigma_x = \sqrt{D(\langle x \rangle)} \text{ or } \sqrt{D(\bar{x})} \tag{2.12}$$

$$\sigma_y = \sqrt{D(\langle y \rangle)} \text{ or } \sqrt{D(\bar{y})} \tag{2.13}$$

Note that variances can be successively rewritten as:

$$\begin{aligned} D(\langle x \rangle) &= \iint (x - \langle x \rangle)^2 \cdot f(x,y) dxdy \\ &= \langle (x - \langle x \rangle)^2 \rangle = \langle x^2 - 2x\langle x \rangle + \langle x \rangle^2 \rangle = \langle x^2 \rangle - 2\langle x \rangle \langle x \rangle + \langle x \rangle^2 \\ &= \langle x^2 \rangle - 2\langle x \rangle^2 + \langle x \rangle^2 = \langle x^2 \rangle - \langle x \rangle^2 \\ &= \iint x^2 \cdot f(x,y) dxdy - \left(\iint x \cdot f(x,y) dxdy \right)^2 \end{aligned} \tag{2.14}$$

and then same for $D(\langle y \rangle)$, and then same for $D(\bar{x})$ and $D(\bar{y})$.

In this framework, the variance allows the introduction of the *covariance* index:

$$\begin{aligned} c_{xy} &= \langle (x - \langle x \rangle)(y - \langle y \rangle) \rangle \\ &= \iint (x - \langle x \rangle)(y - \langle y \rangle) \cdot f(x,y) dxdy = \langle xy - x\langle y \rangle - \langle x \rangle y + \langle x \rangle \langle y \rangle \rangle \\ &= \langle xy \rangle - \langle x \rangle \langle y \rangle - \langle x \rangle \langle y \rangle + \langle x \rangle \langle y \rangle = \langle xy \rangle - \langle x \rangle \langle y \rangle \end{aligned} \tag{2.15}$$

and the same for the discrete random variables.

However, worth observing that if x and y are independent, i.e.,

$$f(x,y) = f(x) \cdot f(y) \tag{2.16}$$

then one has

$$\langle xy\rangle = \iint x\cdot y\cdot f(x)f(y)dxdy = \left(\int x\cdot f(x)dx\right)\left(\int y\cdot f(y)dy\right) = \langle x\rangle\cdot\langle y\rangle \tag{2.17}$$

leaving with vanishing of independent covariance:

$$c_{xy}^{ind} = 0 \tag{2.18}$$

Next, let's turn to the characterization of correlation between the variable *x, y* by the so-called the *correlation coefficient*:

$$r_{xy} = \frac{c_{xy}}{\sigma_x \sigma_y} = \frac{\langle (x-\langle x\rangle)(y-\langle y\rangle)\rangle}{\sqrt{\langle (x-\langle x\rangle)^2\rangle\langle (y-\langle y\rangle)^2\rangle}} \tag{2.19}$$

which for the discrete random variables, i.e., with equal probability appearance

$$p_{ij} = p_i = p_j = \frac{1}{n} \tag{2.20}$$

successively becomes

$$r_{xy}^{Discrete} = \frac{\sum_{i,j}(x_i-\bar{x})(y_i-\bar{y})p_{ij}}{\left(\sum_{i,j}(x_i-\bar{x})^2 p_{ij}\right)\left(\sum_{i,j}(y_i-\bar{y})^2 p_{ij}\right)}$$

$$= \frac{\frac{1}{n}\sum_i (x_i-\bar{x})(y_i-\bar{y})}{\sqrt{\frac{1}{n}\sum_i (x_i-\bar{x})^2 \sum_i (y_i-\bar{y})^2 \frac{1}{n}}} = \frac{\sum_i (x_i-\bar{x})(y_i-\bar{y})}{\sqrt{\sum_i (x_i-\bar{x})^2 \sum_i (y_i-\bar{y})^2}} \tag{2.21}$$

2.2.2 MONO-LINEAR REGRESSION

Next, we like to explore the relation between the correlation factor and the linear regression equation

$$y = ax + b \tag{2.22}$$

To this aim, one may say that the points are of type $M(x_i,y_i)$ arranged on the working table of values:

x	x_1 x_2 ... x_i ...
y	y_1 y_2 ... y_i ...

and with generic representation of the Figure 2.1.

The best least squares procedures minimizes the sum of distance between the points $M(x_i,y_i)$ and $M(x_i,y_x)$, i.e.,

$$\varphi(a,b) = \sum_i d_i^2 \rightarrow \min \tag{2.23}$$

with

$$d_i = y\,\big|_{x=x_i} - y_i = ax_i + b - y_i \tag{2.24}$$

that gives

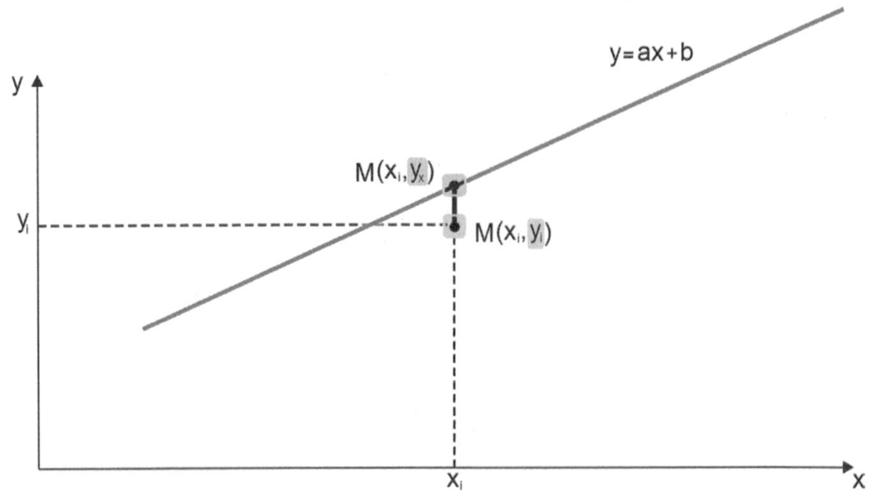

FIGURE 2.1 The regression difference between the computed/predicted value (y_x) and the recorded/observed one (y_i) for a given experiment variable or instance (x_i).

$$\varphi(a,b) = \sum_i (ax_i + b - y_i)^2 \to \min \tag{2.25}$$

or, in analytical terms, by cancelling the first correspondent derivatives

$$\begin{cases} \dfrac{\partial \varphi}{\partial a} = 0 \\ \dfrac{\partial \varphi}{\partial b} = 0 \end{cases} \tag{2.26}$$

The last system successively casts and transforms as

$$\begin{cases} 2\sum_i (ax_i + b - y_i)x_i = 0 \\ 2\sum_i (ax_i + b - y_i) = 0 \end{cases}$$

$$\begin{cases} a\sum_i x_i^2 + b\sum_i x_i = \sum_i x_i y_i \\ a\sum_i x_i + b\sum_{i=1}^{n} 1 = \sum_i y_i \end{cases}$$

$$\begin{cases} a\sum_i x_i^2 + b\sum_i x_i = \sum_i x_i y_i \quad (I) \mid n \\ a\sum_i x_i + \quad bn \quad = \sum_i y_i \quad (II) \mid (-1)\sum_i x_i \end{cases} \tag{2.27}$$

By summation of the results in the system (2.27) one gets

$$a\left[n\sum_i x_i^2 - \left(\sum_i x_i\right)^2\right] = n\sum_i x_i y_i - \sum_i x_i \sum_i y_i$$

so that finding the slope coefficient of the regression

$$a = \frac{n\sum_i x_i y_i - \sum_i x_i \sum_i y_i}{n\sum_i x_i^2 - \left(\sum_i x_i\right)^2} \tag{2.28}$$

With this information, back in the form (II) of the system (2.27) one finds also the free term (the intercept) of the regression (2.22), successively as:

$$b = \frac{1}{n}\left[\sum_i y_i - \sum_i x_i \frac{n\sum_i x_i y_i - \sum_i x_i \sum_i y_i}{n\sum_i x_i^2 - \left(\sum_i x_i\right)^2}\right]$$

$$= \frac{1}{n}\frac{n\sum_i y_i \sum_i x_i^2 - \sum_i y_i \left(\sum_i x_i\right)^2 - n\sum_i x_i \sum_i x_i y_i + \left(\sum_i x_i\right)^2 \sum_i y_i}{n\sum_i x_i^2 - \left(\sum_i x_i\right)^2}$$

$$= \frac{\sum_i y_i \sum_i x_i^2 - \sum_i x_i \sum_i x_i y_i}{n\sum_i x_i^2 - \left(\sum_i x_i\right)^2} \tag{2.29}$$

A worthy comments can be made on the regression equation by observing that

(i) from $II \times \frac{1}{n}$ we get:

$$a\frac{1}{n}\sum_i x_i + \frac{1}{n}\cdot b \cdot n = \frac{1}{n}\sum_i y_i \tag{2.30}$$

that means in fact the form

$$a\bar{x} + b = \bar{y} \tag{2.31}$$

so finding that the regression equation fulfills the arithmetic mean point $M(\bar{x}, \bar{y})$

(ii) one yields the special writing of correlation coefficient (2.19) as:

$$r_{xy} = \frac{c_{xy}}{\sigma_x \sigma_y} = \frac{\langle xy \rangle - \langle x \rangle \langle y \rangle}{\sqrt{\langle x^2 \rangle - \langle x \rangle^2} \sqrt{\langle y^2 \rangle - \langle y \rangle^2}}$$

$$= \frac{\sum_{i,j} x_i y_i p_{ij} - \sum_{i,j} x_i p_{ij} \sum_{i,j} y_i p_{ij}}{\sqrt{\left[\sum_{i,j} x_i^2 p_{ij} - \left(\sum_{i,j} x_i p_{ij}\right)^2\right]\left[\sum_{i,j} y_i^2 p_{ij} - \left(\sum_{i,j} y_i p_{ij}\right)^2\right]}} \tag{2.32}$$

Equation (2.32) can be further transformed by applying the *law of large numbers*: for a sufficiently large number of independent trials the arithmetic mean of the observed values of the random variable converged in probability of its mathematic expectation

$$\wp\left(\left|\frac{1}{n}\sum_i x_i - \bar{x}\right| < \varepsilon\right) > 1 - \delta, \; 0 < \delta < \frac{1}{\varepsilon^2} \cdot \frac{D(x)}{n} \tag{2.33}$$

That is, for large *n*, ($n \to \infty$), we can have:

$$\begin{cases} \langle x \rangle \; or \; \bar{x} \to \frac{1}{n}\sum_i x_i; \; \langle y \rangle \; or \; \bar{y} \to \frac{1}{n}\sum_i y_i \\ \langle x^2 \rangle \to \frac{1}{n}\sum_i x_i^2; \; \langle y^2 \rangle \to \frac{1}{n}\sum_i y_i^2; \; \langle xy \rangle \to \frac{1}{n}\sum_i x_i y_i \end{cases} \tag{2.34}$$

This way we have for the correlation coefficient (2.32) the actual expression

$$r_{xy} = \frac{c_{xy}}{\sigma_x \sigma_y} = \frac{\frac{1}{n}\sum_i x_i y_i - \frac{1}{n^2}\sum_i x_i \sum_i y_i}{\sqrt{\frac{1}{n}\sum_i x_i^2 - \frac{1}{n^2}\left(\sum_i x_i\right)^2} \sqrt{\frac{1}{n}\sum_i y_i^2 - \frac{1}{n^2}\left(\sum_i y_i\right)^2}}$$

$$= \frac{n\sum_i x_i y_i - \sum_i x_i \sum_i y_i}{\underbrace{\sqrt{n\sum_i x_i^2 - \left(\sum_i x_i\right)^2}}_{\sigma_x} \underbrace{\sqrt{\frac{1}{n}\sum_i y_i^2 - \frac{1}{n^2}\left(\sum_i y_i\right)^2}}_{\sigma_y}} \tag{2.35}$$

Accordingly, we may consider the combined factor towards the slope of correlation:

$$r_{xy} \cdot \frac{\sigma_y}{\sigma_x} = \frac{c_{xy}}{\sigma_x \sigma_y} \cdot \frac{\sigma_y}{\sigma_x} = \frac{c_{xy}}{\sigma_x^2}$$

$$= \frac{n\sum_i x_i y_i - \sum_i x_i \sum_i y_i}{n\sum_i x_i^2 - \left(\sum_i x_i\right)^2} = a \tag{2.36}$$

Moreover, if we combine two form of regression equation, we can form the deviation equation

$$\begin{cases} y = ax + b \\ \bar{y} = a\bar{x} + b \end{cases} \Rightarrow y - \bar{y} = a(x - \bar{x}) \tag{2.37}$$

or, in correlation coefficient formulation:

$$y - \bar{y} = r_{xy} \cdot \frac{\sigma_y}{\sigma_x}(x - \bar{x}) \tag{2.38}$$

For practical reasons worth noting that the linear regression (the last equation) is sufficiently possible in the range where

$$\underbrace{\left|r_{xy}\right|}_{\in(0,1)} \sqrt{n-1} \geq 3 \tag{2.39}$$

For instance, one has the specializations:

$$\text{if } n = 10 \Rightarrow \sqrt{n-1} = 3 \;\&\; \forall \; \left|r_{xy}\right| \times 3 < 3 \tag{2.40}$$

$$n = 17 \Rightarrow \sqrt{n-1} = 4 \;\&\; \exists \; \left|r_{xy}\right| \in (0,1) \;|\; \left|r_{xy}\right| \cdot \sqrt{n-1} \geq 3 \tag{2.41}$$

Let's see next the significance of regression equation from the distribution function point of view. Firstly let's introduce *the moments*' definition:

$$\mu_k = \left\langle \left(x - \langle x \rangle\right)^k \right\rangle \; \ldots \text{ of order "}k\text{"} \tag{2.42}$$

supporting the particular cases:

$$\mu_0 = \left\langle (x - \langle x \rangle)^0 \right\rangle = 1 = \int_{-\infty}^{+\infty} f(x) dx \tag{2.43}$$

which equivalents with the normalization condition; then we have the canceling first order moment:

$$\mu_1 = \left\langle (x - \langle x \rangle)^1 \right\rangle = \langle x \rangle - \langle x \rangle = 0 = \int_{-\infty}^{+\infty} (x - \bar{x}) f(x) dx \tag{2.44}$$

while recovering in the second order the dispersion (2.14) definition:

$$\mu_2 = \left\langle (x - \langle x \rangle)^2 \right\rangle = \langle x^2 \rangle - \langle x \rangle^2 = D(x) = \sigma_x^2 = \int f(x)(x - \bar{x})^2 dx \tag{2.45}$$

Now, which is the function $f(x)$ having such properties?

We answer that the so-called Normal (Gaussian) distribution:

$$f(x) = \frac{1}{\sqrt{2\pi}\sigma} e^{-\frac{(x-\bar{x})^2}{2\sigma^2}}, \quad x \in \Re \tag{2.46}$$

has the above properties.

Proof: We have to calculate:

$$\mu_0 = \int_{-\infty}^{+\infty} \frac{1}{\sqrt{2\pi}\sigma} e^{-\frac{(x-\bar{x})^2}{2\sigma^2}} d(x - \bar{x}) = 1 \tag{2.47a}$$

$$\mu_1 = \int_{-\infty}^{+\infty} \frac{1}{\sqrt{2\pi}\sigma} (x - \bar{x}) e^{-\frac{(x-\bar{x})^2}{2\sigma^2}} d(x - \bar{x}) = 0 \tag{2.48a}$$

$$\mu_2 = \int_{-\infty}^{+\infty} \frac{1}{\sqrt{2\pi}\sigma} (x - \bar{x})^2 e^{-\frac{(x-\bar{x})^2}{2\sigma^2}} d(x - \bar{x}) = \sigma^2 \tag{2.49a}$$

To this aim we have to calculate the integral:

$$I_0 = \int_{-\infty}^{+\infty} e^{-ax^2} dx = ? \tag{2.50a}$$

for that we find more advantageously to solve the composed integral

$$I_0^2 = \left(\int_{-\infty}^{+\infty} e^{-ax^2} dx\right)\left(\int_{-\infty}^{+\infty} e^{-ay^2} dy\right) = \iint_{-\infty}^{+\infty} e^{-a(x^2+y^2)} dxdy \tag{2.50b}$$

By performing the polar coordinate transformation (see Figure 2.2)

$$\begin{cases} x^2 + y^2 = r^2 \\ dxdy = rdrd\varphi \end{cases} \tag{2.51}$$

one has the immediate transformations of integral (2.50b)

$$I_0^2 = \int_0^{2\pi}\int_0^{\infty} e^{-ar^2} rdrd\varphi = \left(\int_0^{\infty} e^{-ar^2} rdr\right)\left(\int_0^{2\pi} d\varphi\right)$$

$$= 2\pi\int_0^{\infty}\left(-\frac{1}{2a}\right)d\left(e^{-ar^2}\right) = -\frac{\pi}{a}e^{-ar^2}\Big|_0^{\infty} = \frac{\pi}{a}$$

$$I_0 = \int_{-\infty}^{+\infty} e^{-ax^2} dx = \sqrt{\frac{\pi}{a}} \tag{2.50b}$$

in accordance with the Poisson integrals' recipe – see the Appendices of Volume I of the present five-volume book (Putz, 2016a). Now we can write for the 0th moment (2.47a)

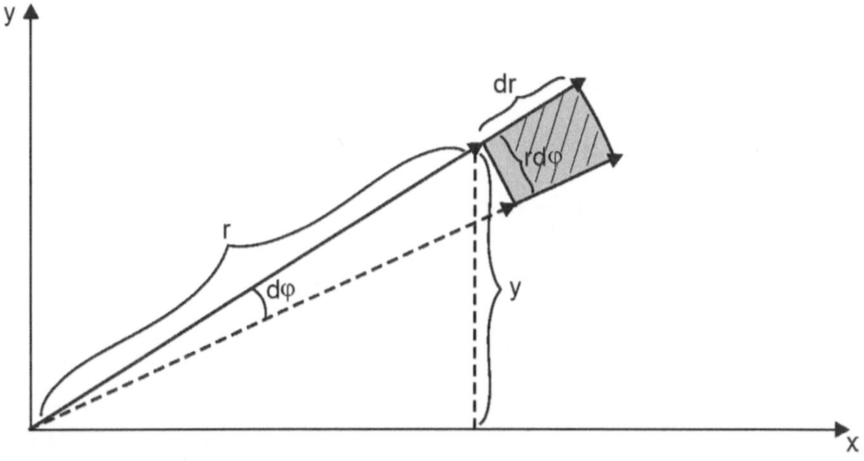

FIGURE 2.2 The construction of the polar coordinates towards covering the 2D space by coupled radial with angular movements (dr, dφ).

$$\mu_0 = \int_{-\infty}^{+\infty} \frac{1}{\sqrt{2\pi}\sigma} e^{-\frac{(x-\bar{x})^2}{2\sigma^2}} d(x-\bar{x}) = \frac{1}{\sqrt{2\pi}\sigma} \cdot \sqrt{\frac{\pi}{\frac{1}{2\sigma^2}}} = \frac{1}{\sqrt{2\pi}\sigma} \cdot \sigma \cdot \sqrt{2\pi} = 1 \tag{2.47b}$$

Then, we can immediately infer the first order computation:

$$I_1 = \int_{-\infty}^{+\infty} e^{-ax^2} dx = -\frac{1}{2a} \int_{-\infty}^{+\infty} d\left(e^{-ax^2}\right) = -\frac{1}{2a} e^{-ax^2}\Big|_{-\infty}^{+\infty} = 0 \tag{2.52}$$

leaving with the expected result – see (2.44):

$$\mu_1 = 0 \tag{2.48b}$$

For the second order we may use the 0th order result by applying the integration by parts

$$I_2 = \int_{-\infty}^{+\infty} x^2 e^{-ax^2} dx = \int_{-\infty}^{+\infty} x\left(xe^{-ax^2}\right)dx = -\frac{1}{2a} \int_{-\infty}^{+\infty} x \frac{d}{dx}\left(e^{-ax^2}\right)dx$$

$$= -\frac{1}{2a}\left[\int_{-\infty}^{+\infty} \frac{d}{dx}\left(xe^{-ax^2}\right)dx - \underbrace{\int_{-\infty}^{+\infty} e^{-ax^2} \frac{dx}{dx} dx}_{I_0}\right] = -\frac{1}{2a}\left(xe^{-ax^2}\right)_{-\infty}^{+\infty} + \frac{1}{2a} \cdot \sqrt{\frac{\pi}{a}}$$

$$= -\frac{1}{2a}\left[\underbrace{\underbrace{\lim_{x\to+\infty} \frac{x}{e^{+ax^2}}}_{0} - \underbrace{\lim_{x\to-\infty} \frac{x}{e^{+ax^2}}}_{0}}_{l'Hospital\ rule}\right] + \frac{1}{2a} \cdot \sqrt{\frac{\pi}{a}} = \frac{1}{2a}\sqrt{\frac{\pi}{a}} \tag{2.53}$$

So we remain to recover the dispersion result of Eq. (2.45) by the second order moment of distribution

$$\mu_2 = \int_{-\infty}^{+\infty} \frac{1}{\sqrt{2\pi}\sigma} (x-\bar{x})^2 e^{-\frac{(x-\bar{x})^2}{2\sigma^2}} d(x-\bar{x})$$

$$= \frac{1}{\sqrt{2\pi}\sigma} \cdot \frac{1}{2 \cdot \frac{1}{2\sigma^2}} \cdot \sqrt{\frac{\pi}{\frac{1}{2\sigma^2}}} = \frac{1}{\sqrt{2\pi}} \cdot \sigma \cdot \sqrt{2\pi} \cdot \sigma = \sigma^2 \tag{2.49b}$$

Next we go to the bi-dimensional normal distribution that, taken assuming the correlation r_{xy} may exist, employed with the general form (see Figure 2.3):

$$f(x,y)=\frac{1}{2\pi\sigma_x\sigma_y\sqrt{1-r_{xy}^2}}\exp\left\{\frac{-1}{2(1-r_{xy}^2)}\left[\frac{(x-\bar{x})^2}{\sigma_x^2}+\frac{(y-\bar{y})^2}{\sigma_y^2}-2r_{xy}\frac{(x-\bar{x})(y-\bar{y})}{\sigma_x\sigma_y}\right]\right\} \tag{2.54}$$

Note that when we have no-correlation

$$r_{xy}=0 \tag{2.55}$$

one gets the factorization of distribution functions by the independent variables:

$$f(x,y)=f(x)\cdot f(y)=\left[\frac{1}{\sqrt{2\pi}\sigma_x}e^{-\frac{(x-\bar{x})^2}{2\sigma_x^2}}\right]\left[\frac{1}{\sqrt{2\pi}\sigma_y}e^{-\frac{(y-\bar{y})^2}{2\sigma_y^2}}\right] \tag{2.56}$$

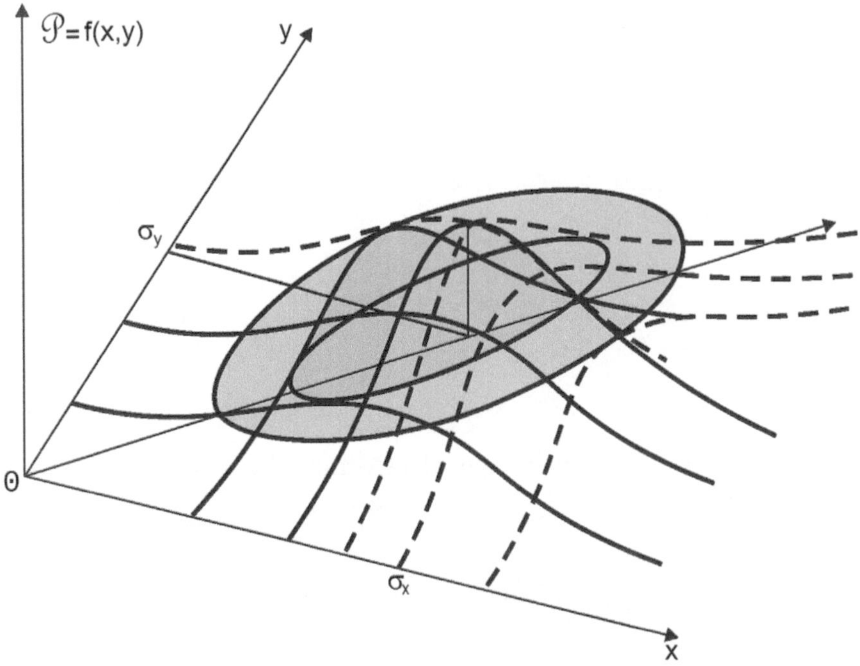

FIGURE 2.3 The 3D representation of the evolving conditional bi-variable distribution probabilities.

However, if $r_{xy} = 1$ then $f(x, y) \to 0$ and all points will be distributed on a line in the (x, y) plane. That is perfect correlation case.

Nevertheless when searching for general $y = y(x)$ dependency with the aid of $f(x, y)$ the *conditional probability* density $g(x|y)$ should be firstly introduced

$$g(y|x) = \frac{f(x, y)}{f(x)} \tag{2.57}$$

We may therefore explicate the conditional probability successively as:

$$g(x|y) = \frac{\dfrac{1}{2\pi\sigma_x\sigma_y\sqrt{1-r_{xy}^2}} e^{\frac{-1}{2(1-r_{xy}^2)}\left[\frac{(x-\bar{x})^2}{\sigma_x^2}+\frac{(y-\bar{y})^2}{\sigma_y^2}-2r_{xy}\frac{(x-\bar{x})(y-\bar{y})}{\sigma_x\sigma_y}\right]}}{\dfrac{1}{\sqrt{2\pi}\sigma_x} e^{-\frac{1}{2\sigma_x^2}(x-\bar{x})^2}}$$

$$= \frac{1}{\sqrt{2\pi}\sigma_y\sqrt{1-r_{xy}^2}} e^{-\frac{1}{2(1-r_{xy}^2)}\left[\frac{(x-\bar{x})^2}{\sigma_x^2}+\frac{(y-\bar{y})^2}{\sigma_y^2}-2r_{xy}\frac{(x-\bar{x})(y-\bar{y})}{\sigma_x\sigma_y}-\frac{2(1-r_{xy}^2)}{2\sigma_x^2}(x-\bar{x})^2\right]}$$

$$= \frac{1}{\sqrt{2\pi}\sigma_y\sqrt{1-r_{xy}^2}} e^{-\frac{1}{2(1-r_{xy}^2)}\left[r_{xy}^2\frac{(x-\bar{x})^2}{\sigma_x^2}+\frac{(y-\bar{y})^2}{\sigma_y^2}-2r_{xy}\frac{(x-\bar{x})(y-\bar{y})}{\sigma_x\sigma_y}\right]}$$

$$g(y|x) = \frac{1}{\sqrt{2\pi}\sigma_y\sqrt{1-r_{xy}^2}} e^{-\frac{1}{2(1-r_{xy}^2)\sigma_y^2}\left[y-\bar{y}-r_{xy}\frac{\sigma_y}{\sigma_x}(x-\bar{x})\right]^2} \tag{2.58}$$

since based on the exponent equivalent writing, by using the preceding (2.38) result

$$\left[\frac{(y-\bar{y})}{\sigma_y} - r_{xy}\frac{(x-\bar{x})}{\sigma_x}\right]^2 = \frac{1}{\sigma_y^2}\left[y-\bar{y}-r_{xy}\frac{\sigma_y}{\sigma_x}(x-\bar{x})\right]^2 \tag{2.59}$$

Now the equation for "y" may be obtained by employing the conditional probability as:

$$y^{calc/comp/predict} = \langle y \rangle_{g(y|x)} = \int y g(y|x) dy$$

$$= \underbrace{\int_{-\infty}^{+\infty} \left[y - \bar{y} - r_{xy} \frac{\sigma_y}{\sigma_x}(x - \bar{x}) \right] g(y|x) d\left(y - \bar{y} - r_{xy} \frac{\sigma_y}{\sigma_x}(x - \bar{x}) \right)}_{=0}$$

$$+ \left[\bar{y} + r_{xy} \frac{\sigma_y}{\sigma_x}(x - \bar{x}) \right] \underbrace{\int_{-\infty}^{+\infty} g(y|x) d\left(y - \bar{y} - r_{xy} \frac{\sigma_y}{\sigma_x}(x - \bar{x}) \right)}_{=1}$$

$$= \bar{y} + r_{xy} \frac{\sigma_y}{\sigma_x}(x - \bar{x}) \tag{2.60}$$

so regaining the regression equation (2.38), see also Figure 2.4.

Next, we like to see also the quantity *sum of residuals conditioned* by correlation with x

$$SR_y = \min \sum_i \left(y_i^{obs} - y_i^{comp} \right)^2$$

$$= \left\langle \left(y_i^{obs} - y_i^{comp} \right)^2 \right\rangle_{g(y|x)} = \left\langle \left[y - \bar{y} - r_{xy} \frac{\sigma_y}{\sigma_x}(x - \bar{x}) \right]^2 \right\rangle_{g(y|x)}$$

$$= \int \left[y - \bar{y} - r_{xy} \frac{\sigma_y}{\sigma_x}(x - \bar{x}) \right]^2 g(y|x) d\left(y - \bar{y} - r_{xy} \frac{\sigma_y}{\sigma_x}(x - \bar{x}) \right)$$

$$= \left(1 - r_{xy}^2\right) \sigma_y^2 \tag{2.61}$$

that equivalently gives

$$r_{xy}^2 = 1 - \frac{SR}{\sigma_y^2} \tag{2.62}$$

or in explicit manner

$$r_{xy} = \sqrt{1 - \frac{\sum_i \left(y_i^{obs} - y_i^{calc} \right)^2}{\sum_i \left(y_i^{obs} - \bar{y} \right)^2}}, \quad \begin{cases} y_i^{calc} = a x_i + b \\ \bar{y} = \frac{1}{n} \sum_i y_i^{calc} \end{cases} \tag{2.63}$$

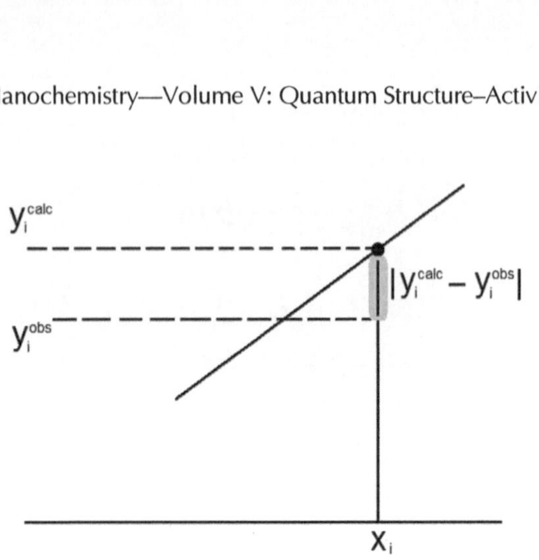

FIGURE 2.4 The detailed difference of Figure 2.1 between the calculated (calc) and the observed (obs) values of a variable's (x_i) effect in an experiment.

that is an alternative (working) form for data-only formulation (2.19)

$$r_{xy} = \frac{c_{xy}}{\sigma_x \cdot \sigma_y} = \frac{\sum_i (x_i - \bar{x})(y_i - \bar{y})}{\sqrt{\sum_i (x_i - \bar{x})^2 \sum_i (y_i - \bar{y})^2}} \tag{2.64}$$

The probability nature according with

$$|r_{xy}| \leq 1 \tag{2.65}$$

is assured by the Cauchy-Schwarz (CS) inequality

$$\left[\sum_i (x_i - \bar{x})(y_i - \bar{y})\right]^2 \leq \left[\sum_i (x_i - \bar{x})^2\right] \cdot \left[\sum_i (y_i - \bar{y})^2\right] \tag{2.66}$$

However, the CS inequality is next proved by introducing the vectorial and scalar product notions:

- Let's consider the $\{x_i\}$, $\{y_i\}$ sets of vectors:

$$\underline{x_1 \ x_2 \ \ldots x_i \ \ldots \ x_n} \rightarrow |x\rangle = |x_1, \ x_2, \ \ldots x_i, \ \ldots \ x_n\rangle \tag{2.67}$$

$$y_1 \ y_2 \ \ldots y_i \ \ldots \ y_n \rightarrow |y\rangle = |y_1, \ y_2, \ \ldots y_i, \ \ldots \ y_n\rangle \tag{2.68}$$

- Then we have the scalar product defined:

$$\langle x|y\rangle = \langle x_1,\ x_2,\ \dots\ x_n|y_1,\ y_2,\ \dots\ y_n\rangle$$
$$= x_1y_1 + x_2y_2 + \dots + x_ny_n = \sum_i x_iy_i \quad (2.69)$$

- If we take the self – scalar product:

$$\langle x|x\rangle = \sum_{i=1}^{n} x_i^2 \quad (2.70)$$

then we can introduce the associate norm:

$$\||x\rangle\| = \sqrt{\langle x|x\rangle} = \sqrt{\sum_{i=1}^{n} x_i^2} \quad (2.71)$$

As an example, take the vector $|r\rangle = |x_1, x_2, x_3\rangle$ in 3D-Euclidian; so we have the decomposition (see Figure 2.5):

$$\langle r|r\rangle = x_1^2 + x_2^2 + x_3^2 \quad (2.72)$$

And then we confirm that the norm

$$\||r\rangle\| = \sqrt{\langle r|r\rangle} = \sqrt{x_1^2 + x_2^2 + x_3^2} \quad (2.73)$$

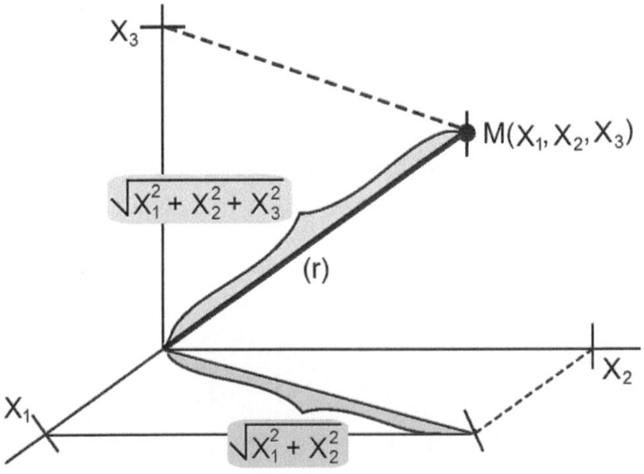

FIGURE 2.5 The vectorial 3D coordinates' decomposing for the "observed length/ radius" of a given experimental/measured point <M (x_1, x_2, x_3)>.

recovers the custom Euclidian measure.

Few observations may provide useful information:

(i) we say that (by the orthogonality condition) we have

$$|x\rangle \perp |y\rangle \text{ if } \langle x|y\rangle = 0 \text{ or if } \sum_i x_i y_i = 0 \tag{2.74}$$

(ii) the distance between two points in the $N-$ space fixed by the vectors "$|x\rangle$" and "$|y\rangle$" is given by Euclidian generalized formula

$$d(|x\rangle,|y\rangle) = \||x\rangle - |y\rangle\| = \|x-y\rangle\| = \sqrt{\langle x-y|x-y\rangle} = \sqrt{\sum_i (x_i - y_i)^2} \tag{2.75}$$

(iii) the angle between two vectors $|x\rangle_N$, $|y\rangle_N$, is defined through generalizing the scalar product formula:

$$\vec{x}\cdot\vec{y} := |\vec{x}|\cdot|\vec{y}|\cdot\cos(\angle\vec{x},\vec{y}) \Rightarrow \cos(\angle\vec{x},\vec{y}) = \frac{\vec{x}\cdot\vec{y}}{|\vec{x}|\cdot|\vec{y}|} \tag{2.76}$$

$$\cos(\angle|x\rangle,|y\rangle) = \frac{\langle x|y\rangle}{\||x\rangle\|\cdot\||y\rangle\|} = \frac{\langle x|y\rangle}{\sqrt{\langle x|x\rangle}\sqrt{\langle y|y\rangle}} = \frac{\sum_i x_i y_i}{\sqrt{\sum_i x_i^2}\sqrt{\sum_i y_i^2}} \tag{2.77}$$

(iv) finally we have the CS inequality (2.66) accordingly rewritten as:

$$|\langle x|y\rangle| \le \||x\rangle\|\cdot\||y\rangle\|,\ \forall\, |x\rangle,|y\rangle \tag{2.78}$$

Proof: for $t \in \Re$ we always have the positive self-scalar product result, i.e.:

$$\begin{aligned} 0 &\le \langle x-ty|x-ty\rangle \\ &= \langle x|x\rangle - 2t\langle x|y\rangle + t^2\langle y|y\rangle \\ &= \||x\rangle\|^2 - 2t\langle x|y\rangle + t^2\||y\rangle\|^2 \end{aligned} \tag{2.79}$$

For the formed second order equation to have maximum one solution one requires its discriminant being zero or negatively ($\Delta \le 0$); so giving the working inequality that proofs the CS condition above:

$$0 \geq \Delta = 4\langle x|y\rangle^2 - 4\||y\rangle\|^2\||x\rangle\|^2 \Leftrightarrow \langle x|y\rangle^2 \leq \||y\rangle\|^2\||x\rangle\|^2 \Rightarrow \text{CS} \quad (2.80)$$

Now, referring to correlation coefficient we can consider the working vectors

$$\begin{cases} |\chi\rangle = |x\rangle - \bar{x} = |x - \bar{x}\rangle \\ |\eta\rangle = |y\rangle - \bar{y} = |y - \bar{y}\rangle \end{cases} \quad (2.81)$$

Thus, we can employ the CS condition (2.78) to proof the sub-unitary hierarchy for the ratio terms in correlation coefficient of Eq. (2.64):

$$\underline{\langle \chi|\eta\rangle^2 \leq \||\chi\rangle\|^2\||\eta\rangle\|^2}$$

$$\langle x-\bar{x}|y-\bar{y}\rangle \leq (\langle x-\bar{x}|x-\bar{x}\rangle)(\langle y-\bar{y}|y-\bar{y}\rangle)$$

$$\left[\sum_i (x_i - \bar{x})(y_i - \bar{y})\right]^2 \leq \left[\sum_i (x_i - \bar{x})^2\right] \cdot \left[\sum_i (y_i - \bar{y})^2\right] \text{(QED)} \quad (2.82)$$

2.2.3 MULTI-LINEAR REGRESSION

Next, we are going to work out the generalized case of finding the best fit of the multivariate correlation

$$y_i = b_0 + b_1 x_{1i} + \ldots + b_k x_{ki} + \ldots b_M x_{Mi} + e_i \quad (2.83)$$

for the values in Table 2.2. Actually, the system of variable (descriptors) – activities is formed as:

$$y_1 = 1b_0 + b_1 x_{11} + b_2 x_{12} + \ldots b_k x_{1k} + \ldots b_M x_{1M} + e_1 \quad (2.84a)$$

$$y_2 = 1b_0 + b_1 x_{21} + b_2 x_{22} + \ldots b_k x_{2k} + \ldots b_M x_{2M} + e_2 \quad (2.84b)$$

$$\vdots$$

$$y_k = 1b_0 + b_1 x_{k1} + b_2 x_{k2} + \ldots b_k x_{kk} + \ldots b_M x_{kM} + e_k \quad (2.84c)$$

$$\vdots$$

$$y_N = 1b_0 + b_1 x_{N1} + b_2 x_{N2} + \ldots b_k x_{Nk} + \ldots b_M x_{NM} + e_N \quad (2.84d)$$

TABLE 2.2 The Custom Table of Variables (x's) and the Records (y's) in a Multi-Regression Analysis of QSAR Type, Corresponding to the Structural Parameters and the Recorded Activities, Respectively

Activity	**Structural parameters/variable**					
y	x_1	x_2	…	x_i	…	x_M
y_1	x_{11}	x_{12}	…	x_{1k}	…	x_{1M}
…	⋮	⋮		⋮		⋮
y_k	x_{k1}	x_{k2}	…	x_{kk}	…	x_{kM}
…	⋮	⋮		⋮		⋮
y_N	x_{N1}	x_{N2}	…	x_{Nk}	…	x_{NM}

The minimization of the residues (errors) in the least square formation:

$$\sum_{i}^{N} e_i^2 = \sum_{i=1}^{N} \left[y_i - \left(b_0 + b_1 x_{i1} + b_2 x_{i2} + \ldots + b_k x_{ik} + \ldots b_M x_{iM} \right) \right]^2 = \min \quad (2.85)$$

leaves with the system:

$$\frac{\partial}{\partial b_k} \left[\sum_{i}^{N} e_i^2 \right] = 0, \quad k = \overline{0, M} \quad (2.86)$$

or in the extended version:

$$2 \sum_{i=1}^{N} x_{ik} \left[y_i - b_0 - b_1 x_{i1} - b_2 x_{i2} - \ldots b_k x_{ik} - \ldots b_M x_{iM} \right] = 0 \quad (2.87)$$

or, better, explicitly as:

$$\frac{\partial \sum_{i=1}^{N} \left[y_i - \left(b_0 + b_1 x_{i1} + b_2 x_{i2} + \ldots + b_k x_{ik} + \ldots b_M x_{iM} \right) \right]^2}{\partial b_0} = 0 \quad (2.88a)$$

$$\frac{\partial \sum_{i=1}^{N} \left[y_i - \left(b_0 + b_1 x_{i1} + b_2 x_{i2} + \ldots + b_k x_{ik} + \ldots b_M x_{iM} \right) \right]^2}{\partial b_1} = 0 \quad (2.88b)$$

$$\vdots$$

$$\frac{\partial \sum_{i=1}^{N}\left[y_i - \left(b_0 + b_1 x_{i1} + b_2 x_{i2} + \ldots + b_k x_{ik} + \ldots b_M x_{iM}\right)\right]^2}{\partial b_k} = 0 \quad (2.88c)$$

$$\vdots$$

$$\frac{\partial \sum_{i=1}^{N}\left[y_i - \left(b_0 + b_1 x_{i1} + b_2 x_{i2} + \ldots + b_k x_{ik} + \ldots b_M x_{iM}\right)\right]^2}{\partial b_M} = 0 \quad (2.88d)$$

That gives:

$$2\sum_{i=1}^{N}\left[y_i - \left(b_0 + b_1 x_{i1} + b_2 x_{i2} + \ldots + b_k x_{ik} + \ldots b_M x_{iM}\right)\right] = 0 \quad (2.89a)$$

$$2\sum_{i=1}^{N}\left[y_i - \left(b_0 + b_1 x_{i1} + b_2 x_{i2} + \ldots + b_k x_{ik} + \ldots b_M x_{iM}\right)\right] x_{i1} = 0 \quad (2.89b)$$

$$\vdots$$

$$2\sum_{i=1}^{N}\left[y_i - \left(b_0 + b_1 x_{i1} + b_2 x_{i2} + \ldots + b_k x_{ik} + \ldots b_M x_{iM}\right)\right] x_{ik} = 0 \quad (2.89c)$$

$$\vdots$$

$$2\sum_{i=1}^{N}\left[y_i - \left(b_0 + b_1 x_{i1} + b_2 x_{i2} + \ldots + b_k x_{ik} + \ldots b_M x_{iM}\right)\right] x_{iM} = 0 \quad (2.89d)$$

$$\Leftrightarrow$$

$$b_0^N + b_1 \sum x_{i1} + \ldots + b_k \sum x_{ik} + \ldots + b_M \sum x_{iM} = \sum_{i=1}^{N} y_i \quad (2.90a)$$

$$b_0 \sum x_{i1} + b_1 \sum x_{i1}^2 + \ldots + b_k \sum x_{ik} x_{i1} + \ldots + b_M \sum x_{iM} x_{i1} = \sum_{i=1}^{N} y_i x_{i1} \quad (2.90b)$$

$$\vdots$$

$$b_0 \sum x_{ik} + b_1 \sum x_{i1} x_{ik} + \ldots + b_k \sum x_{ik}^2 + \ldots + b_M \sum x_{iM} x_{ik} = \sum_{i=1}^{N} y_i x_{ik} \quad (2.90c)$$

$$\vdots$$

$$b_0 \sum x_{iM} + b_1 \sum x_{i1} x_{ik} + \ldots + b_k \sum x_{ik} x_{iM} + \ldots + b_M \sum x_{iM}^2 = \sum_{i=1}^{N} y_i x_{iM} \quad (2.90d)$$

The obtained system can be elegantly solved by employing the matrices' formalism, respectively introduced for:

- independent variable and dependent variable (measures), respectively

$$\hat{x}=\begin{bmatrix}1 & x_{11} & \cdots & x_{1k} & \cdots & x_{1M}\\ 1 & x_{21} & \cdots & x_{2k} & \cdots & x_{2M}\\ \vdots & \vdots & & & & \\ 1 & x_{k1} & \cdots & x_{kk} & \cdots & x_{kM}\\ \vdots & \vdots & & & & \\ 1 & x_{N1} & \cdots & x_{Nk} & \cdots & x_{NM}\end{bmatrix},\ \hat{y}=\begin{bmatrix}y_1\\ y_2\\ \vdots\\ y_k\\ \vdots\\ y_N\end{bmatrix} \tag{2.91}$$

- the parameter vector/correlation vector and errors, respectively

$$\hat{b}=\begin{bmatrix}b_0\\ b_1\\ \vdots\\ b_k\\ \vdots\\ b_M\end{bmatrix}\ \hat{e}=\begin{bmatrix}e_1\\ \vdots\\ e_k\\ \vdots\\ e_N\end{bmatrix} \tag{2.92}$$

Thus the initial system is resumed by the matrix equation

$$\hat{y}=\hat{x}\hat{b}+\hat{e} \tag{2.93}$$

unfolded as

$$\begin{bmatrix}y_1\\ y_2\\ \vdots\\ y_k\\ \vdots\\ y_N\end{bmatrix}=\begin{bmatrix}1 & x_{11} & \cdots & x_{1k} & \cdots & x_{1M}\\ 1 & x_{21} & \cdots & x_{2k} & \cdots & x_{2M}\\ \vdots & \vdots & & & & \\ 1 & x_{k1} & \cdots & x_{kk} & \cdots & x_{kM}\\ \vdots & \vdots & & & & \\ 1 & x_{N1} & \cdots & x_{Nk} & \cdots & x_{NM}\end{bmatrix}\begin{bmatrix}b_0\\ b_1\\ \vdots\\ b_k\\ \vdots\\ b_M\end{bmatrix}+\begin{bmatrix}e_1\\ \vdots\\ e_k\\ \vdots\\ e_N\end{bmatrix} \tag{2.94}$$

So allowing also for the minimized system to be rewritten under the matrix form

$$\underbrace{\begin{bmatrix} 1 & 1 & \cdots & 1 & \cdots & 1 \\ x_{11} & x_{21} & \cdots & x_{k2} & \cdots & x_{N1} \\ \vdots & \vdots & & & & \\ x_{1k} & x_{2k} & \cdots & x_{kk} & \cdots & x_{Nk} \\ \vdots & \vdots & & & & \\ x_{1M} & x_{2M} & \cdots & x_{kM} & \cdots & x_{NM} \end{bmatrix}}_{\hat{x}^T} \overbrace{\begin{bmatrix} 1 & x_{11} & \cdots & x_{1k} & \cdots & x_{1M} \\ 1 & x_{21} & \cdots & x_{2k} & \cdots & x_{2M} \\ \vdots & \vdots & & & & \\ 1 & x_{k1} & \cdots & x_{kk} & \cdots & x_{kM} \\ \vdots & \vdots & & & & \\ 1 & x_{N1} & \cdots & x_{Nk} & \cdots & x_{NM} \end{bmatrix}}^{\hat{x}} \overbrace{\begin{bmatrix} b_0 \\ b_1 \\ \vdots \\ b_k \\ \vdots \\ b_M \end{bmatrix}}^{\hat{b}}$$

$$= \underbrace{\begin{bmatrix} 1 & 1 & \cdots & 1 & \cdots & 1 \\ x_{11} & x_{21} & \cdots & x_{k2} & \cdots & x_{N1} \\ \vdots & \vdots & & & & \\ x_{1k} & x_{2k} & \cdots & x_{kk} & \cdots & x_{Nk} \\ \vdots & \vdots & & & & \\ x_{1M} & x_{2M} & \cdots & x_{kM} & \cdots & x_{NM} \end{bmatrix}}_{\hat{x}^T} \overbrace{\begin{bmatrix} y_1 \\ y_2 \\ \vdots \\ y_k \\ \vdots \\ y_N \end{bmatrix}}^{\hat{y}} \tag{2.95}$$

towards the resumed matrix form

$$\left(\hat{x}^T \cdot \hat{x}\right)\hat{b} = \hat{x}^T \cdot \hat{y} \tag{2.96}$$

By the left multiplication of. eq. (2.96) with the term $\left(\hat{x}^T \cdot \hat{x}\right)^{-1}$ one finds the solution for the estimates:

$$\hat{b} = \underline{\left(\hat{x}^T \cdot \hat{x}\right)^{-1} \cdot \hat{x}^T} \cdot \hat{y} \tag{2.97}$$

known as the Moore-Penrose generalized inverse matrix.

Let's apply this formula for simple linear regression:

$$y_{(i)} = b + ax_{(i)} : \begin{cases} y_1 = b + ax_1 + e_1 \\ y_2 = b + ax_2 + e_2 \\ \vdots \\ y_N = b + ax_N + e_N \end{cases} \tag{2.98}$$

Thus we have the matrices:

$$\hat{y} = \begin{bmatrix} y_1 \\ y_2 \\ \vdots \\ y_N \end{bmatrix}, \ \hat{b} = \begin{bmatrix} b = b_0 \\ a = b_1 \end{bmatrix}, \ \hat{x} = \begin{bmatrix} 1 & x_1 \\ 1 & x_2 \\ \vdots & \\ 1 & x_N \end{bmatrix} \tag{2.99}$$

So we have the composed matrix

$$\hat{A} = \left(\hat{x}^T \cdot \hat{x}\right) = \begin{bmatrix} 1 & 1 & \dots & 1 \\ x_1 & x_2 & \dots & x_N \end{bmatrix} \begin{bmatrix} 1 & x_1 \\ 1 & x_2 \\ \vdots & \\ 1 & x_N \end{bmatrix} = \begin{bmatrix} N & \sum_i^N x_i \\ \sum_i^N x_i & \sum_i^N x_i^2 \end{bmatrix} \tag{2.100}$$

and its determinant

$$\left|\hat{x}^T \cdot \hat{x}\right| = N \sum_i^N x_i^2 - \left(\sum_i^N x_i\right)^2 \tag{2.101}$$

allowing the minors

$$\begin{cases} \widetilde{A}_{11} = \sum_i^N x_i^2; \ \widetilde{A}_{12} = -\sum_i^N x_i; \\ \widetilde{A}_{21} = -\sum_i^N x_i; \ \widetilde{A}_{22} = N; \end{cases} \tag{2.102}$$

entering to the formed associated matrix of minors

$$\hat{A}^* = \begin{bmatrix} \widetilde{A}_{11} & \widetilde{A}_{21} \\ \widetilde{A}_{12} & \widetilde{A}_{22} \end{bmatrix} = \begin{bmatrix} \sum_i x_i^2 & -\sum_i x_i \\ -\sum_i x_i & N \end{bmatrix} \tag{2.103}$$

and finally providing the inverse of the initial matrix (2.100)

$$\hat{A}^{-1} = \left(\hat{x}^T \cdot \hat{x}\right)^{-1} = \frac{\hat{A}^*}{\left|\hat{A}\right|} = \frac{\left(\hat{x}^T \cdot \hat{x}\right)^*}{\left|\hat{x}^T \cdot \hat{x}\right|} \tag{2.104}$$

So leaving with the analytical matrix

$$\left(\hat{x}^T \cdot \hat{x}\right)^{-1} = \begin{bmatrix} \dfrac{\sum_i x_i^2}{N\sum_i x_i^2 - \left(\sum_i x_i\right)^2} & \dfrac{-\sum_i x_i}{N\sum_i x_i^2 - \left(\sum_i x_i\right)^2} \\ \dfrac{-\sum_i x_i}{N\sum_i x_i^2 - \left(\sum_i x_i\right)^2} & \dfrac{N}{N\sum_i x_i^2 - \left(\sum_i x_i\right)^2} \end{bmatrix} \tag{2.105}$$

Along the direct matrix product

$$\hat{x}^T \cdot y = \begin{bmatrix} 1 & 1 & \dots & 1 \\ x_1 & x_2 & \dots & x_N \end{bmatrix} \begin{bmatrix} y_1 \\ y_2 \\ \vdots \\ y_N \end{bmatrix} = \begin{bmatrix} \sum_i^N y_i \\ \sum_i^N x_i y_i \end{bmatrix} \tag{2.106}$$

one can finally involve the Moore-Penrose result (2.97) to produce the matrix solution

$$\hat{b} = \left(\hat{x}^T \cdot \hat{x}\right)^{-1} \cdot \hat{x}^T \cdot y = \begin{bmatrix} \dfrac{\sum_i x_i^2 \sum_i y_i - \sum_i x_i \sum_i x_i y_i}{N\sum_i x_i^2 - \left(\sum_i x_i\right)^2} \\ \dfrac{-\sum x_i \sum y_i + N\sum x_i y_i}{N\sum_i x_i^2 - \left(\sum_i x_i\right)^2} \end{bmatrix} = \begin{bmatrix} b_0 = b \\ b_1 = a \end{bmatrix} \tag{2.107}$$

It finely recovers the basic mono-regression correlation results of Eq. (2.28) and (2.29), thus confirming the validity of the multi-linear matrix formalism reliability; nevertheless, it becomes cumbersome in analytical terms for higher regression forms, for which the actual Spectral-SAR or Spectral-Diagonal SAR successfully overcome, see Chapter 3 of the present volume. For the moment, we go further with the statistical basics analysis.

2.3 INFERENTIAL ANALYSIS OF STATISTICAL SAMPLES

2.3.1 Z-STATISTICS

One considers the descriptive measures for a statistical sample:

- The module: for a variable x that value M_0 of x for which its probability function (or density of probability) is maximum:

$$f(x) = M_0 = \max \tag{2.108}$$

- The median (Figure 2.6): is that value μ for which is equally probably that a random variable to be bigger (on the right sight) or smaller (on the left side) to μ:

$$P(x < \mu) = P(x > \mu) = 0.5 \tag{2.109}$$

For measuring the dispersion of some probability values of a random variable one can use:

- The amplitude

$$t_2 - t_1 : \; P(t_1 \le x \le t_2) = 1 \tag{2.110}$$

- The inter-quartile amplitude

$$Q_{\frac{3}{4}} - Q_{\frac{1}{4}} : \; P\left(Q_{\frac{1}{4}}\right) = \frac{1}{4}; \; P\left(Q_{\frac{3}{4}}\right) = \frac{3}{4} \tag{2.111}$$

More generally, we understand by quartile the solution of equation:

$$f(x) = p \text{ or } p(x) = p \tag{2.112}$$

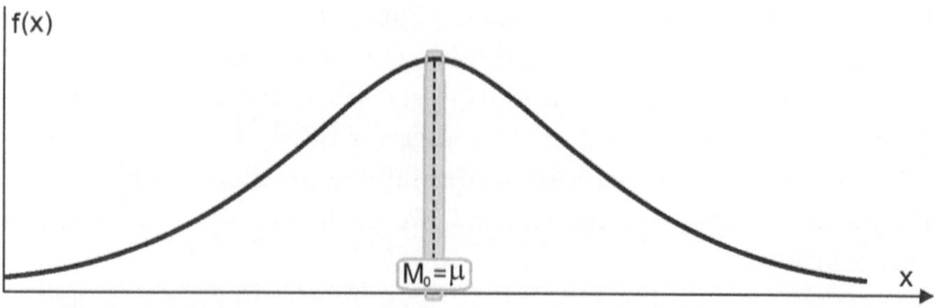

FIGURE 2.6 Illustration of the median value for a normal (Gaussian) distribution.

with $p \in (0,1)$ and $f(x)$ or $p(x)$ – a partition function.

In this condition we have:

$$p = \frac{1}{2} \text{ the quartile is the median itself } \mu \tag{2.113a}$$

$$p = \frac{1}{4} \text{ the quartile is the inferior quartile } Q_{\frac{1}{4}} \tag{2.113b}$$

$$p = \frac{3}{4} \text{ the quartile is the superior quartile } Q_{\frac{3}{4}} \tag{2.113c}$$

$$p = \frac{i}{10},\ i = \overline{1,9} : \text{ the quartiles are named as decile} \tag{2.113d}$$

$$p = \frac{i}{100},\ i = \overline{1,99} : \text{ the quartiles are named as centile} \tag{2.113e}$$

- The significance tests are based on the *null hypothesis*:

$$H_0 = \bar{a} = a_0 \ (\text{e.g., } \bar{x} = \mu) \tag{2.114}$$

If the hypothesis is verified the statistical distribution which made the evaluation of a is true!

- The trusting interval (a_1, a_2) is calculated such that (see Figure 2.7):

$$p(a_1 \le a \le a_2) = 1 - \alpha \tag{2.115}$$

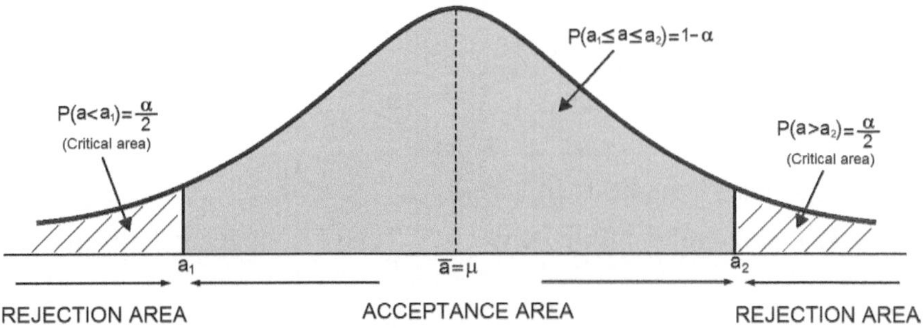

FIGURE 2.7 The main regions of a normal distribution in terms of acceptance and rejection areas respecting a given pre-defined probability confidence interval.

so that:

- the *trusting level* is called as $1-\alpha$ (: 0.9, 0.95, 0.99)
- the *significance level* is called as α (: 0.1, 0.05, 0.01)

- General phases of a significance test are:
  (i) Hypothesis statement containing the null hypothesis (to be memorized and accepted)

$$H_0:\ x = x_0 \tag{2.116}$$

and the alternative hypothesis (to be the rejected hypothesis)

$$H_1:\ x \neq x_0 \tag{2.117}$$

in terms of the x: the variable estimated from the sample (predicted); and x_0: the variable to be expected (the admitted/observed standard)

  (ii) Specifying the significance level α; this represents the possibility to commit an error by rejecting the null hypothesis H_0, rejection caused by sampling errors, in case that the hypothesis is, actually, true.
  (iii) Establishing the critical value (tabulated) $P_{\frac{\alpha}{2}}$ or $f_{\frac{\alpha}{2}}$ or $F_{\frac{\alpha}{2}}$
  (iv) Calculating the statistical value (P, f or F) with the aid of the data provided from sample (the predicted one, $|y^{predict}>$)
  (v) Formulating the acceptation criteria of hypothesis H_0:

$$\text{if: } -P_{\frac{\alpha}{2}} < P < P_{\frac{\alpha}{2}} \tag{2.118}$$

then H_0 is accepted so the P distribution is correct as an *estimation* for the expected values

- else, hypothesis H_0 is rejected

Note that the relationship between F, P and f is, in general, given by:

$$F(x_k) = P(x < x_k) = \int_{-\infty}^{x_k} f(x)dx \tag{2.119}$$

where

- $F(x_k)$: is the repartition function,
- $P(x < x_k)$: the repartition probability
- $\int_{-\infty}^{x_k} f(x)dx$: the repartition density, in relation with repartition function by the direct derivative, ,$f(x) = dF(x)/dx$ and with the graphical representation given in Figure 2.8.

Let's consider the current case of normal distribution (2.46) reloaded as

$$f(x) = \frac{1}{\sigma\sqrt{2\pi}} e^{-\frac{(x-\mu)^2}{2\sigma^2}}, \quad x \in \Re, \quad \int_{-\infty}^{+\infty} f(x)dx = 1 \tag{2.120}$$

with the (already) calculated statistics:

$$\langle x \rangle = \int_{-\infty}^{+\infty} xf(x)dx = \int_{-\infty}^{+\infty} (x - \mu + \mu) f(x)dx = \int_{-\infty}^{+\infty} (x - \mu) f(x)dx + \mu \int_{-\infty}^{+\infty} f(x)dx = \mu \tag{2.121}$$

$$D(x) = \langle (x-\mu)^2 \rangle - \langle (x-\mu) \rangle^2 = \int_{-\infty}^{+\infty} (x-\mu)^2 f(x)dx = \sigma^2 \tag{2.122}$$

so we can say that we deal with, genuinely, by the normal repartition of symmetric Gaussian graphic (Figure 2.9).

There is interest to work with normalized version of the random variable $x = N(\mu, \sigma)$ so that to obtain the normalized Gaussian function:

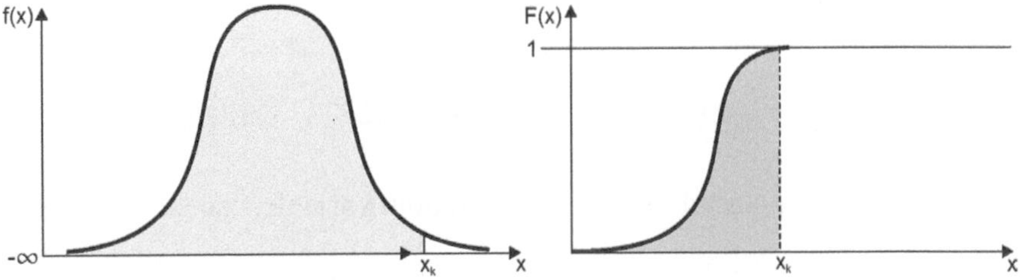

FIGURE 2.8 Total probability (left) and interval probability (right) distributions for a normal Gaussian repartition of a sample of experimental points.

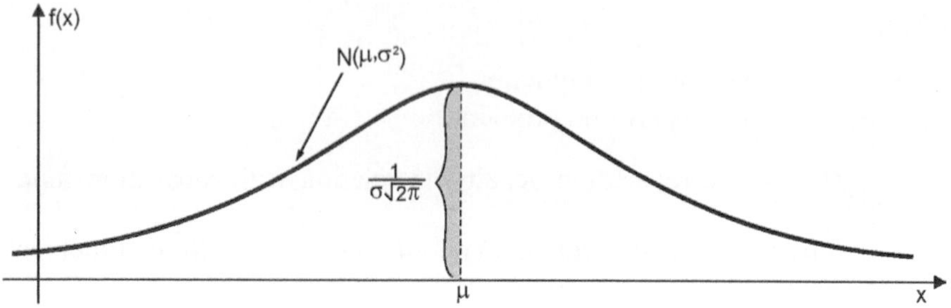

FIGURE 2.9 Normal distribution and its amplitude.

$$z = \frac{x-\mu}{\sigma} = N(0,1) \tag{2.123}$$

Proof:

$$Med\, z = \langle z \rangle = Moment\, no.1$$

$$= \left\langle \frac{x-\mu}{\tau} \right\rangle = \frac{1}{\sigma}\langle x-\mu\rangle = \frac{1}{\sigma}(\langle x\rangle - \mu) = \frac{1}{\sigma}(\mu-\mu) = 0 \tag{2.124}$$

$$D(z) = \langle z^2 \rangle (Moment\ no.2) - \langle z \rangle^2$$

$$= \left\langle \left(\frac{x-\mu}{\sigma}\right)^2 \right\rangle = \frac{1}{\sigma^2}\langle (x-\mu)^2 \rangle = \frac{\sigma^2}{\sigma^2} = 1 \equiv \sigma_z^2,\ \ \sigma_z = 1 \tag{2.125}$$

So the normal distribution goes as

$$N(0,1) = \varphi_{N(0,1)}(t) = \frac{1}{\sqrt{2\pi}} \cdot e^{-\frac{t^2}{2}} \tag{2.126}$$

In these conditions, the repartition function

$$F(x) = P(x < x_k) = \int_{-\infty}^{x_k} f(x)dx \tag{2.127}$$

for the $N(0,1)$ as distribution of a given random variable, can be written as:

$$F_{N(0,1)}(x) = \int_{-\infty}^{x} \varphi_{N(0,1)}(t)dt = \int_{-\infty}^{0} \varphi_{N(0,1)}(t)dt + \int_{0}^{x} \varphi_{N(0,1)}(t)dt \tag{2.128}$$

Now, knowing that

$$1=\int_{-\infty}^{+\infty}\varphi_{N(0,1)}(t)dt=2\int_{0}^{\infty}\varphi_{N(0,1)}(t)dt=2\int_{-\infty}^{0}\varphi_{N(0,1)}(t)dt$$

one gets

$$\int_{0}^{\infty}\varphi_{N(0,1)}(t)dt=\int_{-\infty}^{0}\varphi_{N(0,1)}(t)dt=\frac{1}{2} \quad (2.129)$$

while the function

$$\Phi(x)=\int_{0}^{x}\varphi_{N(0,1)}(t)dt=\frac{1}{\sqrt{2\pi}}\int_{0}^{x}e^{-\frac{t^2}{2}}dt \quad (2.130)$$

is called the *normalized Laplace function*. It associate with the error integral function $Erf(x)$:

$$Erf(x)=\frac{2}{\sqrt{\pi}}\int_{0}^{x}e^{-\frac{t^2}{2}}dt \quad (2.131)$$

by the variable transformation

$$\frac{x}{\sqrt{2}}\to x..."t"\Rightarrow t\to\frac{t}{\sqrt{2}} \quad (2.132)$$

such that to have

$$\frac{1}{2}Erf\left(\frac{x}{\sqrt{2}}\right)=\frac{1}{2}\cdot\frac{2}{\sqrt{\pi}}\cdot\int_{0}^{x/\sqrt{2}}e^{-t^2}dt=\frac{1}{\sqrt{\pi}}\cdot\int_{0}^{x}e^{-\frac{t^2}{2}}\frac{dt}{\sqrt{2}}=\Phi(x);$$

$$\Phi(x)=\frac{1}{2}Erf\left(\frac{x}{\sqrt{2}}\right) \quad (2.133)$$

Altogether, we find that

$$F_{N(0,1)}(x)=\frac{1}{2}+\Phi(x) \quad (2.134)$$

here "x" from $N(0,1)$ being "$\frac{x-\mu}{\sigma}$" from $N(\mu, \sigma)$

Immediately, there results that for a normal random distribution $N(\mu, \sigma)$ we have the repartition function given by:

$$F_{N(\mu,\sigma)}(x) = \frac{1}{2} + \Phi\left(\frac{x-\mu}{\sigma}\right) \tag{2.135}$$

Then, the probability with that the random variable x will take values in the interval a and b for a given distribution will be:

$$P(a < f(x) < b) = F(b) - F(a) \tag{2.136}$$

which, in case of normal distribution will be written

$$\begin{aligned} P(a < N(\mu,\sigma) < b) &= F_{N(\mu,\sigma)}(b) - F_{N(\mu,\sigma)}(a) \\ &= \Phi\left(\frac{b-\mu}{\sigma}\right) - \Phi\left(\frac{a-\mu}{\sigma}\right) \end{aligned} \tag{2.137}$$

Because the intervals came across, one can proof that, for a repartition with a discrete unfold

$$X\begin{pmatrix} x_1 & x_2 & x_3 & \dots & x_n \\ p_1 & p_2 & p_3 & \dots & p_n \end{pmatrix} \tag{2.138}$$

with the known median μ and the dispersion σ, the *Chebyshev inequality* takes place:

$$P(|x-\mu| < \varepsilon) \geq 1 - \frac{\sigma^2}{\varepsilon^2},\ \varepsilon^2 \text{ positive} \tag{2.139}$$

Proof: we write the dispersion explicitly:

$$\sigma^2 = p_1(x_1-\mu)^2 + p_2(x_2-\mu)^2 + \dots + p_i(x_i-\mu)^2 + \dots + p_n(x_n-\mu)^2 \tag{2.140}$$

or, by noting

$$x_i - \mu = d_i \tag{2.141a}$$

one has the equivalent expression

$$\sigma^2 = p_1 d_1^2 + p_2 d_2^2 + \dots + p_i d_i^2 + \dots + p_n d_n^2 \tag{2.142}$$

Supposing now (for simplicity) that the deviation d_i is increasing in the indices order:

$$d_1 < d_2 < \ldots d_i < \ldots < d_n \tag{2.143}$$

we further note that

$$d_i = \varepsilon_i = x_i - \mu \tag{2.141b}$$

Now, when considering the case

$$\varepsilon_j = \begin{cases} \varepsilon, & \forall\, j = \overline{i+1, n} \\ 0, & \forall\, j = \overline{0, i} \end{cases} \tag{2.144}$$

then, certainly we have

$$\sigma^2 \geq \varepsilon^2 \left(p_{i+1} + \ldots + p_n\right) \quad \Big|: \varepsilon^2$$

$$\Leftrightarrow \frac{\sigma^2}{\varepsilon^2} \geq p_{i+1} + p_{i+2} + \ldots + p_n$$

$$1 - \frac{\sigma^2}{\varepsilon^2} \leq \underbrace{1 - \underbrace{\left(p_{i+1} + p_{i+2} + \ldots + p_n\right)}_{P(x_i - \mu = \varepsilon_i \geq \varepsilon)}}_{P(|x-\mu| < \varepsilon)} \text{ (QED)} \tag{2.145}$$

The utility of this inequality consists in evaluating the dispersion domain of values, for example,

$$\varepsilon = k: \; P\left(|x - \mu| < k\sigma\right) \geq 1 - \frac{\sigma^2}{k^2\sigma^2} = 1 - \frac{1}{k^2} \tag{2.146}$$

which, in special case $k = 3$ (3 times the dispersion) we have:

$$P\left(|x - \mu| < 3\sigma\right) \geq 1 - \frac{1}{3^2} = \frac{8}{9} = 0.89\,! \tag{2.147}$$

available for all kind of repartitions! In case of normalized normal distribution we have the representation of the Figure 2.10.

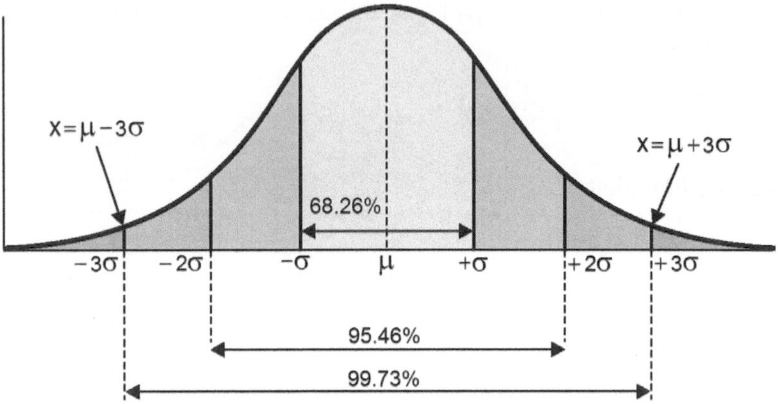

FIGURE 2.10 Representation of the normalized normal (Gaussian) distribution and its probability intervals in terms of its standard deviation respecting the median value.

Let's try to evaluate the next case:

$$P(\mu-3\sigma<N(\mu,\sigma)<\mu+3\sigma)=\Phi\left(\frac{\mu+3\sigma-\mu}{\sigma}\right)-\Phi\left(\frac{\mu-3\sigma-\mu}{\sigma}\right)$$
$$=\Phi(3)-\Phi(-3)$$
$$=2\Phi(3)=2\cdot(0.4987)=0.9974 \qquad (2.148)$$

where we used the odd property of Laplace function:

$$\Phi(-x)=\frac{1}{\sqrt{2\pi}}\int_0^{-x}e^{-\frac{t^2}{2}}dt \underset{\substack{t\to -x\\ -x\to t\ldots\text{"}x\text{"}}}{=} +\frac{1}{\sqrt{2\pi}}\int_0^{x}e^{-\frac{(-t^2)}{2}}d(-t)$$
$$=-\frac{1}{\sqrt{2\pi}}\int_0^{x}e^{-\frac{t^2}{2}}dt$$
$$=-\Phi(x) \qquad (2.149)$$

Furthermore, the Chebyshev inequality justifies the *large numbers law*, which says that the selection median $\bar{x}=\frac{\sum x_i}{n}$ converges in probability to the median value μ (expected) of a random variable of population for a large class of selection ($n\to\infty$)

$$\lim_{n\to\infty} P(|\bar{x}-\mu|<\varepsilon)=1 \quad (2.150)$$

Proof: the Chebyshev inequality recommends that

$$P(|x-\mu|<\varepsilon)\geq 1-\frac{\sigma^2}{\varepsilon^2} \quad (2.151)$$

Now, we look for the phenomenological determination $x \to \bar{x}$: $\sigma \to \sigma_{\bar{x}} = ?$ To this end we calculate the new average:

$$\langle \bar{x} \rangle = \left\langle \frac{\sum_i x_i}{n} \right\rangle == \frac{1}{n}\left\langle \sum_i x_i \right\rangle = \frac{1}{n}\sum_i \langle x_i \rangle = \frac{1}{n}\cdot \sum_i \mu = \frac{n\mu}{n} = \mu \quad \langle \bar{x} \rangle = \mu$$

(2.152)

from where we successively found the new dispersion:

$$\begin{aligned}
\sigma_{\bar{x}}^2 &= \langle \bar{x}^2 \rangle - \langle \bar{x} \rangle^2 \\
&= \left\langle \left(\frac{\sum x_i}{n} \right)^2 \right\rangle - \left\langle \left(\frac{\sum x_i}{n} \right) \right\rangle^2 \\
&= \frac{1}{n^2}\left[\left\langle \left(\sum x_i\right)^2 \right\rangle - \left\langle \left(\sum x_i\right) \right\rangle^2 \right] \\
&= \frac{1}{n^2}\left[\left\langle (x_1+x_2+\ldots+x_n)^2 \right\rangle - \left\langle (x_1+x_2+\ldots+x_n) \right\rangle^2 \right] \\
&= \frac{1}{n^2}\left[\left\langle \sum x_i^2 + 2\sum_{i<j} x_i x_j \right\rangle - \left(\left(\langle x_1 \rangle + \langle x_2 \rangle + \ldots + \langle x_n \rangle \right) \right)^2 \right] \\
&= \frac{1}{n^2}\left[\left(\sum_i \langle x_i^2 \rangle + 2\sum_{i<j} \langle x_i x_j \rangle - \sum_i \langle x_i \rangle^2 - 2\sum_{i<j} \langle x_i \rangle \langle x_j \rangle \right) \right]
\end{aligned} \quad (2.153)$$

For independent variables (measurements), i.e., for $\langle x_i x_j \rangle = \langle x_i \rangle \langle x_j \rangle$, we finally get:

$$\sigma_{\bar{x}}^2 = \frac{1}{n^2}\sum_i \left(\langle x_i^2 \rangle - \langle x_i \rangle^2 \right) = \frac{1}{n^2}\sum_i \sigma_x = \frac{n\sigma_x^2}{n^2} = \frac{\sigma_x^2}{n}$$

$$\sigma_{\bar{x}} = \frac{\sigma_x}{\sqrt{n}} \tag{2.154}$$

This way the Chebyshev inequality becomes:

$$P(|\bar{x} - \mu| < \varepsilon) \geq 1 - \frac{\sigma_{\bar{x}}^2}{\varepsilon^2} = 1 - \frac{\sigma^2}{n\varepsilon^2} \tag{2.155}$$

$$\lim_{n\to\infty} P(|\bar{x} - \mu| < \varepsilon) = 1 \tag{2.156}$$

This result is in accordance with the fact that for large samples we have the median limiting correspondences: $\bar{x} \xrightarrow{n\to\infty} \mu$.

2.3.2 THE SELECTION THEORY

2.3.2.1 From Global (Collectivity) to Local (Sample) Statistics

From a collectivity C we consider a selection, a sample or a pool relatively to collectivity, as a subset of individuals (events) $\varepsilon \subset C$ which are next to be studied from one or more statistical features; be the *volume* the number of individuals/events in collectivity or in sample, respectively. Accordingly, the selection can be:

- non-random, with the realizations:
  - *systemically*, after a specific path (e.g., from 10 to 100);
  - *typically*, when, knowing the previous information referring to collectivity there are considerate individuals with median values close to the median value of whole collectivity;
  - *scientifically*, when the collectivity is bedded (classified) after specific criteria, knowing the individual population for each classification.
- random with the features:
  - *frequently*, when the selected individual or event, after being analyzed, is reentering in collectivity (it has equal probability towards the others to be extracted in collectivities);
  - *non-repeatedly*, when the individual/event once selected is no more "embedded" in collectivity.

Therefore, if upon a selection of n-volume we extract the value (s) x_1, x_2, …, x_n repeatedly, then the random variable

$$x:\begin{vmatrix} x_1 & x_2 & \dots & x_n \\ \frac{1}{n} & \frac{1}{n} & \dots & \frac{1}{n} \end{vmatrix} = \begin{vmatrix} x_i \\ p_i \end{vmatrix} \tag{2.157}$$

is called the *choice variable* (or empirical distribution).

In these conditions we have specific characteristics as:

- *the empirical function of repetition* attached to the previous empirical distribution is defined as:

$$F_n(x) = \frac{n_x}{n} \tag{2.158}$$

with n_x: the number of the values $x_i < x$; $F_n(x)$ expresses the cumulated relative frequency
- *the selection median* representing the average value of empirical distribution:

$$\bar{x}_\varepsilon = \langle x \rangle_n = \sum_i p_i x_i = \sum_i \frac{1}{n} x_i = \frac{1}{n} \sum_{i=1}^{n} x_i \tag{2.159}$$

- *the selection dispersion* of the selection distribution with the form:

$$s_\varepsilon^2 = D_s = \left\langle \left(x - \langle x_n \rangle\right)^2 \right\rangle_n$$
$$= \sum_i p_i (x_i - \bar{x})^2 = \frac{1}{n} \sum_{i=1}^{n} (x_i - \bar{x})^2 \tag{2.160}$$

representing the standard deviation of the sample/choice.

Now, very important, for a statistical C-collectivity we have that

$$\varepsilon(\bar{x}, s) \subset C(\mu_x, \sigma_x) \tag{2.161}$$

with the statistical measures

$$\begin{cases} \langle \bar{x} \rangle = \mu \\ s = \sigma_{\bar{x}} = \dfrac{\sigma_x}{\sqrt{n}} \end{cases} \tag{2.162a}$$

with the representation given in Figure 2.11.

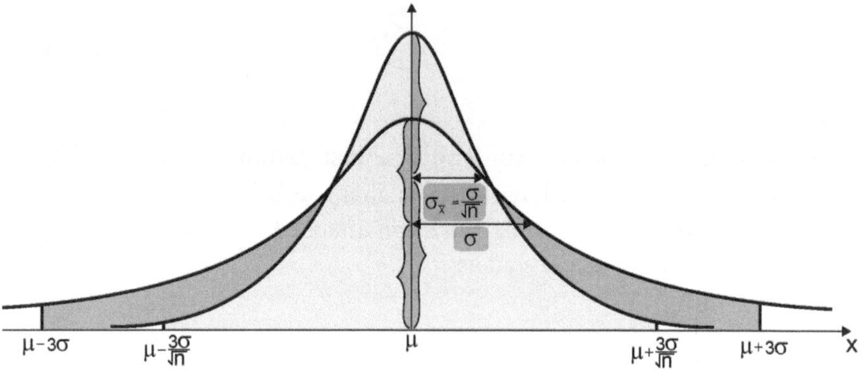

FIGURE 2.11 Illustration of the width effect on the amplitude of the normal distribution, so discriminating between a sample (with a shorten width) from the collectivity volume.

Worth distinguishing between:

- $\varepsilon\left(\mu, s = \dfrac{\sigma}{\sqrt{n}}\right)$ the sample having the same median as the collectivity, but a smaller s-dispersion, by the
- $C(\mu\ \sigma)$ as the general statistical collectivity

The main properties of the median and dispersion in a selection sample are

$$\text{(i)}\ \left\langle s^2 \right\rangle_n \Big|_{\substack{SELECTED \\ SAMPLE\ \varepsilon}} = \frac{n-1}{n}\sigma^2 \Big|_{COLLECTIVITY} \tag{2.162b}$$

Proof:

- We introduce the z random variable with the values

$$z_i = x_i - \mu,\ \ i = \overline{1, n}\Big|_C \tag{2.163}$$

Then:

$$\left\langle z_i \right\rangle_C = \left\langle (x_i - \mu) \right\rangle = \left\langle x_i \right\rangle - \mu = \mu - \mu = 0 \tag{2.164}$$

$$\left\langle z_i^2 \right\rangle_\varepsilon = \left\langle (x_i - \mu)^2 \right\rangle = \left\langle x_i^2 \right\rangle - 2\left\langle x_i \right\rangle \mu + \mu^2 = \left\langle x_i^2 \right\rangle - \mu^2 \Big|_C = \sigma^2 \Big|_C = \sigma_C^2 \tag{2.165}$$

- In addition,

$$\left\langle z_i z_j \right\rangle = \left\langle z_i \right\rangle \left\langle z_j \right\rangle = 0\ for\ j \neq i \tag{2.166}$$

based on the fact that $z_1, z_2, \ldots, z_n$ are independent variables

- Then:

$$\bar{z}_{\varepsilon} = \frac{z_1 + z_2 + \ldots + z_n}{n} = \frac{(x_1 - \mu) + (x_2 - \mu) + \ldots + (x_n - \mu)}{n}$$
$$= \frac{x_1 + x_2 + \ldots + x_n}{n} - \frac{n\mu}{n} \tag{2.167}$$

meaning that

$$\bar{z}_C = \bar{x} - \mu \tag{2.168}$$

from where we can write:

$$x_i - \bar{x} = x_i - \mu + \mu - \bar{x} = \underbrace{x_i - \mu}_{z_i} - \underbrace{(\bar{x} - \mu)}_{\bar{z}} = z_i - \bar{z} \tag{2.169}$$

Therefore, the same applies for the selection dispersion

$$s_{\varepsilon}^2 = \frac{1}{n}\sum_{i=1}^{n}(x_i - \bar{x})^2 = \frac{1}{n}\sum_{i=1}^{n}(z_i - \bar{z})^2 \tag{2.170}$$

In these conditions, we have:

$$\left\langle s^2 \right\rangle = \left\langle \frac{1}{n}\sum_{i=1}^{n}(z_i - \bar{z})^2 \right\rangle = \left\langle \frac{1}{n}\sum_{i=1}^{n}\left(z_i^2 - 2z_i\bar{z} + \bar{z}^2\right) \right\rangle$$
$$= \frac{1}{n}\left\langle \sum_{i=1}^{n} z_1^2 \right\rangle + \left\langle -\frac{1}{n}2\bar{z}\sum_i z_i + \frac{n}{n}\bar{z}^2 \right\rangle$$
$$= \frac{1}{n}\left\langle \sum_{i=1}^{n} z_1^2 \right\rangle - \left\langle \bar{z}^2 \right\rangle$$
$$= \frac{1}{n}\sum_{i=1}^{n}\underbrace{\left\langle z_1^2 \right\rangle}_{\sigma^2} - \left\langle \left(\frac{1}{n}\sum_i z_i\right)^2 \right\rangle$$
$$= \frac{n\sigma^2}{n} - \frac{1}{n^2}\left\langle \left(\sum_i z_i\right)^2 \right\rangle = \sigma^2 - \frac{1}{n^2}\left\langle \sum_i z_i^2 + 2\sum_{i<j} z_i z_j \right\rangle$$

$$= \sigma^2 - \frac{1}{n^2}\sum_i \underbrace{\left\langle z_i^2 \right\rangle}_{\sigma^2} - \frac{1}{n} 2 \sum_{i<j} \underbrace{\left\langle z_i \right\rangle}_{=0} \underbrace{\left\langle z_j \right\rangle}_{=0}$$

$$= \sigma^2 - \frac{n\sigma^2}{n^2} = \frac{n-1}{n}\sigma^2 \tag{2.171}$$

(i) First of all, we demonstrate the relationship:

$$\left\langle s_\varepsilon^2 \right\rangle = \frac{n-1}{n}\sigma_C^2 \tag{2.172}$$

meaning, the dispersion median of selection $\left\langle s_\varepsilon^2 = D_C \right\rangle$ do not tend at the collectivity dispersion $\sigma_C^2 = D_C$, but to the collectivity dispersion minus the term $\frac{\sigma_C^2}{n}$, which introduces the notion of *dispersion biased (displaced) estimate* (of the selection) based on median dispersion of the sample (collectivity).

(ii) Backwards, if one introduces the quantity

$$s_\varepsilon^{*2} = \frac{n}{n-1}s_\varepsilon^2 = \frac{1}{n-1}\sum_{i=1}^{n}(x_i - \bar{x})^2 \tag{2.173}$$

then, obvious, we have:

$$\left\langle s_\varepsilon^{*2} \right\rangle = \sigma_C^2 \tag{2.174}$$

2.3.2.2 The A-Consequence: The Chi-Square Statistics

If the array $\{z_1, z_2, \ldots, z_n\}$ with $z_i = x_i - \mu$ follows a statistics of $N(0, \sigma)$ type, then, because we jointly have the sample-moments

$$\begin{cases} \left\langle z_i \right\rangle = 0 \\ \left\langle z_i^2 \right\rangle = \sigma_C^2 \end{cases} \tag{2.175}$$

then, the array $\left(\frac{x_1 - \mu}{\sigma}, \frac{x_2 - \mu}{\sigma}, \ldots, \frac{x_n - \mu}{\sigma}\right)$ follows the statistics $N(0,1)$. In these conditions, the sum variable of the array

$$\chi_n^2 = \frac{1}{\sigma^2}\sum_{i=1}^{n}(x_i - \mu)^2 \tag{2.176}$$

generates a statistics named as chi-squared type (χ^2) with n degree of liberty for whom the repartition function (called as the Helmert-Pearson function) looks like

$$F_n(x) = \begin{cases} \dfrac{1}{2^{\frac{n}{2}}\Gamma\left(\dfrac{n}{2}\right)}\displaystyle\int_0^x x^{\frac{n}{2}-1} e^{-\frac{x}{2}} dx, & x \geq 0 \\ 0 & , \quad x < 0 \end{cases} \tag{2.177}$$

and it is based on the probability density

$$\varphi_{\chi^2}(x) = \begin{cases} \dfrac{1}{2^{\frac{n}{2}}\Gamma\left(\dfrac{n}{2}\right)} x^{\frac{n}{2}-1} e^{-\frac{x}{2}}, & x \geq 0 \\ 0 & , \quad x < 0 \end{cases} \tag{2.178}$$

It is an asymmetrical function for small degrees of freedom (volumes) of the samples, while becoming more symmetrical with the increasing of n (see Figure 2.12).

In these conditions we have

$$\begin{cases} \left\langle x_{\chi^2} \right\rangle_{\varphi(\chi^2)} = n \\ D(\chi^2) = \left\langle (x_{\chi^2})^2 \right\rangle - \left\langle (x_{\chi^2}) \right\rangle^2 = 2n \end{cases} \tag{2.179}$$

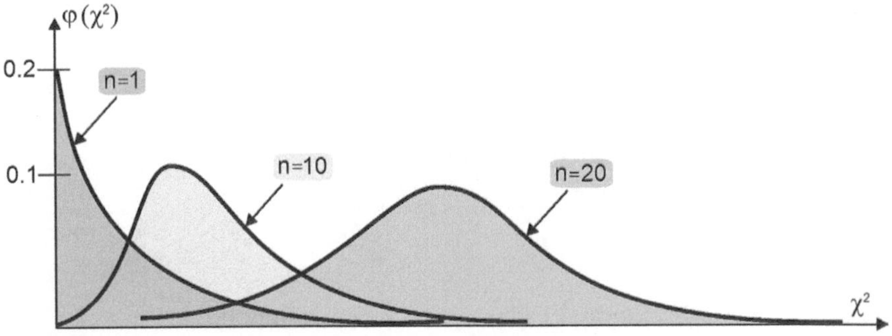

FIGURE 2.12 Illustration of the probability distribution shape hierarchies depending on the asymmetry factor departing the normal repartition of data, see Eq. (2.178).

Proof: before to proceed with the effective the demonstration we need the following adjacent concepts as the *Gamma function of Euler* (see also the Appendices of Volume I of the present five-volume book (Putz, 20016a), although here they are reloaded from another perspective, as the current statistical needs); the starting basic definition of Gamma Euler function is of integral nature and looks like

$$\Gamma(p)=\int_0^\infty x^{p-1}e^{-x}dx,\quad p>0 \tag{2.180}$$

with the properties

$$\begin{cases}\Gamma(p+1)=p\Gamma(p)\\ \Gamma(n+1)=n!\end{cases} \tag{2.181}$$

Therefore, we immediately have

$$\Gamma(p)=\int_0^\infty x^{p-1}e^{-x}dx=\frac{1}{p}\int_0^\infty [\partial(x^p)]e^{-x}dx=\frac{1}{p}\left[\underbrace{x^p\cdot e^{-x}\Big|_0^\infty}_{0}+\int_0^\infty x^p e^{-x}dx\right]=\frac{1}{p}\Gamma(p+1) \tag{2.182}$$

$$\Gamma(p+1)=p\Gamma(p) \tag{2.183a}$$

and by $p\leftrightarrow p-1$ we also have

$$\Gamma(p)=(p-1)\Gamma(p-1) \tag{2.183b}$$

And consecutively one can now evaluate the general term of Gamma function as

$$\Gamma(n+1)=n\cdot(n-1)\cdot\ldots\cdot 2\Gamma(1) \tag{2.183c}$$

where the first factor immediately turns out to be

$$\Gamma(1)=\int_0^\infty e^{-x}dx=-e^{-x}\Big|_o^\infty=1 \tag{2.184}$$

So leaving with the Gamma Euler function fundamental result

$$\Gamma(n+1)=n\cdot(n-1)\cdot\ldots\cdot 2\cdot 1=n! \tag{2.185}$$

Now, returning to the χ^2 statistics we have:

$$\left\langle x_{\chi^2}\right\rangle_{\varphi(\chi^2)} := \left\langle X_{\chi^2}\right\rangle_{\varphi} = \int_0^{\infty} x\cdot\varphi_{\chi^2}(x)dx$$

$$= \int_0^{\infty}\frac{1}{2^{\frac{n}{2}}\Gamma\left(\frac{n}{2}\right)}x^{\frac{n}{2}}e^{-\frac{x}{2}}dx = \frac{1}{2^{\frac{n}{2}}\Gamma\left(\frac{n}{2}\right)}\int_0^{\infty}x^{\frac{n}{2}}e^{-\frac{x}{2}}dx \qquad (2.186)$$

and, by changing the variables

$$\frac{x}{2} = t \Rightarrow \begin{cases} x = 2t \\ dx = 2dt \end{cases} \qquad (2.187)$$

we further have

$$\left\langle X_{\chi^2}\right\rangle_{\varphi} = \frac{1}{2^{\frac{n}{2}}\Gamma\left(\frac{n}{2}\right)}\int_0^{\infty}2^{\frac{n}{2}}t^{\frac{n}{2}}e^{-t}2dt = \frac{2}{\Gamma\left(\frac{n}{2}\right)}\int_0^{\infty}t^{\frac{n}{2}}e^{-t}dt = 2\frac{\Gamma\left(\frac{n}{2}+1\right)}{\Gamma\left(\frac{n}{2}\right)} = n \text{ (QED)}$$

(2.188)

For the dispersion calculus we evaluate the second order momentum:

$$\mu_2 = \left\langle X^2_{\chi^2}\right\rangle = \frac{1}{2^{\frac{n}{2}}\Gamma\left(\frac{n}{2}\right)}\int_0^{\infty}x^{\frac{n}{2}+1}e^{-\frac{x}{2}}dx = \frac{1}{2^{\frac{n}{2}}\Gamma\left(\frac{n}{2}\right)}\cdot(-2)\int_0^{\infty}x^{\frac{n}{2}+1}\left[\partial\left(e^{-\frac{x}{2}}\right)\right]dx$$

$$= \frac{-2}{2^{\frac{n}{2}}\Gamma\left(\frac{n}{2}\right)}\left[x^{\frac{n}{2}+1}e^{-\frac{x}{2}}\Big|_0^{\infty} - \int_0^{\infty}\left(\frac{n}{2}+1\right)x^{\frac{n}{2}}e^{-\frac{x}{2}}dx\right]$$

$$= \frac{2\left(\frac{n}{2}+1\right)}{2^{\frac{n}{2}}\Gamma\left(\frac{n}{2}\right)}\int_0^{\infty}x^{\frac{n}{2}}e^{-\frac{x}{2}}dx = 2n\frac{n+2}{2} = n(n+2)$$

(2.189)

So the dispersion will be

$$D(X_\chi) = \langle X_\chi^2 \rangle - \langle X_\chi \rangle^2 = n(n+2) - n^2 = 2n \text{ (QED)} \qquad (2.190)$$

Now we are in position to formulate the *statistical test significance*:

- A collectivity follows the normal distribution $N(\mu, \sigma)$;
- A sample of collectivity has a median $\bar{x}$ following the normal distribution too, yet of the form $N\left(\mu, \frac{\sigma}{\sqrt{n}}\right)$; while the dispersion of the sample under the form $\frac{n}{\sigma^2} s_\varepsilon^2$ follows a χ^2 distribution with $(n\text{-}1)$freedom degrees.

The last statement will be in next proven.

Consider $z_i = x_i - \mu$ $(i = \overline{1, n})$ independent variables and having the $N(0, \sigma)$ distribution as previous demonstrated; one can introduce their orthogonal transformation as follow:

$$y_i = \sum_{k=1}^{n} a_{ik} z_k, \quad i = \overline{1, n} \qquad (2.191)$$

with a_{ik} the components of an $(n \times n)$ orthogonal matrix $([A^T] \cdot [A] = [I])$ or explicitly written as

$$\sum_{k=1}^{n} a_{ik} a_{jk} = \delta_{ij} = \begin{cases} 1, & i = j \\ 0, & i^{-} = 0 \end{cases} \qquad (2.192)$$

where the first line of elements are all equal with the frequency $\frac{1}{\sqrt{n}}$, for example,

$$A = \begin{bmatrix} \frac{1}{\sqrt{n}} & \frac{1}{\sqrt{n}} & \frac{1}{\sqrt{n}} & \cdots & \frac{1}{\sqrt{n}} \\ \frac{1}{\sqrt{2}} & -\frac{1}{\sqrt{2}} & 0 & \cdots & 0 \\ \frac{1}{\sqrt{6}} & \frac{1}{\sqrt{6}} & -\frac{2}{\sqrt{6}} & \cdots & 0 \\ \vdots & & & & \\ \frac{1}{\sqrt{n(n-1)}} & \frac{1}{\sqrt{n(n-1)}} & \frac{1}{\sqrt{n(n-1)}} & \cdots & \frac{(n-1)}{\sqrt{n(n-1)}} \end{bmatrix} \qquad (2.193)$$

Then

$$\langle y_i \rangle = \sum_{R=1}^{n} a_{ik} \langle z_k \rangle = 0 \tag{2.194}$$

and

$$\langle y_i y_j \rangle = \left\langle \sum_{k=1}^{n} a_{ik} z_k \sum_{r=1}^{n} a_{jr} z_r \right\rangle = \left\langle \sum_{k,r} a_{ik} a_{jr} z_k z_r \right\rangle = \sum_{k=1}^{n} \sum_{r=1}^{n} a_{ik} a_{jr} \langle z_k z_r \rangle$$

$$= \sigma^2 \sum_{k=1}^{n} a_{ik} a_{jk} = \sigma^2 \delta_{ij} = \begin{cases} \sigma^2, & i = j \\ 0, & i \neq j \end{cases}$$

$$\Rightarrow \langle y_i y_j \rangle = \begin{cases} \sigma^2, & i = j \\ 0, & i \neq j \end{cases} \tag{2.195}$$

Note that for

$$\langle z_k z_r \rangle = \sigma^2 \neq 0 \text{ only for } k = r \tag{2.196a}$$

else

$$\langle z_k z_r \rangle = \langle z_k \rangle \langle z_r \rangle = 0 \cdot 0 = 0 \tag{2.196b}$$

In these conditions, $y_{i=\overline{1,n}}$ are uncorrelated two-by-two and although they are independent, y_i has the same statistics as z_i; this way, as based on the general formula of matrix transformation $[y] = A[z]$ there can be written that:

$$\sum_{i=1}^{n} y_i^2 = [y]^T [y] = (A[z])^T (A[z]) = [z]^T A^T A[z] = [z]^T [z] = \sum_{i=1}^{n} z_i^2 \tag{2.197}$$

so

$$\sum_{i=1}^{n} y_i^2 = \sum_{i=1}^{n} z_i^2 \tag{2.198}$$

and therefore one arrives to the same (chi-squared) statistics

$$\frac{1}{\sigma^2} \sum_{i=1}^{n} y_i^2 = \frac{1}{\sigma^2} \sum_{i=1}^{n} z_i^2 = x_{\chi^2, n} \tag{2.199}$$

On the other side

$$y_1 = \sum_{k=1}^{n} a_{1k} z_k = a_{11} z_1 + a_{12} z_2 + \ldots + a_{1n} z_n$$

$$= \frac{z_1 + z_2 + \ldots + z_n}{\sqrt{n}} = \sqrt{n} \frac{z_1 + z_2 + \ldots + z_n}{n}$$

$$= \sqrt{n} \cdot \bar{z} = \sqrt{n}(\bar{x} - \mu) \tag{2.200}$$

which allows the equivalent writing at the dispersion level the successive equations:

$$s_\varepsilon^2 = \frac{1}{n}\sum_i (z_i - \bar{z})^2 = \frac{1}{n}\sum_i \left(z_i^2 + \bar{z}^2 - 2z_i\bar{z}\right) = \frac{1}{n}\sum_i z_i^2 + \frac{n}{n}\bar{z}^2 - 2\bar{z}\frac{1}{n}\sum_i z_i$$

$$= \frac{1}{n}\sum_i z_i^2 - \bar{z}^2 = \frac{1}{n}\sum_i y_i^2 - \frac{y_1^2}{n} = \frac{1}{n}\left(\sum_{i=1} y_i^2 - y_1^2\right) = \frac{1}{n}\sum_{i=2} y_i^2$$

$$\frac{s_\varepsilon^2}{\sigma^2} = \frac{1}{n}\left(\frac{1}{\sigma^2}\sum_{i=2}^{n} y_i^2\right) = \frac{1}{n}\sum_{i=2}^{n}\left(\frac{y_i}{\sigma}\right)^2$$

$$\tag{2.201}$$

When recognizing the χ^2 distribution with (n-1) freedom degree

$$\sum_{i=2}^{n}\left(\frac{y_i}{\sigma}\right)^2 = x_{\chi^2,(n-1)} \tag{2.202}$$

From Eq. (2.201) we get

$$x_{\chi^2,n-1} = n \cdot \frac{s_\varepsilon^2}{\sigma^2} \tag{2.203}$$

or even more

$$x_{\chi^2,n-1} = \frac{(n-1)s_\varepsilon^{*2}}{\sigma^2} \tag{2.204}$$

once performing the replacement

$$s_\varepsilon^2 = n \cdot \frac{n-1}{n} s_\varepsilon^{*2} \tag{2.205}$$

In other words, the statistics $n\frac{s_\varepsilon^2}{\sigma^2}$ and $\frac{(n-1)s_\varepsilon^2}{\sigma^2}$ have the same form as χ^2 statistics with (n-1) degrees of freedom.

2.3.2.3 The B-Consequence: The t-Student Statistics

When the sample volume $\varepsilon\left(\bar{x}, s = \frac{\sigma}{\sqrt{n}}\right)$ is not high, for example, for $n <$ 30, and the collectivity dispersion $C(\mu,\sigma)$ is not known, then it can be formed the random variable

$$t = \frac{\bar{x} - \mu}{\frac{s}{\sqrt{n}}} \tag{2.206}$$

heuristically derived as follows:

$$\left.\frac{x_i - \mu}{\sigma}\right|_{\varepsilon} \rightarrow \left.\frac{\bar{x} - \mu}{s}\right|_{\varepsilon} \rightarrow \frac{\bar{x} - \mu}{\frac{\sigma}{\sqrt{n}}} \xrightarrow[\langle s^{*2}\rangle = \sigma^2]{} \frac{\bar{x}_{\varepsilon} - \mu}{\frac{s^{*}_{\varepsilon}}{\sqrt{n}}} \tag{2.207}$$

The random t variable features a special distribution called as the t-Student, by William Sealy Gosset in 1908, and transforms as:

$$t = \frac{\bar{x} - \mu}{\frac{s^{*}_{\varepsilon}}{\sqrt{n}}} \cdot \frac{\sigma}{\sigma} = \underbrace{\frac{\bar{x}_{\varepsilon} - \mu}{\frac{\sigma}{\sqrt{n}}}}_{(!)not\ z_t\ from N(0,1)} \cdot \frac{1}{\frac{s^{*}_{\varepsilon}}{\sigma}} = \frac{z_t}{\sqrt{\frac{x_{\chi^2, n-1}}{n-1}}} = \sqrt{n-1} \cdot \frac{z_t}{\sqrt{x_{\chi^2, n-1}}} \tag{2.208}$$

$$t = \sqrt{n-1}\frac{N(0,1)}{\chi^2(n-1)} \tag{2.209}$$

This way one is obtaining the t-Student distribution with n–1 degrees of freedom, for which probability function has the form:

$$\varphi_n(t) = \frac{1}{\sqrt{n\pi}} \frac{\Gamma\left(\frac{n}{2}+1\right)}{\Gamma\left(\frac{n}{2}\right)} \frac{1}{\left(1+\frac{t^2}{n}\right)^{\frac{n+1}{2}}}, t \in \Re, n \in N, n > 2 \tag{2.210}$$

having the geometrical representation in the Figure 2.13.

Observe that the t-statistics can be employed in various conditions:

(i) the selection standard deviation (corrected) s^{*}_{ε} is used in estimation of the standard deviation (unknown) of the collectivity population C with σ;

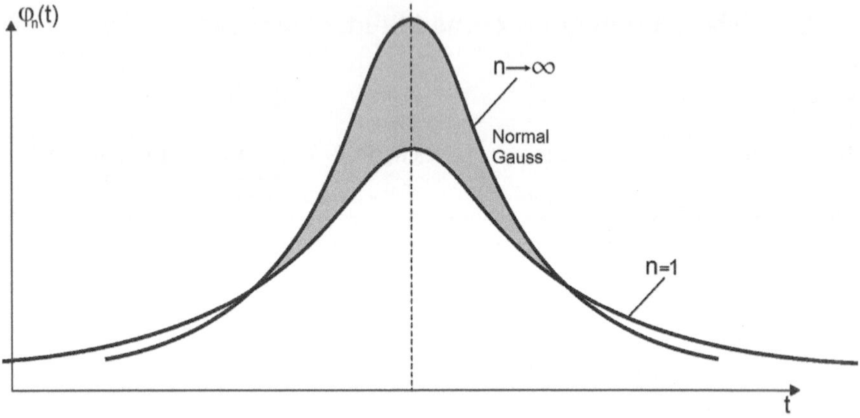

FIGURE 2.13 "Time" evolution/departure of normal (Gaussian) distribution respecting the data sample number (n) dependency: the t-Student distribution.

(ii) the sample volume is small: $2 < n < 30$;

(iii) the established collectivity follow (at the limit) the normal distribution law ($n \to \infty$), see Figure 2.13, a crucial fact in QSAR studies' interpretation.

Next, one would like showing that t-Student repartition has the features

$$\left.\begin{cases} \langle t \rangle = 0 \\ D(t) = \langle t^2 \rangle - \langle t \rangle^2 = \dfrac{n}{n-2} \xrightarrow{n \to \infty} 1 \end{cases}\right| N(0,1) \tag{2.211}$$

Yet, to this aim, we need of another math-intermezzo for the special functions. Actually, we need to show that

$$\Gamma\left(\frac{1}{2}\right) = \sqrt{\pi} \tag{2.212}$$

for which purpose the β-Euler function is introduced

$$\beta(p,q) = \int_0^1 x^{p-1}(1-x)^{q-1}dx \tag{2.213}$$

as well as the Beta–Gamma–Euler identity, see also the Appendices of Volume I of the present five-volume book (Putz, 2016a)

$$\beta(p,q)=\frac{\Gamma(p)\Gamma(q)}{\Gamma(p+q)} \tag{2.214}$$

The relationship (2.214) can be firstly proved after an idea of Dirichlet; accordingly, we can perform the equivalent transformations:

$$\Gamma(p)=\int_0^{\infty} x^{p-1}e^{-x}dx \underset{\substack{x=ty\\dx=tdy}}{=} \int_0^{\infty} t^{p-1}y^{p-1}e^{-ty}tdy = t^p\int_0^{\infty} y^{p-1}e^{-ty}dy$$

$$\Rightarrow \frac{\Gamma(p)}{t^p}=\int_0^{\infty} y^{p-1}e^{-ty}dy \tag{2.215}$$

With the appropriate replacements

$$\begin{cases} p \to p+q \\ t \to t+q \end{cases} \tag{2.216}$$

the last integral identity further becomes

$$\frac{\Gamma(p+q)}{(t+1)^{p+q}}=\int_0^{\infty} y^{p+q-1}e^{-(1+t)y}dy \tag{2.217}$$

It can be then appropriately integrated under the form

$$\Gamma(p+q)\int_0^{\infty}\frac{t^{p-1}}{(t+1)^{p+q}}dt=\int_0^{\infty} t^{p-1}dt\int_0^{\infty} y^{p+q-1}e^{-(1+t)y}dy \tag{2.218}$$

The integral in left hand side of (2.218) is transformed by the variable exchange, as:

$$\theta=\frac{1}{1+t}=\begin{cases} 0, & t=0 \\ 1, & t\to\infty \end{cases} \tag{2.219a}$$

$$\theta+\theta t=t \Leftrightarrow \theta=t(1-\theta) \Leftrightarrow t=\frac{\theta}{1-\theta} \tag{2.219b}$$

$$\& \; dt=\frac{1-\theta+\theta}{(1-\theta)^2}=\frac{1}{(1-\theta)^2}d\theta \;\&\; 1+t=1+\frac{\theta}{1-\theta}=\frac{1}{1-\theta} \tag{2.219c}$$

So that eq. (2.218) is solved by Beta-Euler integral:

$$\int_0^\infty \frac{t^{p-1}}{(t+1)^{p+q}}dt = \int_0^1 \frac{\theta^{p-1}}{(1-\theta)^{p-1}}\cdot(1-\theta)^{p+q}\cdot\frac{1}{(1-\theta)^2}d\theta = \int_0^1 \theta^{p-1}\cdot(1-\theta)^{q-1}d\theta = \beta(p,q) \tag{2.220a}$$

This way, Eq. (2.218) now rewrites as

$$\Gamma(p+q)\beta(p,q) = \int_0^\infty y^{p+q-1}e^{-y}\underline{\underline{\left(\int_0^\infty t^{p-1}e^{-ty}dt\right)}}dy \tag{2.221}$$

With the second integral equivalently transformed

$$\int_0^\infty t^{p-1}e^{-ty}dt = \int_0^\infty \frac{(ty)^{p-1}}{y^{p-1}}\cdot e^{-ty}\cdot\frac{d(ty)}{y} = \frac{1}{y^p}\int_0^\infty (ty)^{p-1}\cdot e^{-ty}d(ty) = \underline{\underline{\frac{\Gamma(p)}{y^p}}} \tag{2.220b}$$

equation (2.221) takes now the compact form

$$\Gamma(p+q)\beta(p,q) = \int_0^\infty y^{p+q-1}e^{-y}\cdot\frac{\Gamma(p)}{y^p}dy = \Gamma(p)\int_0^\infty y^{q-1}e^{-y}dy = \Gamma(p)\Gamma(q) \tag{2.222}$$

so finely proving the Eq. (2.214). This formula does now permit us to write

$$\beta\left(\frac{1}{2},\frac{1}{2}\right) = \frac{\left[\Gamma\left(\frac{1}{2}\right)\right]^2}{\underbrace{\Gamma(1)}_{=1}} = \left[\Gamma\left(\frac{1}{2}\right)\right]^2 \tag{2.223a}$$

On the other hand, with the β-Euler definition (2.213) we have:

$$\beta\left(\frac{1}{2},\frac{1}{2}\right) = \int_0^1 x^{\frac{1}{2}-1}(1-x)^{\frac{1}{2}-1}dx = \int_0^1 x^{-\frac{1}{2}}(1-x)^{-\frac{1}{2}}dx$$

$$= \int_0^1 \frac{dx}{\sqrt{x-x^2}} = \int_0^1 \frac{\partial\left(x-\frac{1}{2}\right)dx}{\sqrt{\frac{1}{4}-\left(x-\frac{1}{2}\right)^2}} = \arcsin\left.\frac{x-\frac{1}{2}}{\frac{1}{2}}\right|_0^1 = \arcsin(2x-1)\Big|_0^1$$

$$= \frac{\pi}{2} - \left(-\frac{\pi}{2}\right) = \pi \tag{2.223b}$$

By comparing Eqs. (2.223a) and (2.223b), we indeed get the Eq. (2.212) result:

$$\Gamma\left(\frac{1}{2}\right) = \sqrt{\pi} \tag{2.223c}$$

With this, we can move forward to determine the t-Student distribution first momentum:

$$\langle t \rangle = \int_{-\infty}^{+\infty} \frac{1}{\sqrt{n\pi}} \frac{\Gamma\left(\frac{n+1}{2}\right)}{\Gamma\left(\frac{n}{2}\right)} \cdot t \cdot \underline{\underline{\left(1+\frac{t^2}{n}\right)^{-\frac{n+1}{2}} dt}} \tag{2.224a}$$

By implementing the changing of variable such as

$$\begin{cases} 1+\frac{t^2}{n} = y \Rightarrow t^2 = n(y-1) \\ t = \sqrt{n} \cdot \sqrt{y-1} = \sqrt{n} \cdot (y-1)^{1/2} \\ dt = \sqrt{n} \cdot \frac{1}{2}(y-1)^{-1/2} \end{cases} \tag{2.225}$$

so that to finally arrive:

$$\langle t \rangle = \frac{1}{\sqrt{n\pi}} \frac{\Gamma\left(\frac{n+1}{2}\right)}{\Gamma\left(\frac{n}{2}\right)} \frac{1}{2} n \cdot 2 \int_0^{\infty} y^{-\frac{n+1}{2}} dy = \frac{n}{\sqrt{n\pi}} \frac{\Gamma\left(\frac{n+1}{2}\right)}{\Gamma\left(\frac{n}{2}\right)} \cdot \left. \frac{y^{-\frac{n+1}{2}+1}}{-\frac{n+1}{2}+1} \right|_0^{\infty} = 0 \text{ (QED)} \tag{2.224b}$$

For the dispersion calculation we need the second order momentum:

$$\mu_2(t) = \langle t^2 \rangle = \frac{1}{\sqrt{n\pi}} \frac{\Gamma\left(\frac{n+1}{2}\right)}{\Gamma\left(\frac{n}{2}\right)} \int_{-\infty}^{\infty} t^2 \left(1+\frac{t^2}{n}\right)^{-\frac{n+1}{2}} dt = 2\frac{1}{\sqrt{n\pi}} \frac{\Gamma\left(\frac{n+1}{2}\right)}{\Gamma\left(\frac{n}{2}\right)} \int_0^{\infty} t^2 \left(1+\frac{t^2}{n}\right)^{-\frac{n+1}{2}} dt \tag{2.226}$$

which by using the variable changing

$$\left\| \begin{array}{l} t^2 = ny \\ t = \sqrt{n} \cdot y^{1/2} \\ dt = \sqrt{n} \cdot \dfrac{1}{2} y^{-1/2} dy \end{array} \right.$$

we can further calculate:

$$\left\langle t^2 \right\rangle = \frac{n^{1+\frac{1}{2}}}{\sqrt{n\pi}} \frac{\Gamma\left(\frac{n+1}{2}\right)}{\Gamma\left(\frac{n}{2}\right)} \int_0^\infty y^{1-t} (1+y)^{-\frac{n+1}{2}} dy = \frac{n}{\sqrt{\pi}} \frac{\Gamma\left(\frac{n+1}{2}\right)}{\Gamma\left(\frac{n}{2}\right)} \underbrace{\int_0^\infty y^{\frac{3}{2}-1} (1+y)^{\frac{n}{2}-1+\frac{3}{2}} dy}_{\beta\left(\frac{3}{2}, \frac{n-2}{2}\right)}$$

$$= \frac{n}{\sqrt{\pi}} \frac{\Gamma\left(\frac{n+1}{2}\right)}{\Gamma\left(\frac{n}{2}\right)} \frac{\Gamma\left(\frac{n-2}{2}\right) \cdot \Gamma\left(\frac{3}{2}\right)}{\Gamma\left(\frac{n+1}{2}\right)} = \frac{n}{\sqrt{\pi}} \frac{\Gamma\left(\frac{n-2}{2}\right)}{\Gamma\left(\frac{n}{2}\right)} \underbrace{\Gamma\left(1+\frac{1}{2}\right)}_{\left(\frac{1}{2}\right)\Gamma\left(\frac{1}{2}\right)=\frac{1}{2}\sqrt{\pi}}$$

$$= \frac{n}{\sqrt{\pi}} \cdot \frac{1}{\frac{n-2}{2}} \cdot \frac{1}{2}\sqrt{\pi} = \frac{n}{n-2} \tag{2.227}$$

From here the t-Student dispersion follows as announced in (2.211):

$$D(t) = \left\langle t^2 \right\rangle - \left\langle t \right\rangle^2 = \frac{n}{n-2} \xrightarrow{n \to \infty} 1 \quad \text{(QED)} \tag{2.228}$$

this way confirming it belongs to the normal distribution; we find again the normal distribution $N(0,1)$.

2.3.2.4 The C-consequence: The Fisher Statistics

When we compare two random samples we can be interested if the medians are equivalent, but most of the time we are not interested if those two samples or selections are equally "grouped" around medians, i.e., whether their dispersions are equivalent or they are significant differently. In this case, neither the Gauss nor the t-Student repartitions help, but the actual matter it is solved by the introduced specific partition called as Fischer-Snedecor (F) statistics. Accordingly, if we have two samples: one of m-volume $\subset N(\mu_1, \sigma_1)$ statistical collectivity, and the other of n-volume $\subset N(\mu_2, \sigma_2)$ statistical collectivity, both with unknown standard deviations but estimated by $\overset{*}{s}_m, \overset{*}{s}_n$, respectively, we can consider the random variable

$$F = \frac{\overset{*}{s}_m}{\overset{*}{s}_n} = \frac{\sigma_1 \chi^2_{m-1}}{\sigma_2 \chi^2_{n-1}} \cdot \frac{n-1}{m-1} \sim \frac{\chi^2_{m-1}}{\chi^2_{n-1}} \tag{2.229}$$

which shows that the F-distribution has a double dependence of $(m - 1)$ and $(n - 1)$ degrees of freedom. If the samples are distinctly characterized but on the same collectivity, then we have

$$\sigma_1 = \sigma_2 = \sigma \quad F = \frac{\chi^2_{m-1}}{\chi^2_{n-1}} \frac{n-1}{m-1} \tag{2.230}$$

Furthermore, if the degrees of freedom rise

$$(m \to \infty, m \to \infty) \text{ then } \begin{cases} \overset{*}{s}_m \to \sigma \\ \overset{*}{s}_n \to \sigma \end{cases} \text{and so } F \to 1 \tag{2.231}$$

this way, the F-distribution is expressed in a way that the biggest dispersion to be on the numerator. Based on the fact that $F \to 1$ for $m,n \to \infty$ there cane be tested if two selection dispersions are estimation of the same general dispersion σ^2. The geometrical interpretation of F-distribution is given in Figure 2.14.

Analytically, F is characterized by a distribution (density) function

$$f_{m,n}(F) = \frac{\Gamma\left(\frac{m+n}{2}\right)}{\Gamma\left(\frac{m}{2}\right)\Gamma\left(\frac{n}{2}\right)} \left(\frac{m}{n}\right)^{\frac{m}{2}} F^{\frac{m}{2}-1} \left(1 + \frac{m}{n} F\right)^{-\frac{m+n}{2}}, \quad F > 0 \tag{2.232}$$

As customarily, we like calculating the moments of Fisher-distribution

$$\begin{cases} \langle F \rangle = ? \\ \& \ D(F) = \langle F^2 \rangle - \langle F \rangle^2 = ? \end{cases} \tag{2.233}$$

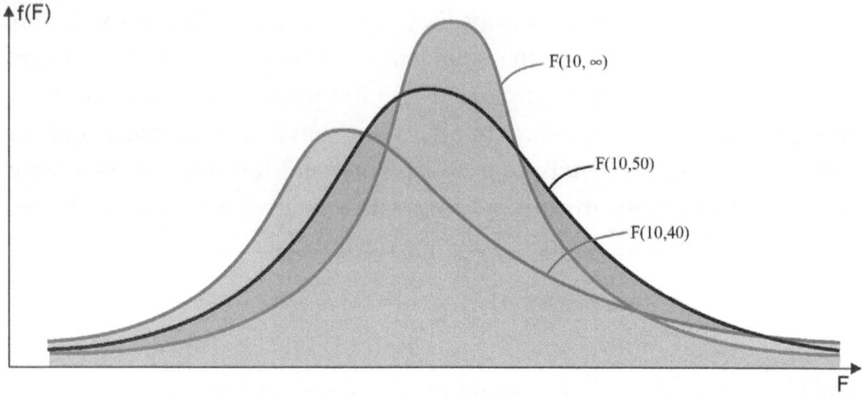

FIGURE 2.14 The Fischer-Snedecor distribution function representations by various degrees of freedom F(m, n) counting for the data numbers of two compared samples' repartitions.

For the first momentum one has:

$$\langle F\rangle = \frac{\Gamma\left(\frac{m+n}{2}\right)}{\Gamma\left(\frac{m}{2}\right)\Gamma\left(\frac{n}{2}\right)}\left(\frac{m}{n}\right)^{\frac{m}{2}}\int_0^\infty F^{\frac{m}{2}}\left(1+\frac{m}{n}F\right)^{-\frac{m+n}{2}}dF \tag{2.234}$$

and, by performing the variable change:

$$\frac{m}{n}F = y \quad \Rightarrow \quad \begin{cases} F = \dfrac{n}{m}y \\ dF = \dfrac{n}{m}dy \end{cases} \tag{2.235}$$

along by using the previous demonstrated equality for Beta-Euler (2.220a), now rewritten as

$$\beta(p,q) = \int_0^1 x^{p-1}(1-x)^{q-1}dx = \int_0^\infty \frac{\theta^{p-1}}{(1+\theta)^{p+q}}d\theta \tag{2.236a}$$

together with the Gamma-Euler properties

$$\left| \begin{array}{l} \Gamma(p+1) = p\Gamma(p) \\ \Gamma(p-1) = \dfrac{1}{p-1}\Gamma(p) \end{array} \right. \tag{2.236b}$$

we obtain:

$$\langle F\rangle = \frac{n}{m}\frac{\Gamma\left(\frac{m+n}{2}\right)}{\Gamma\left(\frac{m}{2}\right)\Gamma\left(\frac{n}{2}\right)}\left(\frac{m}{n}\right)^{\frac{m}{2}}\left(\frac{m}{n}\right)^{\frac{m}{2}}\int_0^{\infty} y^{\frac{m}{2}}(1+y)^{-\frac{m+n}{2}}dy$$

$$= \frac{n}{m}\frac{\Gamma\left(\frac{m+n}{2}\right)}{\Gamma\left(\frac{m}{2}\right)\Gamma\left(\frac{n}{2}\right)}\underbrace{\int_0^{\infty}\frac{y^{\left(1+\frac{m}{2}\right)-1}}{(1+y)^{\left(1+\frac{m}{2}\right)+\left(\frac{n}{2}-1\right)}}dy}_{\underbrace{\beta\left(1+\frac{m}{2},\frac{n}{2}-1\right)}_{\frac{\Gamma\left(1+\frac{m}{2}\right)\Gamma\left(\frac{n}{2}-1\right)}{\Gamma\left(\frac{m}{2}+\frac{n}{2}\right)}}}$$

$$= \frac{n}{m}\cdot\frac{m}{2}\cdot\frac{1}{\frac{n}{2}-1} = \frac{n}{n-2};$$

$$\langle F\rangle = \frac{n}{n-2} \tag{2.237}$$

Next we calculate the second order momentum as:

$$\mu_2(F) = \langle F^2\rangle = \frac{\Gamma\left(\frac{m+n}{2}\right)}{\Gamma\left(\frac{m}{2}\right)\Gamma\left(\frac{n}{2}\right)}\left(\frac{m}{n}\right)^{\frac{m}{2}}\int_0^{\infty} F^{\frac{m}{2}+1}\left(1+\frac{m}{n}F\right)^{-\frac{m+n}{2}}dF$$

$$= \frac{n}{m}\frac{\Gamma\left(\frac{m+n}{2}\right)}{\Gamma\left(\frac{m}{2}\right)\Gamma\left(\frac{n}{2}\right)}\left(\frac{m}{n}\right)^{\frac{m}{2}}\left(\frac{n}{m}\right)^{\frac{m}{2}+1}\int_0^{\infty} y^{\frac{m}{2}+1}(1+y)^{-\frac{m+n}{2}}dy$$

$$= \frac{\Gamma\left(\frac{m+n}{2}\right)}{\Gamma\left(\frac{m}{2}\right)\Gamma\left(\frac{n}{2}\right)}\frac{n^2}{m^2}\int_0^{\infty}\frac{y^{\left(\frac{m}{2}+2\right)-1}}{(1+y)^{\left(\frac{m}{2}+2\right)+\left(\frac{n}{2}-2\right)}}dy$$

$$= \frac{\Gamma\left(\frac{m+n}{2}\right)}{\Gamma\left(\frac{m}{2}\right)\Gamma\left(\frac{n}{2}\right)}\frac{n^2}{m^2}\underbrace{\beta\left(\frac{m}{2}+2,\frac{n}{2}-2\right)}_{\frac{\Gamma\left(\frac{m}{2}+2\right)\Gamma\left(\frac{n}{2}-2\right)}{\Gamma\left(\frac{m}{2}+\frac{n}{2}\right)}}$$

$$=\frac{n^2}{m^2}\frac{\Gamma\left(\frac{m}{2}+2\right)}{\Gamma\left(\frac{m}{2}\right)}\cdot\frac{\Gamma\left(\frac{n}{2}-2\right)}{\Gamma\left(\frac{n}{2}\right)} \tag{2.238a}$$

The final calculation uses the recurrent Gamma-Euler relations, see Eq. (2.183a):

$$\begin{cases}\Gamma\left(\frac{m}{2}+1+1\right)=\left(\frac{m}{2}+1\right)\Gamma\left(\frac{m}{2}+1\right)=\left(\frac{m}{2}+1\right)\frac{m}{2}\Gamma\left(\frac{m}{2}\right)\\ \Gamma\left(\frac{n}{2}-1-1\right)=\frac{1}{\frac{n}{2}-2}\Gamma\left(\frac{n}{2}-1\right)=\frac{1}{\left(\frac{n}{2}-2\right)\left(\frac{n}{2}-1\right)}\Gamma\left(\frac{n}{2}\right)\end{cases} \tag{2.239}$$

to arrive at the Fisher second momentum

$$\left\langle F^2\right\rangle=\frac{n^2}{m^2}\frac{m+2}{2}\cdot\frac{m}{2}\cdot\frac{2}{m-4}\cdot\frac{2}{m-2}=\frac{n^2(m+2)}{m(n-4)(n-2)} \tag{2.238b}$$

and finally to the Fisher statistical dispersion:

$$D(F)=\left\langle F^2\right\rangle-\left\langle F\right\rangle^2$$

$$=\frac{n^2(m+2)}{m(n-4)(n-2)}-\frac{n^2}{(n-2)^2}=\frac{2n^2(m+n-2)}{m(n-2)^2(n-4)} \tag{2.240}$$

2.4 ROBUSTNESS OF CORRELATIONS: CONFIDENCE INTERVALS

2.4.1 *VALIDATION OF STATISTICAL INFORMATION*

2.4.1.1 Synopsis of Statistical Concepts

The most important quantities and qualities in measurements/correlations stand as given in the Table 2.3.

TABLE 2.3 The Median and Dispersion Concepts in "Global"/Collectivity and "Local"/ Sample Statistical Realizations

Statistical Quantity	Global Property	Local Property	General "quantum" statistical Meaning
Median	μ: $\langle x\rangle_c = \mu$	m: $\langle \bar{x}\rangle_{\varepsilon_N \text{ or } n} = \mu$	Localization
Dispersion/Variance	σ: $\langle x^2\rangle - \langle x\rangle^2 = \sigma_x^2 \equiv \sigma$	s: $\langle \bar{x}^2\rangle - \langle \bar{x}\rangle^2 = \sigma_{\bar{x}}^2 = \frac{\sigma_x^2}{n}$	Delocalization
Overall statistical characterization	$C(\mu,\sigma)$	$\varepsilon\left(\mu, s = \frac{\sigma}{\sqrt{n}}\right)$	Connected by $\sigma_C^2 = \frac{n}{n-1}\langle s_\varepsilon^2\rangle$

Accordingly, one may summarize the main statistics characterized so far by: random variable, density probability function, median and dispersion, as following:

- *Z-statistics* is a fundamental statistics mainly driven by the Gaussian distribution (see Section 2.3.1), with:
  - Random variable:

$$z = \frac{x-\mu}{\sigma} \in N(0,1) \tag{2.241a}$$

  - Statistical Function:

$$\varphi(z) = \frac{1}{\sqrt{2\pi}} e^{-\frac{z^2}{2}} \tag{2.241b}$$

  - *Median* (relating the first statistical momentum) along the *dispersion* (relating the second statistical momentum, see Section 2.2.2):

$$\begin{cases} \mu(z) = 0 \\ D(z) = 1 \end{cases} \tag{2.241c}$$

- χ^2-statistics introduces the sample (ε) statistics behavior inside a collectivity (C) statistics, being driven by the displaced statistical information, i.e., at the level of dispersion information, otherwise local-to-global equalized by appropriate factorization:

$$s_\varepsilon \rightarrow \underbrace{s_\varepsilon^{*2}}_{\substack{UN-BIASED\\ESTIMATION}} = \frac{n}{n-1} \times \underbrace{s_\varepsilon^2}_{\substack{BIASED\\ESTIMATION}} \Rightarrow \left\langle s_\varepsilon^{*2} \right\rangle = \sigma_C^2 \tag{2.242}$$

χ^2-statistics works with the main statistical information as (see Section 2.3.2.2):

- Random variable:

$$\chi_{n,n-1}^2 = \sum_{i=1}^{n} z_i^2 = \frac{1}{\sigma^2} \sum_{i=1}^{n} (x_i - \mu)^2 \tag{2.243}$$

- Statistical Function:

$$\varphi\left(x_{\chi_n^2}\right) = \frac{1}{2^{\frac{n}{3}} \Gamma\left(\frac{n}{2}\right)} x_{\chi^2}^{\frac{n}{2}-1} \cdot e^{-\frac{1}{2} x_{\chi_n^2}} \tag{2.244}$$

- Median and dispersion:

$$\begin{cases} \mu\left(x_{\chi^2}\right) = n \\ D\left(x_{\chi^2}\right) = 2n \end{cases} \tag{2.245}$$

- t-statistics (with n–1 degrees of freedom) is mainly designed for small sized samples in a collectivity, and is characterized by the next statistical information (see Section 2.3.2.3):
  - Random variable:

$$t = \frac{\overline{x}_\varepsilon - \mu_C}{\frac{s_\varepsilon}{\sqrt{n}}} \in \sqrt{n-1} \; \frac{N(0,1)}{\chi_{(n-1)}^2} \tag{2.246}$$

  - Statistical function:

$$\varphi_n(t) = \frac{1}{\sqrt{n\pi}} \frac{\Gamma\left(\frac{n+1}{2}\right)}{\Gamma\left(\frac{n}{2}\right)} \frac{1}{\left(1 + \frac{t^2}{n}\right)^{\frac{n+1}{2}}}, t \in \Re, \;\; n \in N, \;\; n > 2 \tag{2.247}$$

- Median and dispersion:

$$\begin{cases} \langle t \rangle = 0 \\ D(t) = \dfrac{n}{n-2} \end{cases} \tag{2.248}$$

- F-statistics is mainly designed for comparing two samples in/of (volumes m and n) a collectivity (see, Section 2.3.2.4) and employs the following statistical information:
  - Random variable:

$$F = \frac{\overset{*}{s}_m}{\overset{*}{s}_n} = \frac{\sigma_{1(m)}\chi^2_{m-1}}{\sigma_{2(n)}\chi^2_{n-1}} \cdot \frac{n-1}{m-1} \underset{\sigma_1=\sigma_2=\sigma}{=} \frac{\chi^2_{m-1}}{\chi^2_{n-1}} \cdot \frac{n-1}{m-1} \tag{2.249}$$

  - Statistical function:

$$\varphi_{m,n}(F) = \frac{\Gamma\left(\dfrac{m+n}{2}\right)}{\Gamma\left(\dfrac{m}{2}\right)\Gamma\left(\dfrac{n}{2}\right)} \left(\frac{m}{n}\right)^{\frac{m}{2}} F^{\frac{m}{2}-1} \left(1 + \frac{m}{n}F\right)^{-\frac{m+n}{2}}, \quad F > 0 \tag{2.250}$$

  - Median and dispersion:

$$\begin{cases} \langle F \rangle = \dfrac{n}{n-2} \\ D(F) = \dfrac{2n^2(m+n-2)}{m(n-2)^2(n-4)} \end{cases} \tag{2.251}$$

However, all these statistical realizations have in common the probability trusting intervals presented in Figure 2.15, where:

- P-value represents the smallest *level of significance* (α) from which the statistic test (no matter which one of the above) can lead us to the *rejection of the null hypothesis* H_0, where the H_0 hypothesis is given by, for example, in the bilateral test, by $P(v_1 \le v \le v_2) = 1 - \alpha$, with a chosen α (confidence) value.
- the null hypothesis H_0 is accepted if the calculated value v is founded between the two *critical values* calculated for those statistics.
- If not, H_0 is rejected.

Next section further discusses on the statistical hypothesis verifications.

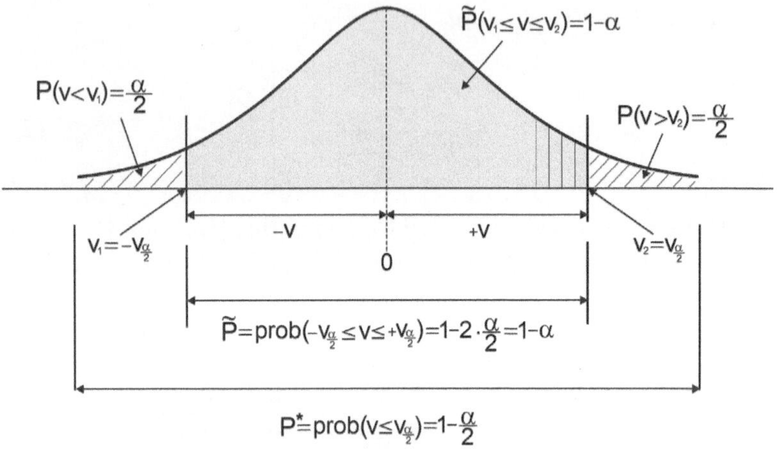

FIGURE 2.15 The main probability distribution in the acceptance (marked) and rejected (hashed) areas of a normal (Gaussian) data repartition.

2.4.1.2 Validation of Statistical Hypothesis for Median Value in Bilateral Case

One likes to test the null-hypothesis by the systematics of the Table 2.4.

Worth noting that accepting H_0 means that "the evidences" delivered from the sample are insufficient for rejection on the significance level considered and not that the hypothesis is necessary true.

Next we are going to present the step in developing a significance test by working examples.

(i) Assume $H_0 = \mu = \mu_0$: we test if the median of the sample ε is the same as the one considered on the collectivity C-level; otherwise the alternative of H_0 is $\mu_0 \neq \mu_0$;

TABLE 2.4 Systematics of the Statistical Hypothesis Verification Cases

Decision	**Situation in collectivity**	
	H_0 true	***H_0 false***
Acceptance of H_0	Correct decision	Second kind error
Rejection of H_0	First kind error	Correct decision

(ii) specify the significance level α;
(iii) choose the right statistics;
(iv) construct the confidence interval based on the Chebyshev inequality (see Section 2.3.1) as applied to median value

$$P\left(\left|\bar{x}-\mu\right|<k\sigma_{\bar{x}}\right)\geq 1-\underbrace{\frac{1}{k^2}}_{\sim\alpha}\Leftrightarrow P\left(-k\sigma_{\bar{x}}\leq\mu-\bar{x}\leq k\sigma_{\bar{x}}\right)\geq 1-\frac{1}{k^2} \tag{2.252}$$

By taking into consideration the fact that k depends on the probability (and backwards) we have:

$$\bar{x}-k_p\sigma_{\bar{x}}\leq\mu\leq\bar{x}+k_p\sigma_{\bar{x}} \tag{2.253}$$

representing *the confidence interval to be tested* according to the statistical hypothesis $H_0 = \mathrm{m} = \mathrm{m}_0$.

In this case the normalized random variable is

$$k_p=\frac{\bar{x}-\mu}{\sigma_{\bar{x}}}=\frac{\bar{x}_\varepsilon-\mu_C}{\dfrac{\sigma_C}{\sqrt{n}}},\ \sigma_{\bar{x}}=s_{\bar{x}}=z\in N(0,1)\ \text{distribution} \tag{2.254}$$

when we know from collectivity the information of $\mu_c = \sigma_c$ while testing $\bar{x}_\varepsilon = \mu_C$.

Accordingly, one rewrites the confidence interval as

$$\bar{x}-z(P)\frac{\sigma}{\sqrt{n}}<\mu<\bar{x}+z(P)\frac{\sigma}{\sqrt{n}} \tag{2.255}$$

covering the μ value with a $P = 1-\alpha$ probability, providing it satisfies the analytical relation (see Figure 2.16):

$$1-\alpha=P\underbrace{\left(\left|z\right|-z_{\frac{\alpha}{2}}\right)}_{-z_{\frac{\alpha}{2}}\leq z\leq +z_{\frac{\alpha}{2}}}=P\left(x<z_{\frac{\alpha}{2}}\right)-P\left(x<-z_{\frac{\alpha}{2}}\right)$$

$$=\int_{-\infty}^{z_{\frac{\alpha}{2}}}\varphi_{N(0,1)}(z)dz-\int_{-\infty}^{-z_{\frac{\alpha}{2}}}\varphi_{N(0,1)}(z)dz=\underbrace{F\left(z_{\frac{\alpha}{2}}\right)}_{:=1-\frac{\alpha}{2}}-\underbrace{F\left(-z_{\frac{\alpha}{2}}\right)}_{:=+\frac{\alpha}{2}}$$

$$= \int_{-\infty}^{0} \varphi_{N(0,1)}(z)dz + \int_{0}^{z_{\frac{\alpha}{2}}} \varphi_{N(0,1)}(z)dz - \left[\int_{-\infty}^{0} \varphi_{N(0,1)}(z)dz - \int_{-z_{\frac{\alpha}{2}}}^{0} \varphi_{N(0,1)}(z)dz \right]$$

$$= \int_{0}^{z_{\frac{\alpha}{2}}} \varphi_{N(0,1)}(z)dz + \underbrace{\int_{-z_{\frac{\alpha}{2}}}^{0} \varphi_{N(0,1)}(z)dz}_{\underbrace{-\int_{0}^{-z_{\frac{\alpha}{2}}} \varphi_{N(0,1)}(z)dz}_{\begin{cases} -dz \xrightarrow{z \to -z} +dz \\ \varphi(-z) = -\varphi(z) \end{cases}}}$$

$$= 2\int_{0}^{z_{\frac{\alpha}{2}}} \varphi_{N(0,1)}(z)dz = 2\Phi\left(z_{\frac{\alpha}{2}}\right) \tag{2.256}$$

$$\Rightarrow \Phi\left(z_{\frac{\alpha}{2}}\right) = \frac{1-\alpha}{2} \tag{2.257}$$

or alternatively

$$\Phi\left(z_{\frac{\alpha}{2}}\right) = \frac{1-\alpha}{2} \Leftarrow 1 - \frac{\alpha}{2} = \int_{-\infty}^{z_{\frac{\alpha}{2}}} \varphi(z)dz = \frac{1}{2} + \Phi\left(z_{\frac{\alpha}{2}}\right) \tag{2.258}$$

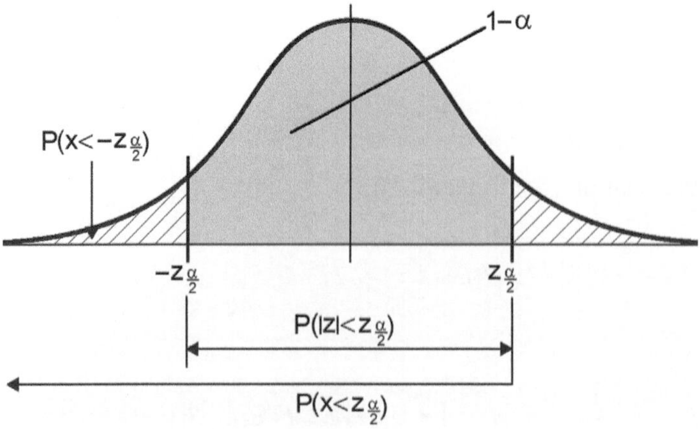

FIGURE 2.16 The main probability distribution quotations for a working normal (Gaussian) repartition of data of the z-statistics.

Now, for the numerical specialization we have:

$$\alpha = 0.06 \Rightarrow 1-\alpha = 0.94 \Rightarrow \frac{1-\alpha}{2} = 0.47 \tag{2.259}$$

So that from Laplace tabulated values one finds that Eq. (2.258) has the particular fulfillment

$$\Phi\underbrace{(0.1808)}_{z_{\frac{\alpha}{2}}} = \underbrace{0.47}_{\frac{1-\alpha}{2}} \Rightarrow z_{\frac{\alpha}{2}} = 0.1808 \tag{2.260}$$

The actual conclusion is that if we practically have

$$z = \left.\frac{\overline{x}_{estimated} - \mu_{0\,tested}}{\frac{\sigma}{\sqrt{n}}}\right|_{\substack{calculated\ when\\ \mu=\mu_0\ is\ hypothetically\ valid}} \in \underbrace{\left[-z_{\frac{\alpha}{2}}, +z_{\frac{\alpha}{2}}\right]}_{[-0.1808, +0.1808]} \tag{2.261}$$

then the hypothesis is accepted; if not, H_0 is rejected.

Before presenting alternative tests about dispersion we have to clarify the calculation law for the dispersion of the sum of two or more independent random variables. To this aim we proceed sequentially as following:

(i) consider $z = x + y$, with $D(x) = \langle x^2\rangle - \langle x\rangle^2$ and $D(y) = \langle y^2\rangle - \langle y\rangle^2$; then we ask for $D(z) = ?$ So we will do successively:

$$\begin{aligned}
\sigma_z^2 &= D(z) = \langle z^2\rangle - \langle z\rangle^2\\
&= \langle (x+y)^2\rangle - \langle x+y\rangle^2\\
&= \langle x^2 + 2xy + y^2\rangle - [\langle x\rangle + \langle y\rangle]^2\\
&= \langle x^2\rangle + 2\langle xy\rangle + \langle y^2\rangle - \langle x\rangle^2 - 2\langle x\rangle\langle y\rangle - \langle y\rangle^2\\
&= D(x) + D(y) = \sigma_x^2 + \sigma_y^2
\end{aligned} \tag{2.262}$$

(ii) in same way we can show that:

$$\text{if } x = a = ct. \Rightarrow \begin{cases}\langle x\rangle = a\\ D(a) = 0\end{cases} \tag{2.263a}$$

$$D(ax) = a^2 D(x), \quad a = ct \tag{2.263b}$$

$$D(x + a) = D(x) \tag{2.263c}$$

$$D(a + bx) = b^2 D(x), \tag{2.263d}$$

$$D(ax + by) = a^2 D(x) + b^2 D(y) \tag{2.263e}$$

$$D(x - y) = D(x) + D(y) \tag{2.264a}$$

meaning overall that

$$\sigma_{(x-y)}^2 = \sigma_x^2 + \sigma_y^2 \ (!) \tag{2.264b}$$

This way one arrives to the *law of propagation of errors*! It can be proved by considering the function $u = f(x, y)$ with x and y random variables; then, for relatively small standard deviations Δx, Δy we have:

$$u + \Delta u = f(x + \Delta x, y + \Delta y) \cong f(x, y) + \left(\frac{\partial u}{\partial x}\right)\Delta x + \left(\frac{\partial u}{\partial y}\right)\Delta y + \ldots$$

$$\Rightarrow \Delta u = \left(\frac{\partial u}{\partial x}\right)\Delta x + \left(\frac{\partial u}{\partial y}\right)\Delta y + \ldots \tag{2.265}$$

Next, by summing on sample and squaring we further obtain (with Δx_i, Δy_i either positive or negative):

$$\sum_{i=1}^{n}(\Delta u_i)^2 = \left(\frac{\partial u}{\partial x}\right)^2 \sum_{i=1}^{n}(\Delta x_i)^2 + 2\left(\frac{\partial u}{\partial x}\right)\left(\frac{\partial u}{\partial y}\right)\underbrace{\sum_{i=1}^{n}\Delta x_i \Delta y_i}_{(>0\&<0)\Rightarrow=0} + \left(\frac{\partial u}{\partial y}\right)^2 \sum_{i=1}^{n}(\Delta y_i)^2 \ \Bigg| \ :n \tag{2.266a}$$

$$\Leftrightarrow \underbrace{\left[\frac{1}{n}\sum_{i=1}^{n}(\Delta u_i)^2\right]}_{\sigma_u^2} = \underbrace{\left[\frac{1}{n}\sum_{i=1}^{n}(\Delta x_i)^2\right]}_{\sigma_x^2}\left(\frac{\partial u}{\partial x}\right)^2 + \underbrace{\left[\frac{1}{n}\sum_{i=1}^{n}(\Delta y_i)^2\right]}_{\sigma_y^2}\left(\frac{\partial u}{\partial y}\right)^2 \tag{2.266b}$$

$$\Rightarrow \sigma_{\bar{u}}^2 = \left(\frac{\partial u}{\partial x}\right)^2 \sigma_{\bar{x}}^2 + \left(\frac{\partial u}{\partial y}\right)^2 \sigma_{\bar{y}}^2 \tag{2.267a}$$

or, in terms of estimations we equivalently have

$$s_{\bar{u}^2}^2 = \left(\frac{\partial u}{\partial x}\right)^2 s_{\bar{x}}^2 + \left(\frac{\partial u}{\partial y}\right)^2 s_{\bar{y}}^2 \tag{2.267b}$$

2.4.1.3 Validation of Statistical Hypothesis for the Difference Between the Median Values

One would like to verify if there is a significant difference between the median values of the two populations, μ_1 and μ_2; testing the null hypothesis

$$H_1 : \mu_1 = \mu_2 \text{ (or } \mu_1 - \mu_2 = 0) \tag{2.268}$$

we distinguish two working cases

(a) *The samples are of big volumes* $(n_1\ n_2 \geq 30)$
Based on the central limit theorem, one can use the *z*-statistics, where

$$z_{calculated} = \frac{(\bar{x}_1 - \bar{x}_2) - (\mu_1 - \mu_2)}{\sigma_{\bar{x}_1 - \bar{x}_2}} \stackrel{H_0}{=} \frac{\bar{x}_1 - \bar{x}_2}{\sigma_{\bar{x}_1 - \bar{x}_2}} \tag{2.269}$$

and where, by further considering the combined errors (see the previous section) we have:

$$\sigma_{\bar{x}_1 - \bar{x}_2} = \sqrt{\sigma_{\bar{x}_1 - \bar{x}_2}^2} = \sqrt{\sigma_{\bar{x}_1}^2 + \sigma_{\bar{x}_2}^2} = \sqrt{\frac{\sigma_1^2}{n_1} + \frac{\sigma_2^2}{n_2}} \tag{2.270}$$

If the dispersion σ_1 & σ_2 are unknown but the samples are big enough one can use their unbiased estimators:

$$s_{\bar{x}_1 - \bar{x}_2} = \sqrt{\frac{s_1^{*2}}{n_1} + \frac{s_2^{*2}}{n_2}} \tag{2.271}$$

with

$$z_{calc} = \frac{\bar{x}_1 - \bar{x}_2}{s_{\bar{x}_1 - \bar{x}_2}} \tag{2.272}$$

For the bilateral test there is checked whether

$$z_{calc} \in \left[-z_{\frac{\alpha}{2}}, +z_{\frac{\alpha}{2}} \right] \text{ with } z_{calc} : \Phi\left(z_{\frac{\alpha}{2}} \right) = \frac{1-\alpha}{2} \tag{2.273}$$

for a certain significance level α.

If valid, that means that H_0 is accepted with a risk degree (of being wrong) ~ α!

(b) *The samples are of small volume* ($n_1, n_2 < 30$)

(i) *The dispersions (σ_1^2 and σ_2^2) are unknown on collectivity*;

In this case the condition of null hypothesis H_0 ($\mu_1 - \mu_2 = 0$) allows the use of the t-Student statistics:

$$t^{calc}_{\bar{x}_1 - \bar{x}_2, k} = \frac{\bar{x}_1 - \bar{x}_2}{s_{\bar{x}_1 - \bar{x}_2}}, \tag{2.274}$$

$$s_{\bar{x}_1 - \bar{x}_2} = \sqrt{\frac{s_1^{*2}}{n_1} + \frac{s_2^{*2}}{n_2}} \tag{2.275}$$

with the number of freedom degrees obtained from the weighted formula

$$k = \frac{1}{\frac{c^2}{k_1} + \frac{(1-c)^2}{k_2}} \xrightarrow{k_1 = k_2 = k_0} k_0 \cdot \frac{1}{1 - 2c^2} \tag{2.276}$$

where the c-term was considered as

$$c = \frac{s_1^{*2}/n_1}{\frac{s_1^{*2}}{n_1} + \frac{s_2^{*2}}{n_2}} \tag{2.277}$$

In our case, $k_1 = n_1 - 1$ and $k_2 = n_2 - 1$ are the freedom degrees of the Student distribution for the separated populations "1" and "2" with unknown σ_1 and σ_2; so we have:

$$c^2 = \frac{\left(s_1^{*2}/n_1\right)^2}{\left(\frac{s_1^{*2}}{n_1} + \frac{s_2^{*2}}{n_2}\right)^2} \Rightarrow (1-c)^2 = \frac{\left(\frac{s_2^{*2}}{n_2}\right)^2}{\left(\frac{s_1^{*2}}{n_1} + \frac{s_2^{*2}}{n_2}\right)^2} \tag{2.278}$$

There is an equal probability that the products will give "+" and "–" values, so to cancel each other (viz. reciprocal orthogonality by null scalar product in a QSAR analysis); this way the resulting degree of freedom for the t-Student distribution for the difference of the unknown medians σ_1 and σ_2 will be

$$k = \frac{\left(\frac{s_1^{*2}}{n_1} + \frac{s_2^{*2}}{n_2}\right)^2}{\frac{1}{n_1 - 1}\left(\frac{s_1^{*2}}{n_1}\right)^2 + \frac{1}{n_2 - 1}\left(\frac{s_2^{*2}}{n_2}\right)^2} \tag{2.279}$$

This "degrees of freedom" factor is to be searched in dedicated statistical tables in order to verify the condition (see Figure 2.17)

$$t_{\bar{x}_1 - \bar{x}_2, k}^{calc} \in \left[-t_{1-\frac{\alpha}{2}}, t_{1-\frac{\alpha}{2}}\right] \tag{2.280}$$

which is equivalent with the probability condition:

$$P\left(t^{calc} \in \left(-t_{1-\frac{\alpha}{2}}, t_{1-\frac{\alpha}{2}}\right) \middle| H_0\right)_{\alpha} = F_k\left(t_{1-\frac{\alpha}{2}}\right) - F_k\left(-t_{1-\frac{\alpha}{2}}\right) = 1 - \alpha \tag{2.281}$$

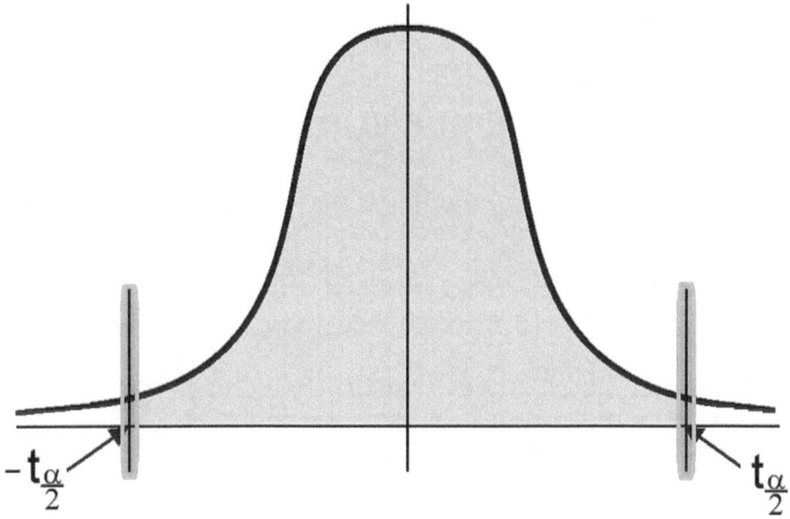

FIGURE 2.17 Marking the working acceptance interval for a normal distribution of data with t-Student distribution.

where

$$F_k(t) = \int_{-\infty}^{t} \varphi_k(x)dx \tag{2.282}$$

Due to the symmetry of the statistical function

$$\varphi(-x) = \varphi(x) \tag{2.283}$$

also the normalization on extended confidence interval remains valid, that is

$$1 = \int_{-\infty}^{+\infty} \varphi(x)dx = 2\int_{0}^{\infty} \varphi(x)dx = 2\int_{-\infty}^{0} \varphi(x)dx \Rightarrow \int_{0}^{\infty} \varphi(x)dx = \int_{-\infty}^{0} \varphi(x)dx = \frac{1}{2} \tag{2.284}$$

resulting the same testing algorithm as in the *z*-statistics case of Section 2.4.1.2, here for *F*(*t*), see Figure 2.18.

Note that in *statistical tables* one searches the $t_{k,\frac{\alpha}{2}}$ for which the fulfillment

$$F_k\left(t_{\frac{\alpha}{2}}\right) - F_k\left(-t_{\frac{\alpha}{2}}\right) = 1 - \alpha \tag{2.285}$$

is provided with the practical expression

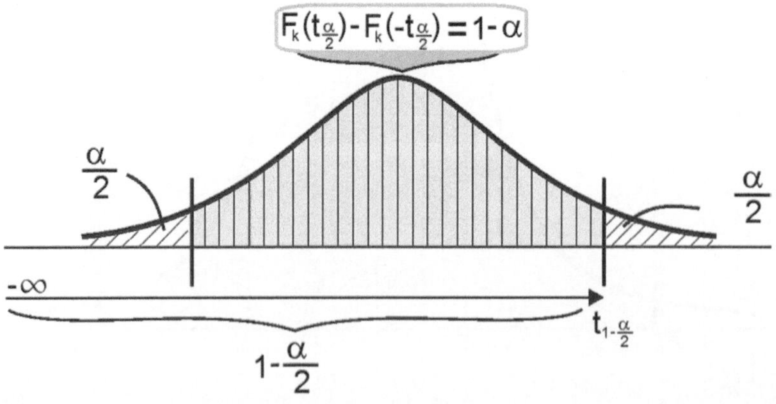

FIGURE 2.18 Synopsis of the probability distribution relevant values and intervals for a normal (Gaussian) distribution of working data repartition with t-Student confidence intervals.

$$F_k\left(t_{1-\frac{\alpha}{2}}\right)=\int_{-\infty}^{t_{1-\frac{\alpha}{2}}}\varphi_k(x)dx=1-\frac{\alpha}{2} \tag{2.286}$$

allowing to find the critical $t_{\frac{\alpha}{2}}$. As an example, consider the numerical case

$$k=10\ \&\ \alpha=0.05\Rightarrow F_k\left(t_{1-\frac{0.05}{2}}\right)=1-0.025=0.975\Rightarrow t_{1-\frac{\alpha}{2}}=2.228 \tag{2.287}$$

so now one can numerically form the confidence interval $\left(-t_{1-\frac{\alpha}{2}},t_{1-\frac{\alpha}{2}}\right)$

(ii) *if the two sets of measurements* (given data to interpret) *correspond to significant equal dispersions on collectivity C* (see the Fisher test below), this means that the measured values differ practically from only one point of view (differ only by the chemist or only by the laboratory (apparatus) by which they were determined, while the rest of experiment protocol is identical)

In these conditions, we have

$$s_{\bar{x}_1-\bar{x}_2}=\sqrt{\frac{s^{*2}}{n_1}+\frac{s^{*2}}{n_2}}=s^*\sqrt{\frac{n_1+n_2}{n_1n_2}} \tag{2.288}$$

with s^* calculated as the "average dispersion" by the weighted expression:

$$s^{*2}=\frac{k_1s_1^{*2}+k_2s_2^{*2}}{k_1+k_2}=\frac{(n_1-1)s_1^{*2}+(n_2-1)s_2^{*2}}{n_1+n_2-2} \tag{2.289}$$

where k_1, k_2 are number of degrees of freedom for the t-Student test applied individually (i.e., separately) to s_1^{*2} and s_2^{*2}.

By introducing the total degree of freedom

$$k=k_1+k_2=(n_1-1)+(n_2-1)=n_1+n_2-2 \tag{2.290}$$

the calculated statistics to be verified by H_0 hypothesis will be:

$$t_{n_1+n_2-2}^{calc}=\frac{\bar{x}_1-\bar{x}_2}{s^{*2}\sqrt{\frac{(n_1+n_2)}{n_1n_2}}}=\frac{\bar{x}_1-\bar{x}_2}{\frac{(n_1-1)s_1^{*2}+(n_2-1)s_2^{*2}}{n_1+n_2-2}\sqrt{\frac{(n_1+n_2)}{n_1n_2}}} \tag{2.291}$$

(iii) *The case of outliers*: if, from the n elements of a sample $(n-1)$ elements seem to be grouped and 1 is "suspicious" the t-Student test between the medians of the two sets $\underline{\{x_1,\ldots,x_{n-1}\}}$ & $\{x_d\}$ is considered

$$\underset{n_1=n-1}{\underline{\underline{\bar{x}_{n-1}=\frac{1}{(n-1)}\sum_{i=1}^{n-1}x_i}}} \quad \& \quad \underset{n_2=1}{\underline{\underline{x_d}}} \tag{2.292}$$

so we actually have to test the statistics:

$$t_{n-2}^{calc}=\frac{\bar{x}_{n-1}-x_d}{\dfrac{(n-1-1)\cdot\underset{s_1^{*2}}{\underline{\underline{\frac{1}{(n-1-1)}\sum_{i=1}^{n}(x_i-\bar{x}_{n-1})^2}}}+(1-1)s_2^{*2}}{(n-2)}\cdot\sqrt{\dfrac{n}{(n-1)\cdot 1}}}$$

$$=\frac{\bar{x}_{n-1}-x_d}{\dfrac{\sum_{i=1}^{n}(x_i-\bar{x}_{n-1})^2}{n-2}\cdot\sqrt{\dfrac{n}{n-1}}} \tag{2.293}$$

Now, if one finds that

$$t_{n-2}^{calc}\notin\left(-t_{1-\frac{\alpha}{2}},t_{1-\frac{\alpha}{2}}\right) \quad , t_{1-\frac{\alpha}{2}}: F_{n-2}\left(t_{\frac{\alpha}{2}}\right)=1-\alpha \tag{2.294}$$

the hypothesis H_0 is rejected so confirming the x_d as outlier, and will be eliminated from working data set for further modeling.

2.4.1.4 Validation of Statistical Hypothesis By Testing the Similitude of Dispersions

Consider two selections of populations of normal repartition for which we do not know σ_1^2 and σ_2^2 but we know the selection estimates s_1^{*2} and s_2^{*2}. Then, for the two selections/populations we can form the statistics:

$$\left.\begin{aligned} "1": \frac{\chi_1^2}{n_1-1}=\frac{s_1^{*2}}{\sigma_1^2} \\ "2": \frac{\chi_2^2}{n_2-1}=\frac{s_2^{*2}}{\sigma_2^2} \end{aligned}\right\} \tag{2.295}$$

which leads to formation of the Fisher statistics (see also the Section 2.3.2.4)

$$F = \frac{s_1^{*2}}{s_2^{*2}} \cdot \frac{\sigma_2^2}{\sigma_1^2} = \frac{\chi_1^2}{\chi_2^2} \cdot \frac{n_2 - 1}{n_1 - 1} \tag{2.296}$$

Now, if we admit the null hypothesis (to be tested) with the form

$$H_0 : \mu_1 = \mu_2 \left(\sigma_1^2 = \sigma_2^2\right) \tag{2.297}$$

we actually have to test the statistics (by calculations)

$$F_{n_1-1,n_2-1}^{calc} = \frac{s_1^{*2}}{s_2^{*2}} \tag{2.298}$$

with the estimates' expressions

$$\begin{cases} s_1^{*2} = \dfrac{1}{n_1 - 1}\sum_{i=1}^{n_1}\left(x_{i1} - \bar{x}_1\right) \\ s_2^{*2} = \dfrac{1}{n_2 - 1}\sum_{i=1}^{n_2}\left(x_{i2} - \bar{x}_2\right) \end{cases} \tag{2.299}$$

If we have the inclusion condition fulfilled under the form

$$F_{n_1-1,n_2-1}^{calc} \in \left(F_{n_1-1,n_2-1}(f_1), F_{n_1-1,n_2-1}(f_2)\right)\Big|_{F(f_1)-F(f_2)=1-\alpha} \tag{2.300}$$

then the H_0 hypothesis is accepted.

Methodologically, one establishes the confidence interval (2.300) through computing the statistics for $F(f_2)$, $F(f_1)$ respecting the samples estimates' dispersions

$$F_{n_1-1,n_2-1}(f_1) \le \frac{s_1^{*2}}{s_2^{*2}} \le F_{n_1-1,n_2-1}(f_2) \tag{2.301}$$

As an example, we successively write for the custom probability case

$$P\left(\frac{s_1^{*2}}{s_2^{*2}} < F_{P,v_1,v_2}\right) = P$$

$$\Leftrightarrow P\left(\frac{s_1^{*2}}{s_2^{*2}} > \frac{1}{F_{P,v_1,v_2}}\right) = 1 - P$$

$$\Leftrightarrow P\left(\frac{s_1^{*2}}{s_2^{*2}} < \frac{1}{F_{1-P,\nu_1,\nu_2}}\right) = P$$

$$\Rightarrow F_{P,\nu_1,\nu_2} = \frac{1}{F_{1-P,\nu_1,\nu_2}}, \begin{cases} \nu_1 = n_1 - 1 \\ \nu_2 = n_2 - 1 \end{cases} \quad (2.302)$$

With Eq. (2.302) one yields the global condition:

$$P\left(F\left(\nu_1,\nu_2,\frac{\alpha}{2}\right) < \frac{s_1^{*2}}{s_2^{*2}} < F\left(\nu_1,\nu_2,1-\frac{\alpha}{2}\right)\right) = 1-\alpha$$

$$\Leftrightarrow P\left(\frac{1}{F\left(\nu_1,\nu_2,1-\frac{\alpha}{2}\right)} < \frac{s_1^{*2}}{s_2^{*2}} < F\left(\nu_1,\nu_2,1-\frac{\alpha}{2}\right)\right) = 1-\alpha$$

$$\Leftrightarrow \begin{Bmatrix} F(f_1) = \dfrac{1}{F_{n_1-1,n_2-1,1-\frac{\alpha}{2}}} \\ F(f_2) = F_{n_1-1,n_2-1,1-\frac{\alpha}{2}} \end{Bmatrix} \quad (2.303)$$

which contains only tabulated values, so with an immediate practical use.

2.4.2 CONFIDENCE INTERVALS

2.4.2.1 Chi-square Statistics

The confidence interval for dispersion is equivalent with testing of null hypothesis $H_0 = \sigma_C^2 = s_\varepsilon^{*2}$ for the unknown dispersion σ^2 of the X-characteristics, which follows the normal law $N(m,\sigma)$ with $m \in \Re$ being the unknown median. While taking into consideration the fact that

$$\chi^2 = (n-1)\cdot\frac{s_\varepsilon^{*2}}{\sigma^2} \xrightarrow{H_0:\, s_\varepsilon^{*2}=\sigma^2} \left(\chi^2\right)^{calc} = n-1 \quad (2.304)$$

we can verify the null hypothesis by requesting the condition:

$$1-\alpha = P\left(\left(\chi^2\right)^{calc} \in \left(\chi_1^2, \chi_2^2\right)\right) = F_{n-1}\left(\chi_2^2\right) - F_{n-1}\left(\chi_1^2\right) \quad (2.305)$$

and we do the appropriate choices:

$$\begin{cases} \chi_1^2 = \chi^2_{n-1,\frac{\alpha}{2}} \rightarrow F(\chi_1^2) = \frac{\alpha}{2} \\ \chi_2^2 = \chi^2_{n-1,1-\frac{\alpha}{2}} \rightarrow F(\chi_2^2) = 1 - \frac{\alpha}{2} \end{cases} \tag{2.306}$$

Now, if is true that

$$\left(\chi^2\right)^{calc} \in \left(\chi^2_{n-1,\frac{\alpha}{2}}, \chi^2_{n-1,1-\frac{\alpha}{2}}\right) \tag{2.307}$$

then the H_0 is accepted, otherwise it is rejected.

Instead, for establishing of the statistical confidence interval, we give up on testing the H_0 hypothesis and we write, in general, that

$$\underline{\chi^2_{n-1,\frac{\alpha}{2}}} < \left(\chi^2\right)^{\substack{certain \\ calculation}} < \underline{\chi^2_{n-1,1-\frac{\alpha}{2}}}$$

$$\Leftrightarrow \chi^2_{n-1,\frac{\alpha}{2}} < (n-1)\cdot\frac{s_\varepsilon^{*2}}{\sigma^2} < \chi^2_{n-1,1-\frac{\alpha}{2}} \left| \cdot \frac{1}{(n-1)\cdot s_\varepsilon^{*2}} \right.$$

$$\Leftrightarrow \frac{\chi^2_{n-1,\frac{\alpha}{2}}}{(n-1)\cdot s_\varepsilon^{*2}} < \frac{1}{\sigma^2} < \frac{\chi^2_{n-1,1-\frac{\alpha}{2}}}{(n-1)\cdot s_\varepsilon^{*2}} \tag{2.308}$$

By reversing the last inequalities reports and by changing the sign we get

$$\frac{(n-1)\cdot s_\varepsilon^{*2}}{\chi^2_{n-1,\frac{\alpha}{2}}} > \sigma^2 > \frac{(n-1)\cdot s_\varepsilon^{*2}}{\chi^2_{n-1,1-\frac{\alpha}{2}}} \tag{2.309}$$

allowing the confidence interval even for σ to be obtained as:

$$\sqrt{\frac{(n-1)\cdot s_\varepsilon^{*2}}{\chi^2_{n-1,1-\frac{\alpha}{2}}}} < \sigma < \sqrt{\frac{(n-1)\cdot s_\varepsilon^{*2}}{\chi^2_{n-1,\frac{\alpha}{2}}}}, \; s_\varepsilon^{*2} = \frac{1}{n-1}\sum_{i=1}^{n}(x_i - \bar{x})^2 \tag{2.310}$$

Remarkably, one can analogously obtain the confidence intervals (after, or by testing the H_0 hypothesis) for the next cases:

(1) The z-repartition, when we determine μ (by knowing σ):

$$z = \frac{|\bar{x} - \mu|}{\frac{\sigma}{\sqrt{n}}} < z_{1-\frac{\alpha}{2}}$$

$$\Rightarrow \bar{x} - \frac{\sigma}{\sqrt{n}} z_{1-\frac{\alpha}{2}} < \mu < \bar{x} + \frac{\sigma}{\sqrt{n}} z_{1-\frac{\alpha}{2}} \tag{2.311}$$

(2) The t-repartition, when we determine μ without knowing σ, and when σ_C is estimated with its unbiased estimator (Figure 2.19)

$$s_\varepsilon^{*2} = \frac{1}{n-1}\sum_{i=1}^{n}(x_i - \bar{x})^2$$

$$t = \frac{|\bar{x} - \mu|}{\frac{s_\varepsilon^{*2}}{\sqrt{n}}} < t_{n-1,1-\frac{\alpha}{2}}$$

$$\Leftrightarrow \bar{x} - t_{n-1,1-\frac{\alpha}{2}} \frac{s_\varepsilon^{*2}}{\sqrt{n}} < \mu < \bar{x} + t_{n-1,1-\frac{\alpha}{2}} \frac{s_\varepsilon^{*2}}{\sqrt{n}} \tag{2.312}$$

(3) The t-repartition for the difference between two medians for two samples of two populations with unknown dispersions σ_1 and σ_2:

$$t = \frac{|(\bar{x}_1 - \bar{x}_2) - (\mu_1 - \mu_2)|}{\sqrt{\frac{s_{1\varepsilon}^{*2}}{n_1} + \frac{s_{2\varepsilon}^{*2}}{n_2}}} < t_{freedom\ deg.,\, 1-\frac{\alpha}{2}}$$

$$\Rightarrow (\bar{x}_1 - \bar{x}_2) - t_{freedom\ deg.,\, 1-\frac{\alpha}{2}} \sqrt{\frac{s_{1\varepsilon}^{*2}}{n_1} + \frac{s_{2\varepsilon}^{*2}}{n_2}} < \underline{\underline{\mu_1 - \mu_1}} < (\bar{x}_1 - \bar{x}_2) + t_{freedom\ deg.,\, 1-\frac{\alpha}{2}} \sqrt{\frac{s_{1\varepsilon}^{*2}}{n_1} + \frac{s_{2\varepsilon}^{*2}}{n_2}} \tag{2.313}$$

which can be specialized depending on our knowledge on dispersions, as follows:

- **Case 1**: $\sigma_1 \neq \sigma_2$ (unknown); with the weighted degree of freedom's recipe:

$$\frac{1}{freedom\ deg.} = \frac{c^2}{freedom\ deg.(1)} + \frac{1-c^2}{freedom\ deg.(2)} \tag{2.314a}$$

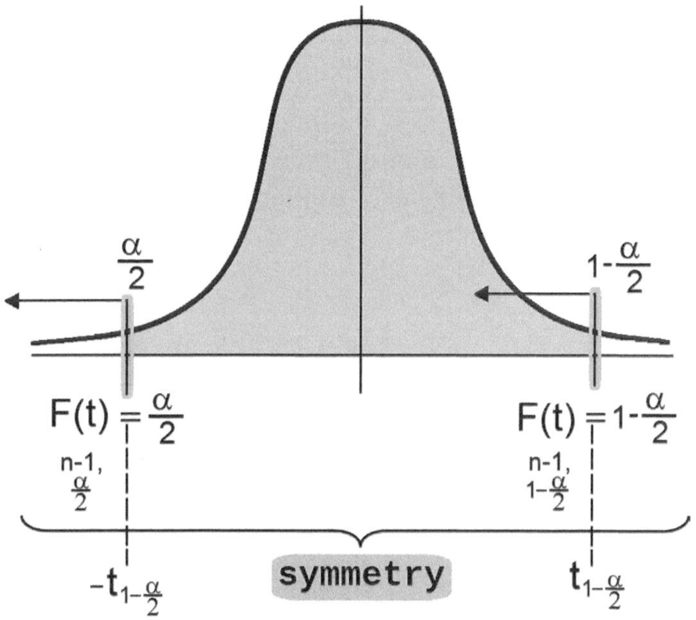

FIGURE 2.19 The symmetrical confidence interval information/formation for a normal (Gaussian) t-Student distribution.

with

$$c = \frac{s_1^{*2}/n_1}{\frac{s_1^{*2}}{n_1} + \frac{s_2^{*2}}{n_2}},\ s_1^{*2} = \frac{1}{n_1 - 1}\sum_{i=1}^{n}(x_{1i} - \bar{x}_1)^2,\ s_2^{*2} = \frac{1}{n_2 - 1}\sum_{i=1}^{n}(x_{2i} - \bar{x}_2)^2 \quad (2.314b)$$

- **Case 2**: $\sigma_1 \cong \sigma_2$ (unknown); with the summed degrees of freedom:

$$\textit{freedom}\ \deg = \textit{freedom}\ \deg(1) + \textit{freedom}\ \deg(2) - 2 \quad (2.315a)$$

and the root squared terms in (2.313) having the actual form

$$\sqrt{\frac{s_1^{*2}}{n_1} + \frac{s_2^{*2}}{n_2}} = s^*\sqrt{\frac{n_1 + n_2}{n_1 n_2}} = \frac{(n_1 - 1)s_1^{*2} - (n_2 - 1)s_2^{*2}}{n_1 + n_2 - 2}\sqrt{\frac{n_1 + n_2}{n_1 n_2}} \quad (2.315b)$$

(4) If σ_1^2 and σ_2^2 are known, the sample is large enough for the dispersions to be identical with that of the collectivity, $n >> 30$, then we go back to the z-statistics:

$$z = \frac{\left|(\bar{x}_1 - \bar{x}_2) - (\mu_1 - \mu_2)\right|}{\sqrt{\frac{\sigma_1^2}{n_1} + \frac{\sigma_2^2}{n_2}}} < z_{1-\frac{\alpha}{2}} \tag{2.316}$$

without having anymore the issue of the degree of freedom; so, by solving (2.316) for the medians' difference confidence interval we are left with the result

$$(\bar{x}_1 - \bar{x}_2) - z_{1-\frac{\alpha}{2}}\sqrt{\frac{\sigma_1^2}{n_1} + \frac{\sigma_2^2}{n_2}} < \underline{\underline{\mu_1 - \mu_1}} < (\bar{x}_1 - \bar{x}_2) + z_{1-\frac{\alpha}{2}}\sqrt{\frac{\sigma_1^2}{n_1} + \frac{\sigma_2^2}{n_2}} \tag{2.317}$$

(5) In case of Fisher repartition with two populations with unknown dispersions σ_1 and σ_2, the asymmetry of the Fisher probability density function can be employed towards successively written the hierarchies:

$$\begin{gathered} F_{\substack{n_1-1 \\ n_2-1}\left|\frac{\alpha}{2}\right.} < F = \frac{s_1^{*2}}{s_2^{*2}} \cdot \frac{\sigma_2^2}{\sigma_1^2} < F_{\substack{n_1-1 \\ n_2-1}\left|1-\frac{\alpha}{2}\right.} \\ \Leftrightarrow \frac{s_2^{*2}}{s_1^{*2}} \frac{1}{F_{\substack{n_1-1 \\ n_2-1}\left|1-\frac{\alpha}{2}\right.}} < \frac{\sigma_2^2}{\sigma_1^2} < \frac{s_2^{*2}}{s_1^{*2}} F_{\substack{n_1-1 \\ n_2-1}\left|1-\frac{\alpha}{2}\right.} \end{gathered} \tag{2.318}$$

They are further solved for the dispersions' ratio σ_2/σ_1 confidence interval:

$$\frac{s_2^*}{s_1^*} \frac{1}{\sqrt{F_{\substack{n_1-1 \\ n_2-1}\left|1-\frac{\alpha}{2}\right.}}} < \frac{\sigma_2}{\sigma_1} < \frac{s_2^*}{s_1^*} \sqrt{F_{\substack{n_1-1 \\ n_2-1}\left|1-\frac{\alpha}{2}\right.}} \tag{2.319}$$

Note that in this case there was preferred the transcription

$$F_{\frac{\alpha}{2}} \to F_{1-\frac{\alpha}{2}} = \frac{1}{F_{\frac{\alpha}{2}}} \tag{2.320}$$

in order to read "at once" the $F_{1-\frac{\alpha}{2}}$ quantile specific to probabilities $1 - \frac{\alpha}{2}$ and $\frac{\alpha}{2}$ (see Figure 2.19).

2.4.2.2 The Chi-Square-Test for Normal Concordance

The typical QSAR (Quantitative Structure-Activity Relationships) correlation has to be evaluated from its robustness point of view (see the OECD-QSAR principles of Chapter 3/Section 3.2.5 and OECD (2004)) in relation with:

- the normal distribution of the activities/residues of correlation;
- calculated statistical parameters
- regression equation's parameters
- robustness of the typical equation, its confidence interval, regarding its prediction capacity.

To this aim, one considers the main working objects:

- *C*: the studied collectivity (a.k.a. the set of molecules with ecotoxicological or pharmacological action);
- *X*: the studied characteristic (the activity/residues);
- *k*: the number of classes of *X* characteristics (the classes of molecules);
- E_i: the event that has a random individuality from collectivity C belongs to the "i" class of the *k*-total;
- p_i: the probability of the E_i event, fulfills the natural conditions

$$p_i = P(E_i),\quad i = \overline{1,k} \quad \Rightarrow \sum_{i=1}^{k} p_i = 1 \tag{2.321}$$

In these conditions, one faces with the actual null hypothesis

$$H_0 : p_i = p_i^{(0)},\quad i = \overline{1,k} \tag{2.322}$$

to be verified by considering a repetitive selection of *n*-volume; in this respect we have the selection data $x_1, x_2, \ldots, x_n$ with the absolute (appearance) frequency for the *X* characteristics classes fulfilling the closure relationship $n = n_1 + n_2 + \ldots, n_k$, with $n_i : i = \overline{1,k}$ counting how many times the E_i event showed up in the considered selection. Note that in the QSAR: $n_i = 1$ means that each molecular activity is an event, while $K = n(\text{or}N)$ denotes the total number of molecules/activities considered in the working data set. Accordingly, one may consider the k-dimensional vectorial association

$$\{n_1, n_2, \ldots, n_k\} \xrightarrow[\substack{\text{the random}\\ k-\text{Dimensional vector}\\ \text{of the selection vector}}]{} \{N_1, N_2, \ldots, N_k\} \quad (2.323)$$

which verifies the multinomial law, which measures the probability for the two values from the $\{N_i\}$ to $\{n_i\}$ sets to be 1-to-1 correspondingly, that is:

$$P(N_1 = n_1, N_2 = n_2, \ldots, N_k = n_k) = \frac{n!}{n_1! n_2! \ldots n_k!} P_1^{n_1} P_2^{n_2} \ldots P_k^{n_k} \quad (2.324)$$

This way, the null hypothesis H_0 is referring to the parameters $p_1 \ldots, p_k$ of the multinomial law. Moreover, one has that the frequency statistics

$$\chi^2 = \sum_{i=1}^{k} \frac{(N_i - np_i)^2}{np_i} \quad (2.325)$$

follows the χ^2-law with $k-1$ degrees of freedom when $n \to \infty$.

Proof: The starting point is the Stirling's formula for the factorial products:

$$n! \cong \sqrt{2\pi n}\, n^n e^{-n}, \; n >> 1 \quad (2.326)$$

which will be firstly analytical justified, see also the Appendices of the Volume I of the present five-volume book's series (Putz, 2016a). To this end, one starts from the Gamma-Euler function $\Gamma(p)$, see also Eq. (2.180):

$$a! = \Gamma(a+1) = \int_0^\infty x^a e^{-x} dx \quad (2.327)$$

which is asymptotically expanded for $a >> 1$ with the saddle-point method, see the Appendix of the Volume II of the present five-volume book's series Putz, 2016b).

Accordingly, one may write

$$\begin{cases} x^a e^{-x} = e^{a \ln x - x} \equiv e^{f(x)} \\ with \;\; f(x) = a \ln x - x \end{cases} \quad (2.328)$$

Then, one may perform the successive derivatives' evaluations

$$\begin{cases} f'(x) = \dfrac{a}{x} - 1 \xrightarrow{x=a} 0 \\ f''(x) = -\dfrac{a}{x^2} \xrightarrow{x=a} -\dfrac{1}{a} \\ \vdots \end{cases} \tag{2.329}$$

so providing the expanded series of the working function, under the form

$$f(x) \cong f(a) + f'(a)(x-a) + f''(a)\frac{(x-a)^2}{2} + \ldots$$

$$= a \ln a - a + 0 - \frac{1}{2a}(x-a)^2 + \ldots \tag{2.330}$$

However, when truncated to the second order expansion in (2.330) we have for (2.327) the expression

$$a! = \Gamma(a+1) \cong \int_0^\infty e^{\left[a \ln a - a - \frac{(x-a)^2}{2a}\right]} dx = e^{a \ln a - a} \underbrace{\int_0^\infty e^{-\frac{(x-a)^2}{2a}} dx}_{\cong \int_{-\infty}^{+\infty} e^{-\frac{(x-a)^2}{2a}} d(x-a)} = \sqrt{2\pi a} a^a e^{-a},\ a >> 1 \tag{2.331}$$

through extending the integration domain to the Poisson integral non-centered/biased domain (Figure 2.20), and applying the associate classical

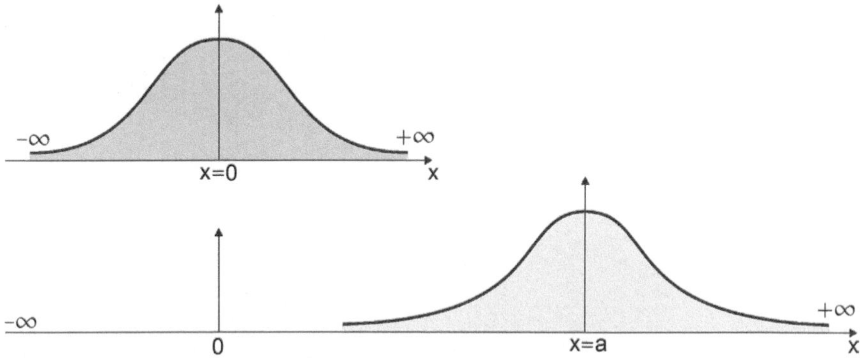

FIGURE 2.20 A centered (up) and the respective ("evolved") non-centered/biased (down) Normal (Gaussian) distribution with their corresponding total (Poisson) area accordingly marked.

integral result, see Eq. (2.50b). For $a = n =$ integer Eq. (2.331) becomes the Stirling formula:

$$n! = \sqrt{2\pi n}\, n^{n} e^{-n} \tag{2.332a}$$

or, usually known under the form:

$$n! = (2\pi)^{1/2} n^{n+\frac{1}{2}} e^{-n} \quad \Big| \ln \tag{2.332b}$$

or as

$$\ln n! = \frac{1}{2}\ln 2\pi + \left(n + \frac{1}{2}\right)\ln n - n \tag{2.332c}$$

Equation (2.332c), for $n >> 1$, it is reduced (by neglecting the accompanied numbers of n) to the fashioned form:

$$\ln n! \cong \ln n - n \tag{2.332d}$$

With this we can go back to the χ^2*-test for concordance*, so that the multinomial probability (2.324) actually transforms as:

$$P(N_1 = n_1, N_2 = n_2, \ldots, N_k = n_k) \approx \frac{\sqrt{2\pi}\, n^{1/2} n^{n} e^{-n}}{\prod_{i=1}^{n}\left(\sqrt{2\pi n_i}\, n_i^{n_i} e^{-n_i}\right)} p_1^{n_1} p_2^{n_2} \ldots p_k^{n_k}$$

$$\approx const \cdot \frac{n^{n+\frac{1}{2}} e^{-n}}{\left(\prod_{i=1}^{k} n_i^{n_i+\frac{1}{2}}\right) e^{-\sum_{i=1}^{k} n_i}} \prod_{i=1}^{k} p_i^{n_i}$$

$$= const \cdot \frac{\left(\prod_{i=1}^{k} n_i^{n_i}\right) n^{\frac{1}{2}}}{\left(\prod_{i=1}^{k} n_i^{n_i+\frac{1}{2}}\right)} \prod_{i=1}^{k} p_i^{n_i}$$

$$= const \cdot \prod_{i=1}^{k} \frac{(np_i)^{n_i} \cdot n^{\frac{1}{2}}}{n_i^{n_i+\frac{1}{2}}}$$

$$\overset{n>>1}{\cong} const^* \cdot \prod_{i=1}^{k} \left(\frac{np_i}{n_i} \right)^{n_i + \frac{1}{2}} \quad \Bigg| \ln \tag{2.333}$$

$$\Rightarrow \ln P \approx \ln const^* + \sum_{i=1}^{k} \left(n_i + \frac{1}{2} \right) \ln \left(\frac{np_i}{n_i} \right) \tag{2.334a}$$

With the appropriate notation:

$$x_i = \frac{n_i - np_i}{\sqrt{np_i}} \Leftrightarrow \left(\frac{np_i}{n_i} \right)^{-1} = 1 + \frac{x_i}{\sqrt{np_i}} \tag{2.335}$$

the result (2.334a) is further arranged as

$$\Rightarrow \ln P = \ln const^* - \sum_{i=1}^{k} \left(n_i + \frac{1}{2} \right) \ln \left(1 + \frac{x_i}{\sqrt{np_i}} \right) \tag{2.334b}$$

and, because $n >> 1$, we can keep the first two terms of the logarithmic expansion:

$$\ln(1+x) \overset{x \to 0}{\cong} \ln(1) + \left(\frac{1}{1+x} \right)_{x=0} \cdot x - \frac{1}{2} \left(\frac{1}{(1+x)^2} \right)_{x=0} \cdot x^2 = x - \frac{1}{2} x^2 \tag{2.336}$$

with

$$x = \frac{x_i}{\sqrt{np_i}} \tag{2.337a}$$

In this line, along the n_i solution from (2.335)

$$n_i = np_i + x_i \sqrt{np_i} \tag{2.337b}$$

one can continue with rewriting the Eq. (2.334a) to get

$$\ln P = \ln const^* - \sum_{i=1}^{k} \left(n_i + \frac{1}{2} \right) \left(\frac{x_i}{\sqrt{np_i}} - \frac{x_i^2}{2np_i} \right)$$

$$= \ln const^* - \sum_{i=1}^{k} \left(np_i + x_i \sqrt{np_i} + \frac{1}{2} \right) \left(\frac{x_i}{\sqrt{np_i}} - \frac{x_i^2}{2np_i} \right) \tag{2.334c}$$

Then, by taking into consideration the fact that $n >> 1$, we can neglect all the terms which contain $\frac{1}{n}$ or $\frac{1}{\sqrt{n}}$, so we remain with

$$\ln P \approx \ln const^* - \sum_{i=1}^{k}\left(x_i\sqrt{np_i} - \frac{x_i^2}{2} + x_i^2 \right)$$

$$= \ln const^* - \sum_{i=1}^{k}\left(x_i\sqrt{np_i} + \frac{x_i^2}{2} \right) \tag{2.334d}$$

Since we recognize that

$$\sum_{i=1}^{k}\left(x_i\sqrt{np_i}\right) \overset{(2.335)}{=} \sum_{i=1}^{k}(n_i - np_i) = \underbrace{\sum_{i=1}^{k} n_i}_{n} - n\underbrace{\sum_{i=1}^{k} p_i}_{1} = n - n = 0 \tag{2.338}$$

we further remain from (2.334d) with the searched result

$$\ln P = \ln const^* - \frac{1}{2}\sum_{i=1}^{k} x_i^2$$

$$\Leftrightarrow P = const^* \exp\left(-\frac{1}{2}\sum_{i=1}^{k} x_i^2 \right) \tag{2.334e}$$

so we get the law that corresponds in the asymptotic ($n \to \infty$) regime to the variables

$$x_i = \left. \frac{n_i - np_i}{\sqrt{np_i}} \right|_{n_i = N_i} \tag{2.335a}$$

and thus, we conclude that the random vector

$$\{x_i\}_{i=\overline{1,k}} = \left(\left\{ \frac{N_i - np_i}{\sqrt{np_i}} \right\} \right)_{i=\overline{1,k}} \tag{2.335b}$$

follows a normal k-dimensional and degenerate law (because every x_i component can be expressed as a linear combination of the other components). This approach results in the square sum of these components which satisfy the normal law for the χ^2-statistics with the number of degrees of freedom given by the components of the sample minus one unit, i.e.

$$\chi^2 = \sum_{i=1}^{k} x_i^2 = \sum_i \frac{(N_i - np_i)^2}{np_i} \tag{2.339}$$

thus it is a χ^2_{k-1} law.

In this framework, for the given significance level $\alpha \in (0, 1)$ the quantile $\chi^2_{k-1,\,1-\alpha}$ can be determined accordingly:

$$1-\alpha = P\left(\chi^2 \le \chi^2_{k-1,\,1-\alpha} \mid H_0\right) = F_{k-1}\left(\chi^2_{k-1,\,1-\alpha}\right) \tag{2.340}$$

meaning that

$$F_{k-1}\left(\chi^2_{k-1,\,1-\alpha}\right) = 1-\alpha \Rightarrow \text{critical } \chi^2_{k-1,\,1-\alpha} \tag{2.341}$$

So, if one has:

$$\chi^2_{calculated|H_0} = \sum_{i=1}^{k} \frac{n_i - np_i^{(0)}}{np_i^{(0)}} < \chi^2_{k-1,\,1-\alpha} \tag{2.342}$$

the H_0 is admitted with the risk degree (of being wrong) ~ α.

This mechanism of analysis can be extended to the non-parametrical testing regarding the concordance between the probability law which is exhibited by the characteristic X, *say to be denoted by* F – and the tested or proposed one or to be validated – *say* F_0. In these conditions the test's steps will be:

- the null hypothesis is sill $H_0 : F = F_0$ (what we are testing)
- the domain of X's values, say (a, b), is coarsened in order to extract the "k" classes by the points:

$$a_0 = a < a_1 < \ldots < \text{b} = a_k \tag{2.343}$$

- the parameters $p_i = p_i^{(0)}$ will be written with the tested function $F=F_0$ on the intermediate intervals, this way:

$$p_i^{(0)} = P(a_{i-1} < x \le a_i) = F_0(a_i) - F(a_{i-1}) \tag{2.344}$$

- one fixes the random variable with the (2.344) result:

$$\chi^2_{calc} = \sum_{i=1}^{k} \frac{\left[n_i - np_i^{(0)}\right]^2}{np_i^{(0)}}, \quad i = 1, k \tag{2.345}$$

- finally, the null hypothesis is tested by fulfilling of the formed confidence hierarchy

$$\chi^2_{calc} < \chi^2_{k-1,\,1-\alpha} \Rightarrow \begin{cases} YES : \exists\, H_0 \\ NO : \nexists\, H_0 \end{cases} \tag{2.346}$$

2.4.2.3 The Chi-Square Test for the Parametrical Concordances and Likelihood

One may consider the F_0 function as being depending by the unknown parameters:

$$F_0 = F_0(x;\, \theta_1,\, \theta_2,\ldots,\theta_s) \tag{2.347}$$

In this case, supplementary confidence testing steps appears as:

(1) the unknown parameters are estimated to be $\{\hat{\theta}_1,\hat{\theta}_2,\ldots,\hat{\theta}_s\}$ by using the selection data by employing the *maximum likelihood criterion* (see below);

(2) the hypothesis $H_0 : F = F_0$ means now that

$$p_i^{(0)} = p\left(a_{i-1} < x \le a_i \middle| H_0\right) = F_0\left(a_i;\, \hat{\theta}_1,\hat{\theta}_2,\ldots,\hat{\theta}_s\right) - F\left(a_{i-1};\, \hat{\theta}_1,\hat{\theta}_2,\ldots,\hat{\theta}_s\right),\quad i = \overline{1,k} \tag{2.348}$$

(3) the χ^2-law, which has to be applied, is considered now with k – s – 1 degrees of freedom.

Now, let's discuss *the maximum likelihood criterion*.

Consider the X-characteristic with the probability function $f(n;\tilde{\theta})$ with $\tilde{\theta} \in A \subset \Re^p$ a p-dimensional vector; then we name *the maximum likelihood estimator* for the θ parameter the statistics $\hat{\theta} = \hat{\theta}\{x_1, x_2,\ldots,x_n\}$ for which the maximum of the *likelihood* function is obtained as the total probability with product of independents probabilities

$$g\left(x_1, x_2,\ldots,x_n,\tilde{\theta}\right) = \prod_{i=1}^{n} f\left(x_k;\tilde{\theta}\right) \tag{2.349}$$

This means that the optimum (maximum) condition is applied to the system

$$\left\{ \frac{\partial}{\partial \theta_k} g\left(x_1, x_2,\ldots,x_n,\tilde{\theta}\right) = 0 \right. ,\ k = \overline{1,p} \tag{2.350}$$

eventually equivalent with the system of equations obtained by logarithm of the previous form:

$$\left\{\frac{\partial}{\partial\theta_k}\ln g\left(x_1, x_2, \ldots, x_n, \tilde{\theta}\right) = 0 \quad, k = \overline{1,p}\right. \tag{2.351}$$

with the practical form:

$$\left\{\sum_{i=1}^{n}\frac{\partial}{\partial\theta_k}\ln f\left(x_i, \tilde{\theta}\right) = 0 \quad, k = \overline{1,p}\right. \tag{2.352}$$

called as the set of equations of maximum likelihood.

As a working example we will take the normal distribution $N(\mu, \sigma)$, the X-characteristic following the well-known law:

$$\begin{cases}\langle x\rangle = \mu; \quad D(x) = \sigma \\ f(x;\mu,\sigma) = \dfrac{1}{\sigma\sqrt{2\pi}}e^{-\frac{(x-\mu)^2}{2\sigma^2}}\end{cases} \tag{2.353}$$

For it we like to establish the estimators of the maximum likelihood for median μ and dispersion σ, as a basal parameters of the normal distribution function. To this aim we state the likelihood by writing

$$\ln f = -\ln\sqrt{2\pi} - \ln\sigma - \frac{(x-\mu)^2}{2\sigma^2} \tag{2.354}$$

from which we formulate the respective derivatives:

$$\begin{cases}\dfrac{\partial}{\partial\mu}\ln f(x;\mu,\sigma) = \dfrac{x-\mu}{\sigma^2} \\ \dfrac{\partial}{\partial\sigma}\ln f(x;\mu,\sigma) = -\dfrac{1}{\sigma} + \dfrac{(x-\mu)^2}{\sigma^3}\end{cases} \tag{2.355}$$

towards forming the system with the optimum equations as:

$$\begin{cases}0 = \dfrac{\partial\ln g}{\partial\mu} = \displaystyle\sum_{i=1}^{n}\dfrac{\ln f(x_i;\mu,\sigma)}{\partial\mu} = \sum_{i=1}^{n}\dfrac{x_i-\mu}{\sigma^2} \\ 0 = \dfrac{\partial\ln g}{\partial\sigma} = \displaystyle\sum_{i=1}^{n}\dfrac{\ln f(x_i;\mu,\sigma)}{\partial\sigma} = \sum_{i=1}^{n}\left[-\dfrac{1}{\sigma} + \dfrac{(x_i-\mu)^2}{\sigma^3}\right]\end{cases} \tag{2.356}$$

from where one remains with the results:

$$\begin{cases}\sum_{i=1}^{n}(x_i-\mu)=0\\ \sum_{i=1}^{n}\left[-\sigma^2+(x_i-\mu)^2\right]=0\end{cases} \tag{2.357}$$

It has to be solved for the estimators of maximum likelihood, i.e., for the μ^* parameter:

$$\sum_{i=1}^{n}x_i-\mu^*\sum_{i=1}^{n}1=0\Rightarrow\mu^*=\frac{1}{n}\sum_{i=1}^{n}x_i=\bar{x} \tag{2.358}$$

and respectively for the maximum likelihood σ^* parameter from the second equation of (2.357):

$$-\sigma^{*2}\sum_{i=1}^{n}1+\sum_{i=1}^{n}\left(x_i-\mu^*\right)^2\Rightarrow\sigma^*=\sqrt{\frac{1}{n}\sum_{i=1}^{n}\left(x_i-\mu^*\right)^2} \tag{2.359}$$

offering therefore the practical formulas for the selection data $\{x_i\}_{i=\overline{1,n}}$.

Returning to the χ^2 test of selection data concordance, while testing the concordance with the normal distribution:

$$H_0:\ F=F_0(x;\mu,\sigma)=\frac{1}{\sigma\sqrt{2\pi}}\int_{-\infty}^{x}e^{-\frac{(t-\mu)^2}{2\sigma^2}}dt\ ,t\in\Re \tag{2.360}$$

we have the following test's steps:

- for the values' domain of the X-characteristic one considers the coarsening of the $\Re$ domain:

$$-\infty=a_0<a_1<a_2<\ldots<a_{k-1}<a_k=+\infty \tag{2.361}$$

- the respective probabilities are obtained by the custom evaluations:

$$H_0:\ p_i=p_i^{(0)}=F_0\left(a_i;\mu^*,\sigma^*\right)-F_0\left(a_{i-1};\mu^*,\sigma^*\right)\ ,\ i=\overline{1,k} \tag{2.362}$$

which, in a particular, are written as:

$$p_1^{(0)}=F_0\left(a_1;\mu^*,\sigma^*\right)-F_0\left(-\infty;\mu^*,\sigma^*\right)$$

$$= \int_{-\infty}^{a_1} \varphi_{N(\mu^*,\sigma^*)}(t)dt - \underbrace{\int_{-\infty}^{-\infty} \varphi_{N(\mu^*,\sigma^*)}(t)dt}_{=0}$$

$$= \int_{-\infty}^{0} \varphi_{N(\mu^*,\sigma^*)}(t)dt + \int_{0}^{a_1} \varphi_{N(\mu^*,\sigma^*)}(t)dt$$

$$= \frac{1}{2} + \Phi\left(\frac{a_i - \mu^*}{\sigma^*}\right) \tag{2.363}$$

- in the similar way for a random $p_i^{(0)}$ with $i = \overline{2, k-1}$ we have:

$$p_i^{(0)} = F_0(a_i; \mu^*, \sigma^*) - F_0(a_{i-1}; \mu^*, \sigma^*)$$

$$= \left(\int_{-\infty}^{0} \varphi_{N(\mu^*,\sigma^*)}(t)dt + \int_{0}^{a_i} \varphi_{N(\mu^*,\sigma^*)}(t)dt\right) - \left(\int_{-\infty}^{0} \varphi_{N(\mu^*,\sigma^*)}(t)dt + \int_{0}^{a_{i-1}} \varphi_{N(\mu^*,\sigma^*)}(t)dt\right)$$

$$= \Phi\left(\frac{a_i - \mu^*}{\sigma^*}\right) - \Phi\left(\frac{a_{i-1} - \mu^*}{\sigma^*}\right), \quad i = \overline{2, k-1} \tag{2.364}$$

- finally, for $i = k$ we have:

$$p_k^{(0)} = F_0(+\infty; \mu^*, \sigma^*) - F_0(a_{k-1}; \mu^*, \sigma^*)$$

$$= \underbrace{\int_{-\infty}^{+\infty} \varphi_{N(\mu^*,\sigma^*)}(t)dt}_{=1} - \left(\int_{-\infty}^{0} \varphi_{N(\mu^*,\sigma^*)}(t)dt + \int_{0}^{a_{k-1}} \varphi_{N(\mu^*,\sigma^*)}(t)dt\right)$$

$$= \frac{1}{2} - \Phi\left(\frac{a_{k-1} - \mu^*}{\sigma^*}\right) \tag{2.365}$$

- with all the results (2.363)–(2.365) now we can calculate the critical value for chi-square variable

$$\chi^2_{calc} = \sum_{i=1}^{k} \left(\frac{n_i - np_i^{(0)}}{np_i^{(0)}}\right)^2 \underset{\substack{n_i = 1 \\ \sum n_i = n \\ k = n}}{\overset{QSAR}{=}} \sum_{i=1}^{n} \left(\frac{1 - np_i^{(0)}}{np_i^{(0)}}\right)^2 \tag{2.366}$$

which is compared with the quantile $\chi^2_{k-3,1-\alpha}$ for the hypothesis verification

$$\exists\ H_0 \Rightarrow \chi^2_{calc} < \chi^2_{k-3,1-\alpha} \ \Big|\ F_{k-3}\left(\chi^2_{k-3,1-\alpha}\right) = 1-\alpha \tag{2.367}$$

where the degrees of freedom are now as:

$$k - s - 1 = k - 2 - 1 = k - 3 \tag{2.368}$$

In a QSAR analysis we will have n – 3 with n – the number of compounds selected for correlation. If H_0 is accepted, the selection with the X-characteristics (the structure descriptors and their activities in QSAR – see Chapter 3) follows the normal law $N(m^*, s^*)$.

Note that since

$$\sum_k n_i = n = k \tag{2.369}$$

every single chemical compound (in a QSAR study) goes as just once, while the number of classes (k) equalize the number of compounds (n or N) with known/observed or predicted activities.

2.4.3 ROBUSTNESS OF REGRESSION: MULTI-REGRESSION ANALYSIS OF VARIANCE (ANOVA)

Consider the regression equations cases

$$y_i^P = a + bx \text{ – uni-parametric } (M = 1) \tag{2.370a}$$

$$y^P_{\binom{i}{M}} = b_0 + \sum_{j=1}^{M} b_j \cdot x_{ij} \text{ – multi-parametric } (M > 1) \tag{2.370b}$$

Then, the dispersion σ_0^2 of the entire population of the y values is estimated based on the $y^{o(bserved)}$ selected observations respecting those calculated by regression model $y^{p(redicted)}$:

$$\sigma_0^2\Big|_{C(y)} \rightarrow s_0^{*2} = \frac{\sum_{i=1}^{n=N}\left(y_i^o - y_i^P\right)^2}{n-1(-M)} \tag{2.371}$$

However, since the regression parameters *"b"* are calculated through employing the y^{obs} values, the additional constraint (-*M*) appears in dispersion estimation (2.371). The unbiased estimation $\sqrt{s_0^{*2}}$ is also-called as the *standard error of estimation* (SEE) and it is calculated with the formula

$$\sqrt{s_0^{*2}} = SEE_{\sigma_0(y^o)} = \sqrt{\frac{\sum_{i=1}^{n} e_i^2}{n-1-M}} = \sqrt{\frac{\sum_{i=1}^{n}\left[y_i^o - b_0 - \sum_{j=1}^{M} b_j x_{ij}\right]^2}{n-1-M}}$$

$$= \sqrt{\frac{\sum_{i=1}^{n}\left[y_i^o - b_0 - \sum_{j=1}^{M} b_j x_{ij}\right]\left[y_i^o - b_0 - \sum_{j=1}^{M} b_j x_{ij}\right]}{n-1-M}}$$

$$= \sqrt{\frac{\sum_{i=1}^{n} y_i^o\left(y_i^o - b_0 - \sum_{j=1}^{M} b_j x_{ij}\right) - b_0\sum_{i=1}^{n}\left[y_i^o - b_0 - \sum_{j=1}^{M} b_j x_{ij}\right] - \sum_{j=1}^{M} b_j \sum_{i=1}^{n} x_{ij}\left[y_i^o - b_0 - \sum_{j=1}^{M} b_j x_{ij}\right]}{n-1-M}} \tag{2.372}$$

Because the summing terms two and three from the right member of Eq. (2.372) are nulls, due to the fact that the sums $\sum_{i=1}^{n}\bullet$ are equal with the zero-sums of the partial derivatives from the multivariate regressions, see the system (2.89), we finally obtain the expression:

$$\sqrt{s_0^{*2}} = SEE_{\sigma_0(y^o)} = \sqrt{\frac{\sum_{i=1}^{n}(y_i^o)^2 - b_0\sum_{i=1}^{n} y_i^o - \sum_{j=1}^{M} b_j \sum_{i=1}^{n} y_i^o x_{ij}}{n-1-M}} \tag{2.373}$$

On the other hand, in order to estimate the dispersion of the regression model we take into consideration the first equation in the multivariate regression system (2.90), now with $n = N$ (n: the selection volume and N: the number of molecules), to further transform it as:

$$b_0 n + b_1\sum_{i=1}^{n} x_{i1} + \ldots + b_k\sum_{i=1}^{n} x_{ik} + \ldots + b_M\sum_{i=1}^{n} x_{iM} = \sum_{i=1}^{n} y_i \quad \Big| : \frac{1}{n}$$

$$\Rightarrow \frac{1}{n}\sum_{i=1}^{n} y_i = b_0 + b_1\frac{1}{n}\sum_{i=1}^{n} x_{i1} + \ldots + b_k\frac{1}{n}\sum_{i=1}^{n} x_{ik} + \ldots + b_M\frac{1}{n}\sum_{i=1}^{n} x_{iM} \tag{2.374}$$

meaning that

$$\bar{y}^o = b_0 + b_1\bar{x}_1 + \ldots + b_k\bar{x}_k + \ldots + b_M\bar{x}_M$$

$$\Rightarrow \bar{y}^o = b_0 + \sum_{j=1}^{M} b_j\bar{x}_j \tag{2.375}$$

Then, if we extract from the Eq. (2.375) the free term value

$$b_0 = \bar{y}^o - \sum_{j=1}^{M} b_j\bar{x}_j \tag{2.376}$$

and it is introduced it the expression or predicted model

$$y^P = b_0 + \sum_{j=1}^{M} b_j x_j \tag{2.377}$$

for a collection of (working) points

$$(x_1 = x_{01}; x_2 = x_{02}, \ldots, x_M = x_{0M}) \tag{2.378}$$

we actually obtain the predicted-observed relationship:

$$y_0^P = \bar{y}^o - \sum_{j=1}^{M} b_j\bar{x}_j + \sum_{j=1}^{M} b_j x_{0j}$$

$$= \bar{y}^o + \sum_{j=1}^{M} b_j\left(x_{0j} - \bar{x}_j\right) \tag{2.379a}$$

At the level of dispersion the Eq. (2.379a) yields

$$D\left(y_0^P\right) = D\left(\bar{y}^o\right) + D\left(\sum_{j=1}^{M} b_j\left(x_{0j} - \bar{x}_j\right)\right)$$

$$= D\left(\bar{y}^o\right) + D\left(b_j\right)\sum_{j=1}^{M}\left(x_{0j} - \bar{x}_j\right)^2 \tag{2.379b}$$

on this point we have to evaluate $D\left(\bar{y}^o\right)$ & $D\left(b_j\right)$; to this end the next steps will be unfolded:

(a) Because $\bar{y}^o$ is the general median of the n observations, while σ_0^2 stays for the general dispersion of the population, we simply have the estimate-to-collectivity connection

$$\underbrace{\sigma^2_{\bar{y}^o}}_{\substack{selection\\dispersion}} = \underbrace{\frac{\sigma_0^2}{n} = \frac{s_0^{*2}}{n}}_{\substack{population\\dispersion}} \tag{2.380}$$

(b) In order to calculate the $D(b_j)$ we have to firstly generalize the $(b_j)_{j=\overline{1,M}}$ expression from the multivariate analysis and we have successively, the reloading of Eqs. (2.60) and (2.64), then combined with Eqs. (2.28) and (2.36) (considering also the interchange between the "a" and "b" notation for the correlation slope):

$$b_j = \left(r_{x_j y^o}\right)\frac{\sigma_{y^o}}{\sigma_{x_j}} = \frac{c_{x_j y^o}}{\sigma_{y^o}\sigma_{x_j}} \cdot \frac{\sigma_{y^o}}{\sigma_{x_j}} = \frac{c_{x_j y^o}}{\sigma^2_{x_j}}$$

$$= \frac{n\sum_{i=1}^{n} x_{ij} y_i - \left(\sum_{i=1}^{n} x_{ij}\right)\left(\sum_{i=1}^{n} y_i^o\right)}{n\sum_{i=1}^{n} x_{ij}^2 - \left(\sum_{i=1}^{n} x_{ij}\right)^2}$$

$$\overset{\sum_i x_{ij} = n\bar{x}_j}{=} \frac{n\sum_{i=1}^{n} x_{ij} y_i - n\bar{x}_j \sum_{i=1}^{n} y_i}{n\sum_{i=1}^{n} x_{ij} - n^2\bar{x}_j^2}$$

$$= \frac{\sum_{i=1}^{n} x_{ij} y_i - \bar{x}_j \sum_{i=1}^{n} y_i}{\sum_{i=1}^{n} x_{ij}^2 - 2n\bar{x}_j^2 + n\bar{x}_j^2} = \frac{\sum_{i=1}^{n} (x_{ij} - \bar{x}_j) y_i}{\sum_{i=1}^{n} x_{ij}^2 - 2\bar{x}_j \sum_{i=1}^{n} x_{ij} + n\bar{x}_j^2}$$

$$= \frac{\sum_{i=1}^{n} (x_{ij} - \bar{x}_j) y_i}{\sum_{i=1}^{n} (x_{ij}^2 - 2x_{ij}\bar{x}_j + \bar{x}_j^2)} = \frac{\sum_{i=1}^{n} (x_{ij} - \bar{x}_j) y_i}{\sum_{i=1}^{n} (x_{ij} - \bar{x}_j)^2}$$

$$= \frac{(x_{1j} - \bar{x}_j)}{\sum_{i=1}^{n} (x_{ij} - \bar{x}_j)^2} y_1 + \frac{(x_{2j} - \bar{x}_j)}{\sum_{i=1}^{n} (x_{ij} - \bar{x}_j)^2} y_2 + \ldots$$

$$\Rightarrow b_j = \sum_{i=1}^{n} c_{ij} y_i^o \tag{2.381a}$$

with

$$c_{ij} = \frac{(x_{1j} - \bar{x}_j)}{\sum_{i=1}^{n}(x_{ij} - \bar{x}_j)^2} \quad , j = \overline{1, M} \tag{2.381b}$$

Now we are in position to calculate the dispersion of $\{b_j\}_{j=\overline{1,M}}$:

$$D(b_j) = \sigma_{b_j}^2 = D\left(\sum_{i=1}^{n} c_{ij} y_i^o\right)$$

$$= \sum_{i=1}^{n} c_{ij}^2 D(y_i^o) = \underbrace{D(y^o)}_{=\sigma_0^2 \text{ or, more precisely } = s_0^{*2}} \sum_{i=1}^{n} c_{ij}^2 \tag{2.382a}$$

So, we can equivalently and successively write that

$$D(b_j)\Big|_{estimated} = s_{b_j}^{*2}\Big|_{unbiased}$$

$$= s_0^{*2} \cdot \sum_{i=1}^{n} \frac{(x_{ij} - \bar{x}_j)^2}{\left[\sum_{i=1}^{n}(x_{ij} - \bar{x}_j)^2\right]^2} = s_0^{*2} \frac{\sum_{i=1}^{n}(x_{ij} - \bar{x}_j)^2}{\left[\sum_{i=1}^{n}(x_{ij} - \bar{x}_j)^2\right]^2}$$

$$= s_0^{*2} \frac{1}{\sum_{i=1}^{n}(x_{ij} - \bar{x}_j)^2} = \frac{1}{n-1-M} \frac{\sum_{i=1}^{n}(y_i^o - y_i^p)^2}{\sum_{i=1}^{n}(x_{ij} - \bar{x}_j)^2} \tag{2.382b}$$

Back in calculating $D(\bar{y}^o)$ by (2.379b) we have at the *unbiased* estimation level (for which 1:1 analysis substitutes the sample with collectivity dispersion σ):

$$s_{y_0^p}^{*2} = \left(\frac{s_0^{*2}}{n}\right) + \sum_{j=1}^{M}(x_{ij} - \bar{x}_j)^2 \left(\frac{s_0^{*2}}{\sum_{i=1}^{n}(x_{ij} - \bar{x}_j)^2}\right)$$

$$= s_0^{*2}\left[\frac{1}{n} + \sum_{j=1}^{M} \frac{(x_{0j} - \bar{x}_j)^2}{\sum_{i=1}^{n}(x_{ij} - \bar{x}_j)^2}\right] \tag{2.383}$$

Now, if we want to obtain the dispersion of the individuals observed values (in a given point) from the dispersion of individual predicted model, then, on the observed dispersion one should *add also the error dispersion* (of predicted vs. observed values):

$$s^{*2}_{y_0^{obs}} = s^{*2}_{y_0^p} + SSE^2_{\sigma_0(y^{obs})} = s^{*2}_{y_0^p} + s_0^{*2} = s_0^{*2}\left[1 + \frac{1}{n} + \sum_{j=1}^{M} \frac{\left(x_{0j} - \bar{x}_j\right)^2}{\sum_{i=1}^{n}\left(x_{ij} - \bar{x}_j\right)^2}\right] \quad (2.384)$$

With this, one can immediately obtain the estimated value for b_0 by observing that:

$$b_0 = y^p\Big|_{\substack{x_{0j}=0,\\ \forall j=\overline{1,M}}} \Rightarrow \begin{cases} \sigma^2_{b_0} = \sigma^2_{y^p_{(x_{0j}=0,\forall j)}} \\ s^{*2}_{b_0} = s_0^{*2}\left[\dfrac{1}{n} + \displaystyle\sum_{j=1}^{M} \dfrac{\bar{x}_j^2}{\sum_{i=1}^{n}\left(x_{ij} - \bar{x}_j\right)^2}\right] \end{cases} \quad (2.385)$$

In these conditions we can move forward to evaluate the confidence intervals regarding the regression. For the confidence intervals regarding the coefficients:

$$\{b_0, b_1, \ldots, b_k, \ldots, b_M\} \to b_k : k = \overline{0, M} \quad (2.386)$$

we have, upon the regression analysis, the specific results as the values of parameters $b_{k=\overline{0,M}}$ and the values for the estimates $s^{*2}_{b_0}, s^{*2}_{b_{j=\overline{1,M}}}$; thus, one can form the random normalized variable:

$$\frac{b_{k=\overline{0,M}} - \beta_{b_{k=\overline{0,M}}}}{s^{*}_{b_{k=\overline{0,M}}}} = t_{n-1-M} \quad (2.387)$$

representing a t-Student distribution with $n - 1 - m$ degrees of freedom (just because of the presence of this restricting factor in the expression of s_0^{*2} referring to the y^{obs} population). Then, from the inequality condition

$$\frac{b_k - \beta_{b_k}}{s^{*2}_{b_k}} < t_{n-1-M, 1-\frac{\alpha}{2}}, k = \overline{0, M} \quad (2.388)$$

the confidence interval results as:

$$\beta_k = b_k \pm t_{n-1-M,\, 1-\frac{\alpha}{2}} \times s_{b_k}^{*2} \quad ,k = \overline{0,M} \tag{2.389}$$

to be specialized for each regression coefficient's correspondence $b_0 \leftrightarrow s_{b_0}^*, b_j \leftrightarrow s_{b_j}^*$ with the critical value of the confidence interval found such that specializing the statistical tables with the confidence constrain:

$$F\left(t_{n-1-M,\, 1-\frac{\alpha}{2}}\right) = 1-\alpha \Rightarrow t_{n-1-M,\, 1-\frac{\alpha}{2}} \tag{2.390}$$

Regarding the significance of the regression coefficients $\beta_{j=\overline{1,M}}$, they features the magnitude in changing of the dependent variable y when the independent variables x_{ij} rise with one unit, while maintaining unchanged the remaining variables. This way, the coefficient β_j is validated for the allied null hypothesis test, i.e., $H_0 : \beta_j = 0,\ j = \overline{1,M}$ must be rejected. Accordingly, one calculates

$$t_{n-1-M}^{calc(j)} = \frac{b_j - 0}{s_{b_j}^*} = \frac{b_j}{s_{b_j}^*} \quad ,j = \overline{1,M} \tag{2.391}$$

and consequently one verifies if it belongs in the above confidence interval (2.389), i.e.

$$t_{n-1-M}^{calc(j)} \in \left[b_j - t_{n-1-M,\, 1-\frac{\alpha}{2}} s_{b_j}^*, b_j + t_{n-1-M,\, 1-\frac{\alpha}{2}} s_{b_j}^* \right] \tag{2.392}$$

If the answer is *yes* then H_0 accepted so β_j is invalidated and x_j should be removed from correlation, and the regression analysis should be re-done without it; If the answer is *no*, the above hypothesis H_0 is rejected so β_j is validated (as $\neq 0$) and therefore also x_j is validated as significant for the correlation. Overall, one may build the Table 2.5 of global parameters in a multivariate regression.

Finally, arise the problem of evaluating the reducing (error) of variance of the tested model respecting the "absolute" (ideally, yet unknown) one, which exactly correlates with the observed data. To this end the dispersion/

TABLE 2.5 The Confidence and Significance Analysis for the Multiregression Variables of Eq. (2.370b)

Regression variable	Correlation coefficient	Standard deviation	t-Student variable	Confidence interval	Significance test
Constant	b_0	$s^*_{b_0}$	$t^{calc(0)}_{n-1-M}$	$b_0 - t_{n-1-M, 1-\frac{\alpha}{2}} s^*_{b_0}$	$\exists H_0 \rightarrow \nexists b_0$
⋮	⋮	⋮	⋮	⋮	⋮
$x_{j=\overline{1,M}}$	b_j	$s^*_{b_j}$	$t^{calc(j)}_{n-1-M}$	$b_j - t_{n-1-M, 1-\frac{\alpha}{2}} s^*_{b_j}$	$\exists H_0 \rightarrow \nexists x_j$

variance analysis (ANOVA) is employed on a multifactorial manner, as next describing. One may assume that the n detections (observed or measured activities) are grouped in $r = M + 1$ classes based on the grouping regression factors, see Table 2.6.

In this case the null H_0 hypothesis says that: if the predicted models features equal medians, there are no significant differences between the parameters types in multivariate correlation, i.e., along the (const./free term, b_0), (b_1, x_1),…,(b_M, x_M), with the observed activity.

Therefore, since the H_0 is infirmed as the medians of multiple models based on different structural factors give a more significant grouping information; this way, consolidating the robustness of the working model of grouping, $\bar{y}_t$ with t=1, …, r, …, r=M+1 are used for comparing the

TABLE 2.6 The Basic Structure of a QSAR Table, As Provided By the Multivariate Regression Analysis

y^{obs}	$b_0 1$	$b_0 x_1$	…	$b_j x_j$	…	$b_M x_M$
y_1^{O}	y_1^{pO}	y_1^{p1}		y_1^{pj}		y_1^{pM}
y_2^{O}	y_2^{pO}	y_2^{p1}		y_2^{pj}		y_2^{pM}
⋮	⋮	⋮	…	⋮	…	⋮
y_k^{O}	y_k^{pO}	y_k^{p1}		y_k^{pj}		y_k^{pM}
⋮	⋮	⋮		⋮		⋮
y_n^{O}	y_n^{pO}	y_n^{p1}		y_n^{pj}		y_n^{pM}

*See Eq. (2.370b) and the subsequent Chapter 3.

variances between the medians (among the r groups of data) with the residual variance of the groups (or results).

Therefore, one uses the working notations

$$\bar{y}_t = \frac{1}{n_t}\sum_{i=1}^{n_t} y_{i,t}, \quad t = \overline{1,r} \tag{2.393a}$$

giving the median of the studied class of data modeled by $\sim y^{predicted}_{regression}$; along

$$\bar{y} = \frac{1}{n}\sum_{t=1}^{r}\sum_{i=1}^{n_t} y_{i,t} \tag{2.393b}$$

giving the median over all observations (data) from all classes; being both quantities of (2.393a, b) associated with the observed i-value of the t-class:

$$y_{i,t} \; ---\rightarrow \bar{y}_t ---\rightarrow \bar{y} \tag{2.394}$$

Therefore, the dispersion analysis appropriately decomposes/expands the square sum of the $y_{i,t}$ deviation from the general median $\bar{y}$ in components corresponding to a real or hypothetical source of statistical modeling (including errors or residuals) of the predicted medians:

$$\begin{aligned}
S_{\substack{ABSOLUTE \\ (observed)}\left\}\left\{\substack{no-MODELING \\ no-grouping}\right.\right.} &= \sum_{t=1}^{r}\sum_{i=1}^{n_t}(y_{i,t} - \bar{y})^2 \\
&= \sum_{t=1}^{r}\sum_{i=1}^{n_t}\left[(y_{i,t} - \bar{y}_t) + (\bar{y}_t - \bar{y})\right]^2 \\
&= \sum_{t=1}^{r}\sum_{i=1}^{n_t}(y_{i,t} - \bar{y}_t)^2 + \sum_{t=1}^{r}\sum_{i=1}^{n_t}(\bar{y}_t - \bar{y})^2 + 2\sum_{t=1}^{r}\sum_{i=1}^{n_t}\left[(y_{i,t} - \bar{y}_t) + (\bar{y}_t - \bar{y})\right]
\end{aligned} \tag{2.395}$$

which can be better understood with the help of further notations and identifications:

$$\sum_{t=1}^{r}\sum_{i=1}^{n_t}(y_{i,t} - \bar{y}_t)^2 = S_{\substack{RESIDUALS \\ (errors)}} \tag{2.396a}$$

$$\sum_{t=1}^{r}\sum_{i=1}^{n_t}(\bar{y}_t - \bar{y})^2 = S_{\substack{MODEL \\ (predicted)}} \tag{2.396b}$$

While noting that the third term in Eq. (2.395) identically vanishes

$$\sum_{t=1}^{r}\sum_{i=1}^{n_t}\left[(y_{i,t}-\bar{y}_t)+(\bar{y}_t-\bar{y})\right]$$

$$=\sum_{t=1}^{r}\left[(\bar{y}_t-\bar{y})\sum_{i=1}^{n_t}(y_{i,t}-\bar{y}_t)\right]=\sum_{t=1}^{r}(\bar{y}_t-\bar{y})\left[\underbrace{\sum_{i=1}^{n_t}y_{i,t}-n_t\bar{y}_t}_{=0}\right]=0 \quad (2.396c)$$

one remains with the worthy result

$$S_{obs}=S_{pred}+S_{error} \quad (2.397)$$

However, at the variance or unbiased dispersions we have the following picture of freedom degrees:

$$\left.\begin{array}{l} S_{obs}:n-1 \\ S_{pred}:r-1=(M+1)-1=M \end{array}\right\}\Rightarrow S_{err}:n-r=n-M-1 \quad (2.398)$$

In terms of a QSAR analysis we therefore have the respective synopsis of ANOVA:

$$s_{obs}^{*2}=\frac{1}{n-1}\sum_{i=1}^{n}\left(y_i^{obs}-\bar{y}\right)^2,\quad \bar{y}=\frac{1}{n}\sum_{i=1}^{n}y_i^{obs} \quad (2.399)$$

$$s_{pred}^{*2}=\frac{1}{M}\sum_{i=1}^{n}\left(y_i^{pred}-\bar{y}\right)^2 \quad (2.400)$$

$$s_{error}^{*2}=\frac{1}{n-M-1}\sum_{i=1}^{n}\left(y_i^{obs}-y_i^{pred}\right)^2 \quad (2.401)$$

This information contributes in forming the major statistical test of confidence intervals by the major above statistics:

(i) Through employing the *Fisher's ratio* of calculated critical value

$$F_{M,\,n-M-1}^{calc}=\frac{s_{pred}^{*2}}{s_{error}^{*2}}=\frac{\frac{1}{M}\sum_{i=1}^{n}\left(y_i^{pred}-\bar{y}\right)^2}{\frac{1}{n-M-1}\sum_{i=1}^{n}\left(y_i^{obs}-y_i^{pred}\right)^2} \quad (2.402)$$

towards comparing it with the *tabulated* value $F_{tab,M,\,n-M-1}\big|_{1-\alpha}$ and accordingly forming the testing cases of confidence and robustness of correlation itself:

$$F^{calc}_{\substack{M,\\ n-M-1}} > F^{tab}_{\substack{M,\\ n-M-1}}\Big|_{1-\alpha} \quad H_0 \text{ is rejected and regression is } \textit{significant} \tag{2.403}$$

against

$$F^{calc}_{\substack{M,\\ n-M-1}} \cong 4F^{tab}_{\substack{M,\\ n-M-1}}\Big|_{1-\alpha} \quad H_0 \text{ is rejected and regression is } \textit{robust} \tag{2.404}$$

(ii) The *adjusted correlation coefficient* (or the *explained variance* EV) because it contains the dependence of the degrees of freedom of the QSAR (predicted) model, beyond the standard correlation coefficient information:

$$EV = r^2_{adjusted} = 1 - \frac{s^{*2}_{error}}{s^{*2}_{obs}} = 1 - \frac{n-1}{n-M-1}\frac{\sum_{i=1}^{n}\left(y_i^{obs} - y_i^{pred}\right)^2}{\sum_{i=1}^{n}\left(y_i^{obs} - \bar{y}\right)^2} \tag{2.405}$$

with the role in highlighting of the residual factors' significance. Actually, the obtained value for r^2_{adj} (or simply called as r^2) can be relatively tested to its robustness by two methods:

(a) The case with r not to high on a big statistical sample is solved by the Fisher test with the random variable:

$$f = \frac{n-1-(M+\sigma)}{M+\sigma}\cdot\frac{r^2}{1-r^2}, \text{ with } \sigma = \begin{cases}1, & b_0 \neq 0\\ 0, & b_0 = 0\end{cases} \tag{2.406}$$

such that the hypothesis $H_0 : r = 0$ is associated with f^{calc}:

$$f^{calc} = \frac{n-1-(M+\sigma)}{M+\sigma}\cdot\frac{r^2_{calc}}{1-r^2_{calc}} \tag{2.407}$$

This way, one finds that

$$f^{calc} > F^{tab}_{\substack{M+\sigma,\\ n-1-(M+\sigma)}}\Big|_{1-\alpha} \Rightarrow H_0 \text{ is rejected} \Rightarrow r \neq 0 \text{ and significant;} \tag{2.408}$$

(b) The model with big r yet on a small sample is tested for its reliability through using the t-Student type testing variable

$$t_{calc} = \sqrt{n-1-(M+\alpha)} \cdot \sqrt{\frac{r^2_{calc}}{1-r^2_{calc}}} \tag{2.409}$$

and by means of fulfillment of the critical hierarchy, for instance as $t_{calc} > t^{tab}_{n-1-(M+\sigma)|1-\alpha} \Rightarrow H_0$ is rejected and $\Delta r \neq 0$ and significant.

(iii) In the final, one recognizes that:

$$s_0^{*2} = SEE = s^{*2}_{error} = \frac{1}{n-M-1}\sum_{i=1}^{n}\left(y_i^{obs} - y_i^{pred}\right)^2 \tag{2.410}$$

being it the last statistic indicator of the robustness of a QSAR analysis, while it is validated through the H_0 test applied to the χ^2_{n-M-1} distribution, namely:

$$H_0 : s_0^{*2} = \sigma_0^2 \Big|_C \Rightarrow \left(\chi^2\right)^{calc} = n - M - 1 \tag{2.411}$$

This way, whenever occurs the critical inclusion as

$$\underbrace{\chi^{2(tab)}_{n-M-1;\,\frac{\alpha}{2}}}_{F(\bullet)=\frac{\alpha}{2}} < \chi^{2(calc)} < \underbrace{\chi^{2(tab)}_{n-M-1;\,1-\frac{\alpha}{2}}}_{F(\bullet)=1-\frac{\alpha}{2}} \tag{2.412}$$

one concludes that H_0 is true and s_0^{*2} is significant. Alternately, the *SEE* value is judged respecting the $\left(y^{obs}_{max} - y^{obs}_{min}\right)$ domain; for example *SEE* = 0.3 is considered excellent over 10–15 units but unsatisfactory on just few units interval.

All-in-all, a QSAR regression can be represented in a synthetically-analytical form as:

$$y = \left(b_0 \pm s^*_{b_0} \cdot t_{n-1-M,\,1-\frac{\alpha}{2}}\right) + \sum_{j=1}^{M}\left(b_j \pm s^*_{b_j} \cdot t_{n-1-M,\,1-\frac{\alpha}{2}}\right) x_M \tag{2.413}$$

jointly reported along the statistical indicators

$$N = n;\ r^2\left(or\ r^2_{adj}\right);\ SEE;\ F^{calc}_{M,\,n-M-1} \tag{2.414}$$

in the case the input activities and the identified errors (by the residues $e_i = y_i^{obs} - y_i^{pred}$) follow a normal distribution (eventually tested with the χ^2 concordances' contingency); if this is not the case, further non-parametrical correlations are in view – yet this will be let for other time to be discussed.

2.5 CONCLUSION

The main lessons to be kept for the further theoretical and practical investigations of the correlation analysis for the chemical-biological complex interaction by the mathematical tools of statistics were presented with the aim to prepare the framework in which the huge and or diverse amount of experimental information may be quantitatively organized in comprehensive equations with a predictive value the quantitative structure-activity (or property) relationships QSA(P)Rs. This method seems to offer the best key for unifying the chemical (and biological) interaction into single model content (Hansch et al., 1996; Ogihara, 2003); written in the multi-linear form of causes $|X\rangle$ (in the Dirac bra-ket notation of quantum states) resulted in the predicted effect (Putz, 2012a, b)

$$|Y\rangle = \sum_i p_i |X_i\rangle \tag{2.415}$$

The QSA(P)R equation (2.415) has the justification in the quantum superposition principle (Dirac 1944), thus providing the appropriate framework in searching for new "natural laws" by various statistical means for computing the coefficients of this expansions (the p 's) so that the error of the predicted to the recorded effects $|Y\rangle$ is minimum.

On other side, depending on the way of considering the stated optimization centered on the molecular structure and various types of descriptors in hand the current methods can be conceptually grouped into the following main directions (Putz & Lacrămă, 2007; Putz & Putz, 2011; Putz, 2012a-b), see also the Figure 2.21:

(i) *Classical QSA(P)R* (Liwo et al., 1992; Crippen et al., 1993; Kubinyi, 1994b; Lhuguenot, 1995; Schmidli, 1997; Kier & Hall,

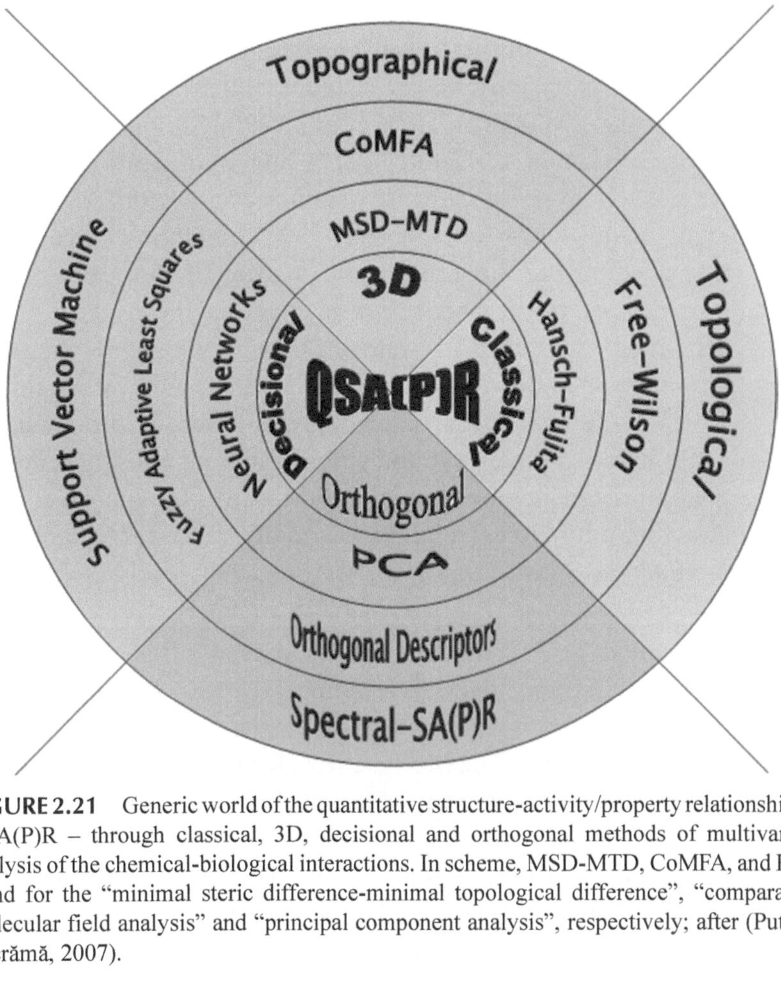

FIGURE 2.21 Generic world of the quantitative structure-activity/property relationships – QSA(P)R – through classical, 3D, decisional and orthogonal methods of multivariate analysis of the chemical-biological interactions. In scheme, MSD-MTD, CoMFA, and PCA stand for the "minimal steric difference-minimal topological difference", "comparative molecular field analysis" and "principal component analysis", respectively; after (Putz & Lacrămă, 2007).

1986; Balaban et al., 1993): assumes as descriptors the structural indices that directly reflect the electronic structures of the tested chemical compounds, for example, factors describing the lipophilicity (e.g., LogP, surfaces), electronic effects (e.g., Hammett constants, polarization, localization of charges), and steric effects (e.g., Taft indices, Verloop indices, topological indices, molecular mass, total energy at optimized molecular geometry);

(ii) *3-dimensional 3D-QSA(P)R*: specific for quantifying the biological activity in the ligand-receptor interactions, is characterized by entry indices; for instance, the minimal topological difference – MTD (Balaban et al., 1990; Simon et al., 1994; Duda-Seiman et al. 2006, 2007) and comparative molecular field analysis – CoMFA (Cramer et al., 1988, 1993; Sun et al., 2006; Duda-Seiman et al., 2011) methods closely take into account the bioactive conformation of the receptor, the topology of the ligand series as well as their steric fit, in accordance with the "key-into-lock" principle, while the topographical schemes (Randić et al., 1990; Randić & Razinger, 1995) make use of the graph representation of the chemical compounds, replacing in the associated connectivity matrices the optimized stereochemical indices. A visible increase in the structure-activity correlation is usually recorded when these methods are used (Navia & Peattie, 1993; Perkins & Dean, 1993; Lemmen & Lengauer, 1997);

(iii) *Decisional QSA(P)R*: contains an arsenal of heuristic methods of classifying data, developing GA, i.e., neural networks (Manallack & Livingstone, 1992, 1994), fuzzy methodologies (Moriguchi et al., 1992; Moriguchi & Hirono, 1995; Duchowicz & Castro, 2008), or support vector machine for learning (Vapnik, 1982, 1998; Schölkpof et al., 1999; Schölkpof & Smola, 2002; Mattera et al., 1999), in order to find optimal solutions for combinatorial problems; they offer the advantage of providing a quick estimation regarding the quality of correlation we should expect from the data and furnishing several best regression models to decide upon; moreover, the decisional analysis can be made in high-dimensional space always giving a solution by standard algorithm (Marchant & Combes, 1996; Mangasarian & Musicant, 1999);

(iv) *Orthogonal QSA(P)R*: addresses the heart of QSAR analysis, namely the orthogonal problem; statistically, the orthogonal property was attributed to those descriptors whose values form a basis set that pose little inter-correlation factors; in practice, data reduction techniques such as PCA (Sutter et al., 1992; Cash & Breen, 1992) describe biological activity or chemical properties through a fewer number of independent (orthogonal) descriptors giving a regression equation on these principal components; unfortunately, even combined with PLS cross-validation technique to produce higher predictive QSAR models, the main drawback still remains since they provide scarce possibility to interpret the obtained models (Nendza & Wenzel, 1993; Hemmateenejad et al., 2006); further progress was made by producing an orthogonal space by transforming the original basis set of descriptors in an orthogonal one by searching inter-regression equations between them (Randić, 1991a), followed, eventually, by their reciprocal subtractions (Randić, 1991a-b); unfortunately, this method was found to give, in almost all cases, the same correlation and statistical factors as those provided by regressions with original basis set of descriptors (Amić et al., 1995; Lučić et al., 1995a-c; Šoškić et al., 1996; Klein et al., 1997; Ivanciuc et al., 2000; Fernandez et al., 2004), moreover, producing a QSAR equation in the orthogonal space where the orthogonal descriptors have little interpretation against the real ones; the situation was recently improved by the new SPECial TRace of ALgebraic Structure-Activity Relationship (SPECTRAL-SAR, or Spectral-SAR or SSAR) method, where, due to the alternative algebraic solution of the problem of Eq. (2.415) one can better understand the mechanics of the molecular interaction by the so-called *least-spectral path principle* and the associate *spectral maps* that enable many computational ecotoxicological applications and interpretations being recently reported (Putz & Lacrămă, 2007; Putz & Putz, 2011; Putz, 2012a-b).

It is the last framework for QSAR the one that will be the extensive subject of the forthcoming Chapter 3, in relation with both the present Chapters 1 and 2; this because it employs the fundamental feature of the quantum theory: the orthogonality between descriptors – the chemical structural indices or characters that combines in a superposition manner (viz. the multi-regression), see Eq. (2.415), to produce the interaction with the biological site towards the recorded (viz. observed) effect; This way the Quantum-SAR (Qu-SAR) is rooted in orthogonal QSAR and on its variants, as will be unfolding in the sequel.

KEYWORDS

- **bilateral case**
- **chi-square statistics**
- **confidence intervals**
- **dispersions**
- **Fisher statistics**
- **median values**
- **multi-regression analysis of variance (ANOVA)**
- **regression**
- **robustness of correlations**
- **statistical correlation analysis**
- **statistical samples**
- **t-Student statistics**
- **Z-statistics**

REFERENCES

AUTHOR'S MAIN REFERENCES

Duda-Seiman, C., Duda-Seiman, D., Dragoş, D., Medeleanu, M., Careja, V., Putz, M. V., Lacrămă, A.-M., Chiriac, A., Nuţiu, R., Ciubotariu, D. (2006). Design of anti-HIV ligands by means of minimal topological difference (MTD) method. *Int. J. Mol. Sci.* 7(11), 537–555 (DOI: 10.3390/i7110537).

Duda-Seiman, C., Duda-Seiman, D., Putz, M. V., Ciubotariu, D. (2007). QSAR modeling of anti-HIV with HEPT derivatives. *Digest, J. Nanomater. Biostruct.* 2(2), 207–219.

Duda-Seiman, D., Avram, S., Mancas, S., Careja, V., Duda-Seiman, C., Putz, M. V., Ciubotariu, D. (2011). MTD-CoMSIA modeling of HMG-CoA reductase inhibitors. *J. Serb. Chem. Soc.* 76(1), 85–99 (DOI: 10.2298/JSC100601019D).

Lacrămă, A. M., Putz, M. V., Ostafe, V. (2008). Designing a spectral structure-activity ecotoxico-logistical battery. Putz, M. V. (Ed.) *Advances in Quantum Chemical Bonding Structures*, Transworld Research Network, Kerala, Chapter 16, pp. 389–419.

Putz, M. V. (2016a). *Quantum Nanochemistry. A Fully Integrated Approach: Vol. I. Quantum Theory and Observability.* Apple Academic Press & CRC Press, Toronto-New Jersey, Canada-USA.

Putz, M. V. (2016b). *Quantum Nanochemistry. A Fully Integrated Approach: Vol. II. Quantum Atoms and Periodicity.* Apple Academic Press & CRC Press, Toronto-New Jersey, Canada-USA.

Putz, M. V. (2012a). *Chemical Orthogonal Spaces*, Mathematical Chemistry Monographs Vol. 14, University of Kragujevac.

Putz, M. V., Ed. (2012b). *QSAR & SPECTRAL-SAR in Computational Ecotoxicology*, Apple Academics, Toronto.

Putz, M. V., Putz, A. M. (2011). Timisoara Spectral – Structure Activity Relationship (Spectral-SAR) algorithm: from statistical and algebraic fundamentals to quantum consequences. In Putz, M. V. (Ed.) *Quantum Frontiers of Atoms and Molecules*, NOVA Science Publishers, Inc., New York, Chapter 21, pp. 539–580.

Putz, M. V., Lacrămă, A. M. (2007). Introducing Spectral Structure Activity Relationship (S-SAR) analysis. Application to ecotoxicology, *Int. J. Mol. Sci.* 8(5), 363–391 (DOI: 10.3390/i8050363).

SPECIFIC REFERENCES

Amić, D., Davidović-Amić, D., Trinajstić, N. (1995,) Calculation of retention times of anthocyanins with orthogonalized topological indices. *J. Chem. Inf. Comput. Sci.* 35, 136–139.

Anderson, T. W. (1958). *An Introduction to Multivariate Statistical Methods*, Wiley, New York.

Balaban, A. T., Chiriac, A., Motoc, I., Simon, Z. (1980). *Steric Fit in QSAR*, Springer, Berlin (Lecture Notes in Chemistry Series).

Balaban, A. T., Motoc, I., Bonchev, D., Mekenyan, O. (1983). Topological indices for structure-activity correlations. *Top. Curr. Chem.* 114, 21–55.

Benigni, R., Bossa, C., Netzeva, T. I., Worth, A. P. (2007). Collection and evaluation of [(Q)SAR] models for mutagenicity and carcinogenicity, European Commission – Joint Research Centre, Ispra. Available online: http://ecb.jrc.it/qsar/publications/, accessed January 2009.

Box, G. E. P., Hunter, W. G., Hunter, J. S. (1978). *Statistics for Experimenters*, John-Wiley, New York.

Cash, G. G., Breen, J. J. (1992). Principal component analysis and spatial correlation: environmental analytical software tools. *Chemosphere* 24, 1607–1623.

Chatterjee, S., Hadi, A. S., Price, B. (2000). *Regression Analysis by Examples*, 3rd Ed., John-Wiley, New-York.

Cramer, R. D. III; DePriest, S. A., Patterson, D. E., Hecht, P. (1993). *The developing practice of comparative molecular field analysis*; In: H. Kubinyi (Ed.), *3D QSAR in Drug Design. Theory, Methods and Applications,* Escom, Leiden, pp. 443–485.

Cramer, R. D. III; Patterson, D. E., Bunce, J. D. (1988). Comparative molecular field analysis (CoMFA). 1. Effect shape on binding of steroids to carrier proteins. *J. Am. Chem. Soc.* 110, 5959–5967.

Crippen, G. M., Bradley, M. P., Richardson, W. W. (1993). Why are binding-site models more complicated than molecules? *Perspectives in Drug Discovery and Design* 1, 321–328.

Cronin, M. T. D., Aptula, A. O., Duffy, J. C., Netzeva, T. I., Rowe, P. H., Valkova, I.V; Schultz, T. W. (2002). Comparative assessment of methods to develop QSARs for the prediction of toxicity of phenols to Tetrahymena pyriformis. *Chemosphere* 49, 1201–1221.

Cronin, M. T. D., Netzeva, T. I., Dearden, J. C., Edwards, R., Worgan, A. D. P. (2004). Assessment and modeling of the toxicity of organic chemicals to Chlorella Vulgaris: development of a novel database. *Chem. Res. Toxicol.* 17, 545–554.

Cronin, M. T. D., Schultz, T. W. (2003). Pitfalls in QSAR. *J. Mol. Struct. (Theochem)* 622, 39–51.

Cronin, M. T. D., Worth, A. P. (2008). (Q)SARs for predicting effects relating to reproductive toxicity. *QSAR Comb. Sci.* 27, 91–100.

Dirac, P. A. M. (1944). *The Principles of Quantum Mechanics*, Oxford University Press, Oxford.

Duchowicz, P. R., Castro, E. A. (2008). *The Order Theory in QSPR-QSAR Studies*; In: Mathematical Chemistry Monographs, University of Kragujevac, Kragujevac.

EC (2003). Proposal for a Regulation of the European Parliament and of the Council concerning the Registration, Evaluation, Authorization and Restriction of Chemicals (REACH), establishing a European Chemicals Agency and amending directive 1999/45/ EC and Regulation (EC) {on Persistent Organic Pollutants}, Brussels, Belgium.

EU-COM (2008). *Green Paper* on the management of bio-waste in the European Union, COM(2008). 811 final, Brussels, 3.12.2008 (http://ec.europa.eu/green-papers/index_en.htm)

Fernandez, F. M., Duchowicz, P. R., Castro, E. A. (2004). About orthogonal descriptors in QSPR/QSAR theories. *MATCH Commun. Math. Comput. Chem.* 51, 39–57.

Gallegos Saliner, A., Patlewicz, G., Worth, A. P. (2008). A review of (Q)SAR models for skin and eye irritation and corrosion. *QSAR Comb. Sci.* 27, 49–59.

Green, J. R., Margerison, D. (1978). *Statistical Treatment of Experimental Data*; Elsevier: New York.

Haegawa, K., Kimura, T., Fanatsu, K. (1999). GA strategy for variable selection in QSAR Studies: Enhancement of comparative molecular binding energy analysis by GA-based PLS method. *Quant. Struct.-Act. Relat.* 18, 262–272.

Hansch, C., Hoekman, D., Gao, H. (1996). Comparative QSAR: toward a deeper understanding of chemicobiological interactions. *Chem. Rev.* 96, 1045–1075.

Hemmateenejad, B., Miri, R., Jafarpour, M., Tabarzad, M., Foroumadi, A. (2006). Multiple linear regression and principal component analysis-based prediction of the anti-tuberculosis activity of some 2-aryl-1,3,4-thiadiazole derivatives. *QSAR Comb. Sci.* 25, 56–66.

Hulzebos, E. M., Posthumus, R. (2003). (Q)SARs: gatekeepers against risk on chemicals? *SAR QSAR Environ. Res.*14(4), 285–316.

Hulzebos, E. M., Sijm, D., Traas, T., Posthumus, R., Maslankiewicz, L. (2005). Validity and validation of expert (Q)SAR systems. SAR and QSAR in Environmental Research. *SAR QSAR Environ. Res.* 16(4), 385–401.

Ivanciuc, O., Taraviras, S. L., Cabrol-Bass, D. (2000). Quasi-orthogonal basis sets of molecular graph descriptors as chemical diversity measure. *J. Chem. Inf. Comput. Sci.* 40, 126–134.

Kier, L. B., Hall, L. H. (1986). *Molecular Connectivity in Structure-Activity Analysis*, Research Studies Press, Letchworth.

Klein, D. J., Randić, M., Babić, D., Lučić, B., Nikolić, S., Trinajstić, N. (1997). Hierarchical orthogonalization of descriptors. *Int. J. Quantum Chem.* 63, 215–222.

Kubinyi, H. (1994a). Variable selection in QSAR studies. 1. An evolutionary algorithm. *Quant. Struct.-Act. Relat.* 13, 285–294.

Kubinyi, H. (1994b). Der Schlüssel zum Schloß, I. Grundlagen der Arzneimittelwirkung. *Pharmazie in unserer Zeit* 23(3), 158–168.

Kubinyi, H. (1996,) Evolutionary variable selection in regression and PLS analysis. *J. Chemometr.* 10, 119–133.

Lemmen, C., Lengauer, T. (1997). Time-efficient flexible superposition of medium-sized molecules. *J. Comput.-Aided Mol. Design* 11, 357–368.

Lhuguenot, J.-C. (1995). Relation quantitative structure-activité (QSAR), une méthode mal reconnue car trop souvent mal utilisée. *Ann. Fals. Exp. Chim.* 88, 293–310.

Liwo, A., Tarnowska, M., Grzonka, Z., Tempczyk, A. (1992). Modified Free-Wilson method for the analysis of biological activity data. *Computers Chem.* 16, 1–9.

Lucic, B., Trinajstic, N. (1999). Multivariate regression outperforms several robust architectures of neural networks in QSAR modeling. *J. Chem. Inf. Comput. Sci.* 39, 121–132.

Lučić, B., Nikolić, S., Trinajstić, N., Juretić, D. (1995a). The structure-property models can be improved using the orthogonalized descriptors. *J. Chem. Inf. Comput. Sci.* 35, 532–538.

Lučić, B., Nikolić, S., Trinajstić, N., Juretić, D., Jurić, A. (1995c). A Novel QSPR approach to physicochemical properties of the α-amino acids. *Croatica Chem. Acta* 68, 435–450.

Lučić, B., Nikolić, S., Trinajstić, N., Jurić, A., Mihalić, Z. (1995b). A structure-property study of the solubility of aliphatic alcohols in water. *Croatica Chem. Acta* 68, 417–434.

Manallack, D. T., Livingstone, D. J. (1992). Artificial neural networks: application and chance effects for QSAR data analysis. *Med. Chem. Res.* 2, 181–190.

Manallack, D. T., Livingstone, D.J (1994). Limitations of functional-link nets as applied to QSAR data analysis. *Quant. Struct-Act. Relat.* 13, 18–21.

Mangasarian, O. L., Musicant, D. R. (1999). Successive over relaxation for support vector machines. *IEEE Trans. Neural Networks* 10, 1032–1036.

Marchant, C. A., Combes, R. D. (1996). Artificial intelligence: the use of computer methods in the prediction of metabolism and toxicity; In: Ford, M. G., Greenwood, R. (Eds.) *Bioactive Compound Design: Possibilities for Industrial Use*, G. T. Brooks and, R. Franke BIOS Scientific Publishers Limited.

Mattera, D., Palmieri, F., Haykin, S. (1999). Simple and robust methods for support vector expansions. *IEEE Trans. Neural Networks* 10, 1038–1047.

Moriguchi, I., Hirono, S. (1995). Fuzzy adaptive least squares and its use in quantitative structure-activity relationships; In: Fujita, T. (Ed.) *QSAR and Drug Design – New Developments and Applications*, Elsevier Science, B. V.

Moriguchi, I., Hirono, S., Matsushita, Y., Liu, Q., Nakagome, I. (1992). Fuzzy adaptive least squares applied to structure-activity and structure-toxicity correlations. *Chem. Pharm. Bull.* 40, 930–934.

Navia, M. A., Peattie, D. A. (1993). Structure-based drug design: applications in immunopharmacology and immunosuppression. *Immunology Today* 14, 296–301.
Nendza, M., Wenzel, A. (1993). Statistical approach to chemicals classification. *Environ. Toxicol. Chem. Supplement,* 1459–1470.
Netzeva, T., Pavan, M., Worth, A. P. (2008). Review of (quantitative) structure-activity relationship for acute aquatic toxicity. *QSAR Comb. Sci.* 27, 77–90.
OECD (2004). Report from the Expert Group on (Quantitative) Structure-Activity Relationships [(Q)SARs] on the Principles for the Validation of (Q)SARs, Series on Testing and Assessment, No. 49, OECD, Paris; 2004. pp. 206 (http://www.oecd.org/document/30/0,2340, en_2649_34365_1916638_1_1_1_1, 00.html, accessed 3 March 2011).
Ogihara, N. (2003). Drawing out drugs. *Mod. Drug Discovery* 6 (9), 28–32.
Patlewicz, G., Aptula, A., Roberts, D. W., Uriarte, E. (2008). A mini-review of available skin sensitization (Q)SARs/Expert systems. *QSAR Comb. Sci.* 27, 60–76.
Pavan, M., Netzeva, T. I., Worth, A. P. (2006). Validation of a QSAR model for acute toxicity. *SAR QSAR Environ. Res.* 17(2), 147–171.
Pavan, M., Netzeva, T., Worth, A. P. (2008a). Review of literature based quantitative structure-activity relationship models for bioconcentration. *QSAR Comb. Sci.* 27, 21–31.
Pavan, M., Worth, A. P. (2008b). Review of estimation models for biodegradation. *QSAR Comb. Sci.* 27, 32–40.
Perkins, T. D. J., Dean, P. M. (1993). An exploration of a novel strategy for superposing several flexible molecules. *J. Comput.-Aided Mol. Design* 7, 155–172.
Randić, M. (1991a). Resolution of ambiguities in structure-property studies by use of orthogonal descriptors. *J. Chem. Inf. Comput. Sci.* 31, 311–320.
Randić, M. (1991b). Orthogonal molecular descriptors. *New, J. Chem.* 15, 517–525.
Randić, M., Jerman-Blazic, B., Trinajstić, N. (1990). Development of 3-dimensional molecular descriptors. *Comput. Chem.* 14, 237–246.
Randić, M., Razinger, M. (1995). Molecular topographic indices. *J. Chem. Inf. Comput. Sci.* 35, 140–147.
Schmidli, H. (1997). Multivariate prediction for QSAR. *Chemometrics and Intelligent Laboratory Systems* 37, 125–134.
Schultz, T. W., Netzeva, T. I., Cronin, M. T. D. (2003). Selection of data sets for QSARs: analyses of Tetrahymena toxicity from aromatic compounds. *SAR QSAR Environ. Res.* 14(1), 59–81.
Schölkpof, B., Burges, C. J. C., Smola, A. J. (Eds.) (1999). *Advances in Kernel Methods. Support Vector Learning,* MIT Press, Cambridge.
Schölkpof, B., Smola, A. J. (2002). *Learning with Kernels,* MIT Press, Cambridge.
Seyfel, J. K. (1985). *QSAR and Strategies in the Design of Bioactive Compounds*, VCH Weinheim, New York.
Shorter, J. (1973). *Correlation Analysis in Organic Chemistry: An Introduction to Linear Free Energy Relationships*, Oxford Univ. Press, London.
Simon Z; Chiriac, A., Holban, S., Ciubotariu, D., Mihalas, G. I. (1984). *Minimum Steric Difference. The MTD Method for QSAR Studies*, Res. Studies Press (Wiley), Letchworth.

So, S. S., Karpuls, M. (1996). Evolutionary optimization in quantitative structure-activity relationship: An application of genetic neural network. *J. Med. Chem.* 39, 1521–1530.

Sun, J., Chen, H. F., Xia, H. R., Yao, J. H., Fan, B. T. (2006,) Comparative study of factor Xa inhibitors using molecular docking/SVM/HQSAR/3D-QSAR methods. *QSAR Comb. Sci.* 25, 25–45.

Sutter, J. M., Kalivas, J. H., Lang, P. K. (1992). Which principal components to utilize for principal component regression. *J. Chemometrics* 6, 217–225.

Teko, I. V., Alessandro, V. A. E. P., Livingston, D. J. (1996). Neutral network studies. 2. Variable selection. *J. Chem. Inf. Comput. Sci.* 36, 794–803.

Topliss, J. (1983). *Quantitative Structure-Activity Relationships of Drugs*, Academic Press: New York.

Tsakovska, I., Lessigiarska, I., Netzeva, T., Worth, A. P. (2008). A mini review of mammalian toxicity (Q)SAR models. *QSAR Comb. Sci.* 27, 41–48.

Vapnik, V. N. (1982). *Estimation of Dependencies Based on Empirical Data*, Springer-Verlag, Berlin.

Vapnik, V. N. (1998). *Statistical Learning Theory*, John Wiley & Sons, New York.

Worth, A. P., Balls, M. (2002). Alternative (non-animal) methods for chemicals testing: current status and future prospects. A report prepared by ECVAM and the ECVAM Working Group on Chemicals. *ATLA* 30(1), 1–125.

Worth, A. P., Bassan, A., Gallegos Saliner, A., Netzeva, T. I., Patlewicz, G., Pavan, M., Tsakovska, I., Vracko, M. (2005). The characterization of quantitative structure-activity relationships: Preliminary guidance. European Commission – Joint Research Centre: Ispra; Available online: http://ecb.jrc.it/qsar/publications/, accessed January 2009.

Worth, A. P., Cronin, M. T. D. (2003). The use of discriminant analysis, logistic regression and classification tree analysis in the development of classification models for human health effects. *J. Mol. Struct. (Theochem)* 622, 97–111.

Zhao, V. H., Cronin, M. T. D., Dearden, J. C. (1998). Quantitative structure-activity relationships of chemicals acting by non-polar narcosis – theoretical considerations. *Quant. Struct.-Act. Relat.* 17, 131–138.

Zheng, W., Tropsha, A. Novel variable selection quantitative structure-property relationship approach based on the k-nearest neighbor principle. (2000). *J. Chem. Inf. Comput. Sci.* 40, 185–194.

Šoškić, M., Plavšić, D., Trinajstić, N. (1996). Link between orthogonal and standard multiple linear regression models. *J. Chem. Inf. Comput. Sci.* 36, 829–832.

FURTHER READINGS

Draper, N. R., Smith, H. (1998). *Applied Regression Analysis* (3rd ed.), John Wiley.

Freedman, D. A. (2005). *Statistical Models: Theory and Practice*, Cambridge University Press.

Good, P. I., Hardin, J. W. (2009). *Common Errors in Statistics (And How to Avoid Them)* (3rd ed.), Hoboken, Wiley, New Jersey.

Pearson, E. S. (1939). 'Student' as statistician. *Biometrika* 30(3/4), 210–250.

Pearson, E. S. (1990). *'Student, 'A Statistical Biography of William Sealy Gosset,* Edited and Augmented by Plackett, R. L. with the Assistance of, G. Barnard, Oxford: University Press.

Sen, A., Srivastava, M. (2011). *Regression Analysis—Theory, Methods, and Applications*, Springer-Verlag, Berlin.

CHAPTER 3

CHEMICAL ORTHOGONAL SPACES FOR STRUCTURE-ACTIVITY RELATIONSHIP (COS-SAR)

CONTENTS

ABSTRACT

With the present-day interest in correlating chemical structure with biological activity the quantitative structure-activity relationships (QSARs) is here presented under a plethora of novel, fresh and fruitful picture of regression analysis aiming to closely approach the quantum interpretation of data and of ligand-receptor interaction by means of systematic orthogonal and scalar (dot) product of either molecular (chemicals or toxicants) descriptors between them and with the observed (recorded, measured) activities. The resulted Spectral-, Diagonal-Spectral-, Projective-, Catastrophe-, Residual-, SMILES (simplified molecular-input line-entry system)-, Topo-Reactive and Logistic-SARs may be conceptually and computationally considered as realization forms of the general Quantum-SAR (Qu-SAR) which widely employs the present data as a whole vectors, to be associated in principle with the eigen-states in quantum Hilbert space, while opening the way for assigning a sort of wave function or wave packet for the congeneric active molecular series rather than for a single molecule as used to be; this way the specific interaction may be eventually modeled by structure (intrinsic)-metabolic (extrinsic) quantum rather quantitative correlation picture so further allowing in establishing the specific "quantum paths" (or what is customarily known as the mechanistically map of ligand-receptor bonding) for a given or designed chemical-biological interaction.

3.1 INTRODUCTION

Paradoxically, the main problem for QSAR resides not in performing the correlation itself but setting the variable selection for it; the mathematical counterpart for such problem is known as the "factor indeterminacy" (Spearman, 1927; Wilson, 1928; Wilson & Hilferty, 1931; Wilson & Worcester, 1939; Steiger & Schonemann, 1978) and affirms that the same degree of correlation may be reached with in principle an infinity of latent variable combinations. Fortunately, in physical-chemical there are a limited (although many enough) indicators to be considered with a clear-cut meaning in molecular structure that allows for rationale of reactivity and bindings (Topliss & Costello, 1972; Topliss & Edwards, 1979).

Therefore, although undoubtedly useful, the "official" trend in employing QSAR methods is to classify, over-classify and validate through (external or molecular test set) prediction, a gap between the molecular computed orderings and the associate mechanistic role in bio-/eco-activity assessment remains as large as the QSAR strategy has not turned into a versatile tool in identifying the inter-molecular role in receptor binding sites through recorded activities by means of structurally selected common variables; that is to use QSAR information for internal mechanistic predictions among training molecules to see their inter-relation respecting the whole class of observed activities employed for a specific correlation. Such an approach will also be helpful for checking the chemical domain spanned by training molecules – a feature of the paramount importance also for further external tests.

The modern *in silico* (computational) chemical analysis respecting the bio- activity and availability of analogues substances, potentially beneficial or detrimental for specific interaction in organs and organisms, faces with a paradoxical dichotomy: if searching for the best correlation useful for *prediction* of specific molecular bio- or eco- activity QSAR models involving un-interpretable many latent variables may be produced, while always remaining the question of correlation factor indeterminacy (i.e., the assumed descriptors can be at any time replaced with other producing at least the same correlation performances); instead, when restricting the analysis to search for molecular design and mechanisms throughout performing SARs by means of special structural indicators for a given class of relevant molecules, arises the price of limiting the use of generated models for further prediction.

The present chapter aims filling this gap by deepening the modeling of inter-molecular activity through extending the main concepts around the so-called Spectral-SAR (Putz, 2012a, b), developed the fully algebraic version of traditional statistically optimized QSAR picture, targeting the quantification of the competition between molecular inter-activity and inter-endpoints records. As such, the present review was mainly oriented in presenting and developing the second (Q)SAR facet by rationalized the recent introduced notion of spectral-path-linking-endpoints and the associate least action principle to spectral path quantification, in terms of the best fitted molecules, along the contained computed models, by means of the introduced q(uantum)-SAR factor within the generally called Quantum-SAR (QuaSAR) methodology.

On the other side, the so-called *green chemistry* stands as a priority field of research which is approached by the research programs of United States and European Commission as well. It has the goal of characterization, prediction and the control of the chemical structures acting as toxicants on organisms and environment. The main reason for such research links the economical, ecological and public health issues in a general paradigm: *method* → *data* → *information* → *knowledge* → *use*. Within this epistemological chain *the method* relates the involved procedure in obtaining the experimental data and is regulated by the chemical-physical and biological scientific laws; *the data* represent the chemicals and their toxic or carcinogenic values; *information* refers to elaboration of models through the recorded data; the *knowledge* means the prediction or the final model of the molecular action mechanisms; *the use* is defined by the legal boundaries for the toxic values or classes of chemicals admitted.

In this context, the actual Spectral-to-Quantum SAR project propose an advanced study based on the epistemological bulk data-information-knowledge of the chemicals used in green chemistry in order to asses: a specific model of quantum characterization of concerned active substances at the bio-, eco- and pharmaco-logic levels through unitary formulation of the atomic-molecular indices for the effector-receptor binding degree potential of the logistic type (including the temporal dependency); a computational consistent model aiming to minimize the residual recorded activities in the experiments studying the enzymic, ionic liquid, antagonists and allosteric inhibition interactions. The methodology allows pattering both the controlling as well as the design of new compounds for synthesis this way eventually covering also the method-and-use segments of the economical-social life in XXI.

3.2 SPECTRAL REGRESSIONS' MODELS ON CHEMICAL ORTHOGONAL SPACE

3.2.1 INTRODUCING SPECTRAL-SAR

The key concept in SAR discussion regards the independence of the considered structural parameters in Table 3.1. As a consequence we may further employ this feature to quantify the basic SAR through an orthogonal space. The idea is to transform the columns of structural data of Table 3.1 into an abstract orthogonal space, Figure 3.1, where necessarily all predictor variables are independent, solve the SAR problem there and then referring the result to the initial data by means of a coordinate transformation ; the present discussion mainly follows (Putz & Lacrămă, 2007).

TABLE 3.1 The Vectorial Descriptors in a Spectral-SAR Analysis*

Activity	*Structural predictor variables*					
$\lvert A\rangle = \lvert Y_{OBS(ERVED)}\rangle$	$\lvert X_0\rangle$	$\lvert X_1\rangle$	...	$\lvert X_k\rangle$	...	$\lvert X_M\rangle$
$A_1 = y_{1\text{-}OBS}$	1	x_{11}	...	x_{1k}	...	x_{1M}
$A_2 = y_{2\text{-}OBS}$	1	x_{21}	...	x_{2k}	...	x_{2M}
⋮	⋮	⋮	⋮	⋮	⋮	⋮
$A_N = y_{N\text{-}OBS}$	1	x_{N1}	...	x_{Nk}	...	x_{NM}

*Putz & Lacrămă (2007); Putz & Putz (Lacrămă) (2008).

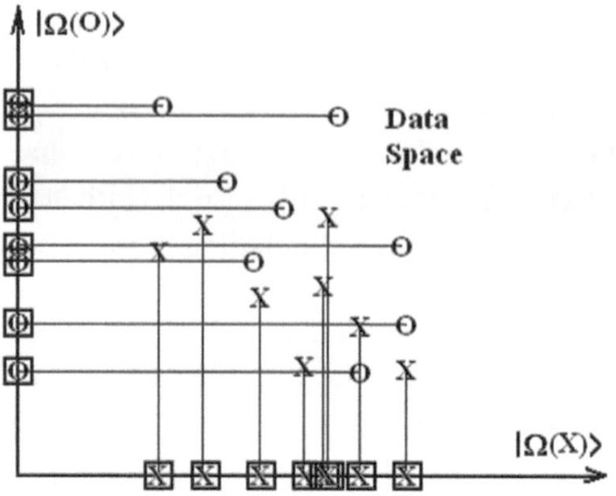

FIGURE 3.1 Generic mapping of data space containing the vectorial sets {|X⟩, |0⟩} into orthogonal basis {|Ω(X)⟩, |Ω(O)⟩} (Putz & Lacrămă, 2007).

The idea is to transform the columns of structural data of Table 3.1 into an abstract orthogonal space, where necessarily all predictor variables are independent, see Figure 3.1; solve the SAR problem there and then referring the result to the initial data by means of a coordinate transformation. The analytical procedure is unfolded in simple three steps.

Since QSAR models aims correlations between concerned molecular structures and measured (or otherwise evaluated) activity, appears naturally that the *structure* part of the problem to be accommodated within the quantum theory and of its formalisms. In fact, there are few quantum characters that we are using within the present approach (Chicu & Putz, 2009; Putz, 2012a, b):

- Any molecular structural state (dynamical, since undergoes interaction with organism) may be represented by a $|ket\rangle$ *state vector*, in an abstract space of allowed states within Hilbert space, following the $\langle bra|ket\rangle$ Dirac formalism; such states are to be here represented by any reliable molecular index, or, in particular in our study by hidophobicity $|LogP\rangle$, polarizability $|POL\rangle$, total optimized energy $|E_{tot}\rangle$, just to name only the so-called Hansch parameters, usually employed for accounting the diffusion, electrostatic and steric effects for molecules acting within organisms' cells, respectively. Actually, the columns of Table 3.1 are now considered as vectors in data space, then we are looking for the "spectral" decomposition of the activity vector $|Y\rangle$ upon the considered basis of the structural vectors $\{|X_0\rangle, |X_1\rangle, ..., |X_k\rangle, ..., |X_M\rangle\}$ based on the quantum superposition principle applied to indicators of the structural molecular causes, i.e., with inherent quantum nature.
- The (quantum) *superposition principle* that assures that sum combinations of molecular states map on other resulting molecular state, here interpreted as bio-, eco- or toxico- logical activity, e.g.

$$|Y\rangle = |Y_0\rangle + C_{LogP}|LogP\rangle + C_{POL}|POL\rangle + ... \tag{3.1}$$

  with $|Y_0\rangle$ meaning the free or unperturbed activity (when all other influences are absent).
- The *orthogonalization feature* of quantum states, a crucial condition for that the superimposed molecular states generates other molecular state (here quantified as molecular-linking organism activity); analytically, the orthogonalization condition is represented by the $\langle bra|ket\rangle$ scalar product of two envisaged states (molecular indices) whom value if it is evaluated to be zero, $\langle bra|ket\rangle = 0$, then the states are said orthogonal and molecular descriptors independent, therefore suitable to be added as states in resulted activity state and as molecular indices in activity correlation. Further details on scalar product and related properties are given in Volume I of this five-volume work (Putz, 2016a), while one may shortly remembered by the *quantum* generalized scalar product throughout the basic rule:

$$\langle \Psi_l | \Psi_k \rangle = \sum_{i=1}^{N} \psi_{il}\psi_{ik} = \langle \Psi_k | \Psi_l \rangle \tag{3.2}$$

giving out a real number from two arbitrary N-dimensional vectors (wave functions' states)

$$|\Psi_l\rangle = |\psi_{1l} \quad \psi_{2l} \quad \dots \quad \psi_{Nl}\rangle \text{ and } |\Psi_k\rangle = |\psi_{1k} \quad \psi_{2k} \quad \dots \quad \psi_{Nk}\rangle$$

In what follows the Spectral-SAR correlation method is resumed by the analytical procedure unfolded in three fundamental steps (I, II, and III) (Putz & Lacrămă, 2007).

I. Given a set of N molecules being studies against biological activity they produce by means of their $M-$ structural indicators, all input information (the states) may be vectorial expressed by the columns of the Table 3.1 and correlated upon equation

$$\begin{aligned} |Y_{OBS(ERVED)}\rangle &= b_0|X_0\rangle + b_1|X_1\rangle + \dots + b_k|X_k\rangle + \dots + b_M|X_M\rangle \\ &\qquad + |prediction \quad error\rangle \\ &= |Y_{PRED(ICTED)}\rangle + |prediction \quad error\rangle \end{aligned} \tag{3.3}$$

which the unity vector $|X_0\rangle = |1\,1 \cdots 1_N\rangle$ added to account for the free term.

In order equation (3.3) to represent a reliable model of the given activities, the molecular states (indices) assumed should constitute an orthogonal set, having this constraint a quantum mechanically fundament, as above described. However, unlike other important studies addressing this problem, the present employed Spectral-SAR assumes the prediction error vector in Eq. (3.3) as being from beginning orthogonal on all others, since it cannot be considered input data as the others,

$$\langle Y_{PRED} | prediction \quad error\rangle = 0 \tag{3.4a}$$

being not known *apriori* any correlation is made. Moreover, from Eqs. (3.3) and (3.4a) there follows that the prediction error vector has to be orthogonal on all other descriptor states of predicted activity.

$$\langle X_{i=\overline{0,M}} | prediction \quad error\rangle = 0 \tag{3.4b}$$

for consistency of the present vectorial (quantum formalized by means of $|ket\rangle$ states) approach. In other terms, conditions (3.4a) and (3.4b) confirm the form (3.3) in the sense that prediction vector and the prediction activity $|Y_{PRED}\rangle$ (with all its sub-intended states $|X_{i=\overline{0,M}}\rangle$) belongs to disjoint (thus orthogonal) Hilbert spaces; or even more, one can say that the Hilbert space of the observed activity $|Y_{OBS}\rangle$ may be decomposed into a predicted and error independent Hilbert sub-spaces of states.

Therefore within *Spectral-SAR* (SPEcial Computing TRace of Algebraic SAR) procedure the very beginning step in orthogonalization is prediction vector orthogonalization to prediction activity and of its predictor states, while the remaining orthogonalization algorithm do not search for optimizing the minimization of errors, but for producing the ideal correlation between $|Y_{PRED}\rangle$ and the given descriptors $|X_{i=\overline{0,M}}\rangle$.

II. To achieve the minimal errors in Eq. (3.3) the transformation of the data basis $\{|X_0\rangle, |X_1\rangle, \ldots, |X_k\rangle, \ldots, |X_M\rangle\}$ into an orthogonal one, say $\{|\Omega_0\rangle, |\Omega_1\rangle, \ldots, |\Omega_k\rangle, \ldots, |\Omega_M\rangle\}$, is now considered (see Figure 3.2 for the basic idea). In this respect the consecrated Gram-Schmidt procedure is

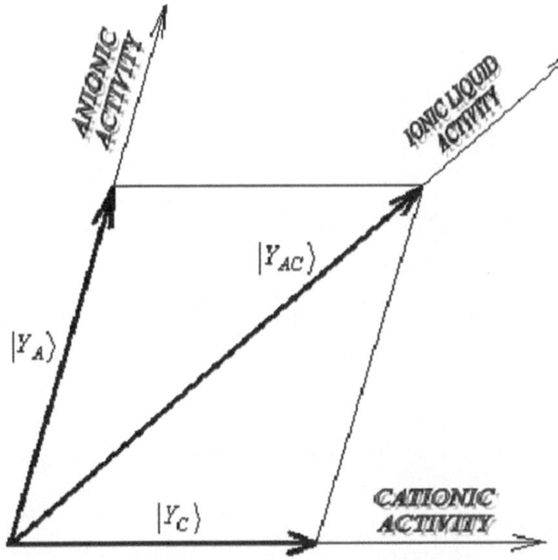

FIGURE 3.2 The vectorial composition of anionic and cationic sub-system activities in the ionic liquid global system, leading with the internal angle formation, with the trigonometric expression of Eq. (3.83) (Lacrămă et al., 2007; Putz, 2012b).

employed. It is worth noting that this procedure is well known in quantum chemistry when searching for an orthogonal basis for an orthogonal basis set in atomic and molecular wave function spectral decomposition (Daudel et al., 1983). This way, the Gram-Schmidt orthogonalization algorithm is applied through constructing the orthogonal set of descriptors by means of the consecrated iteration (Fadeeva, 1959; Steen, 1973; Siegmund-Schultze, 1986):

Choose

$$\left|\Omega_0\right\rangle = \left|X_0\right\rangle \tag{3.5a}$$

i. Then, by picking $\left|X_1\right\rangle$ as the next vector to be transformed, one can write that:

$$\left|\Omega_1\right\rangle = \left|X_1\right\rangle - r_0^1\left|\Omega_0\right\rangle,\ r_0^1 = \frac{\left\langle X_1 \middle| \Omega_0\right\rangle}{\left\langle \Omega_0 \middle| \Omega_0\right\rangle} \tag{3.5b}$$

so that $\left\langle \Omega_0 \middle| \Omega_1\right\rangle = 0$ assuring so far that $\left|\Omega_0\right\rangle$ and $\left|\Omega_1\right\rangle$ are orthogonal.

ii. Next, repeating steps (i) and (ii) above until the vectors $|\Omega_0\rangle$, $|\Omega_1\rangle,\ldots, |\Omega_{k-1}\rangle$ are orthogonally constructed, we can, for instance, further transform the vector $|X_k\rangle$ into:

$$\left|\Omega_k\right\rangle = \left|X_k\right\rangle - \sum_{i=0}^{k-1} r_i^k\left|\Omega_i\right\rangle,\ r_i^k = \frac{\left\langle X_k \middle| \Omega_i\right\rangle}{\left\langle \Omega_i \middle| \Omega_i\right\rangle},\ k = \overline{1, M}; \tag{3.5c}$$

so that the vector $\left|\Omega_k\right\rangle$ is orthogonal on all previous ones.

iii. Step (iii) is repeated and extended until the last orthogonal predictor vector $\left|\Omega_M\right\rangle$ is obtained.

Therefore, grounded on the Gram-Schmidt recipe the starting predictor vectorial basis

$$\left\{\left|X_0\right\rangle, \left|X_1\right\rangle, \ldots, \left|X_k\right\rangle, \ldots, \left|X_M\right\rangle\right\}$$

is replaced with the orthogonal one

$$\left\{\left|\Omega_0\right\rangle, \left|\Omega_1\right\rangle, \ldots, \left|\Omega_k\right\rangle, \ldots, \left|\Omega_M\right\rangle\right\}$$

by appropriately subtracting from the original vectors the non-wished non-orthogonal contributions. Note that the above procedure holds for any arbitrary order of original vectors to be orthogonalized. Within the constructed orthogonal space, the vector activity $|Y\rangle$ achieves true spectral decomposition form (Putz & Lacrămă, 2007):

$$|Y_{PRED}\rangle = \omega_0|\Omega_0\rangle + \omega_1|\Omega_1\rangle + \ldots + \omega_k|\Omega_k\rangle + \ldots + \omega_M|\Omega_M\rangle \quad (3.6)$$

Note that the residual vector in Eq. (3.3) has disappeared in Eq. (3.6) since it has no structural meaning in the abstracted orthogonal basis. Or, alternatively, one can say that in the abstract orthogonal space the residual vector $|prediction \quad error\rangle$ was identified with the vector with all components zero $|0,0,...,0\rangle$ that is always perpendicular with all other vectors of orthogonal basis. This way, the Gram-Schmidt algorithm, by its specific orthogonal recursive rules, absorbs or transforms the minimization condition of errors in Eq. (3.3) to simple identification with the origin of the orthogonal space of data. At this point, since there is no residual vector remaining in Eq. (3.6) one can consider that the SAR problem is in principle solved once the new coefficients in Eq. (3.6) (ω_0, ω_1,..., ω_k,...,ω_M) are determined. These new coefficients can be immediately deduced based on the orthogonal peculiarities of the spectral decomposition (3.6) grounded on the fact that $\langle\Omega_k|\Omega_l\rangle = 0,\ k \neq l$, a condition assured by the very nature of the vectors from the constructed orthogonal basis. As such, each coefficient comes out as the scalar product of its specific predictor vector with the activity vector (3.6) is performed (Putz & Lacrămă, 2007):

$$\omega_k = \frac{\langle\Omega_k|Y\rangle}{\langle\Omega_k|\Omega_k\rangle},\ k = \overline{0, M} \quad (3.7)$$

With coefficients given by expressions of type (3.7) the spectral expansion of the activity vector into an orthogonal basis (3.6) is completed. Yet, this does not mean that we have found the coefficients that directly link the activity with the predictor vectors as Eq. (3.3) demands.

III. Remarkably, while the studies dedicated to orthogonal problem usually stops at this stage the Spectral-SAR uses it to provide the solution for the original searched correlation, Eq. (3.3) with the error vector orthogonal on the predicted activity and on all its predictor states of Table 3.1. This can be adequately achieved through rearranging Eqs. (3.5) and (3.6) so that the system of all descriptors of Table 3.1 to be written in terms of orthogonal descriptors (Putz & Lacrămă, 2007):

$$\begin{cases} |Y_{PRED}\rangle = \omega_0|\Omega_0\rangle + \omega_1|\Omega_1\rangle + ... + \omega_k|\Omega_k\rangle + ... + \omega_M|\Omega_M\rangle \\ |X_0\rangle = 1\cdot|\Omega_0\rangle + 0\cdot|\Omega_1\rangle + ... + 0\cdot|\Omega_k\rangle + ... + 0\cdot|\Omega_M\rangle \\ |X_1\rangle = r_0^1|\Omega_0\rangle + 1\cdot|\Omega_1\rangle + ... + 0\cdot|\Omega_k\rangle + ... + 0\cdot|\Omega_M\rangle \\ \cdots\cdots\cdots \\ |X_k\rangle = r_0^k|\Omega_0\rangle + r_1^k|\Omega_1\rangle + ... + 1\cdot|\Omega_k\rangle + ... + 0\cdot|\Omega_M\rangle \\ \cdots\cdots\cdots \\ |X_M\rangle = r_0^M|\Omega_0\rangle + r_1^M|\Omega_1\rangle + ... + r_k^M|\Omega_k\rangle + ... + 1\cdot|\Omega_M\rangle \end{cases} \quad (3.8)$$

The system (3.8) has no trivial (orthogonal) solution if and only if the associated extended determinant vanishes; this condition introduces the Spectral-SAR determinant and of its equation (Putz & Lacrămă, 2007):

$$\begin{vmatrix} |Y_{PRED}\rangle & \omega_0 & \omega_1 & \cdots & \omega_k & \cdots & \omega_M \\ |X_0\rangle & 1 & 0 & \cdots & 0 & \cdots & 0 \\ |X_1\rangle & r_0^1 & 1 & \cdots & 0 & \cdots & 0 \\ \vdots & \vdots & \vdots & \vdots & & \vdots & \\ |X_k\rangle & r_0^k & r_1^k & \cdots & 1 & \cdots & 0 \\ \vdots & \vdots & \vdots & \vdots & & \vdots & \\ |X_M\rangle & r_0^M & r_1^M & \cdots & r_k^M & \cdots & 1 \end{vmatrix} = 0 \quad (3.9)$$

If the determinant of Eq. (3.9) is expanded on it first column, and the result rearranged so that to have $|Y_{PRED}\rangle$ on left side and the rest of states/indicators on the right side the searched QSAR solution of the initial problem of Eq. (3.3) is obtained as Spectral-SAR vectorial expansion (from where the "spectral" name is justified as well) with the error vector already absorbed in the orthogonalization procedure.

In fact Spectral-SAR procedure uses the double conversion passages: one forward from the given problem of Eq. (3.3) to the orthogonal one of

Eq. (3.6) in which the error vector is orthogonally "dissolved"; and the reverse one, back from the orthogonal to the real descriptors throughout the system (3.8), leaving with the determinant (3.9) to be expanded as the QSAR solution.

The result is that now QSAR/Spectral-SAR equation is delivered directly by the determinant (3.9) and not through matrices products as in statistical Pearson approach, see Section 2.1, while furnishing directly the Spectral-SAR correlation equation and not only the parameters of multi-variate correlation (Anderson, 1958; Box et al., 1978; Green & Margerison, 1978; Topliss, 1983; Seyfel, 1985; Draper & Smith, 1998; Chatterjee et al., 2000). Moreover, the Spectral-SAR algorithm is *invariant* also to the order of descriptors chosen in orthogonalization procedure, providing equivalent determinants just with rearranged lines, a matter that was not previously achieved by other orthogonalization techniques (Amić et al., 1995; Lučić et al., 1995; Šoškić et al., 1996; Klein et al., 1997).

Remarkably, apart from being conceptually new through considering the spectral (orthogonal) expansion of the input data space (of both activity and descriptors) throughout the system (3.8), the present method also has the computational advantage of being simpler than the classical "standard" statistical way of treating SAR problem previously exposed. That because, one has nothing to do with computations of matrix of the coefficients (2.92) or (2.97), this being a quite involving and time consuming procedure for higher dimensional systems. Instead, one can write directly the Spectral-SAR solution (equation) as the expansion of a (M+2)-dimensional determinant of Eq. (3.9) whose components are the activity and structural vectors involving the Gram-Schmidt and the spectral decomposition coefficients, r_i^k and ω_k, respectively (Putz & Lacrămă, 2007).

However, although different from the mathematical procedure, both standard- and spectral-SAR give similar results due to the theorem that states that (Fadeeva, 1959): *if the matrix* X, as that from (2.91), *with dimension* $N\times(M+1)$, $N>M+1$, *has linear independent columns*, i.e., they are orthogonal as in the spectral approach, *then there exists an unique matrix* [Q] *of dimension* $N\times(M+1)$ *with orthogonal columns and a triangular matrix* [R] *of dimension* $(M+1)\times(M+1)$ *with the elements of the principal diagonal equal with* 1, as identified in the first small determinant in Eq. (3.9), *so that the matrix* [*X*] *can be factorized as*

$$[X]=[Q][R] \tag{3.10}$$

When combining Eq. (3.10) with the optimal Eq. (2.96) one can get, after straight algebraic rules, that the [*B*] vector of estimates takes the form

$$[B]=\left([Q]^T[Q]\right)^{-1}[Q]^T[Y] \tag{3.11}$$

in close agreement with previous normal one, see Eq. (2.97). However, by comparison of matrices $[X]^T[X]$ and $[Q]^T[Q]$ of Eqs. (2.97) and (3.11), respectively, there is clear that the last case certainly furnishes a diagonal form which for sure is easier to handle (i.e., to take its inverse) when searching for the vector [*B*] of SAR coefficients.

However, worth being convinced by the equivalence of the present Spectral method with the standard statistical one by specializing the general problem (3.3) to the linear case (Putz & Putz (Lacrămă), 2008)

$$|Y_{PRED}\rangle=b_0|X_0\rangle+b_1|X_1\rangle \tag{3.12}$$

and to check whether this is unfolded through the Spectral-equation (3.9) as providing the parameters of linear regression given by Eqs. (2.98). In this respect, actually, we deal with the particular equation (Putz & Putz, 2011)

$$0=\begin{vmatrix}|Y_{PRED}\rangle & \omega_0 & \omega_1\\ |X_0\rangle & 1 & 0\\ |X_1\rangle & r_0^1 & 1\end{vmatrix}=|Y_{PRED}\rangle\begin{vmatrix}1 & 0\\ r_0^1 & 1\end{vmatrix}-|X_0\rangle\begin{vmatrix}\omega_0 & \omega_1\\ r_0^1 & 1\end{vmatrix}+|X_1\rangle\begin{vmatrix}\omega_0 & \omega_1\\ 1 & 0\end{vmatrix} \tag{3.13}$$

which is immediately rearranged as

$$|Y_{PRED}\rangle=\underbrace{\left(\omega_0-r_0^1\omega_1\right)}_{b}|X_0\rangle+\underbrace{\omega_1}_{a}|X_1\rangle \tag{3.14}$$

so that to identify the actual with the previous linear coefficients of Eq. (2.98):

$$a=\omega_1, b=\omega_0-r_0^1\omega_1 \tag{3.15}$$

Going to evaluate the expressions of (3.15) within the Spectral-SAR algorithm, there is instructive to identify form Table 3.1 only the relevant actual variables, with convenient denotation of instantaneous structural ones as the columns:

$\|Y_{PRED}\rangle$	$\|X_0\rangle$	$\|X_1\rangle$
y_1	1	x_1
y_2	1	x_2
$\vdots$	$\vdots$	$\vdots$
y_N	1	x_N

Other working tools are the zero-th and the first orthogonal vectors, accordingly considered and computed respectively as (Putz & Putz, 2011)

$$|\Omega_0\rangle = |1\,1 \cdots 1_N\rangle \tag{3.16a}$$

$$\begin{aligned}|\Omega_1\rangle &= |X_1\rangle - r_0^1|\Omega_0\rangle\\ &= |x_1\ x_2\ \ldots\ x_N\rangle - \frac{1}{N}\sum_{i=1}^{N} x_i|1\,1\,1\ \ldots\ 1\rangle = \left|x_1 - \frac{1}{N}\sum_{i=1}^{N} x_i\ \ldots\ x_N - \frac{1}{N}\sum_{i=1}^{N} x_i\right\rangle\end{aligned} \tag{3.16b}$$

with the help of coefficient

$$r_0^1 = \frac{\langle X_1|\Omega_0\rangle}{\langle\Omega_0|\Omega_0\rangle} = \frac{1}{N}\sum_{i=1}^{N} x_i \tag{3.17}$$

specialized from the general definition (3.5b).

In the same manner, the other specific Spectral coefficients from the general orthogonal recipe (3.6) are now for linear regression computed as the zero-th order contribution (Putz & Putz, 2011)

$$\omega_0 = \frac{\langle\Omega_0|Y\rangle}{\langle\Omega_0|\Omega_0\rangle} = \frac{1}{N}\sum_i y_i \tag{3.18}$$

while the first orthogonal one recovers precisely the previous linear slope of Eq. (2.98) and (2.107):

$$
\begin{aligned}
\omega_1 &= \frac{\langle \Omega_1 | Y \rangle}{\langle \Omega_1 | \Omega_1 \rangle} \\
&= \frac{\left\langle x_1 - N^{-1}\sum x_i \; ... \; x_N - N^{-1}\sum x_i \middle| y_1 \; ... \; y_N \right\rangle}{\sum_i \left(x_i - N^{-1}\sum_i x_i \right)^2} \\
&= \frac{\sum_i y_i \left(x_i - N^{-1}\sum_i x_i \right)}{\sum_i \left(x_i - N^{-1}\sum_i x_i \right)^2} = \frac{\sum_i y_i x_i - N^{-1}\left(\sum_i y_i\right)\left(\sum_i x_i\right)}{\sum_i \left[x_i^2 + N^{-2}\left(\sum_i x_i\right)^2 - 2N^{-1} x_i \sum_i x_i \right]} \\
&= \frac{N\sum_i y_i x_i - \left(\sum_i y_i\right)\left(\sum_i x_i\right)}{N\sum_i x_i^2 - \left(\sum_i x_i\right)^2} \equiv a
\end{aligned}
\tag{3.19}
$$

as prescribed by the correspondence of (3.15). Additionally, also its companion free term coefficient of relationship (3.14) may be now straightly evaluated as

$$
\begin{aligned}
b &\equiv \omega_0 - r_0^1 \omega_1 \\
&= \frac{1}{N}\sum_i y_i - \frac{1}{N}\left(\sum_i x_i\right) \frac{N\sum_i y_i x_i - \left(\sum_i y_i\right)\left(\sum_i x_i\right)}{N\sum_i x_i^2 - \left(\sum_i x_i\right)^2} \\
&= \frac{\left(\sum_i y_i\right)\left(\sum_i x_i\right)^2 - \left(\sum_i x_i\right)\left(\sum_i y_i x_i\right)}{N\sum_i x_i^2 - \left(\sum_i x_i\right)^2}
\end{aligned}
\tag{3.20}
$$

as well successfully regaining the previously computed linear free terms counterpart of Eqs. (2.98) and (2.107), yet by means of variational statistical (optimization of errors' squares summation) procedure (Putz & Putz, 2011).

With this there is clear that the present Spectral-SAR algebraic SAR methodology not only recovers in great details the standard statistical QSAR routine but also generalizes to a great analyticity extent towards better assessment of mechanistically ordering and influences in practical eco- and bio- logical applications.

3.2.2 SPECTRAL-DIAGONAL-SAR

The present approach makes a step further, thus going beyond the Gram-Schmidt orthogonalization procedure to employing the basic ideas of (Löwdin, 1950, 1970) symmetrical orthogonalization, already applied for atomic orbital's basis in molecular orbital computation (in the light of the above required quantum chemical superposition principle implying the orthogonalization of the combined components); the present discussion follows (Putz, 2013a).

Given an N-dimensional space, with a coordinate unfold for a (quantum) vector

$$|\phi_I\rangle = |\phi_{I1}, \phi_{I2}, \ldots, \phi_{IN}\rangle \tag{3.21}$$

one may consider its transformation (Putz, 2013a)

$$|\phi_I'\rangle = \hat{S}^{-1/2} e^{-i\frac{I}{N}\phi} |\phi_I\rangle \tag{3.22}$$

with the help of (operator) matrix $\hat{S}$ having the components defined by the dot (scalar) product in the allied Banach-Hilbert space,

$$S_{Ij} = \langle \phi_I | \phi_j \rangle = \sum_{q=1}^{N} \phi_{Iq}\phi_{jq} \tag{3.23}$$

Eq. (3.23) is susceptible to the diagonalization by means of the Löwdin orthogonal transformation

$$\hat{S}_{Diag(onal)} = \hat{U}^{+}\hat{S}\hat{U} \tag{3.24}$$

throughout the unitary matrix $\hat{U}$, at its turn fulfilling the consecrated conditions:

Its adjoined form equals its transpose form:

$$\hat{U}^{+} = \hat{U}^{T} \tag{3.25}$$

- It is "normalizated" through the (operatorial) matrix product

$$\hat{U}^{+}\hat{U} = \hat{U}\hat{U}^{+} = \hat{1} \tag{3.26}$$

This way, the square root and then the inverse of square root of transformation operator in Eq. (3.24) may be evaluated from Eq. (3.25) to look like (Putz, 2013a):

$$\hat{S}^{1/2} = \hat{U}^{+}\hat{S}_{Diag}^{1/2}\hat{U} \tag{3.27a}$$

$$\hat{S}^{-1/2} = \left(\hat{S}^{1/2}\right)^{-1} = \hat{U}^{+}\left(\hat{S}_{Diag}^{1/2}\right)^{-1}\hat{U} = \hat{U}^{+}\hat{S}_{Diag}^{-1/2}\hat{U} \tag{3.27b}$$

since recognizing (by construction) the successive identities

$$\hat{S}^{-1/2}\hat{S}^{1/2} = \hat{U}^{+}\hat{S}_{Diag}^{-1/2}\underbrace{\hat{U}\hat{U}^{+}}_{\substack{=1,\\ by\,Eq.(3.26)}}\hat{S}_{Diag}^{1/2}\hat{U} = \hat{U}^{+}\underbrace{\hat{S}_{Diag}^{-1/2}\hat{S}_{Diag}^{1/2}}_{\substack{=1,\\ by\,Eq.(3.27b)}}\hat{U} = \hat{U}^{+}\hat{U} \underset{\substack{by\\ Eq.(3.26)}}{=} \hat{1} \tag{3.28a}$$

along the other obvious ones

$$\hat{S}^{1/2}\hat{S}^{1/2} = \hat{U}^{+}\hat{S}_{Diag}^{1/2}\underbrace{\hat{U}\hat{U}^{+}}_{\substack{=1,\\ by\,Eq.(3.26)}}\hat{S}_{Diag}^{1/2}\hat{U} = \hat{U}^{+}\hat{S}_{Diag}\hat{U} \underset{\substack{by\\ Eq.(3.24)}}{=} \hat{S} \tag{3.28b}$$

$$\hat{S}^{-1/2}\hat{S}^{-1/2} = \hat{U}^{+}\hat{S}_{Diag}^{-1/2}\underbrace{\hat{U}\hat{U}^{+}}_{\substack{=1,\\ by\,Eq.(3.26)}}\hat{S}_{Diag}^{-1/2}\hat{U} = \hat{U}^{+}\hat{S}_{Diag}^{-1}\hat{U} \underset{\substack{by\\ Eq.(3.27b)}}{=} \hat{S}^{-1} \tag{3.28c}$$

These inter-relationship are most useful for turning a non-orthonormalized vectorial system, say $\{\phi_l\}_{l=\overline{1,M}}$ (description of a given reality or interaction, here focused on QSP(A)R), see Eq. (3.23)

$$\langle\phi_l|\phi_j\rangle = S_{lj} \neq \begin{cases}1 ... l = j\\ 0 ... l \neq j\end{cases} \tag{3.29a}$$

into a general ortho-normalization relationship between the "directions 0-l and 0-j" by the gauge transformed space of Eq. (3.22) type, by noticing that (Putz, 2013a)

$$\langle \phi_l ' | \phi_j ' \rangle = \left(\frac{1}{2\pi} \int_0^{2\pi} e^{-i\frac{j-l}{N}\phi} d\phi \right) \langle \phi_l | \left(\hat{U}^+ \hat{S}_{Diag}^{-1/2} \hat{U} \right) \left(\hat{U}^+ \hat{S}_{Diag}^{-1/2} \hat{U} \right) | \phi_j \rangle$$
$$= \delta_{lj} \langle \phi_l | \phi_j \rangle \left(\hat{U}^+ \hat{S}_{Diag}^{-1/2} \hat{U} \hat{U}^+ \hat{S}_{Diag}^{-1/2} \hat{U} \right)_{lj} = \delta_{lj} S_{lj} \left(\hat{U}^+ \hat{S}_{Diag}^{-1} \hat{U} \right)_{lj}$$
$$= \delta_{lj} S_{lj} \left(\hat{S}^{-1} \right)_{lj} = \delta_{lj} S_{lj} S_{lj}^{-1} = \delta_{lj} = \begin{cases} 1 ... l = j \\ 0 ... l \neq j \end{cases} \tag{3.29b}$$

while recognizing the Kronecker representation by the definite integral describing the rotation in the complex plane for the difference in the gauge exponents for the sates $|\phi_l '\rangle$ and $|\phi_j '\rangle$ (Wilfred, 2003; Hassani, 2008)

$$\delta_{lj} = \frac{1}{2\pi} \int_0^{2\pi} e^{-i\frac{j-l}{N}\phi} d\phi \tag{3.30}$$

Note that the variable ϕ firstly appearing in transformation of Eq. (3.22) plays the gauge role in this type of transformation, so that assuring the invariance or scaling parameter specific to the descriptor's states $|\phi_l\rangle_{l=\overline{1,M}}$; for instance, it may be related with a quantum constant (Planck, Rydberg, or the fine structure).

However, there remains the big challenge providing a physicochemical property/biological activity correlation to structural (inner) atomic or molecular states' indices, by using the actual ortho-normalization framework in such way to be independent of the unitary transformation itself, i.e., for acquiring the desiderated universality of QSP(A)R method. Such exciting perspective is in the next section approached.

However, when dealing with QSP(A)R problems, one has to find the analytic expansion linking (correlating) the observed property/activity in a N-dimensional space of toxicants

$$|Y\rangle = |Y_1, Y_2, ..., Y_N\rangle \tag{3.31a}$$

with the structural M-causes/descriptors

$$|X_1\rangle = |X_{11}, X_{12}, ..., X_{1N}\rangle$$
$$...$$
$$|X_M\rangle = |X_{M1}, X_{M2}, ..., X_{MN}\rangle \tag{3.32a}$$

In other words, one looks for having explicitly the dependence (assuming no error in prediction – the optimum case), abstracted from Eq. (3.3)

$$|Y\rangle = b_0|X_0\rangle + b_1|X_1\rangle + ... + b_M|X_M\rangle \tag{3.31b}$$

where the "zero" term associates with the unity vector, as previously noted

$$|X_0\rangle = |1\rangle_N = |1,1,...,1\rangle_{N-components} \tag{3.32b}$$

However, since there is no guarantee that in general the descriptors are independent, or truly orthogonal, their choice is somehow limitative if one seeks for "pure" orthogonal states to be correlated with observed property/activity of their superposition or coupling; instead, for the general case one can employ the above diagonal recipe with the so-called Spectral decomposition of Y into the pure orthogonal states, associated with the original ones by the transformation of Eq. (3.22) type,

$$|Y\rangle = \omega_0|X_0'\rangle + \omega_1|X_1'\rangle + ... + \omega_M|X_M'\rangle \tag{3.33}$$

Equation (3.33) has the advantage of allowing the immediate yield of coefficients by successive formations of scalar products since the descriptors crossing combinations vanish according with the rule of Eq. (3.29b), see also (3.7):

$$\omega_0 = \langle X_0'|Y\rangle,$$
$$\omega_1 = \langle X_1'|Y\rangle,$$
$$...$$
$$\omega_M = \langle X_M'|Y\rangle \tag{3.34}$$

However, the expansion (3.33) may be considered along the diagonalization transformations of Eq. (3.22) so that one may arrange them into the system (Putz, 2013a)

$$\begin{cases} |Y\rangle = \omega_0 |X_0'\rangle + \omega_1 |X_1'\rangle + ... + \omega_M |X_M'\rangle \\ |X_0\rangle = \hat{S}^{1/2} |X_0'\rangle + 0 \cdot |X_1'\rangle + ... + 0 \cdot |X_M'\rangle \\ |X_1\rangle = 0 \cdot |X_0'\rangle + \hat{S}^{1/2} |X_1'\rangle + ... + 0 \cdot |X_M'\rangle \\ ... \\ |X_M\rangle = 0 \cdot |X_0'\rangle + 0 \cdot |X_1'\rangle + ... + \hat{S}^{1/2} |X_M'\rangle \end{cases} \quad (3.35)$$

Note that the exponential (rotational) factor appearing in Eq. (3.22) disappeared in Eq. (3.35) since working within the statistical context in which the multidimensional chemical space is quite large (N>>1) and in any case one has the number of cases (observed toxicants) in much larger set than the number of descriptors (N >> M); this way, this yet disputed condition appears naturally for consistently/uniformly composing the (M+2) system of Eq. (3.35). Nevertheless, it has non-trivial solution only if the associated determinant vanishes, i.e., producing the actual Spectral-Diagonal equation (Putz, 2013a):

$$\begin{vmatrix} |Y\rangle & \omega_0 & \omega_1 & \cdots & \omega_M \\ |X_0\rangle & \hat{S}^{1/2} & 0 & \cdots & 0 \\ |X_1\rangle & 0 & \hat{S}^{1/2} & \cdots & 0 \\ \vdots & \vdots & \vdots & \ddots & \vdots \\ |X_M\rangle & 0 & 0 & \cdots & \hat{S}^{1/2} \end{vmatrix} = 0 \quad (3.36)$$

Remarkably, the actual spectral equation significantly differs by that based on Gram-Schmidt orthogonalization algorithm; being simpler just in the sense it is diagonal and thus providing a direct evaluation when explicated upon the first line and column, as follows (Putz, 2013a):

$$0 = |Y\rangle \begin{vmatrix} \hat{S}^{1/2} & 0 & \cdots & 0 \\ 0 & \hat{S}^{1/2} & \cdots & 0 \\ \vdots & \vdots & \ddots & \vdots \\ 0 & 0 & \cdots & \hat{S}^{1/2} \end{vmatrix}_{M+1} - |X_0\rangle \begin{vmatrix} \omega_0 & \omega_1 & \cdots & \omega_M \\ 0 & \hat{S}^{1/2} & \cdots & 0 \\ \vdots & \vdots & \ddots & \vdots \\ 0 & 0 & \cdots & \hat{S}^{1/2} \end{vmatrix}_{M+1}$$

$$+|X_1\rangle\begin{vmatrix}\omega_0 & \omega_1 & \cdots & \omega_M\\ \hat{S}^{1/2} & 0 & \cdots & 0\\ \vdots & \vdots & \ddots & \vdots\\ 0 & 0 & \cdots & \hat{S}^{1/2}\end{vmatrix}_{M+1} -\ldots+|X_M\rangle\begin{vmatrix}\omega_0 & \omega_1 & \cdots & \omega_{M-1} & \omega_M\\ \hat{S}^{1/2} & 0 & \cdots & 0 & 0\\ 0 & \hat{S}^{1/2} & \vdots & 0 & 0\\ \vdots & \vdots & \ddots & \vdots & \vdots\\ 0 & 0 & \cdots & \hat{S}^{1/2} & 0\end{vmatrix}_{M+1} \tag{3.37a}$$

Eq. (3.37a) further develops by reducing the determinants to those obtained by considering the single non-zero line pivots that give:

$$0=\left(\hat{S}^{1/2}\right)^{M+1}|Y\rangle-\omega_0\left(\hat{S}^{1/2}\right)^{M}|X_0\rangle-\omega_1\left(\hat{S}^{1/2}\right)^{M}|X_1\rangle-\ldots-\omega_M\left(\hat{S}^{1/2}\right)^{M}|X_M\rangle \tag{3.37b}$$

The last equation may even more be rearranged to formally yield (Putz, 2013a)

$$|Y\rangle=\omega_0\hat{S}^{-1/2}|X_0\rangle+\omega_1\hat{S}^{-1/2}|X_1\rangle+\ldots+\omega_M\hat{S}^{-1/2}|X_M\rangle \tag{3.37c}$$

Eq. (3.37c) fully accords with the Eq. (3.33) combined with transformation of Eq. (3.22) under statistical joint conditions $N>>1$, $N>>M$; Still, Eq. (3.37c) may be once more simplified when twice combined with general expression for ω's of Eq. (3.34):

$$|Y\rangle=\sum_{k=1}^{M}\omega_k\hat{S}^{-1/2}|X_k\rangle=\sum_{k=1}^{M}\hat{S}^{-1}\langle X_k|Y\rangle|X_k\rangle=\sum_{k=1}^{M}\hat{S}^{-1/2}\langle X_k|Y\rangle|X_k'\rangle \tag{3.38a}$$

Thus we remain with workable expansion in ortho-normal chemical space; accordingly, one may use also the conjugate Hilbert property/action vector of Eq. (3.38a)

$$\langle Y|=\sum_{k=1}^{M}\langle Y|X_k\rangle\langle X_k'|\hat{S}^{-1/2} \tag{3.38b}$$

This way, one finally makes the scalar product of the last two quantities, Eqs. (3.38a) and (3.38b), by formal necessity, to obtain (Putz, 2013a):

$$\langle Y|Y\rangle = \sum_{l,j=1}^{M} \langle X_l|Y\rangle\langle Y|X_j\rangle\langle X_j{}'|\hat{S}^{-1/2}\hat{S}^{-1/2}|X_l{}'\rangle$$

$$= \sum_{l=j=k=1}^{M} |\langle X_k|Y\rangle|^2 \langle X_k{}'|\hat{S}^{-1}|X_k{}'\rangle = \sum_{k=1}^{M} |\langle X_k|Y\rangle|^2 \left[\hat{S}^{-1}\right]_{kk} \quad (3.39)$$

Now, when remember that the transformation matrix is common for all QSP(A)R descriptors as revealed by Eq. (3.37c), i.e.,

$$\left[\hat{S}^{-1}\right]_{kk} = ct. \quad (3.40)$$

one can find out its invariant expression from Eq. (3.39) with the global value

$$\left[\hat{S}^{-1}\right] = \frac{\langle Y|Y\rangle}{\sum_{k=1}^{M} |\langle X_k|Y\rangle|^2} \quad (3.41)$$

With Eq. (3.41), the Eq. (3.38a) can be accordingly specialized so furnishing the actual Spectral-Diagonal QSP(A)R expression (Putz, 2013a):

$$|Y\rangle = \sum_{k=1}^{M} \left[\hat{S}^{-1}\right] \langle X_k|Y\rangle |X_k\rangle$$

$$= \sum_{k=1}^{M} \frac{\langle Y|Y\rangle\langle X_k|Y\rangle}{\sum_{k=1}^{M} |\langle X_k|Y\rangle|^2} |X_k\rangle \quad (3.42)$$

This is an analytic expression for the full correlation itself, while its performances against various numerical data, and in comparison with the consecrated QSP(A)R analysis, is let for future works and communications; However, we can more notices that when Eq. (3.42) is limited to the mono-linear correlation of Eq. (3.12). Then, the involved coefficients are identified from Eq. (3.12) with the actual Diagonal-SPECTRAL forms (Putz, 2013a):

$$b_0 = \frac{\langle Y|Y\rangle\langle X_0|Y\rangle}{\sum_{k=0,1}\left|\langle X_k|Y\rangle\right|^2} = \frac{\left(\sum_{q=1}^{N} Y_q\right)\left(\sum_{q=1}^{N} Y_q^2\right)}{\left(\sum_{q=1}^{N} Y_q\right)^2 + \left(\sum_{q=1}^{N} X_{1q}Y_q\right)^2} \tag{3.43a}$$

$$b_1 = \frac{\langle Y|Y\rangle\langle X_1|Y\rangle}{\sum_{k=0,1}\left|\langle X_k|Y\rangle\right|^2} = \frac{\left(\sum_{q=1}^{N} X_q Y_q\right)\left(\sum_{q=1}^{N} Y_q^2\right)}{\left(\sum_{q=1}^{N} Y_q\right)^2 + \left(\sum_{q=1}^{N} X_{1q}Y_q\right)^2} \tag{3.43b}$$

with clear different shape than the usual statistical or algebraically ones which do no depend on the number of descriptors used, while displaying ratios of differences as in Eq. (2.107):

$$b_0 \;...\; b = \frac{\left(\sum_{q=1}^{N} X_{1q}^2\right)\left(\sum_{q=1}^{N} Y_q\right) - \left(\sum_{q=1}^{N} X_{1q}\right)\left(\sum_{q=1}^{N} X_{1q}Y_q\right)}{N\left(\sum_{q=1}^{N} X_{1q}^2\right) - \left(\sum_{q=1}^{N} X_{1q}\right)^2} \tag{3.44a}$$

$$b_1 \;...\; a = \frac{N\left(\sum_{q=1}^{N} X_{1q}Y_q\right) - \left(\sum_{q=1}^{N} Y_q\right)\left(\sum_{q=1}^{N} X_{1q}\right)}{N\left(\sum_{q=1}^{N} X_{1q}^2\right) - \left(\sum_{q=1}^{N} X_{1q}\right)^2} \tag{3.44b}$$

However, their eventual inter-relationship and hierarchical performances should be established on concrete application and on the general analytical level when the statistical approach is replaced by the algebraic one – for more versatility in formal expressions' manipulations.

The actual approach complements the SPECTRAL-SAR QSAR modeling through employing the N-dimensional chemical space of as such chemical compounds analyzed by M-descriptors in a quantum superposition (vectorial) problem, or the so-called qSAR; the results are quite general and requires the inherent statistical working condition N>>M; at least at the analytical level the obtained diagonal-spectral equation display in the first order of development (i.e., for linear QSP(A)R) formally different coefficients driving the solution of modeling the N-chemical series by M-descriptors in orthogonal space (Putz, 2013a).

The fist tests for numerically testing the reliability of the present spectral-diagonal QSP(A)R approach respecting the Gram-Schmidt Spectral – SAR method of Section 3.2.2 is going to test the correlation performance (Dudaş & Putz, 2014), see Section 3.3.7.2, towards consecrating the actual Spectral-Diagonal SAR as a viable algebraic alternative for both the statistical QSAR and Spectral-SAR while further exploration should reveal the application Quantum-SAR fields when this diagonal receipe prevails (by its inner quantum nature).

3.2.3 ALGEBRAIC CORRELATION FACTOR

Let's explore in next whether the present Spectral regression gives the opportunity in defining another correlation index, beyond the standard statistical one given by Eq. (2.35) (Chicu & Putz, 2009; Putz, 2012a, b).

One starts with the simple connection between the observed, predicted and error vectors of Eq. (3.3), however specialized on their instantaneous entries; the present discussion follows (Putz & Putz, 2011):

$$Y_{i-OBS} = Y_{i-PRED} + pe_i \tag{3.45}$$

where "*pe*" stays here as abbreviation for "prediction error." Then, by means of squaring relation (3.45)

$$Y_{i-OBS}^2 = Y_{i-PRED}^2 + pe_i^2 + 2Y_{i-PRED} \cdot pe_i \tag{3.46}$$

and summing for all working N-molecules (of Table 3.1),

$$\sum_{i=1}^{N} Y_{i-OBS}^2 = \sum_{i=1}^{N} Y_{i-PRED}^2 + \sum_{i=1}^{N} pe_i^2 + 2\sum_{i=1}^{N} Y_{i-PRED} \cdot pe_i \tag{3.47}$$

the last relation simplifies to:

$$\sum_{i=1}^{N} Y_{i-OBS}^2 = \sum_{i=1}^{N} Y_{i-PRED}^2 + \sum_{i=1}^{N} pe_i^2 \tag{3.48}$$

based on applying of scalar product definition (3.29a) and of prediction error orthogonalization condition (3.4a) upon the last term of Eq. (3.47), i.e.,

$$\sum_{i=1}^{N} Y_{i-PRED} \cdot pe_i = \langle Y_{PRED} | pe \rangle = 0 \tag{3.49}$$

Now, substituting the prediction error values of Eq. (3.45) into remaining expression (3.48) one firstly gets (Putz & Putz, 2011):

$$\sum_{i=1}^{N} Y_{i-OBS}^2 = \sum_{i=1}^{N} Y_{i-PRED}^2 + \sum_{i=1}^{N} (Y_{i-OBS} - Y_{i-PRED})^2 \tag{3.50}$$

or the equivalent identity

$$\sum_{i=1}^{N} Y_{i-PRED}^2 = \sum_{i=1}^{N} Y_{i-OBS} \cdot Y_{i-PRED} \tag{3.51}$$

which further rewrites, recalling the norm and scalar product definitions of Eqs. (3.94)–(3.96), respectively, as:

$$\| Y_{PRED} \rangle \|^2 = \langle Y_{OBS} | Y_{PRED} \rangle \tag{3.52}$$

Finally, the Cauchy-Schwarz form (2.66) is employed on the right side term of (3.52), noting that the observed and predicted activities are of the same nature for a given molecule – i.e., either both positive or both negative – thus providing their scalar product as positively defined; with these, the relation (3.52) immediately reads as the inequality:

$$\| Y_{PRED} \rangle \|^2 \leq \| Y_{OBS} \rangle \| \cdot \| Y_{PRED} \rangle \| \tag{3.53}$$

leaving with the predicted-observed norms' hierarchy

$$\| Y_{PRED} \rangle \| \leq \| Y_{OBS} \rangle \| \tag{3.54}$$

that guarantees the *consistent probability definition* while introducing *algebraic correlation factor* with the form (Putz & Putz, 2011):

$$RA \equiv r_{ALGEBRAIC} = \frac{\| Y_{PRED} \rangle \|}{\| Y_{OBS} \rangle \|} \leq 1 \tag{3.55}$$

Nevertheless, there remains to compare this new correlation factor, written in algebraically manner as the ration of predicted – to – observed norms of investigated molecular activity or of their effects, with the fashioned statistical counterpart given by Eq. (2.63); this issue will be addressed in what follows.

3.2.4 ALGEBRAIC VS. STATISTIC CORRELATIONS

BanaterAnsatz on the algebraic Spectral-SAR correlation: *For any QSAR analysis, once considering the measured/observed and computed/predicted activity data as the vectors* $|Y_{OBS}\rangle$ *and* $|Y_{PRED}\rangle$ *with the associate norms through the scalar products of Eqs. (2.69)–(2.71), the algebraic norm order* (3.55) *valid in defining the algebraic correlation factor (3.54), sets also the hierarchy at the levels of correlations factors in a sense that the algebraic one of always exceed the standard correlation factor (2.63)*; the present discussion follows (Putz et al., 2008a; Putz & Putz, 2011; Putz, 2012a, b):

$$r_{S-SAR}^{ALGEBRAIC} \geq r_{QSAR}^{STATISTIC} \tag{3.56}$$

Proof: by straight algebraic translation the condition (3.56) firstly it rewrites as:

$$\frac{\langle Y_{PRED} | Y_{PRED} \rangle}{\langle Y_{OBS} | Y_{OBS} \rangle} \geq 1 - \frac{\langle Y_{OBS} - Y_{PRED} | Y_{OBS} - Y_{PRED} \rangle}{\langle Y_{OBS} - \overline{Y}_{OBS} | Y_{OBS} - \overline{Y}_{OBS} \rangle} \tag{3.57}$$

where we have introduced the averaged observed activity

$$\overline{Y}_{OBS} = \frac{1}{N} \sum_{i=1}^{N} y_{i-OBS} \tag{3.58}$$

and its associate N-dimensional vector (state in Hilbert space):

$$|\overline{Y}_{OBS}\rangle = \left(\frac{1}{N} \sum_{i=1}^{N} y_{i-OBS} \right) |1 1 \ldots 1_N\rangle \tag{3.59}$$

Note that the inequality (3.57) becomes equality in the case of perfect identity between observed and predicted activity values, i.e., perfect correlation, the case in which the second term of the right hand side vanishes while that of the left hand side become unity. For all other non-perfect correlations strict inequality holds and this will be considered in next, for the equivalent expression (Putz et al., 2008a; Putz & Putz, 2011)

$$\langle Y_{PRED} | Y_{PRED} \rangle \langle Y_{OBS} - \overline{Y}_{OBS} | Y_{OBS} - \overline{Y}_{OBS} \rangle$$
$$> \langle Y_{OBS} | Y_{OBS} \rangle \left[\langle Y_{OBS} - \overline{Y}_{OBS} | Y_{OBS} - \overline{Y}_{OBS} \rangle - \langle Y_{OBS} - Y_{PRED} | Y_{OBS} - Y_{PRED} \rangle \right] \tag{3.60a}$$

which may be further rearranged as

$$\left[\langle Y_{PRED} | Y_{PRED} \rangle - \langle Y_{OBS} | Y_{OBS} \rangle \right] \left[\langle Y_{OBS} | Y_{OBS} \rangle - 2 \langle Y_{OBS} | \overline{Y}_{OBS} \rangle + \langle \overline{Y}_{OBS} | \overline{Y}_{OBS} \rangle \right]$$
$$+ \langle Y_{OBS} | Y_{OBS} \rangle \left[\langle Y_{OBS} | Y_{OBS} \rangle - 2 \langle Y_{OBS} | Y_{PRED} \rangle + \langle Y_{PRED} | Y_{PRED} \rangle \right] > 0 \tag{3.60b}$$

At this point, after obvious simplifications and factorization may easily recognize and employ both the identities (3.52) and (3.54), specific to algebraic correlation,

$$2 \langle Y_{PRED} | Y_{PRED} \rangle \langle Y_{OBS} | Y_{OBS} \rangle - 2 \langle Y_{OBS} | Y_{OBS} \rangle \underbrace{\langle Y_{OBS} | Y_{PRED} \rangle}_{\langle Y_{PRED} | Y_{PRED} \rangle}$$
$$+ \underbrace{\left[\langle Y_{OBS} | Y_{OBS} \rangle - \langle Y_{PRED} | Y_{PRED} \rangle \right]}_{\geq 0} \left[2 \langle Y_{OBS} | \overline{Y}_{OBS} \rangle - \langle \overline{Y}_{OBS} | \overline{Y}_{OBS} \rangle \right] > 0 \tag{3.60c}$$

the simplified expression is obtained

$$2 \langle Y_{OBS} | \overline{Y}_{OBS} \rangle > \langle \overline{Y}_{OBS} | \overline{Y}_{OBS} \rangle \tag{3.61a}$$

that finally is analytically explicated with the aid of introduced vector (3.59) of the average activity to the unfolded scalar ordered products (Putz et al., 2008a; Putz & Putz, 2011)

$$2\sum_{i=1}^{N}\left(y_{i-OBS}\frac{1}{N}\sum_{i=1}^{N}y_{i-OBS}\right)>\sum_{i=1}^{N}\left(\frac{1}{N}\sum_{i=1}^{N}y_{i-OBS}\right)\left(\frac{1}{N}\sum_{i=1}^{N}y_{i-OBS}\right) \quad (3.61b)$$

leaving with the equivalent strict inequality

$$\frac{2}{N}\left(\sum_{i=1}^{N}y_{i-OBS}\right)^2>\frac{1}{N}\left(\sum_{i=1}^{N}y_{i-OBS}\right)^2 \quad (3.61c)$$

fully satisfied by the natural ordering as $2 > 1$.

Therefore, there was proofed both the (qualitative) simplicity and the (quantitative) superiority of algebraic correlation factor. Many applications proof these statements also on dedicated molecular-biological or molecular-ecotoxicological cases. For instance, the specialization on modern bi-component molecular system, as concerned for the *ionic liquids* (IL) toxicological actions – is in next explicated by the general projective QSAR form.

3.2.5 PROJECTIVE QSAR

When utilizing the analytical model of QSAR/Spectral-SAR for environmental interactions, one should consider the framework of the principles both at the general and applied levels. As such, regarding the general principles or green chemistry and engineering, they are provided in Table 3.2 to ensure that they can be readily compared (Putz, 2010a).

Note that the fundamental principles constitute the background or the general framework or desiderate that is eventually supported by the associate engineering principle; the green engineering principle is primarily based on "minimizing" or "maximizing" the time, space, energy and costs, and it is constituted either in an economical enterprise and an extension of the main principles of nature or in terms of optimizing mass-energy and time-space. From this point forward, the most basic physical and chemical principles are observed as those acting on each process, system, or state to be created, maintained or modified.

However, while restraining the analysis to the specific interactions between chemical structures and biological species, the Organization for

TABLE 3.2 The Twelve Principles of Green Chemistry and Engineering*

No.	*Principle of Green Chemistry*	*Principle of Green Engineering*
1.	Prevention of waste that must cleaned afterwards	Prevention rather than treatment
2	Inherently safer chemistry for accident prevention such as releasing, explosions, and fires	Inherent rather than circumstantial processes and components to prevent hazard
3.	Atom economy in maximizing the incorporation of all material used	Conserving complexity of embedded entropy for minimizing the recycling process
4.	Less hazardous chemical systems should be designed with little or no toxicity	Design for commercial "afterlife" through their nontoxic availability
5.	Designing safer chemicals to minimize their toxicity	Durability rather than immortality because whatever compound should be degradable
6.	Safer solvents and auxiliars (separation agents)	Integrate material and energy flows allowing interconnectivity in components
7.	Designing for energy efficiency while synthetic methods should be conducted at ambient temperature and pressure whenever possible.	Maximizing efficiency in producing products through minimizing mass, energy, space, and time consumption
8.	Use of renewable raw materials and feedstocks rather than depleting them	Design for separation and purification operations should maximize recycling
9.	Reducing derivatives as those modifying physical-chemical processes because they are virtually converted into waste	Minimizing material diversity in multicomponent products towards promoting easiest disassembly process
10.	The use of catalytic rather than stoichiometric reagents is desirable for maintaining control over the selectivity	Output-pulled of reaction products rather than input-pushed reactants as additional starting material
11.	Design for degradation targeting biodegradability and not persistent components in environment	Renewable rather than depleting of material and energy inputs
12.	Real time analysis for pollution prevention by means of in-process monitoring and analytical methodologies	Meet need while minimizing the excess of unnecessary capacities or capabilities for bio-physicochemical systems

*Ritter (1989); Anastas & Zimmerman (2003); Anastas et al. (2009).

Economic Co-operation and Development (OECD) advanced a set of standard principles for the validation and for regulatory purposes of the (quantitative) structure-activity relationship models (OECD, 2004, 2005, 2006, 2007):

- QSAR-1: a defined endpoint
- QSAR-2: an unambiguous algorithm
- QSAR-3: a defined domain of applicability
- QSAR-4: appropriate measures of goodness-of–fit, robustness and predictivity
- QSAR-5: a mechanistic interpretation, if possible

Within this context, the present QSAR-Spectra-SAR (QSAR-SSAR) approach "responds" to these OECD-QSAR principles by the present Spectral-SAR ecotoxicological principles' realization, the present discussion mainly follows (Putz et al., 2011a), with special reference to projective systems/ionic liquids (Principle 3):

- *ECOTOX-SSAR Principle 1* may be assured by: the "length" of the predicted/measured (eco)biological action follows the self-scalar product rule of the computed endpoint activity

$$\left\| |Y\rangle^{ENDPOINT\,(MEASURED\,/\,PREDICTED)} \right\| = \sqrt{\sum_{i=1}^{N} \left(y_i^2\right)^{MEASURED\,/\,PREDICTED}} \quad (3.62)$$

- *ECOTOX-SSAR Principle 2* may be assured by: the "orthogonality" of assumed molecular factors that correlate with eco- and bio-effects is assured by the spectral decomposition of the associate activity respecting them, see Eq. (3.3), and the orthogonal-real space transformation given by the Spectral-determinant, Eq. (3.9), giving in fact the searched structure-activity relationship model;
- *ECOTOX-SSAR Principle 3* may be assured by: the method of considering the structural parameters and the activities with which they should be correlated, for specific (target) class of molecules; for instance, the consecrated Hansch quantitative structure activity relationships (QSARs) generally prescribes the activity expansion under the generic minimal but meaningful form:

$$A = b_0 + b_1\begin{pmatrix} hydrophobic \\ parameter \end{pmatrix} + b_2\begin{pmatrix} electronic \\ parameter \end{pmatrix} + b_3\begin{pmatrix} steric \\ parameter \end{pmatrix} \tag{3.63}$$

which provides sufficient information about the transport, electronic affinity and specific interaction at the molecular level, respectively; whereas the hydrophobicity index, LogP, describes, at best, the quality of molecular transport through cellular membranes, see Eq. (3.1). For the electronic and steric contributions, many structural parameters may be considered (Lacrămă et al., 2008; Putz et al., 2009a); among them, the polarizability (POL) measures the inductive electronic effect that reflects the long range or van der Waals bonding, whereas for the steric component, the total energy (E_{TOT}) is assumed to be the representative index because it is calculated at the optimum molecular geometry at which the stereo-specificity is included. These parameters have been demonstrated to be quite reliable in modeling the ecotoxicological interactions (Putz & Lacrămă, 2007; Chicu & Putz, 2009a; Putz, 2012b), for example, as they are used in multy-systems in general and in those composed by cation-anion (CA) complexes, as there is the case of IL applications. To this aim, when information about the eco-biological influence of complex-system (AC/IL) is desired, the particular anionic-cationic structure has to be properly considered because almost all structural information about ionic liquids is based on the superposition of the separate anionic and cationic contributions. In this situation, one deals with the so-called *projective QSAR* when two different additive models for modeling anionic-cationic interaction can be considered.

The first projective model is based on the vectorial summation of the produced anionic and cationic biological effects. In other words, this so-called |1+> model is constructed from the superposition of the anionic (subscripted with A) and cationic (subscripted with C) activities, and can be formally represented as (Lacrămă et al., 2007; Putz, 2012b; Putz & Putz, 2013a):

$$|Y_{AC}\rangle^{1+} = |1+\rangle = |Y_A\rangle + |Y_C\rangle = \hat{O}_{S\text{-}SAR}\left[g(\{|X_A\rangle\}) + g(\{|X_C\rangle\})\right] \tag{3.64}$$

with Hansch combinations

$$\left\{X_{A,C}\rangle\right\} = \left\{LogP_{A,C}, POL_{A,C}, E_{TOT(A,C)}\right\} \tag{3.65}$$

Practically, with the |1+> model, the SPECTRAL-SAR procedure is separately performed for the anionic and cationic subsystems, and it is subsequently summed in the resulting AC/IL-activity.

The second projective SPECTRAL-SAR model can be advanced here when the additive stage is considered at the incipient stage of the SPECTRAL-SAR operator $\hat{O}_{S-SAR}$ such that the considered Hansch factors, for instance, are first combined to produce the anionic-cationic indices that are further used to produce the spectral mechanistic map of the concerned interaction, producing the so-called |0+> model (Putz et al., 2007; Putz et al., 2010; Putz, 2012b; Putz & Putz, 2013a):

$$|Y_{AC}\rangle^{0+} = \hat{O}_{S-SAR}|0+\rangle = \hat{O}_{S-SAR}\, f(\{X_A\rangle\},\{X_C\rangle\}) \tag{3.66}$$

with the Hansch specification of the spectral vectors as follows.

- When considering the hidrophobicity with the model $|0+\rangle$ or 0^{+} the Spectral-SAR is projected on separate AC components:

$$\log P_A = e_{P_A} \Rightarrow P_A = e^{e_{PA}} = e^{\log P_A} \tag{3.67a}$$

$$\log P_C = e_{P_C} \Rightarrow P_C = e^{e_{PC}} = e^{\log P_C} \tag{3.67b}$$

  with the composed recipe

$$P_{AC} = P_A + P_C = e^{\log P_A} + e^{\log P_C} \tag{3.68}$$

  so that providing the AC-complex hidrophobicity

$$f(LogP_A, LogP_C) \equiv LogP_{AC} = \log\left(e^{LogP_A} + e^{LogP_C}\right) \in \left\{X_{1AC}\rangle\right\} \tag{3.69}$$

- Next, for evaluating the composed polarizability and total energy in the electrostatic AC-complex one should relay on the basic principles of electrostatic, starting from the Gauss electric flux law linking the total (source) charge with the electrostatic field created through a closed surface

$$\oint_{\Sigma} \vec{E} d\vec{S} = \frac{1}{\varepsilon_0} q_{total} \tag{3.70}$$

With the help of which the Columbus law springs out for a constant electrostatic field on a spherical surface distribution at the given radius (r) from the charge point (q_{total})

$$\begin{cases} S = 4\pi r^2 \\ E = ct \end{cases} \quad E\oint_{\Sigma} dS = \frac{Q}{\varepsilon_0} \Leftrightarrow E \cdot 4\pi r^2 = \frac{Q}{\varepsilon_0} \Rightarrow E = \frac{Q}{4\pi\varepsilon_0 r^2} \tag{3.71}$$

The electrostatic field immediately relates with the electrostatic force

$$\vec{F} = q \cdot \vec{E} = \frac{1}{4\pi\varepsilon_0} \frac{q \cdot Q}{r^3} \vec{r} \tag{3.72}$$

and with the electrostatic potential:

$$U = \int_1^2 \vec{E} \cdot d\vec{l} = \Delta\varphi \tag{3.73a}$$

which may be further written successively as

$$\Delta\varphi = \varphi_1 - \varphi_2 = \frac{1}{4\pi\varepsilon_0} \frac{q}{r_1} - \frac{1}{4\pi\varepsilon_0} \frac{q}{r_2} = \frac{q}{4\pi\varepsilon_0} \left(\frac{1}{r_1} - \frac{1}{r_2} \right) = \frac{q}{4\pi\varepsilon_0} \cdot \frac{r_2 - r_1}{r_1 r_2}$$

$$\cong \frac{1}{4\pi\varepsilon_0} \frac{ql \cos\theta}{r^2} = \frac{1}{4\pi\varepsilon_0} \frac{d \cos\theta}{r^2} \tag{3.73b}$$

when one employed the general expressions for the electrostatic potential

$$\varphi = \frac{1}{4\pi\varepsilon_0} \frac{\vec{d} \cdot \vec{r}}{r^3} \tag{3.74}$$

and for the dipole moment

$$\vec{d} \cdot \vec{r} = d \cdot r \cdot \cos\theta \tag{3.75}$$

However, from Eqs. (3.73a) and (3.74) there follows that in general one has the electrostatic field-potential relationship

$$\vec{E}\cdot\vec{r}=\varphi \quad \vec{E}\cdot\vec{r}=\frac{1}{4\pi\varepsilon_0}\frac{\vec{d}\cdot\vec{r}}{r^3} \tag{3.76}$$

leaving with the general electrostatic field-dipole connection too

$$\vec{d}=4\pi\varepsilon_0 r^3\cdot\vec{E}=\alpha\cdot\vec{E} \tag{3.77}$$

From Eq. (3.77) one recognizes the polarizability basic formulation

$$\alpha=4\pi\varepsilon_0 r^3 \Leftrightarrow POL=ct\cdot R^3,\ ct=4\pi\varepsilon_0 \tag{3.78}$$

allowing further AC-complex polarizability formation as based on the anionic-cationic summation of the associate radii of action

$$R_{AC}=R_A+R_C\cong ct^{-1/3}\left(POL_A^{1/3}+POL_B^{1/3}\right)$$

which corresponds to the actual projective polarizability formulation for the AC (IL) complex

$$f(POL_A,POL_C)\equiv POL_{AC}=\left(POL_A^{1/3}+POL_C^{1/3}\right)^3\in\{\langle X_{2AC}\rangle\}[\text{Å}^3] \tag{3.79}$$

- On the same analysis line, when one goes to evaluate the total energy of the AC complex, one has to consider also the electrostatic interaction energy which superimposes to the individual radicalic (atomic or molecular fragments) of A and C sysbsystems, to get

$$E_{AC}=E_A+E_C-\frac{q_Aq_C}{(4\pi\varepsilon_0)R_{AC}}$$
$$\cong E_A+E_C-\frac{q_Aq_C}{(4\pi\varepsilon_0)ct^{1/3}POL_{AC}^{1/3}} \tag{3.80}$$

Thus producing the working projective form for the total energy produced by the the AC (IL) complex

$$f(E_A,E_C)\equiv E_{AC}=E_A+E_C-627.71\frac{q_Aq_C}{POL_{AC}^{1/3}}\in\{\langle X_{3AC}\rangle\}\ [\text{kcal/mol}] \tag{3.81}$$

where one implemented the unitary electrostatic AC conditions $q_Aq_C=(-1)(+1)=-1$ along the atomic units transformation

$$\frac{1}{4\pi\varepsilon_0 ct^{1/3}} = 1at = 627.71\frac{kcal}{Mol} \tag{3.82}$$

Having exposed the two projective forms of complex-related QSARs, the open issue addresses whether the |0+> & |1+> states leave with the same results or what aspects of the SPECTRAL-SAR operator $\hat{O}_{S\text{-}SAR}$ might differ in the projective (multy-systems/subsystems) ecotoxicity, a matter that is solved by computing the so-called projective internal angle between the anion-cationic activity vectors, Figure 3.2, with $y_{iA}, y_{iC}, i = \overline{1,N}$ components following the prescription (Lacrămă et al., 2007; Putz et al., 2007; Putz et al., 2010; Putz, 2012b; Putz & Putz, 2013a):

$$\cos\theta_{AC} = \frac{\langle Y_C | Y_A \rangle}{\| |Y_C\rangle \| \| |Y_A\rangle \|} = \frac{\sum_{i=1}^{N} y_{iC} y_{iA}}{\sqrt{\sum_{i=1}^{N} y_{iC}^2 \sum_{i=1}^{N} y_{iA}^2}} \begin{cases} \geq 0.707107 \ldots |0+\rangle \; MODEL \\ < 0.707107 \ldots |1+\rangle \; MODEL \end{cases} \tag{3.83}$$

Accordingly, in general, one has the operators and vectorial unfold for projective Spectral-SAR/QSAR corresponding, respectively to the specifically combing of the structural causes

$$|0+\rangle S-SAR \begin{cases} |x_{AC}\rangle_i = f(|x_A\rangle_i, |x_C\rangle_i) \\ \& \; |y_{AC}^P\rangle = \sum_i b_i |x_{AC}\rangle_i \end{cases} \Bigg| \; \cos\theta_{AC} \cong 1 \tag{3.84}$$

and to the structural effects' or recorded bio-actions' superposition

$$|1+\rangle S-SAR \begin{cases} |y_A^P\rangle = \sum_i b_i^A |x_A\rangle_i \\ |y_C^P\rangle = \sum_i b_i^C |x_C\rangle_i \\ |y_{AC}^P\rangle = |y_A^P\rangle + |y_C^P\rangle \end{cases} \Bigg| \; \cos\theta_{AC} << 1 \tag{3.85}$$

- *ECOTOX-SSAR Principle 4* is assured by the "intensity" of the chemical-eco-bio-interaction is determined by the ratio of the expected to measured activity norms

$$RA \equiv r_{S-SAR}^{ALGEBRAIC} = \sqrt{\frac{\sum_{i=1}^{N} y_{i-PRED}^{2}}{\sum_{i=1}^{N} y_{i-OBS}^{2}}} = \frac{\left\| Y_{PRED} \right\rangle\|}{\left\| Y_{OBS} \right\rangle\|} \leq 1 \tag{3.86}$$

as a counterpart or along to the classical statistical correlation factor (Putz & Putz, 2011; Putz et al., 2008a, 2009a; Putz, 2012a-b)

$$R \equiv r_{QSAR}^{STATISTIC} = \sqrt{1 - \frac{\sum_{i=1}^{N} (y_{i-OBS} - y_{i-PRED})^2}{\sum_{i=1}^{N} \left(y_{i-OBS} - \frac{1}{N} \sum_{i=1}^{N} y_{i-OBS} \right)^2}} \tag{3.87}$$

- *ECOTOX-SSAR Principle 5* may be assured by the "selection" of the manifested chemical-eco-(bio-)binding parallels the minimum distances of paths (Putz M.V., Lacrămă, 2007; Putz, 2012a-b)

$$\delta[A, B] = 0 \tag{3.88}$$

connecting all possible endpoints in the norm-correlation hyperspace

$$[A, B] = \sqrt{\left(\left\| Y_B \right\rangle\| - \left\| Y_A \right\rangle\| \right)^2 + \left(r_B^{STATISTIC/ALGEBRAIC} - r_A^{STATISTIC/ALGEBRAIC} \right)^2} \tag{3.89}$$

In this manner, the "validation" of the obtained mechanistic picture is achieved by requiring that the influential minimum paths are numbered by the cardinal of the input structural factors set such that, excepting that the final endpoint that is always considered as the final evolution target, all other endpoints are activated one time and one time only.

3.2.6 NON-LINEAR CATASTROPHE QSAR

Put differently the above QSAR-OECD/ECOTOX principles express the essence of the chemical modeling of biological effects while relaying (Husserl-Russell) knowledge phenomenology in a more general manner (Putz et al., 2011a):

- QSAR-1: why does one do modeling?
- QSAR-2: how does one do modeling?
- QSAR-3: with what tools do I model?
- QSAR-4: how reliable is what I modeled?
- QSAR-5: what knowledge did the model provide?

Therefore, although the backbone of QSAR modeling is based on equation (3.3), one should be aware that it represents, despite the innumerable extant studies, only one type of model-the multi-linear type. It is therefore worth refreshing QSAR studies by exploring other ways of combining the structural parameters that cause the observed biological activity. However, although it is clear that non-linear QSAR is the next generation of correlations, one should not search arbitrarily or randomly while having at hand a well-designed theory of non-linear modeling of natural phenomena: Thom's catastrophe theory, the basic assumptions and main working tools of which are next presented.

The foreground of the Catastrophe Theory lies on expressing the Taylor series associated to a smooth function $\eta(c,x)$, $(c,x) = (c_1,\ldots,c_k,x_1,\ldots,x_m)$, say in its origin $(c,x) = 0$ under the form; the present discussion follows (Putz et al., 2011b)

$$\eta(c,x) = j^s\eta(c,x) + tayl \tag{3.90}$$

viewed as the summation of the so-called *s-jet* or *s-current*

$$j^s f(c,x) = \sum_{r=0}^{s} \frac{1}{r!} D^r\eta\Big|_0 x^r \tag{3.91}$$

and of its tail generically called here as "*tayl*." However, in modeling the natural phenomena, unlike the regular (like planets orbits) or continuous ones (with small perturbations included) where the truncation to the *s-jet* works fine, many of registered events display sudden (or "catastrophic") characters, like earthquakes, population growth, or cancer spreading, thus highly requiring for counting of the Taylor tail as well; such need was elegantly resumed by C. E. Zeeman, one of the pioneers of Catastrophe Theory (Zeeman, 1976), by "allowing the tail of the Taylor series to wag the dog." When the *tayl* part is becoming important it shapes as the quadratic type dependency on the control × behavior joint space where the original function was defined:

$$tayl \propto g^2(c,x) \tag{3.92}$$

This is due to the celebrated Morse's bifurcation lemma (Morse, 1931) around the so-called critical points of the original function, $\eta(c,x)$, where it actually equivalents the original function with the family of function (Putz et al., 2011b)

$$\eta(c,x) \to \tilde{\eta}(g_1(c,x),..., g_s(c,x)) \pm g_{s+1}^2 \pm ... \pm g_m^2 \tag{3.93}$$

Here s-stays also the co-rank of the Hessian of $\eta(c,x)$ in the point $(c.x) = 0$. The main question that arises hereby is to try to identify the so-called local types of function in a k-parametric (control space) family of functions, or, even more, being given a function to identify in its neighborhood the family it belongs to. The solution to this problem was furnished by Thom (1973) and then by Arnold (1976), by using the powerful concepts of co-dimension and structural tranversality, such that the resulted classification theorem formulates the seven elementary so-called catastrophe function of Table 3.3 as governing all the natural phenomena where the co-dimension is no greater than 4 (four). To better understand that this is indeed covering quite general plethora of natural dynamic systems (with complicated local/turning/singular points modeling sudden changes), enough recalling the heuristic example of the co-dimension for England-Scotland frontier, for instance, that is always equal to 1 (one) no mater one represents the frontier as a line (the road along it), as bidimensional (the road through it on the Earth), as tridimensional (the road through it by plain), or as 4-D (in relativistic vision when the space-time cone is considered as well along it) (Poston & Stewart, 1978). It is this co-dimension that controls so powerfully the reduction of all possible power expansions of smooth functions to those seven presented on Table 3.3; there, one sees the co-dimension number is always equal with the number of parameters from the control space appearing in the Thom polynomials; they, in fact, represent families of functions, i.e., controlling large classes of functions that drive open systems in similar (local) ways. In taxonomical (or algebraically) terms, it is said that although not all functions are typical (or elementary) their families are typical as families; In analytically terms, as all minima through origin look the same (there are said to be typical, and typical like the Morse minima of generalized parabola $x_1^2+...+x_m^2$, eventually after re-parameterization) likewise any transverse path through any non-Morse function that can be found within a family of

TABLE 3.3 Thom's Classification of Elementary Catastrophes*

Name	Co-dimension	Co-rank	Universal unfolding	Parametric representation
Fold	1	1	x^3+ux	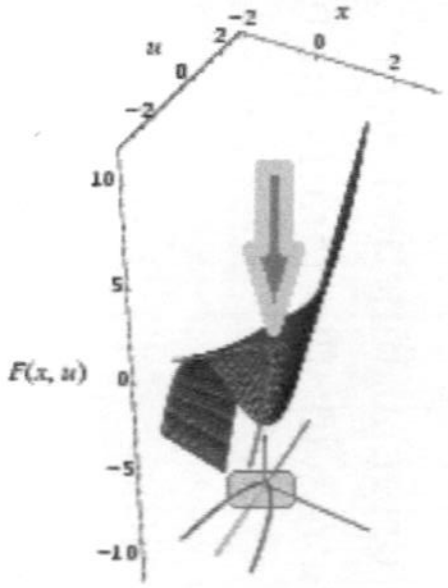
Cusp	2	1	$x^4=ux^2+vx$	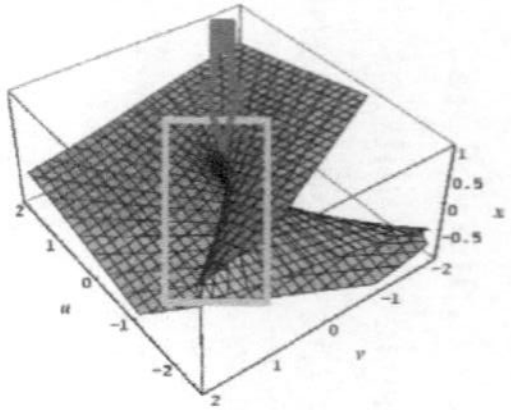
Swallow tail	3	1	$x^5+ux^3+vx^2+wx$	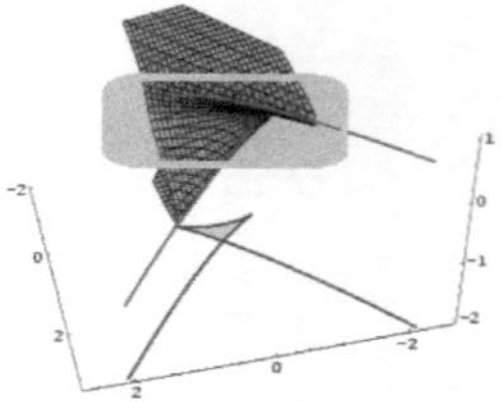

TABLE 3.3 Continued

Name	Co-dimension	Co-rank	Universal unfolding	Parametric representation
Butterfly	4	1	$x^6 + ux^4 + vx^3 + wx^2 + tx$	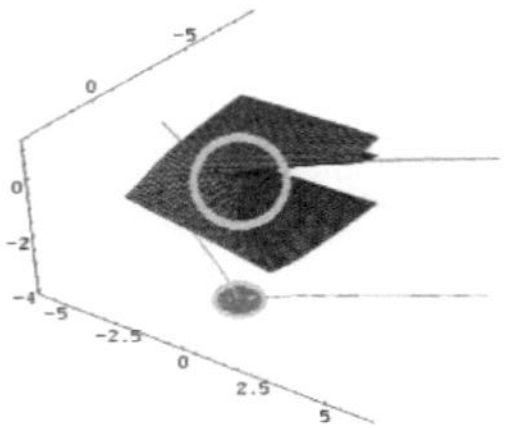
Hyperbolic umbilic	3	2	$x^3 + y^3 + uxy + vx + wy$	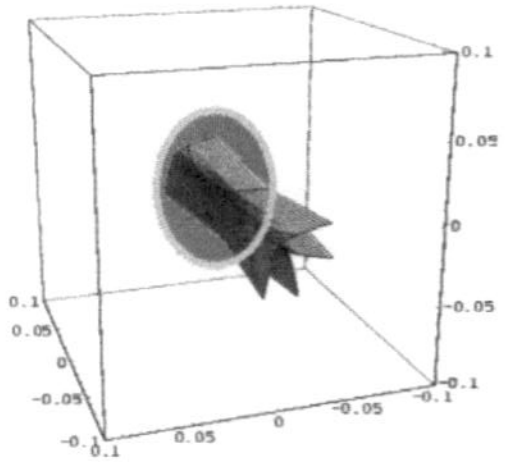
Elliptic umbilic	3	2	$x^3 - xy^2 + u(x^2 + y^2) + vx + wy$	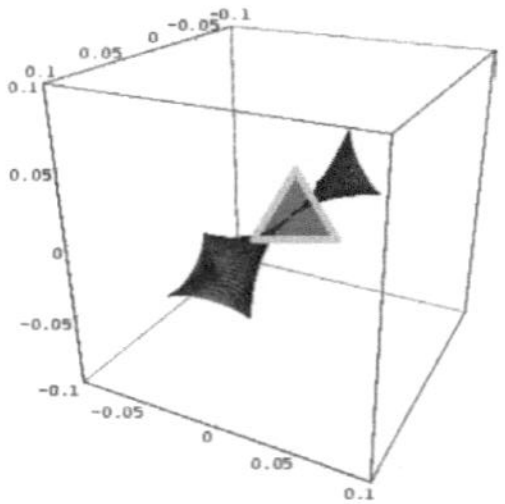

TABLE 3.3 Continued

Name	Co-dimension	Co-rank	Universal unfolding	Parametric representation
Parabolic umbilic	4	2	$x^2y + y^4 + ux^2 + vy^2 + wx + ty$	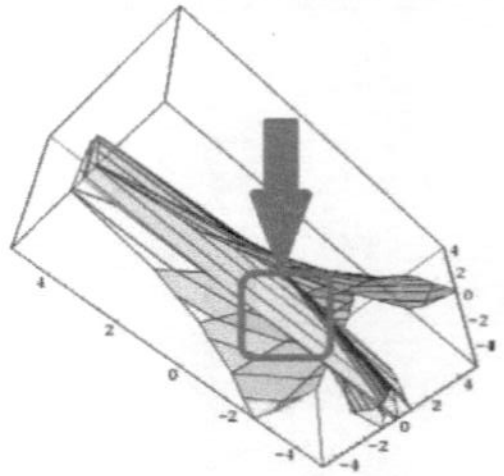

*After Putz (2009a, 2012a); Weisstein (2011).

finite functions looks the same as all other transverse paths in the family (those of Table 3.3). Even more, the co-rank of those functions (as the co-rank of their Hessian on the critical/singular/turning points) fixes also the minimum of variables that function can be reduced to; for example, if a function of 2011 variables has a critical point of co-rank equal 1, the actual function to be studied is of only 1 variable! This makes the Catastrophe Theory extremely interesting for being applied on QSAR studies, where the available structural variables are listed on hundred pages (Todeschini & Consonni, 2000), while in fact one searches for modeling functions that enter natural classes or family of functions with an universal character—as the Thom polynomials are—and therefore aiming to work with appropriate functions with considerable lower number of variables/structural descriptors, see Tables 3.3 and 3.4 (Putz et al., 2011b).

René Thom's catastrophe theory basically describes how, for a given system, continuous action on the *control space* (C^k), parameterized by C_k's, provides a sudden change in its *behavior space* (I^m), described by x_m variables through stable singularities of the smooth map (Krokidis et al., 1997; Putz, 2009a, 2012a)

$$\eta(c_k, x_m): C^k \times I^m \rightarrow \Re \quad (3.94)$$

with $\eta(c_k, x_m)$ called the *generic potential* of the system. Therefore, catastrophes are given by the set of *critical points* (c_k, x_m) for which the field gradient of the generic potential vanishes

$$M^{k \times m} = \left\{ (c_k, x_m) \in C^k \times I^m \middle| \nabla_{x_m} \eta(c_k, x_m) = 0 \right\} \quad (3.95)$$

or, more rigorously: a catastrophe is a singularity of the map $M^{k'm} \rightarrow C^k$.

Next, depending on the number of parameters in space C^k (also-called the *co-dimension*, *k*) and on the number of variables in space I^m (also-called the *co-rank*, *m*), Thom classified the generic potentials (or maps) given by Eq. (3.94) as seven unfolding elementary (in the sense of universal) catastrophes, i.e., providing the multi-variable (with the co-rank up to two) and multi-parametrical (with the co-dimension up to four) polynomials listed in Table 3.3. Going to the higher derivatives of the generic potential (the fields), the control parameter c_k* for which the Laplacian of the generic potential vanishes (Putz, 2009a, 2012a; Putz et al., 2011b)

$$\Delta_x \eta(c_k^*, x_m) = 0 \tag{3.96}$$

gives the *bifurcation point*. Consequently, the set of control parameters $c^\#$ for which the Laplacian of a critical point is non-zero defines the *domain of stability* of the critical point. It is clear now that small perturbations of $\eta(c^*, x)$ bring the system from one domain of stability to another; otherwise, the system is located within a *domain of structural stability*.

Remarkably, the cases described above correspond to the equilibrium limit of the dynamical (non-equilibrium) evolution of an open system

$$F\left(c_k; t; \eta(c_k; x_m); \frac{\partial \eta(c_k; x_m)}{\partial t}, \ldots\right) = 0 \tag{3.97}$$

where the behavior space is further parameterized by the temporal paths $x_m(c_k, t)$. The connection with equilibrium is recovered through the stationary time regime imposed on the critical points. In this way, the set of points giving a critical point in the stationary $t \to +\infty$ regime (the so-called ω-*limit*) corresponds to *an attractor*, and it forms a *basin*, whereas the stationary regime $t \to -\infty$ (the so-called *α-limit*) describes *a repellor*. In this way, the catastrophe polynomials may be regarded either as an asymptotic solution of a dynamical evolutionary system or as a steady state solution allowing the quasi-equilibrium of the ligand-receptor or inhibitor-organism interactions to be described. However, in complex binding systems with multiple evolutionary phases, e.g., the HIV-1 life cycle, the possibility of "linking" the various classes of catastrophes themselves may provide a striking analytical approach to the dynamics and mutational sensitivity of the studied interaction that starts with the actual catastrophe-QSAR method. Nevertheless, the natural consequence on statistical (Pearson) correlation behavior should be also analyzed, as following.

Since the transformation of the original smooth function into catastrophe one involves the Morse parabolic polynomials contribution, see Eq. (3.93), one may employ this recipe to consider the ordinary QSAR predicted activity, say Y^{QSAR}, and of its transformation into the Catastrophe-QSAR one, say $Y^{\Gamma/QSAR}$, through the Gaussian mapping (Putz et al., 2011b)

$$Y_{i=\overline{1,N}}^{\Gamma/QSAR}(X_1,...X_M)=Y_{i=\overline{1,N}}^{QSAR}(X_1,...X_M)\pm Y_{i=\overline{1,N}}^{M/\Gamma/QSAR}(X_2,...X_M) \quad (3.98)$$

with

$$Y_{i=\overline{1,N}}^{M/\Gamma/QSAR}(X_2,...X_M)=\sum_{j=2}^{M}\frac{1}{\sqrt{\pi}}\exp\left(-\frac{X_j^2}{4\sigma_{i=\overline{1,N}}}\right) \quad (3.99)$$

while referring to the running-indices assumed in Table 3.1. The form (3.98) with (3.99) recovers the original QSAR predicted function/value when all dispersions over all structural variables vanish

$$Y_{i=\overline{1,N}}^{\Gamma/QSAR}(X_1,...X_M)\xrightarrow{\sigma_{i=\overline{1,N}}\to 0}Y_{i=\overline{1,N}}^{QSAR}(X_1,...X_M) \quad (3.100)$$

thus motivating the actual generalization for treating the natural non-zero dispersive phenomena. On the other side, for higher dispersive values of structural variables (i.e., when their domains of applicability eventually overlap and promote interactions, i.e., the appearance of cross products in Tables 3.3 and 3.4) it produces the second order development

$$Y_{i=\overline{1,N}}^{\Gamma/QSAR}(X_1,...X_M)\xrightarrow{\sigma_{i=\overline{1,N}}\to\infty}Y_{i=\overline{1,N}}^{QSAR}(X_1,...X_M)\pm\widetilde{X}_2^2\pm...\pm\widetilde{X}_M^2 \quad (3.101)$$

under appropriate transformations $\widetilde{X}_{j=\overline{2,M}}=\widetilde{X}_{j=\overline{2,M}}\left(X_{j=\overline{1,M}}\right)$. However, one can see that in the Catastrophe Theory's language the first function of the right hand side in Eq. (3.101) stays for the 1-jet for the function Y^{QSAR}, while the hole expression (3.101) having the Hessian co-rank of order 2 is in full consistence with the maximum co-rank universal unfolding for the polynomials of Table 3.3. Next, one likes to check for the effect the Gaussian development (aka the catastrophe transformation) of Eq. (3.98) has on the statistical (Pearson) statistical coefficient respecting the QSAR value

$$R_0=\sqrt{1-\frac{1}{\sigma_A}\sum_{i=1}^{N}\left(A_i-Y_i^{QSAR}\right)^2}\ ,\ \sigma_A=\sum_{i=1}^{N}\left(A_i-\overline{A}\right)^2 \quad (3.102)$$

For the sake of clarity we will chose only one sign on (3.98), while the result will not depend on it, and successively obtain (Putz et al., 2011b)

$$R_\Gamma = \sqrt{1 - \frac{1}{\sigma_A}\sum_{i=1}^{N}\left(A_i - Y_i^{\Gamma/QSAR}\right)^2}$$

$$= \sqrt{R_0^2 - \frac{1}{\sigma_A}\sum_{i=1}^{N}\left(Y_i^{M/\Gamma/QSAR}\right)^2 + \frac{2}{\sigma_A}\sum_{i=1}^{N}\left(A_i - Y_i^{QSAR}\right)\left(Y_i^{M/\Gamma/QSAR}\right)}$$

$$\leq \sqrt{R_0^2 - \frac{1}{\sigma_A}\sum_{i=1}^{N}\left(Y_i^{M/\Gamma/QSAR}\right)^2 + 2\sqrt{\frac{1}{\sigma_A}\sum_{i=1}^{N}\left(A_i - Y_i^{QSAR}\right)^2}\sqrt{\frac{1}{\sigma_A}\sum_{i=1}^{N}\left(Y_i^{M/\Gamma/QSAR}\right)^2}} \tag{3.103}$$

where in the last relation the Cauchy-Schwarz inequality (2.66) was used under the actual working form:

$$\sum_{i=1}^{n} u_i v_i \leq \sqrt{\sum_{i=1}^{n} u_i^2}\sqrt{\sum_{i=1}^{n} v_i^2} \tag{3.104}$$

Next, in order to draw results that do not depend either on M-the number of structure variables nor on N-the number of chemicals/molecules involved in a custom QSAR study, one assumes dealing with the same dispersion of the observed activity as well as for each descriptor (the so-called homogeneous assumption, $\sigma = \sigma_A = \sigma_i, \forall i = \overline{1, N}$) likely to be valid when dealing with great number of structural descriptors; this way, one actually performs the asymptotic limit $M \to \infty$ on (3.99) for all $i = \overline{1, N}$ and recognizes the Poisson integral result

$$Y^{\infty/\Gamma/QSAR} = \int_0^\infty \frac{1}{\sqrt{\pi}} \exp\left(-\frac{X^2}{4\sigma}\right) dX = \sqrt{\sigma} \tag{3.105}$$

Accordingly, the inequality (3.103) now reads

$$R_{\infty/\Gamma}^2 \leq R_0^2 - N + 2\sqrt{N\left(1 - R_0^2\right)} \tag{3.106}$$

It may be rearranged upon the second order equation in N-chemicals' space

$$-N + 2\sqrt{\left(1 - R_0^2\right)}\sqrt{N} + \left(R_0^2 - R_{\infty/\Gamma}^2\right) \geq 0 \tag{3.107}$$

whose universal fulfillment leads with the condition

$$R_{\infty/r} \geq 1 \tag{3.108}$$

Since the result (3.108) was obtained within asymptotic conditions regarding the number of structural descriptors and homogeneous dispersion against recorded activity, it can be naturally asserted to its minimum as

$$R_{M/r} \rightarrow 1 \geq \forall R_{\alpha} \tag{3.109}$$

thus heuristically proving the superiority for the catastrophe-QSAR modeling over the fashioned QSAR, therefore further motivating the present approach. As numerical illustration of the general prescription of inequality (3.109) the present application confirms it by all one-to-one (i.e., catastrophe-QSAR *vs.* simple QSAR) results to be reported in the forthcoming Section 3.3.4.

Aiming to construct a QSAR rationale from the elementary catastrophes, the next steps are implemented (Putz et al., 2011b; Putz, 2012a):

(i) Assuming the vectorial form of activities and of associated QSARs are according to Table 3.3, the Table 3.4 showing catastrophe-QSAR is thereby formed.

(ii) Determine the norms for each model, see also (3.52)

$$\left\| Y \rangle \right\| = \sqrt{\langle Y | Y \rangle} = \sqrt{\sum_{i=1}^{N} y_i^2} \tag{3.110}$$

(iii) Calculate the algebraic correlation factor for each model, see also (3.55) (Putz & Putz, 2011; Putz, 2012a-b)

$$R_{ALG} = \frac{\left\| Y \rangle \right\|}{\left\| A \rangle \right\|} = \sqrt{\frac{\sum_{i=1}^{N} y_i^2}{\sum_{i=1}^{N} A_i^2}} \tag{3.111}$$

(iv) Calculate the so-called "*statistical relative power*" index for each model with each set of descriptors

$$\Pi = \sqrt{r^2 + t^2 + f^2} \tag{3.112}$$

TABLE 3.4 Algebraic Realization of Thom's Elementary Catastrophes as Uni- and Bi-Nonlinear QSARs*

Model	QSAR Equation
Group I: with one descriptor only, $\lvert X_1\rangle$	
QSAR-(I)	$\lvert Y_I\rangle = a_0\lvert 1\rangle + a_{11}\lvert X_1\rangle$
Fold	$\lvert Y_F\rangle = f_0\lvert 1\rangle + f_{11}\lvert X_1\rangle + f_{13}\lvert X_1^3\rangle$
Cusp	$\lvert Y_C\rangle = c_0\lvert 1\rangle + c_{11}\lvert X_1\rangle + c_{12}\lvert X_1^2\rangle + c_{14}\lvert X_1^4\rangle$
Swallow tail	$\lvert Y_{ST}\rangle = s_0\lvert 1\rangle + s_{11}\lvert X_1\rangle + s_{12}\lvert X_1^2\rangle + s_{13}\lvert X_1^3\rangle + s_{15}\lvert X_1^5\rangle$
Butterfly	$\lvert Y_B\rangle = b_0\lvert 1\rangle + b_{11}\lvert X_1\rangle + b_{12}\lvert X_1^2\rangle + b_{13}\lvert X_1^3\rangle + b_{14}\lvert X_1^4\rangle + b_{16}\lvert X_1^6\rangle$
Group II: with two descriptors, $\lvert X_1\rangle, \lvert X_2\rangle$	
QSAR- (II)	$\lvert Y_{II}\rangle = q_0\lvert 1\rangle + q_{11}\lvert X_1\rangle + q_{21}\lvert X_2\rangle$
Hyperbolic umbilic	$\lvert Y_{HU}\rangle = h_0\lvert 1\rangle + h_{11}\lvert X_1\rangle + h_{21}\lvert X_2\rangle + h_{1121}\lvert X_1X_2\rangle + h_{13}\lvert X_1^3\rangle + h_{23}\lvert X_2^3\rangle$
Elliptic umbilic	$\lvert Y_{EU}\rangle = e_0\lvert 1\rangle + e_{11}\lvert X_1\rangle + e_{21}\lvert X_2\rangle + e_{12}\lvert X_1^2\rangle + e_{22}\lvert X_2^2\rangle + e_{1122}\lvert X_1X_2^2\rangle + e_{13}\lvert X_1^3\rangle$
Parabolic umbilic	$\lvert Y_{PU}\rangle = p_0\lvert 1\rangle + p_{11}\lvert X_1\rangle + p_{21}\lvert X_2\rangle + p_{12}\lvert X_1^2\rangle + p_{22}\lvert X_2^2\rangle + p_{1221}\lvert X_1^2X_2\rangle + p_{24}\lvert X_2^4\rangle$

*The systematics of the sub-indices indicates consecutive coupled pairs, where each pair is interpreted as: the index of a structural factor followed by its power (Putz et al., 2011b).

where the components are defined as follows:

- relative index of correlation:

$$r = \frac{R_{ALG}}{R_{Pearson}} \tag{3.113}$$

relative index for Student's *t*-test

$$t = \frac{t_{Computed}}{t_{\substack{Tabulated \\ (1-\alpha=0.99; \\ N-M-2)}}} \tag{3.114}$$

relative index for Fisher's test

$$f = \frac{F_{Computed}}{F_{\substack{Tabulated \\ (1-\alpha=0.99; \\ M,N-M-1)}}} \tag{3.115}$$

(v) Determine the generalized Euclidian distances between corresponding type-I and type-II models employing different descriptors

$$\Delta\Pi = \sqrt{(r-r')^2 + (t-t')^2 + (f-f')^2} \tag{3.116}$$

and establish formal matrices for the models' differences for single descriptors, respectively

$$\Delta^2\Pi_{I(X_1,X_2)} = \left|\Delta\Pi_{I(X_1)} - \Delta\Pi_{I(X_2)}\right| \tag{3.117}$$

where

$$\Delta\Pi_{I(X=X_1 \vee X_2)} = \begin{pmatrix} QSAR_{I(X)} - F_{(X)} & QSAR_{I(X)} - C_{(x)} & QSAR_{I(X)} - ST_{(X)} & QSAR_{I(X)} - B_{(X)} \\ & F_{(X)} - C_{(X)} & F_{(X)} - ST_{(X)} & F_{(X)} - B_{(X)} \\ & & C_{(X)} - ST_{(X)} & C_{(X)} - B_{(X)} \\ & & & ST_{(X)} - B_{(X)} \end{pmatrix} \tag{3.118}$$

and for pair descriptors

$$\Delta\Pi_{II(X_1 \wedge X_2)} = \begin{pmatrix} QSAR_{II(X_1,X_2)} - HU_{(X_1,X_2)} & QSAR_{II(X_1,X_2)} - EU_{(X_1,X_2)} & QSAR_{II(X_1,X_2)} - PU_{(X_1,X_2)} \\ & HU_{(X_1,X_2)} - EU_{(X_1,X_2)} & HU_{(X_1,X_2)} - PU_{(X_1,X_2)} \\ & & EU_{(X_1,X_2)} - PU_{(X_1,X_2)} \end{pmatrix} \quad (3.119)$$

(vi) Identify all minimum paths across all differences $\Delta\Pi_{I(X_1 \vee X_2)}$, $\Delta^2\Pi_{I(X_1,X_2)}$ and $\Delta\Pi_{II(X_1 \wedge X_2)}$ for a given set of descriptors (X_1, X_2)

$$\begin{cases} \delta\{\Delta\Pi_{I(X)}\} = 0 \\ \delta\{\Delta^2\Pi_{I(X_1 \vee X_2)}\} = 0 \\ \delta\{\Delta\Pi_{II(X_1 \wedge X_2)}\} = 0 \end{cases} \quad (3.120)$$

The combination of descriptors that fulfills this system provides the *molecular mechanism* of the interaction. The correlation models involved are ordered according to their relative statistical power within the same molecular mechanism, thereby providing the *best models*. Because pair-descriptors are primarily involved in the present analysis, one can consider the first two such "waves" and their best correlation models up to the second order minimum paths, as in Eq. (3.120).

(vii) For selected correlation models, in either structure-driven or molecular mechanistic "waves," one employs them to compute the associated predicted activities for test molecules and to provide the statistics regarding the observed activity. If the obtained relative statistical power is close to those characteristics for the trial set of molecules, then these models may be validated for the specific eco-, bio-, or pharmacological problem. Moreover, further insight will be provided by the analysis of the catastrophe shape of the models involved and discussed accordingly.

3.2.7 QSAR BY QUANTUM AMPLITUDE (QUA-SAR)

One considers the main point of a QSAR, i.e., given a set of N-molecules one can chose to correlate their observed activities $A_{i=\overline{1,N}}$ with M-selected structural indicators in as many combinations as; the present discussion follows (Putz et al., 2009b)

$$C = \sum_{k=1}^{M} C_M^k \, , \; C_M^k = \frac{M!}{k!(M-k)!} \tag{3.121}$$

linked by different endpoint paths, as many as:

$$K = \prod_{k=1}^{M} C_M^k \tag{3.122}$$

indexing the numbers of paths built from connected distinct models with orders (dimension of correlation) from k=1 to k=M.

Basically, for each of the C-combinations a correlation (endpoint) QSAR equation is determined, say $Y_{l=\overline{1,C}} = \{y_{il}\}_{l=\overline{1,C}}^{i=\overline{1,N}}$, containing all computed activities for all considered N-molecules within the l-selected correlation. Now, the Spectral-SAR version of QSAR analysis computes these activities in a complete non-statistical way, i.e., by assuming the vectors for both observed (activities) and unobserved (latent variables) quantities while furnishing their correlation throughout a specific S-SAR determinant obtained from the transformation matrix between the orthogonal (desirable) and oblique (input) correlations. Yet, besides producing essentially the same results as the statistical least-square fit of residues the S-SAR method introduces new concepts; the present discussion follows (Putz et al., 2009b; Putz, 2012b):

- *endpoint spectral norm*, reloading (3.110) as

$$\left\| Y_l \right\rangle \| = \sqrt{\langle Y_l | Y_l \rangle} = \sqrt{\sum_{i=1}^{N} y_{il}^2} \, , \; l = \overline{1,C} \tag{3.123}$$

  gaining the possibility of the unique assignment of a number to a specific type of correlation, i.e., performing a sort of resumed quantification of the models (Putz & Lacrămă, 2007; Putz, 2012a-b);

- *algebraic correlation factor*, reloading (3.111) as

$$R_{ALG,l} = \frac{\||Y_l\rangle\|}{\||A\rangle\|} = \sqrt{\frac{\sum_{i=1}^{N} y_{il}^2}{\sum_{i=1}^{N} A_i^2}}, \quad l = \overline{1,C}, \tag{3.124}$$

viewed as the ratio of the spectral norm of the predicted activity to that of the measured one, giving the measure of the overall (or summed up) potency of the computed activities respecting the observed one rather than the local (individual) molecular distribution of activities around the mean statistical yields; thus, it is a specific measure of the molecular selection under study, always with a superior value to that yielded from statistical approach (Putz et al., 2008a), however preserving the same hierarchy in a shrink (less dispersive) manner being therefore better suited for intra-training set molecular analysis;

- *spectral path*, with the distance defined in the Euclidian sense, reloading (3.89) as

$$[l,l'] = \sqrt{\left(\||Y_l\rangle\| - \||Y_{l'}\rangle\|\right)^2 + (R_l - R_{l'})^2}, \ \forall (l,l') = \overline{1,C} \tag{3.125}$$

allows in defining complex information as path distances in norm-correlation space with norms computed from Eq. (3.123) while correlation free to be considered either from statistical (local) or algebraically (global) – Eq. (3.124) approaches; note that as far as computed activity Y_l corresponds to the measured activity A_l defined as logarithm of inverse of 50%-effect concentration (EC50), see below, both modulus of Y_l vectors and R values have no units so assuring the consistency of the Eq. (3.125).

- *least spectral path principle*, reloading (3.88) as

$$\delta[l_1,...l_k...,l_M] = 0; \ l_1,....,l_k,...,l_M : ENDPOINTS \tag{3.126}$$

provides a practical tool in deciding the dominant $\{\alpha,....\}$ hierarchies along the paths constructed by linking all possible k-models (i.e., models with k correlation factors) from Eq. (3.121) combinations selected one time each on a formed path – generating the so-called "M-endpoints containing ergodic path on K-paths assembly" of Eq. (3.122). However, the implementation of the principle (3.126) is recursively performed through selecting the least distance computed upon systematically application of Eq. (3.125) on ergodic paths; if, by instance, two paths are equal there is selected that one containing the first two models with shorter norm difference in accordance with the natural least action; the procedure is repeated until all C-models where connected on shortest paths; there was already conjectured that only the first M-shortest paths (called as $\alpha,\ldots,\alpha_M$) are enough to be considered for a comprehensive (and self-consistent) mechanistic analysis (Putz & Lacrămă, 2007; Lacrămă et al., 2007; Putz et al., 2007; Lacrămă et al., 2008; Putz & Putz, 2008; Putz et al., 2010; Putz, 2012a-b).

Nevertheless, for present purpose another two quantities are here introduced, namely (Putz et al., 2009b; Putz, 2012b):

- *inter-endpoint norm difference (IEND)*,

$$\Delta Y_{l\|l'} = \left\||Y_{l'}\rangle\right\| - \left\||Y_{l}\rangle\right\|,\ (l,l') \in \{\alpha_1,\ldots,\alpha_M\} \tag{3.127}$$

  that accounts for norm differences of the models lying on the M-shortest spectral paths linking M- from the C-models of Eq. (3.121);

- *inter-endpoint molecular activity difference (IEMAD)*,

$$\Delta A_{i\|j}^{l\|l'} = A_j^{l'} - A_i^{l} = \ln\frac{1}{(EC_{50})_j^{l'}} - \ln\frac{1}{(EC_{50})_i^{l}} = \ln\frac{(EC_{50})_i^{l}}{(EC_{50})_j^{l'}} \tag{3.128}$$

  is considered from activity difference between the fittest molecules (i, j), in the sense of minimum residues, for the models (l, l') belonging to the shortest paths $\alpha_1,\ldots,\alpha_M$ for which the inter-endpoint norm difference is given by Eq. (3.127).

This way, we can interpret the two fittest molecules (i, j) as reciprocally activated by the models (l, l') through the spectral path whom they belong; putted in analytical terms, the difference between quantities of Eqs. (3.127) and (3.128) may assure the "jump" or *transition activity* that turns the effect of *i* molecule on that of *j* molecule across the least spectral (here revealed as metabolization) path connecting the models *l* and *l'*(Putz et al., 2009b; Putz, 2012b):

$$\ln \frac{1}{q_{i|j}^{l|l'}} \equiv \Delta Y_{l|l'} - \Delta A_{i|j}^{l|l'} \tag{3.129}$$

Note that if we rearrange Eq. (3.129) in terms of 50-effect concentrations of Eq. (3.128) one gets the wave-like form of molecular EC_{50} inter-molecular transformation:

$$\left(EC_{50}\right)_i^l = \left(EC_{50}\right)_j^{l'} q_{i|j}^{l|l'} \exp\left(i\Delta Y_{l|l'}\right) \tag{3.130}$$

providing the analytic continuation in the complex plane for the *IEND* of Eq. (3.127) was assumed, i.e., $\Delta Y_{l|l'} \rightarrow i\Delta Y_{l|l'}$ outside the factor $q_{i|j}^{l|l'}$. Remark that although the differences in Eqs. (3.127) and (3.128) were consider mathematically, along the "arrow" *i-to-j*, the "quantum transformation" from Eq. (3.130) suggests that the bio-chemical-physical equivalence (metabolization) of the concentration effects evolves from *j-to-i*, revealing a typical quantum behavior with the factor $q_{i|j}^{l|l'}$ playing the propagator role as the quantum kernels in path integral formulation of quantum mechanics, see Volume I of the present multi-volumes work, see (Putz, 2016a) and (Dittrich & Reuter, 1992).

Equation (3.130) stands as the present "quantum"-SAR equation since (Putz et al., 2009b; Putz, 2012b):

- involves *the wave-type* expression of molecular effect of concentration, however, for special selected molecules (the fittest out of the *C*-models) and for special selected paths (the least for the *M*-ergodic assembly), being *M* and *C* related by Eq. (3.121);
- provides the *specific transition* or specific transformation of the effect of a certain molecule into the effect of another special molecule out from the *N*-trained molecules, paralleling the phenomenology of consecrated quantum transitions;
- has the amplitude of transformation driven by the so-called *quantum-SAR factor* of an exponential form

$$q_{i|j}^{l|l'} = \exp\left(\Delta A_{i|j}^{l|l'} - \Delta Y_{l|l'}\right) \tag{3.131}$$

defining the specific quantum-SAR wave;

- allows the *identity*

$$(EC_{50})_i^{l} = (EC_{50})_i^{l} \tag{3.132}$$

when the reverse effects is considered

$$(EC_{50})_j^{l'} = (EC_{50})_i^{l} \frac{1}{q_{i|j}^{l|l'}} \exp\left(-i\Delta Y_{l|l'}\right) \tag{3.133}$$

and substituted in the direct one (3.130), as absorption and emissions stand as reciprocal quantum effects;

- has a "phase" with unity norm, in the same manner as ordinary quantum wave functions, allowing the inter-molecular *"real" quantum-SAR transformation*

$$\left|(EC_{50})_i^{l}\right| = q_{i|j}^{l|l'} \cdot \left|(EC_{50})_j^{l'}\right| \tag{3.134}$$

exclusively regulated by the quantum-SAR factor of Eq. (3.131), in the same fashion as quantum tunneling is characterized by the transmission coefficient;

- when *multiple transformations* take place across paths with multiple linked models, say (*l*, *l*, ' *l*"), the inter-molecular transformation *i*→*j*→*t* is characterized by the overall quantum-SAR factor (3.131) written as product of intermediary ones

$$q_{i|t}^{l|l''} = q_{i|j}^{l|l'} \cdot q_{j|t}^{l'|l''} \tag{3.135}$$

due to the two-equivalent ways the $(EC_{50})_i^{l}$ effect may be described directly from *t* or intermediated by *j* molecular effect transformations, respectively:

$$\left|(EC_{50})_i^{l}\right| = q_{i|t}^{l|l''} \cdot \left|(EC_{50})_t^{l''}\right| = q_{i|j}^{l|l'} \cdot \left|(EC_{50})_j^{l'}\right| = q_{i|j}^{l|l'} \cdot \left(q_{j|t}^{l'|l''} \cdot \left|(EC_{50})_t^{l''}\right|\right) \tag{3.136}$$

in the same way as the quantum propagators behave along quantum paths, see Volume I of the present multi-volumes work (Putz, 2016a), and (Dittrich & Reuter, 1992); certainly, such contraction scheme may be generalized for least paths connecting the M-contained k-endpoints giving an overall quantum-SAR ("metabolization power") factor as:

$$q_{i_1|i_M}^{l_1|l_M} = \prod_{w=2}^{M} q_{i_{w-1}|i_w}^{l_{w-1}|l_w} \tag{3.137}$$

- Eq. (3.130) supports the *self-transformation* as well, with the driven qua-SAR factor given by

$$q_{i|j=i}^{l|l'} = \exp\left(-\Delta Y_{l|l'}\right) \tag{3.138}$$

during its evolution along the least paths when the same molecule ($i=j$) is metabolized by activating certain structural features ($l \neq l'$) though specific indicators (variables) in correlation (bindings with receptor site); this case resembles the stationary quantum case according which even isolated (or with free motion), the molecular structures suffer dynamical wave-corpuscular or fluctuant transformation along their quantum paths;

With the present Qua-SAR methodology one can appropriately identify the molecular pairs that drive certain bio-/eco- activities against given receptor by means of selected descriptors in a "wave"- or "quantum" mechanistic formal way. The ultimate goal will be the computation of quantum-SAR factors along the least paths of actions that give the potential information of the conversion power of the fittest molecules in their specific bindings.

3.2.8 FROM RESIDUAL TO STRUCTURAL ALERTS' QSAR (RES/SA-SAR)

Assuming there is a structure–activity multi-linear correlation problem with the parameters and observed endpoint set as $\left(\{X_i\}_{i=\overline{1,M}}, A\right)$, the standard QSAR corresponds to the ordinary regression equation producing the following computed activity; the present discussion follows (Putz & Putz, 2011; Putz, 2011a):

$$f^0\left(\{X_i\}_{i=\overline{1,M}}\right) = Y^0 = a_0 + \sum_{i=1}^{M} b_{0i} X_i \tag{3.139}$$

However, in carcinogenic modeling, it is difficult to find a proper set of structural parameters with significant correlation to the observed activity, especially when considering compounds having highly diverse molecular structures (i.e., being non-congeners) yet producing similar carcinogenic endpoints. Even by applying the available commercial or academic software to compute thousands of structural parameters and their non-linear combinations (Tarko et al., 2005), the obtained significant correlation relies on structural parameters or combinations thereof with little physical or chemical meaning. This makes QSAR analysis an artifact outside of reality. Such studies may not include the hydrophobic feature (LogP) within the correlation equation (Tarko & Putz, 2012), which has less physico-chemical meaning, especially with respect to cellular toxicity.

In such circumstances, it is preferable to test the *induced influence* of a given set of structural parameters with established significance over the cancer genotoxicity correlation, see Eq. (3.139). Hypothetically, this shows the direct, scarce correlation with the observed activity. The *residual correlation* follows (Putz, 2011a):

$$f^1\left(A-Y^0\right)=Y^1=a_1+b_1\left(A-Y^0\right) \quad (3.140)$$

From this point forward, one may use the various *residual-QSAR* (*res-QSAR*) models to obtain the correlation equation of the computed activity in terms of the original structural parameters.

***Self-Consistent res-QSAR Model*:** One may insert Eq. (3.139) into equation (3.140), while preserving the observed activity by the rule of computed activity:

$$Y_{SC}=a_1-b_1a_0+b_1A-b_1\sum_{i=1}^{M}b_{0i}X_i \quad (3.141)$$

This model has the conceptual advantage of containing looping or self-consistent QSAR information that is in line with the recursive evolution of cancer at the cellular level. It has also an apparent weakness in that it requires prior knowledge of the observed activity, even for the untested compounds or those that are designed in silico. However, such a drawback may now be avoided with the advent of unified databases with the aid of software to presumptively assess the "observed" activity of any common molecular-species couples (OECD Toolbox, 2012).

***Asymptotic res-QSAR Model*:** The obtained residual-QSAR matches were assumed with the observed activity,

$$A = Y^1 \tag{3.142}$$

yielding the following *asymptotic residual-model* from Eqs. (3.139) and (3.140):

$$Y_A = \frac{1}{1-b_1}\left[a_1 - b_1 a_0 - b_1 \sum_{i=1}^{M} b_{0i} X_i\right] \tag{3.143}$$

This model illustrates the residual QSAR method to amplify asymptotically the computed toxicity towards the observed carcinogenicity (Figure 3.3). This considers the limitation of no use when considering the case of $b_1 \to 1$, which produces the asymptotic (infinite) expressed activity $Y_A \to \infty$ with residual correlation. This difficult computation can be removed by reconsidering the residual equation (3.140) within different computational activity frameworks that are suited to assess the carcinogenic molecular mechanisms.

***Factor res-QSAR Model*:** If the observed, computational activity is proportionality confirmed by the following residual correlation factor,

$$R_1 Y^1 = A \tag{3.144}$$

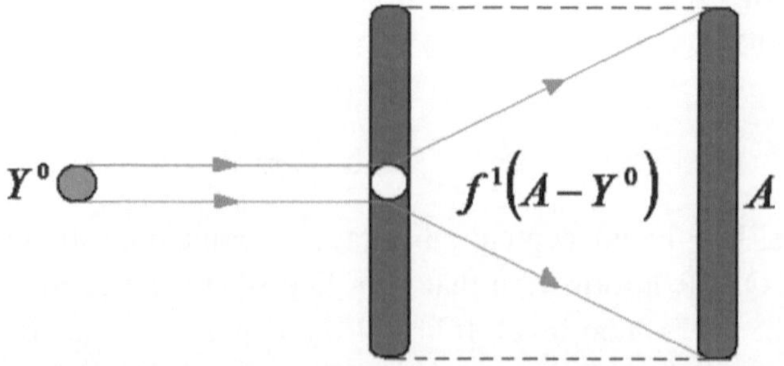

FIGURE 3.3 Representation of the residual-QSAR algorithm from a given computed activity (Y^0) to the observed one (A) through the "diffracting" process of the residual $A - Y^0$ activity (Putz, 2011a).

then Eq. (3.143) can be modified to the following workable model

$$Y_{F_1} = \frac{1}{1-R_1 b_1}\left[a_1 - b_1 a_0 - b_1 \sum_{i=1}^{M} b_{0i} X_i\right] \tag{3.145}$$

This model will eventually "diverge" when the residual correlation factor approaches unity ($R_1 \to 1$), along with the asymptotic condition, $b_1 \to 1$, noting the same asymptotic feature of this model as its ancestor, Eq. (3.143). This model is still identical to that obtained from replacing the residual factor with its complement, $R_1 \to 1-R_1$, because of the scale multiplication operation with the same correlation efficiency.

***Averaged res-QSAR Model*:** When the presence of the observed activity dependency is replaced by its average within the self-consistent equation, see Eq. (3.141), over the entire N-molecular series, the averaged residual-QSAR model is changed to the following:

$$Y_{AV} = a_1 - b_1 a_0 + b_1 \langle A \rangle - b_1 \sum_{i=1}^{M} b_{0i} X_i \tag{3.146}$$

where the average activity may be computed either as a simple statistical mean

$$\langle A \rangle \to \bar{A} = N^{-1} \sum\nolimits_{i=1}^{N} A_i \tag{3.147}$$

or as the interpolation function, $A = f_A(N)$, which is averaged as the integral

$$\tilde{A} = N^{-1} \int_1^N f_A(N) dN \tag{3.148}$$

Conceptually, the residual QSAR features correlation performances complementary to the direct QSAR analysis. This is effective in assessing the molecular phenomenology of cancer genotoxicity, as the direct structural parameters show little correlation. In addition, they apparently have no direct influence on observed activity, and slow-acting carcinogenesis does not have a significant, direct influence on physicochemical, structural parameters. However, for congeneric molecular species, significant direct correlation is expected, with low residual-QSAR influence as its statistical-information complement. Therefore, the present residual-QSAR approach is best suited for non-congeneric compounds, such as those involved in genotoxic carcinogenesis (Putz, 2011a).

Next, going to generalize the residual correlations one firstly resumed it respecting the direct QSAR with which it is combined; actually, for a given parameter set and an observed endpoint set, $\left(\{X_i\}_{i=\overline{1,M}}, A\right)$, the direct QSAR can be written as

$$A^M = a_0 + \sum_{i=1}^{M} b_{0i} X_i \tag{3.149}$$

while residual analysis gives

$$ARA^M = a_1 + b_1\left(A - A^M\right) \tag{3.150}$$

Here the superscript "*M*" refers to either the full molecule or the structural parameters for each molecule under study. Equations (3.149) and (3.150), with the assumption that the obtained residual-QSAR matches the observed activity; the present discussion also follows (Putz et al., 2011c),

$$A = ARA^M \tag{3.151}$$

providing the "*asymptotic residual QSAR*", see above Eq. (3.145)

$$ARA^M = \frac{1}{1-b_1}\left[a_1 - b_1 a_0 - b_1 \sum_{i=1}^{M} b_{0i} X_i\right] \tag{3.152}$$

However, since Eq. (3.152) is modulated by the parameters $b_1 \to 1$, it is invariantly obtained with the same form throughout this procedure regardless of the structural parameters or form of the direct QSAR, due to its asymptotic/divergent (catastrophic): while such behavior is common in cancer modeling (of asymptotically growing activity), the revealed conceptual limitation drawbacks in assessing the model of interest.

The aim is then to avoid such drawbacks by making use of the properties of the *structural alerts* (SA). It specifically employs the physicochemical properties of these alerts to build associated QSARs and residual QSAR counterparts to build a multi-regression model of the molecular mechanism.

The conceptual structural alert-QSAR algorithm is shown qualitatively in Figure 3.4. This algorithm provides activity predictions either by considering the full molecular structures or the substructures of the structural alerts.

The algorithm assumes a structure–activity multi-linear correlation problem using the structural alert (SA) parameters and observed endpoint set: $\left(\left\{X_i^{SA}\right\}_{i=\overline{1,M}}, A\right)$. The associated *alert-QSARs* corresponding to specific regressions over subsets of the structural parameters may be computed, for instance, for $(1 = m_1 < \ldots < m_i < m_{i+1} < \ldots < m_{j-1} < m_j < \ldots < m_M = M)$ (Putz et al., 2011c):

$$A_{(m_i,m_j)}^{SA} = a_{0(m_i,m_j)}^{SA} + \sum_{k=m_i}^{m_j} b_{0k}^{SA} X_k^{SA} \tag{3.153}$$

The residuals associated with Eq. (3.153) are

$$RA_{(m_i,m_j)}^{SA} = A - A_{(m_i,m_j)}^{SA} \tag{3.154}$$

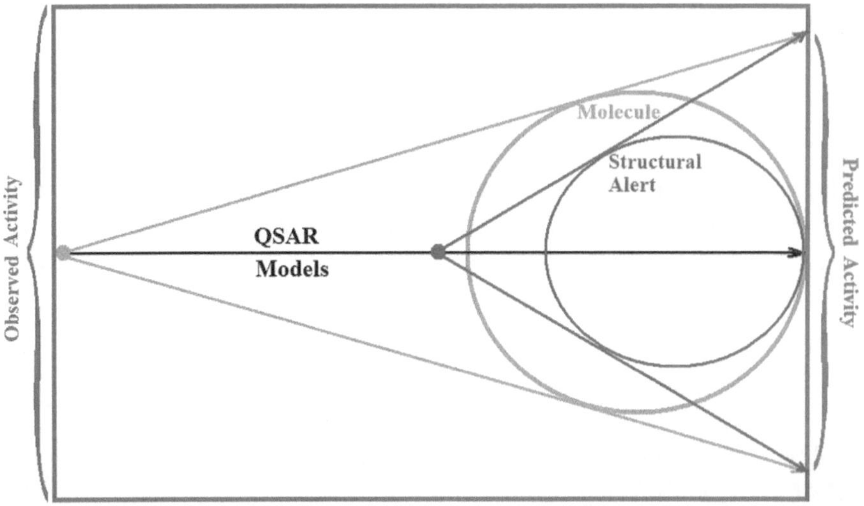

FIGURE 3.4 The *alert-QSAR method* uses structural alerts to assemble a molecular fragment QSAR model that has predictive power similar to that of full molecular modeling (Putz et al., 2011c).

Equations (3.153) and (3.154) give the formed residual-alert-QSAR equation

$$ARA^{SA}_{\{(m_i,m_j)\}} = a^{SA}_{1\{(m_i,m_j)\}} + \sum_{m_i=1}^{m_j} \sum_{m_j=m_i+1}^{m_k \le M} b^{SA}_{1(m_i,m_j)} RA^{SA}_{(m_i,m_j)} \quad (3.155)$$

Combining this expression with the activity matching condition of Eq. (3.151), here rewritten in actual terms,

$$A = ARA^{SA}_{\{(m_i,m_j)\}} \quad (3.156)$$

yields the structural alert residual correlation (Putz et al., 2011c)

$$ARA^{SA}_{\{(m_i,m_j)\}} = \frac{1}{1 - \sum_{m_i=1}^{m_j} \sum_{m_j=m_i+1}^{m_k \le M} b^{SA}_{1(m_i,m_j)}} \left[a^{SA}_{1\{(m_i,m_j)\}} - \sum_{m_i=1}^{m_j} \sum_{m_j=m_i+1}^{m_k \le M} b^{SA}_{1(m_i,m_j)} \left(a^{SA}_{0(m_i,m_j)} + \sum_{k=m_i}^{m_j} b^{SA}_{0k} X^{SA}_k \right) \right] \quad (3.157)$$

Equation (3.157) improves upon Eq. (3.152) and therefore also on Eq. (3.145) by using structural alert information instead of molecular information, avoiding singularities in the denominator, i.e.

$$1 \neq \sum_{m_i=1}^{m_j} \sum_{m_j=m_i+1}^{m_k \le M} b^{SA}_{1(m_i,m_j)} \quad (3.158)$$

This gives a residual-alert QSAR, a self-consistent correlation equation for the observed activity based on structural alert-predicted activities and residuals thereof. The method is to illustrated side-by-side with the previous residual – QSAR on representative toxicological carcinogenic series among the exposed studies in Section 3.3.6.

3.2.9 QSAR BY SMILES' STRUCTURE AND CHEMICAL REACTIVITY PRINCIPLES

Mechanistic QSAR stays as the fifth and the final of the consecrated OECD-QSAR principles a structure-activity/property correlation study should fulfill, alongside the QSAR-1 (Putz et al., 2011a): a defined endpoint, QSAR-2: an unambiguous algorithm, QSAR-3: a defined domain of applicability, and QSAR-4: appropriate measures of goodness-of–fit, robustness and predictivity (see Section 3.2.5). It appears therefore as the final goal of any QSAR rationale; it is nevertheless grounded on the precedent principles: starting from appropriate scrutiny of employed molecules such that the applicability domain to be well defined; then, the defined algorithm involves appropriateness in descriptors' proper meaning such that leaving with meaningful interpretation towards insight in chemical-biological mechanism of interaction.

For the screening side, the present work advances an original idea: actually, it uses the 2D-to-1D representation of a genuine molecule to treat it as a sort of "transition state/meta-stable" structure favoring the binding with the receptor site; it looks like a "fractalic chain" and has some of its genuine chemical bonds broken so that being preparing to travel through cellular walls till docking with active sites of the target receptors (see Figure 3.5); the present discussion follows (Putz & Dudaş, 2013a-b).

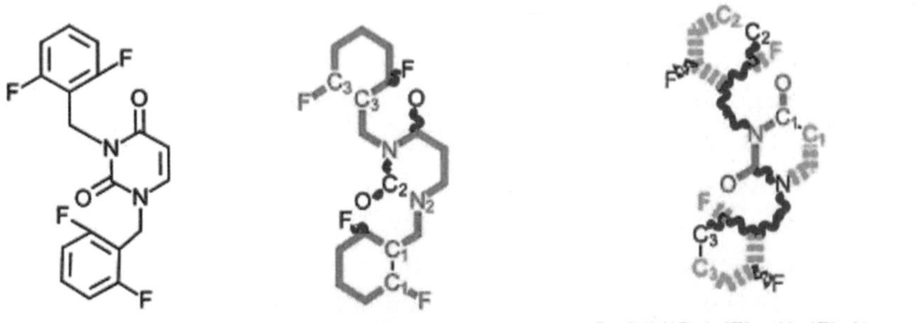

FIGURE 3.5 The anti-HIV molecule of 1,3-Bis-(2,6-difluoro-benzyl)-1H-pyrimidine-2,4-dione (left side) along the two SMILES generated codes as obtained by two different conceptual algorithms: the first one (middle) based on the longest SMILES molecular chain (LoSMoC) along the principal chain ( ) with few branches ( ); and the second one (in right) based on brancing SMILES (BraS) containg a shorter (or shortest) principal chain with brances of superior orders (,). The alternative unequivocally InChi and InChiKet codes were given for completeness.

Fortunately, such molecular description exists under simplified molecular-input line-entry system (SMILES) form; initially developed as a computational (ASCII strings) entry it has certain features making it an ideal candidate for present purpose (Weininger, 1988; Weininger et al., 1989; Ertl, 2010; Drefahl, 2011):

- It is based on principles of molecular graph theory, yet also including specification of configuration at tetrahedral centers and double bond geometry that were not included in the connectivity information;
- Allows for rigorous partial specification of chirality, specification of isotopes, etc.;
- It is unique for each structure, although dependent on the canonicalization algorithm used to generate it, i.e., it forms a surjective relationship with molecule but not a bijective one.

According to OECD guidance, "the intent of QSAR Principle 1 (*defined endpoint*) is to ensure clarity in the endpoint being predicted by a given model, since a given endpoint could be determined by different experimental protocols and under different experimental conditions. It is therefore important to identify the experimental system that is being modeled by the (Q)SAR." Note that, for instance, when the actual endpoint is considered as the inhibitory effect predicted by a series of 1,3-disubstituted uracil-based anti-HIV compounds on reverse transcriptase (Garg et al., 1999; Mehellou & De Clercq, 2010; Esposito et al., 2012; Quashie et al., 2012) in the highly active antiretroviral therapy (HAART) (Kaufmann & Cooper, 2000; De Clercq, 2009a; Krausslich & Bartenschlager, 2010), the concerned anti-HIV effect arises, in principle, through the same binding mechanism as binding/breaking DNA, i.e., through a group of non-necessarily similar structures, giving rise to the following updating QSAR end-point approaches (Benigni et al., 2007, 2008; Lessigiariska et al., 2010):

- *(Eco-) toxicological studies*, having various end-points (such as inhibition, activation, death, sterility, irritations, *etc.*) yet produced by a group of similar molecules, i.e., the case of *congeneric studies*; and
- *carcinogenic studies*, having essentially the same end-point as the exacerbated apoptosis that in principle diffuses in the organism no matter what the initial trigger point is, and may be initiated by highly structurally diverse molecules, being therefore classified as *non-congeneric studies*.

While the first case above is usually treated by ordinary (or direct) QSAR approaches, the second category is less frequently treated with the central QSAR dogma of congenericity. It therefore requires special approaches, such as the previously described residual-QSAR study (Putz, 2011a) (see also Section 3.2.8 as well as the case studies in Section 3.3). This relies on the fact that if no direct high correlation can be found, then there is a high probability that the action is residual, complementary or indirect (Tarko & Putz, 2012). For this point one considers the working molecules under study the most likely form producing the considered end-point, namely the anti-HIV activity produced, fro instance by uracil-based pyrimidines (Maruyama et al., 2003; Chemical Identifier Resolver beta 4, 2013), along two aspects of their SMILES structure, namely (Putz & Dudaş, 2013a-b):

- the longest SMILES molecular chain (LoSMoC), see Figures 3.5 and 3.6, when bonds are breaking on aromatic rings and moieties such that the resulting molecule displays a sort of 2D form of the original molecule along the "fractalic" chain, assumed to be the first stage in intermediary molecular defolding targeting the receptor. The maximum SMILES chains in LoSMoC are presumably responsible for best transport/transduction of ligand molecules through cellular (lipidic) walls, after which they may be released with a modified structure due to their further ionization resulting from interactions with cellular layers; accordingly, another SMILES form is generated and considered next, namely:
- the Branching SMILES (BraS), see Figures 3.5 and 3.6, representing the second phase of molecular defolding and providing ligand bond breakages such that many "bays" are formed, yet with consistent "arms" linking the short molecular "skeleton" aiming to favor the binding with a receptor in its pockets. Accordingly, the branching is not necessary in the same points of molecules through a series, but the maximum branching combined with equilibrium of branches is to be obtained in the final BraS. For instance, a long branch adjacent to a short one will not make a strong enough "anchor" to bind in a receptor pocket; therefore, the branching principle is to have the anchor-clefs balanced among themselves.

However, one should note that the fact that the most drugs are ionized once immersed in a biological medium is in accordance with the present

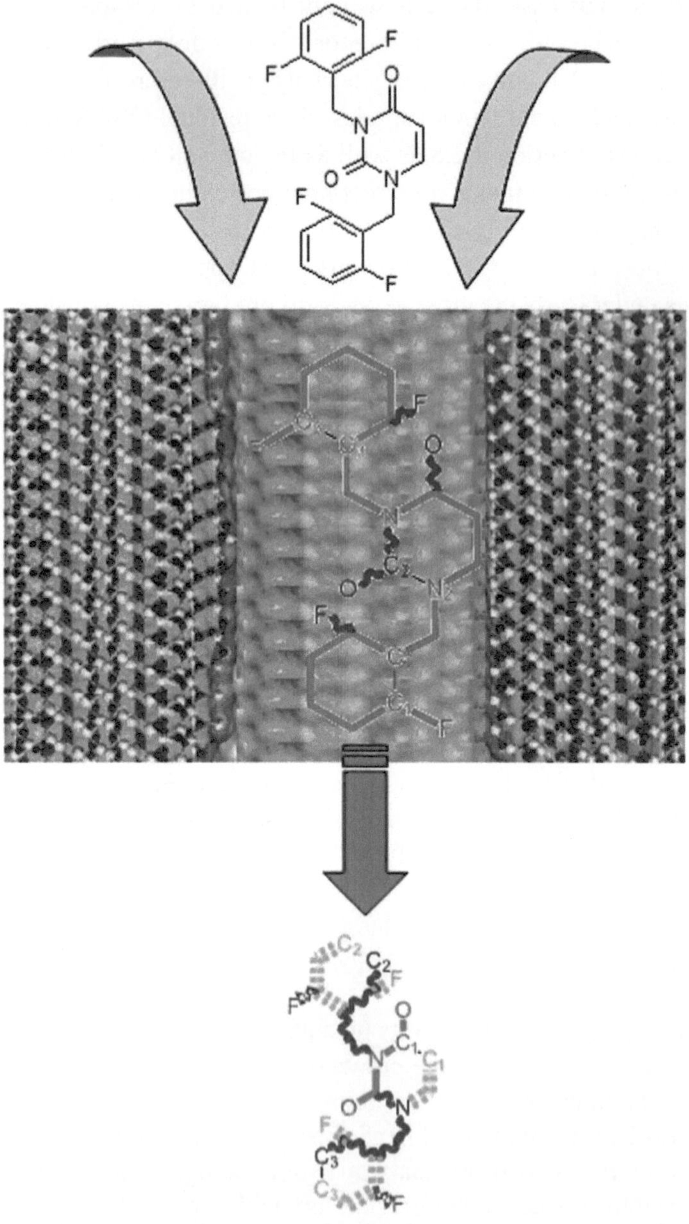

FIGURE 3.6 Illustration of the cellular wall's penetration, by the LoSMoC molecule towards hitting the receptor site under the BraS form of Figure 3.5; adapted from Refs. (Heng et al., 2004; Putz & Dudaş, 2013a-b).

two-steps of SMILES conformations, since in each of them more nucleophilic compounds are considered due to the successive bonding breaking and the loss of pairs of electrons as the unfolding goes from the original to the LoSMoC to the BraS configuration. These SMILES metabolic intermediates acting as nucleophilic active sides are confirmed at least for fused and non-fused diazines (Moldoveanu et al., 2009), among which are also those based on pyrimidines, which have already demonstrated antiviral and anti-HIV activity (De Clercq, 2005; Muhanji & Hunter, 2007; Gammon et al., 2008; Fan et al., 2009) and antiinflammatory effects in general (Butnariu et al., 2007; Butnariu, R.; Mangalagiu, 2009; Balan et al., 2009).

However, towards consistently describing the chemical-biological interaction by QSAR modeling in general and by SMILES-QSAR in special, worth shortly remembering that the chemical bonding and reactivity were at the forefront of modern chemistry in the last century, described through various qualitative theories (*viz.* (Lewis', 1916) theory of atoms and molecules or the resonance theory (Wheland, 1944) of Pauling (& Wheland, 1933; & Sherman, 1933; & Wilson, 1935)) as well as through quantitative ones [e.g., (Heitler & London, 1927) homopolar theory, (Hückel, 1931a-b) and extended Hückel heteropolar theories (Hoffmann, 1963), or the Bader-Gillespie Atoms in Molecules – AIM (Bader, 1990, 1994, 1998) and Valence Shell Electron Pair Repulsion – VSEPR formulations (Gillespie & Nyholm, 1957; Gillespie & Hargittai, 1991), just to name a few], before finally being united within the *conceptual* Density Functional Theory (Kohn et al., 1996; Putz, 2012c) leading to the recent bonding-by-reactivity scenario within the so- called *chemical orthogonal space* (Putz, 2012b-c, 2013b-c) of electronegativity (Parr et al., 1978; Putz, 2008a, 2011) and chemical hardness (Pearson, 1997; Putz, 2008a, 2011a). Accordingly, the next step was made when chemical-biology binding interactions and binding were considered as a superior phenomenological level of ordinary chemical bonding. To treat it, however, the descriptors' orthogonality feature turns out to be of prime importance so that the quantitative structure-activity relationship QSAR approach, see the Section 3.1.1, while incorporating it, establishes itself as the current paradigm in modeling biological activity. Eventually it may fully employ the fundamental chemical reactivity concepts such as the electronegativity and chemical hardness along their second generation of descriptors such as chemical power (Putz & Putz, 2013b; Putz & Dudaş, 2013a) and electrophilicity (Parr et al., 1999), and their associated variational principles, while assuming a given (parabolic) electronic total energy vs. number of electrons $E = E(N)$ shape dependency (Ayers & Parr, 2000; Putz, 2011b, 2012c), see also the related discussions on Volume II (Putz, 2016b) for atoms and in Volume III (Putz, 2016c) for molecules of the present five-volumes' work.

Moreover, according to the OECD guidance, the intent of QSAR-Principle 2 (unambiguous algorithm) is to ensure transparency in the predictive algorithm. In order to achieve this aim one needs reliable descriptors with physicochemical relevance. In this regard, the QSAR modeling in general and the present SMILES-QSAR approach in special employs the so-called *chemical orthogonal space*–COS of chemical bonding (Putz, 2012a-b, 2013b-c), which is based on the main chemical reactivity indices and the principles of electronegativity (χ) and chemical hardness (η), alongside their related quantities such as chemical power index (π) and electrophilicity (ω), see also the related atomic and reactivity indices discussions on Volumes II and III (Putz, 2016a-b), respectively, of the present five-volume work. Their short description follows with the aim of better understanding the QSAR- based mechanism to be then applied on particular series (Putz & Dudaş, 2013b).

3.2.9.1 Electronegativity and Its Minimum Principle

Electronegativity is viewed as an instantaneous variation of total (or valence) energy for a neutral or charged system (Parr & Yang, 1989):

$$\chi \equiv -\left(\frac{\partial E_N}{\partial N}\right)_{V(\mathbf{r})} \tag{3.159}$$

It may be also be related to frontier electronic behavior by performing the central finite difference development of Eq. (3.159) in terms of ionization potential (IP) and electronic affinity (EA), thus facilitating further connection with the highest occupied molecular orbitals (HOMO) and lowest unoccupied molecular orbitals (LUMO), respectively, according to Koopmans' (1934) frozen spin orbitals' theorem (see Figure 3.7):

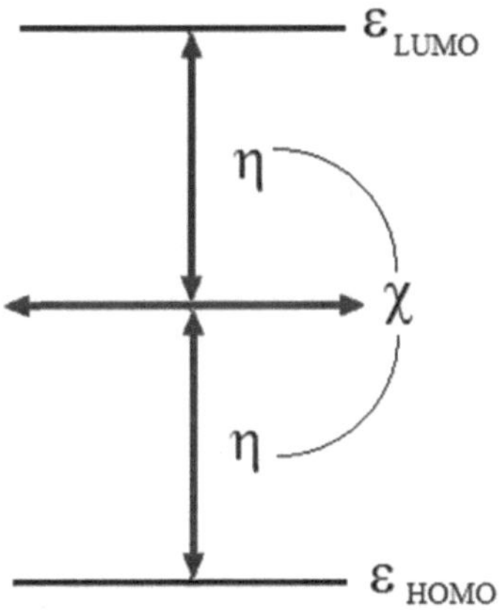

FIGURE 3.7 Electronegativity (χ) and chemical hardness (η) as phenomenological orthogonal chemical reactivity indices related to HOMO and LUMO levels (Putz, 2012a, 2013b; Putz et al., 2013a).

$$\chi_{FD} \cong \frac{(E_{N_0-1} - E_{N_0}) + (E_{N_0} - E_{N_0+1})}{2} \equiv \frac{IP + EA}{2} \cong -\frac{\varepsilon_{LUMO} + \varepsilon_{HOMO}}{2} \tag{3.160}$$

As such, in the course of a chemical reaction, or in chemical reactivity in general, electronegativity basically assures HOMO-LUMO energetic stabilization around its half-gap value; this is sustained by its inner definition by Eq. (3.159) which identifies it with the negative of the chemical potential of a system, as according to Parr et al. (1978), by the natural thermodynamic law of two fluids in contact its complex evolves towards equalization of the individual chemical potentials into a global one, while this principle, for the electronic fluid systems, was already consecrated from solid state physics (Mortier et al., 1985), in Chemistry it was coined by the so-called electronegativity equalization (EE) principle ($\Delta\chi = 0$), as originally stated by Sanderson under the assumption that "*for molecules in their fundamental state, the electronegativities of different electronic regions in the molecule – are equal*" (Sanderson, 1988); however, its variational form was recently clarified within the context of the double variational procedure (Putz, 2011d), specific to chemical systems:

$$\delta\chi \leq 0 \tag{3.161}$$

under the minimum electronegativity principle stating that: "a chemical reaction is promoted so as to minimize further charge transfer between atoms-in-molecules or between molecular fragments within a complex" (Tachibana, 1987; Tachibana & Parr, 1992; Tachibana et al., 1999). Nevertheless it was firstly formulated by Parr and Yang as under the maximum form favoring chemical reactivity (Parr & Yang, 1984; Yang et al., 1984): "*given two different sites with generally similar disposition for reacting with a given reagent, the reagent prefers the one which on the reagent's approach is associated with the maximum response of the system's electronegativity. In short,* $\Delta\chi \geq 0$ *is good for reactivity (n. a.).*" Yet, for assessing the chemical stability the reverse form of the latter idea will be considered, from where the minimum electronegativity principle $\Delta\chi \leq 0$ immediately results. However, in order to not conflict with the equality of electronegativity, this principle should be seen as a quantum fluctuation remnant effects in system upon the EE was consumed, i.e., it needs to be minimized so that the system reaches stable equilibrium (Putz, 2003), see also the Volume III of the present five-volume series (Putz, 2016c).

3.2.9.2 Chemical Hardness and Its Maximum Principle

Chemical hardness is viewed as the instantaneous electronegativity change with charge (Pearson, 1997):

$$\eta \equiv -\frac{1}{2}\left(\frac{\partial \chi}{\partial N}\right)_{V(\mathbf{r})} \tag{3.162}$$

It also supports the Koopmans' frozen spin orbitals reformulation at the level of molecular frontier, see Volume I of the present five-volume series (Putz, 2016a), i.e., there where chemical reactivity takes place, through the expression (see Figure 3.7) (Putz, 2011b):

$$\eta_{FD} = \frac{1}{2}\left(\frac{\partial^2 E_N}{\partial N^2}\right)_{V(r)} \cong \frac{E_{N_0+1} - 2E_{N_0} + E_{N_0-1}}{2} = \frac{IP - EA}{2} \cong \frac{\varepsilon_{LUMO} - \varepsilon_{HOMO}}{2} \tag{3.163}$$

At this point, while comparing Eqs. (3.160) and (3.163), it is clear that the electronegativity and chemical hardness may be viewed as the basis for an orthogonal space $\{\chi,\eta|\chi \perp \eta\}$ for chemical reactivity analysis since the conceptual and practical differences noted between the energetic level characterizing the "experimental" electronegativity and the energetic gap characterizing the "experimental" chemical hardness, respectively (Putz, 2009b, 2010b, 2012a, 2013b-c).

Like electronegativity, chemical hardness also supports two types of equations accompanying the chemical reactions and transformations. The first one promoting equalization of chemical hardness $\Delta\eta = 0$ of the atoms in a molecule or between molecular fragments in a complex or between adducts in a chemical bond refers to the so-called the hard and soft acids and bases (HSAB) principle (Chattaraj & Schleyer, 1994; Chattaraj & Maiti, 2003; Putz et al., 2004); it was initially formulated by Pearson and says that "*the species with a high chemical hardness prefer the coordination with species that are high in their chemical hardness, and the species with low softness (the inverse of the chemical hardness) will prefer reactions with species that are low in their softness, respectively*" (Pearson, 1990). This leads to numerous applications in both inorganic and organic chemistry, since it practically reshapes the basic Lewis and Brönsted qualitative theories of acids and bases (Pearson, 1985) into a rigorous orbital-based rule of chemical reactivity and bonding quantification.

Nevertheless, being of a quantum nature, chemical hardness inherently contains fluctuations leading to the inequality or variational form of its evolution towards bonding stabilization; as such, within the abovementioned double-variational variational formalism the actual maximum hardness principle is advanced (Chattaraj et al., 1991, 1995; Putz, 2008b):

$$\delta\eta \geq 0 \tag{3.164}$$

stating that the charge transfer during a chemical reaction or binding continues until the resulted bonded complex acquires maximum stability through hardness, i.e., maximizing the HOMO-LUMO energetic gap thus impeding further electronic transitions (Ayers & Parr, 2000). It was originally based on the Pearson observation according which "*there seems to be a rule of nature that molecules (or the many-electronic systems in general, n. a.) arrange themselves (in their ground or valence states, n.a.) to be as hard as possible*" (Pearson, 1985); it also leads to the practical application merely through its inverse formulation, the chemical softness is in turn related with the polarizability features of a system, i.e., as an observable quantity rooted in the quantum structure of the system, so that the minimum polarization principle was actually tested for various chemical systems (Mineva et al., 1998), for example, to rotational barriers accounting for conformational properties and thus with the steric effects (Torrent-Sucarrat & Solà, 2006), such that the actual chemical hardness variational principle of Eq. (3.164) is also indirectly validated.

3.2.9.3 Chemical Power and Its Double Minimum Principle

Since noting the opposition of electronegativity and chemical hardness, i.e., being the former associated with the tendency of the system to attract electrons and the latter with the tendency to inhibit the coordination and with the system stability, one may introduce the concept of chemical power, as the dynamic charge of atoms in a molecule, between molecular fragments or between adducts in a chemical bond, through the basic definition (Putz & Putz, 2013):

$$\pi = \frac{\chi}{2\eta} \tag{3.165}$$

Equation (3.165) gives us a sort of "reduced" or "normalized" electronegativity when its inertial hardness also counts. Moreover, for establishing a quantitative meaning one considers the Cartesian system where the coordinates are the hardness (on abscise) and electronegativity (on ordinate), see Figure 3.8(a); in this framework there follows that:

$$\pi = \frac{1}{2}\frac{\chi_A}{\eta_B} = \frac{1}{2}\tan(\theta_A) \cong -\Delta N_A \tag{3.166}$$

The last identity in Eq. (3.166) follows from chemical hardness-to-electronegativity definition (3.162) and allows the practical interpretation

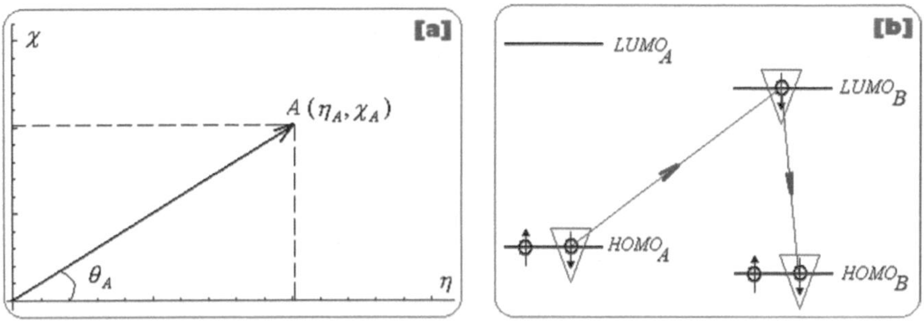

FIGURE 3.8 (a) Orthogonal hardness-electronegativity ($\eta - \chi$) representation for an electronic system with coordinate A (η_A, χ_A); (b) the "ABB" mechanism of frontier chemical reactivity driven by chemical power in A-B bonding complex (Putz & Dudaş, 2013b).

of chemical power in the chemical reactivity and bonding realm, providing the electronic charge transfer released by the adduct "A" when in bonding in an "A-B" complex, see Figure 3.8(b). Accordingly, the original frontier orbital $HOMO_A$ is minimized to the $HOMO_B$ in bonding, through the intermediate $LUMO_B$. In variational terms, the chemical power index is associated with minimizing HOMOs in bonding by means of charge transfer without spin changing:

$$\delta\pi \leq 0 \tag{3.167}$$

While principle (3.167) is consistent with principles relating minimum electronegativity and inverse of maximum of chemical hardness, it also emphasizes the necessity of the double variational principle when combined with Eq. (3.166), i.e., the released charge transfer of A in bonding is minimized so as to fit with the HOMO of bonding; in other terms, LUMO/HOMOA and LUMO/HOMOB levels also tend to equalize in bonding thus jointly fulfilling the conditions of equalization of electronegativity and chemical hardness.

3.2.9.4 Electrophilicity and Its Triple Minimum Principle

Electrophilicity (Parr et al., 1999), further allows coupling of chemical power index with electronegativity to provide the energetic information of activation towards charge tunneling of the potential between adducts (Putz & Putz, 2013; Putz & Dudaş, 2013a-b):

$$\omega = \chi \times \pi = \frac{\chi^2}{2\eta} \tag{3.168}$$

Electrophilicity actually accounts for energy consumed by a system for manifesting its chemical power in a chemical orthogonal space see Figure 3.9(a), essentially complementing it in bonding by electron transfer through tunneling between the bonding adducts, having the parent LUMO as an intermediate state, see Figure 3.9(b) as "orthogonal/complementary" to that of Figure 3.8(b).

As a mixed reactive index electrophilicity was developed to characterize the electrophilic/ nucleophilic action of charge transfer through accepting/ donating electrons, in modeling a variety of physical-chemical phenomena such as site selectivity (Pérez et al., 2002; Chamorro et al., 2003), molecular vibrations and rotation (Parthasarathi et al., 2005a), intramolecular and intermolecular reactivity patterns (Domingo et al., 2002a-b), solvent and external field effects (Pérez et al., 2001; Meneses et al., 2006; Parthasarathi et al., 2003a) as well as biological activity and toxicity

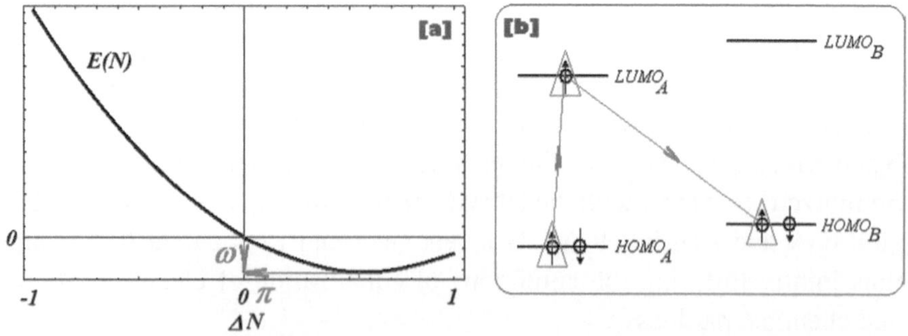

FIGURE 3.9 (a) Orthogonal representation of chemical power-electrophilicity ($\pi - \omega$) scheme for a parabolic form of total energy respecting the number of electrons for an elementary reactivity (accepting-donating) range; (b) the "AAB" mechanism of frontier chemical reactivity driven by electrophilicity in a A-B bonding complex as the complementary/orthogonal one respecting the "ABB" counterpart of Figure 3.8(b) (Putz & Dudaş, 2013b).

(Parthasarathi et al., 2003b, 2004, 2005b, 2006a-b; Rong et al., 2007; Roy et al., 2007; Putz & Putz, 2013; Putz & Dudaş, 2013a-b).

However, electrophilicity involves even stronger than the chemical power the double minimum character (through squaring of electronegativity and of its principle), which corresponds to charge penetration of the A-B energetic barrier towards fulfilling electronic pairing in a bonded complex. In practical circumstances, electrophilicity drives the electronic jump from HOMOA to LUMOA then relaxes to HOMOB in an "A-B" bond complex thus covering the "AAB" pathway in chemical reactivity and bonding, see Figure 3.9(b); in this case it minimizes the LUMOB-HOMOB gap, as the inverse of chemical hardness as promoted by Eq. (3.168), that nevertheless leaves the bonding complex in an activated state which competes with minimization of electronegativity (through pairing) which tends to stabilizes the structure. For this reason the overall variational principle of electrophilicity assumes its minimization form:

$$\delta\omega \leq 0 \tag{3.169}$$

yet whether this is a characteristic of a reactive or stabilized bonding system remains an open issue and should be assessed for each case under study.

These reactivity indices and principles are suited for analyzing the molecular interaction mechanism for a bonding complex chemically formed in a chemical-biological interaction. However, the resulting scheme with the "$\chi + \eta$" actions show the two systems "bonded" and the bonding electrons in the LUMO state for each adduct ligand and receptor. Therefore, a coupled L-R "anti-bonding" or transition state for the L-R interaction is generated, not a stabilized L-R chemical bond (see Figure 3.10a). This observation is the principal motivation and frontier orbital molecular illustration that suggests electronegativity and chemical hardness do not exhaust the chemical bond potential in a ligand-receptor system in its transition state without electronic relaxation. Thus, further description is necessary for such a chemical bonding scheme through structural action with related chemical descriptors to produce a stable ligand-receptor bond associated with either electronic (pair) transfer (ionic interaction) or common electronic (pair) overlapping (covalent) ; the present discussion also follows (Putz et al. 2013a).

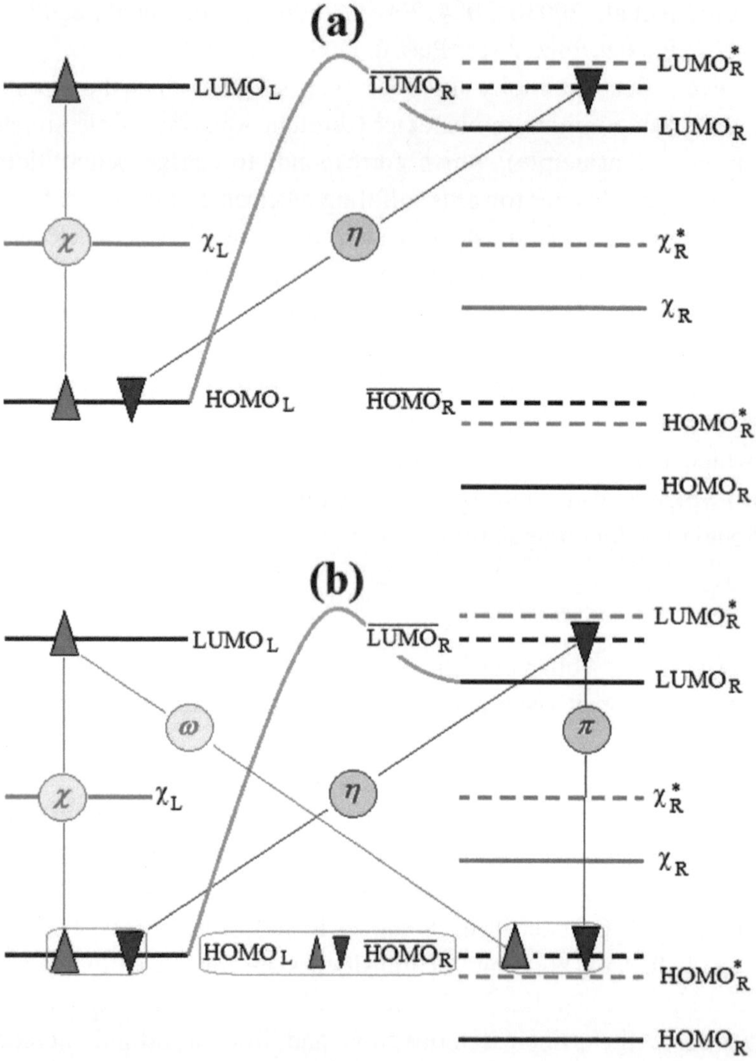

FIGURE 3.10 (a) The incomplete ligand (L)-receptor (R) interaction illustrates the two spin-electrons at the HOMO and LUMO levels through electronegativity (χ) and chemical hardness (η), which equalize the HOMO-LUMO mid-levels and gaps, respectively (see text); (b) a complete reaction scheme for (a) with the effect of electrophilicity (ω) and chemical power (π) on stabilizing the pairing transfer between ligand and receptor (i.e., completing the ligand-receptor bond using 4 chemical reactivity indices and principles). The representation herein is an illustration; additional variations for this bonding scheme may include L in place of R or with a different initial hierarchy for the L- and R- species' HOMO and LUMO levels, respectively (see text for details), after (Putz et al., 2013a).

Therefore, the chemical bonding mechanism should include the remaining two structural actions that are driven by the following.

- Chemical power (π) as representing the dynamic of maximum charge for atoms in a molecule; therefore, Eq. (3.166) indicates a role in relaxing the previous "maximum chemical hardness" through a quantum L-R-barrier transition into an electronic transfer from the LUMO-receptor into a newly "created" HOMO level in the receptor; the newly created bar-HOMO-receptor, see Figure 3.10(b) "aligns" with the corresponding (original) site on the ligand to complete the HOMO-receptor-ligand bonding; along
- Electrophilicity which is generated after the effect from chemical power; by definition, it is second order kind of chemical reactivity index, see Eq. (3.168). Thus, it contributes by restoring the above electronegativity and, in turn, extending through the L-R barrier potential between the involved adducts. Eventually, this property yields the LUMO-ligand electron at the newly created HOMO-receptor-ligand level, which completes the bonding scheme, as revealed by Figure 3.10(b).

Alltogether, the underlying notion is that the complete chemical bonding scheme through ionic or covalent means may be completed when chemical reactivity is driven by electronegativity, chemical hardness, chemical power and electrophilicity, in the following order (called as "*the chemical reactivity hierarchy dogma*"), see the chemical reactivity algorithm of Volume III/Chapter 3 of the present five-volumes' book (Putz, 2016c):

$$\chi \rightarrow \eta \rightarrow \pi \rightarrow \omega \tag{3.170}$$

Note that in the Figure 3.10 the order in Eq. (3.179) should not depend on whether L and R change electronic roles or other relative energetic HOMO and LUMO positions, althougheach particular situation may have its own orbital diagrams; however the chemcail reactivity dogma (3.170) may be largely disproved by particular chemical-biological interaction, due to complex binding mechanism which may inverse some specific quantum causes, as may be checked by the case studies exposed in Section 3.3.7, being such behavior characteristic to Qu-SAR treatment for such bio-molecular interactions.

Specifically, returning to SMILES analyses as combined with QSAR approach by variational principles of the involved structural parameters, according with the "OECD-QSAR fifth commandment" the mechanistic QSAR based picture of chemical-biological SMILES' interaction assumes the natural and computational *variational principles* in various forms, eventually resumed as (Putz & Dudaş, 2013a-b):

- By considering SMILES counterpart of envisaged molecules and making the screening classes based on the *longest* SMILES molecular chain (LoSMoC) or on branching SMILES (BraS) in various lengths of main chains and complexity of branches;
- By considering QSAR orthogonal descriptors with associate *min-max* prescription in chemical reactivity (as exposed for the consecrated chemical reactivity descriptors, to which other may be added, as related to solubility, topology, etc.);
- By selecting the *highest* QSAR correlation among SMILES screening based compounds;
- By employing the transitivity of the QSAR descriptors, as a manifestation of the *close up* of the molecular physicochemical properties, in extracting from endpoint's chains the structural causal hierarchy;
- By ordering the multi-descriptor dependencies with the help of *ordering* of local/mono-linear correlations, especially for establishing the triggering chemical causes of biological action;
- By recognizing the restrained number of relevant correlation models on endpoints' paths through assuring the complete coverage of the considered Banach chemcial space of chemical reactivity descriptors by *smallest* number of highly rated correlations' dependencies;
- By performing the *least* path-length search along all possible correlation' combinations built on the selected models from previous step, for instance through the search along the formed paths along endpoints' (mono-*I*, bi-*II*, and many/multi-*M*) correlations, in accordance with Eqs. (3.88) and (3.126)

$$0 = \delta\left\{|Y_{Ii}\rangle, |Y_{IIi}\rangle, |Y_{IIIi}\rangle, \ldots |Y_{Mi}\rangle\right\}_{i-PATH}$$
$$= \left\{\sum_{\theta=I,II,\ldots}\left(R_{\theta i} - R_{(\theta+I)i}\right)^2 + \left(R_{(M-1)i} - R_{Mi}\right)^2\right\}_{i-PATH}^{1/2} \tag{3.171}$$

All together, these steps provide 7-fold variations ("*longest*", "*min-max*", "*highest*", "*close up*", "*ordering*", "*smallest*", and the "*least*") from a given pool of molecules with a recorded biological activity towards establishing the chemical mechanism of structural causes targeting the observed effect.

However, worth noting that, computationally, for BraS molecular structures, since more ionized (bonding breaking) structures, naturally after passing the lipidic cellular layers targeting the biological receptor (Figure 3.6), one should consider more HOMO and LUMO states, beyond the first order of frontier orbitals used in "custom" chemical reactivity calculations, see Eqs. (3.160) and (3.163) for electronegativity and chemical hardness, respectively. Therefore the so-called "branching" effect of SMILES structures should be accordingly reflected at the energetic level too; fortunately, within Koopmans' approximation, such formulations exist up to the third order of compact finite differences, see Volume I of the present five-volume work (Putz, 2016a) when they look like (Putz, 2011b-c, 2012a):

$$\chi_{CFD} = -\left[a_1\left(1-\alpha_1\right)+\frac{1}{2}b_1+\frac{1}{3}c_1\right]\frac{\varepsilon_{HOMO(1)}+\varepsilon_{LUMO(1)}}{2}$$
$$-\left[b_1+\frac{2}{3}c_1-2a_1\left(\alpha_1+\beta_1\right)\right]\frac{\varepsilon_{HOMO(2)}+\varepsilon_{LUMO(2)}}{4}$$
$$-\left(c_1-3a_1\beta_1\right)\frac{\varepsilon_{HOMO(3)}+\varepsilon_{LUMO(3)}}{6} \quad (3.172)$$

$$\eta_{CFD} = \left[a_2\left(1-\alpha_2+2\beta_2\right)+\frac{1}{4}b_2+\frac{1}{9}c_2\right]\frac{\varepsilon_{LUMO(1)}-\varepsilon_{HOMO(1)}}{2}$$
$$+\left[\frac{1}{2}b_2+\frac{2}{9}c_2+2a_2\left(\beta_2-\alpha_2\right)\right]\frac{\varepsilon_{LUMO(2)}-\varepsilon_{HOMO(2)}}{4}$$
$$+\left[\frac{1}{3}c_2-3a_2\beta_2\right]\frac{\varepsilon_{LUMO(3)}-\varepsilon_{HOMO(3)}}{6} \quad (3.173)$$

in terms of lowest unoccupied and highest occupied molecular orbital energies ε_{LUMO} and ε_{HOMO}, respectively, with the parameters given in Table 3.5.

TABLE 3.5 Numerical Parameters for the Compact Finite Second (2C)-, Fourth (4C)- and Sixth (6C)-Order Central Differences; Standard Padé (SP) Schemes; Sixth (6T)- and Eight (8T)-Order Tridiagonal Schemes; Eighth (8P)- and Tenth (10P)-Order Pentadiagonal Schemes Up To Spectral-Like Resolution (SLR) Schemes for the Electronegativity and Chemical Hardness*

Scheme	Electronegativity					Chemical Hardness				
	a_1	b_1	c_1	α_1	β_1	a_2	b_2	C_2	α_2	β_2
2C	1	0	0	0	0	1	0	0	0	0
4C	$\frac{4}{3}$	$-\frac{1}{3}$	0	0	0	$\frac{4}{3}$	$-\frac{1}{3}$	0	0	0
6C	$\frac{3}{2}$	$-\frac{3}{5}$	$\frac{1}{10}$	0	0	$\frac{12}{11}$	$\frac{3}{11}$	0	$\frac{2}{11}$	0
SP	$\frac{5}{3}$	$\frac{1}{3}$	0	$\frac{1}{2}$	0	$\frac{6}{5}$	0	0	$\frac{1}{10}$	0
6T	$\frac{14}{9}$	$\frac{1}{9}$	0	$\frac{1}{3}$	0	$\frac{3}{2}$	$-\frac{3}{5}$	$\frac{1}{5}$	0	0
8T	$\frac{19}{12}$	$\frac{1}{6}$	0	$\frac{3}{8}$	0	$\frac{147}{152}$	$\frac{51}{95}$	$-\frac{23}{760}$	$\frac{9}{38}$	0
8P	$\frac{40}{27}$	$\frac{25}{54}$	0	$\frac{4}{9}$	$\frac{1}{36}$	$\frac{320}{393}$	$\frac{310}{393}$	0	$\frac{344}{1179}$	$\frac{23}{2358}$
10P	$\frac{17}{12}$	$\frac{101}{150}$	$\frac{1}{100}$	$\frac{1}{2}$	$\frac{1}{20}$	$\frac{1065}{1798}$	$\frac{1038}{899}$	$\frac{79}{1798}$	$\frac{334}{899}$	$\frac{43}{1798}$
SLR	1.303	0.994	0.038	0.577	0.09	0.216	1.723	0.177	0.502	0.056

*Putz (2011b-c, 2012a).

Actually, for treating the BraS molecules the most complex finite difference scheme of computation is used for basic chemical reactivity indices, i.e., the spectral-like-resolution (SLR) numerics of Table 3.5 (Lele, 1992), to specialize the Eqs. (3.172) and (3.173) to the working ones (Putz, 2011b-c, 2012a]:

$$\chi_{CFD}^{SLR} = -1.06084\frac{\varepsilon_{HOMO(1)} + \varepsilon_{LUMO(1)}}{2} + 0.718869\frac{\varepsilon_{HOMO(2)} + \varepsilon_{LUMO(2)}}{4} + 0.31381\frac{\varepsilon_{HOMO(3)} + \varepsilon_{LUMO(3)}}{6} \quad (3.174)$$

$$\eta_{CFD}^{SLR} = 0.582177\frac{\varepsilon_{LUMO(1)} - \varepsilon_{HOMO(1)}}{2} + 0.708161\frac{\varepsilon_{LUMO(2)} - \varepsilon_{HOMO(2)}}{4} + 0.022712\frac{\varepsilon_{LUMO(3)} - \varepsilon_{HOMO(3)}}{6} \quad (3.175)$$

The analytical descriptors of Eqs. (3.174) and (3.175) greatly help in considering the chain and branching modeling of actual molecules as being differentiated for LoSMoC and BraS intermediates also at the level of frontier chemical reactivity. Concluding,

- we consider only first orders (2C of Table 3.5) of χ & η for the LoSMoC molecules;
- we consider all three orders of HOMO and LUMO for the BraS molecules under SLR scheme of Table 3.5 for χ & η.

Note that with the values of χ and η they are further implemented in π and ω of Eqs. (3.165) and (3.168), for SMILES molecules in the same way as for genuine (as in gas phase environment) to provide the respective LoSMoC and BraS results to be then – altogether modeled by the SMILES-QSAR methodology.

Nevertheless the whole plethora of finite difference schemes of Table 3.5 may be used for electronegativity and chemical hardness while employing a different kind of QSAR, in a topo-reactive fashion as is in the next described.

3.2.10 TOPO-REACTIVE QSAR

3.2.10.1 A Historical Perspective on Topology and Link to Chemical Graphs

Beside quantum-mechanical analysis also the so-called topological or "graph theory" assessment of the molecular structure there was consecrated throughout a rich and fascinating history. It begins with the Kant's each-day walks on the Königsberg bridges in the mid of XVIII, see Figure 3.11-left, when he rise the problem whether it is possible to stroll across each bridge only once while returning to the starting point. Fortunately, although the great philosopher left the problem as unsolvable one it was soon accomplished by the (Euler, 1736). The latter had attributed to each land area a point (A, B, C, and D in Figure 3.11) and to each bridge a line linking two points (a, b, c, d, e, f, and g in Figure 3.11) thus obtaining a (multi)graph. Very soon he realized that the Kant's problem has a solution only when a connected graph is also "eulerian", i.e., each vertex has an even number of lines starting from it – in other words, having an even degree. There follows that the Königsberg bridges were no so economically designed since an eulerian path is to be attained only when the line "g" would be between point "B and C" instead of "A and D", as easily observed from the Figure 3.11 – middle and right, respectively; the present discussion follows (Putz et al., 2008b).

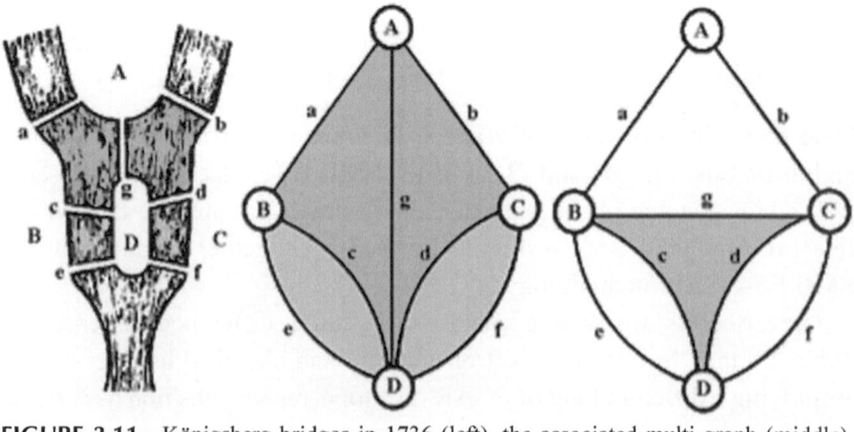

FIGURE 3.11 Königsberg bridges in 1736 (left), the associated multi-graph (middle), and its eulerian version (right); adapted from Refs. (Atanasiu, 1971; Putz et al., 2008b).

However, the graphs were rediscovered with other several occasions: in physics by Kirchhoff's (1842) studies on electrical networks and, in chemistry, by the seminal works of Couper, Butlerov, and Kekulé when founding the modern structure theory of chemical compounds.

Aiming to apply the features of the graph's representation of chemical formulas on predicting the number of isomers in organic chemistry the enumeration problem was successively solved by the pioneers of graph theory in chemistry, among which worth quoting some key contributions paving the topo-chemcial field (Cayley, 1857, 1874, 1875, 1881; Crum-Brown, 1864; Blair & Henze, 1932; Henze & Blair, 1933, 1934; Coffman et al., 1933; Pólya, 1937; Senior, 1955; Balaban & Harary, 1971).

Basically, the use of graph theory in chemistry may be seen as another representation of quantum chemistry, as a simplification version of it. As most of molecular structures may be seen as having two main electronic components: one is giving the σ-electron framework that creates the effective field in which the π-electrons – the other component – move, the simplification may be achieved since these two components may be seen as independents. Although having no rigorous justification such context enables the path descriptions of atoms-in-molecule and, nevertheless, was recorded with an impressing success in chemical bond description.

Practically, the correspondence of quantum and chemical terms with those used in graph theory, see Table 3.6, make possible the developing of the mathematical chemistry still preserving its observable interpretation.

Without going into details (Kier & Hall, 1986), for the background notions, let us denote a graph by the set $G(V, E)$ with the sub-sets $V{=}V(G)$ and $E{=}E(G)$ of the vertices and edges, respectively. In these conditions the number of elements of the vertices set will be assigned as $N = |V(G)|$ while the edges set will have the dimension $M = |E(G)|$. Since the degree $\mathbf{deg}_i$ of a vertex v_i is further defined as the number of its adjacency vertices we may as well introduce $\mathbf{Deg} = \mathbf{Deg}(\boldsymbol{G})$ as the degree vector containing all the degrees of the existent vertices of the G graph.

This way, the common representation of the structural formulas through the graph characteristics is constructing specific matrices for them, some of them with clear physical interpretation while other being less intuitive, however, still most helpful in QSAR correlations (Putz et al., 2008b).

TABLE 3.6 THE Face-To-Face Quantum Chemical and Mathematical Terms*

Quantum Chemical Terms		Mathematical Terms
Atom	↔	*Vertex*
Bond	↔	*Edge*
Structural formula	↔	*Graph*
Acyclic molecule	↔	*Tree graph*
Alternant molecule	↔	*Bipartite graph*
Atomic valence	↔	*Vertex degree*
n-polyene	↔	*Chain of n vertices*
n-annulene	↔	*Cycle of n vertices*
Number of rings	↔	*Cyclomatic number*
Secular polynomial	↔	*Characteristic polynomial*
Topological matrix	↔	*Adjacency matrix*
Level of energy	↔	*Eigenvalues*
Non-bonding level	↔	*Zero eigenvalue*
Bonding level	↔	*Positive eigenvalue*
Anti-bonding level	↔	*Negative eigenvalue*
Hückel theory	↔	*Spectral theory*

*After Rouvray (1976).

3.2.10.2 Common Construct of Topological Indices

One of the most famous stands the adjacency matrix **A** = **A**(G) whose elements are defined as (Kier & Hall, 1986):

$$A_{ij} = \begin{cases} 1, & \dots\ i \neq j \wedge e_{ij} \in E(G) \\ 0, & \dots\ i = j \vee e_{ij} \notin E(G) \end{cases} \quad (3.176)$$

The interpretation is simple: the sum of elements on the i-row or on the i-column of matrix **A**(G) gives the i-vertex degree $\mathbf{deg}_i$.

The degree set of a graph can be analytically represented by the so-called Chi-matrix, $\chi = \chi(G)$, symmetrically defined as (Randić, 1992):

$$\chi_{ij} = \begin{cases} (\mathbf{deg}_i \mathbf{deg}_j)^{1/2}, & \dots\ e_{ij} \in E(G) \\ 0, & \dots\ in\ rest \end{cases} \quad (3.177)$$

thus counting of the connectivity degree of the concerned structural formula; the present discussion follows (Putz et al., 2008b).

Regarding the paths information of a molecular graph the symmetrical distance matrix **D** = **D**(G) can be as well considered with elements (Mihalić et al., 1992):

$$D_{ij} = \begin{cases} d_{ij}, & \dots\ i \neq j \\ 0, & \dots\ i = j \end{cases} \quad (3.178)$$

with d_{ij} representing the distance between the vertices v_i and v_j. When the elements of rows or columns are summed up results the sum of the bonds in the studied molecule, this way making from the distance matrix a vital ingredient of the molecular topological indices.

For instance, the first proposed index for thermodynamic correlations of the acyclic hydrocarbons that used the distance matrix was the so-called (Wiener, 1947) index **WiI**, defined by Hosoya, see (Hosoya, 1971; Hosoya & Murakami, 1975; Mohar & Pisanski, 1988):

$$W(G) = \sum_{i=1}^{N} \sum_{j=i}^{N} D_{ij} \quad (3.179)$$

Actually, the calculated Wiener index of the molecule the average topological atom distance (half of the sum of all atom distances) in the molecule.

On a chemical graph $G(N)$ with N atoms $W(N)$ represents an invariant of the graph arising from the half-summation of the minimum distances d_{ij} between all pairs of $G(N)$ vertices (Wiener, 1947; Cataldo et al., 2010; Cataldo et al., 2011a; Iranmanesh et al., 2012):

$$W(N) = ½\, \Sigma_{ij}\, d_{ij},\ d_{ii}=0,\ i, j = 1,2, \dots, N-1, N \quad (3.180)$$

Entries d_{ij} constitute the NxN distance matrix of the graph $\hat{D} = [d_{ij}]$. Matrix $\hat{D}$ is symmetric by definition, being $d_{ij} = d_{ij}$. In the present study in particular, N corresponds not only to the number of PAH carbon atoms, but also to the number N_π of available π-electrons in each molecule as well. The Wiener index, Eq. (3.180), measures the average *topological compactness* of the molecular graphs and, when a minimum-principle is imposed on it, $W(N)$ promptly selects stable systems among possible candidates as demonstrated by recent studies on fullerenes isomers or defective graphenic planes (see different applications of topological modeling (TM) methods given in the recent extended report (Ori & D'Mello, 1992). Indicating with $M(N)$ the graph *diameter* (e.g., the largest distance in the graph) and with b_{ik} the number of k-neighbors of the i-atom, the contribution w_i to (3.180) is:

$$w_i = \tfrac{1}{2} \Sigma_k k b_{ik} \quad k = 1, 2, \ldots, M\text{-}1, M \tag{3.181}$$

with $W(N) = \Sigma_i w_i$ and 1-N= $\Sigma_k b_{ik}$ for each graph vertex v_i. Note that hexagonal networks as nanotori, fullerenes, graphenic layers, schwarzites feature connectivity b_{i1}=3 for each node; in case of aromatic molecules (e.g., polycyclic aromatic hydrocarbons, PAHs), their atoms show a double possibility b_{i1}=2 or b_{i1}=3, the lowest value being valid for atoms lying along the molecular boundary.

Very close with the Wiener number is the Szeged index (**SzI**) calculated by simply counting the number of nodes on both sides of each edge (and sum these counts). The Szeged index extends the Wiener number for cyclic graphs by counting those nodes only, which are nearer to the given side of the edge than to the other. Unreachable nodes (for disconnected graphs) or those at equidistance position to both sides of the given edge are not taken into account.

In the same simple distance- (path) or degree-based category we may recall the Harary index (**HaI**) Harary (1969), defined as half-sum of the off-diagonal elements of the reciprocal molecular distance matrix of the molecule, as well as the Platt index (**PaI**) computed as the sum of the edge degrees of a molecular graph (Platt, 1952).

Next, going to the connectivity class of indices, by employing the distance sum (Rouvray, 1986):

$$DS_i = \sum_{j=1}^{N} D_{ij} \tag{3.182}$$

for the vertices v_i and v_j, Balaban has introduced his connected Balaban index (**BaI**) (Balaban, 1982),

$$J = \frac{M}{\mu+1}\sum_{E(G)}\left(DS_i DS_j\right)^{-1/2} \tag{3.183}$$

In last relation the factor $M/(\mu + 1)$ represents the normalization of the graph edges' number to the variation of the edges' number with the cyclicity through the cyclometer number:

$$\mu = M - N + 1 \tag{3.184}$$

representing the number of cycles in a graph.

However, worth noting that the Balaban index displays a quite low degeneracy, i.e., the uniqueness degree its value for a given structural formula, while for an infinite linear molecule tends to the number $\pi = 3.14159$.

In terms of the connectivity, the Balaban index gives, in fact, the Balaban distance connectivity of the molecule, which is the average distance sum connectivity.

In the same manner, the Randić index (**RaI**) or molecular connectivity index can be introduced as the harmonic sum of the geometric means of the node degrees for each edge (Randić, 1975). It is close related with the identification number derived from the paths of a molecular graph (Randić, 1984a, b, 1986):

$$ID(G) = N + \sum_{p} f^{*}(p) \tag{3.185}$$

for all paths "p" of the graph, where

$$f^{*}(p) = \sum_{i,j}^{M} f(e_{ij}) \tag{3.186}$$

contains the read edge-defined functions:

$$f(e_{ij}) = \left(\mathbf{deg}_i \mathbf{deg}_j\right)^{-1/2} \tag{3.187}$$

These topological indices were successfully employed in many QSPR studies and are currently assumed as ones of the most used indicators of the chemical structure. However, in the present study, beside their individual evaluation and implementation a sort of super-topological indices will be employed in a form revealed in next section.

3.2.10.3 Introducing SuTIs: Super-Topological Indices

Since the considered Wiener (**WiI**), Szeged (**SzI**), Harary (**HaI**), Platt (**PaI**), Balaban (**BaI**), and Randic (**RaI**) topological indices may be seen as classified in two groups, (**WiI, SzI, HaI, PaI**) and (**BaI, RaI**), according with their way of calculation based on *simple* or *connectivity* formulas of distances and degrees, see the Eqs. (3.179), (3.183), and (3.185)–(3.187), respectively, we may further combined in the simple and connectivity indices, *SI* and *CI*, respectively, for instance through formulas; the present discussion follows (Putz et al., 2008b):

$$SI_1 = \frac{\mathbf{WiI}}{\mathbf{PaI}},\ SI_2 = \frac{\mathbf{SzI}}{\mathbf{HaI}},\ CI_1 = \mathbf{BaI},\ CI_2 = \mathbf{RaI} \tag{3.188}$$

Note that in Eq. (3.188) in the simple indices class the normalization of the Wiener and Szeged indices were considered normalized to the Platt and Harary indices, respectively, in the view of producing numerical results that are comparable with those specific to the indices of connectivity class. This way the problem of dominance of CI class over *SI* class (due to the products involved by connectivity formulas) is avoided an the "competition" between the influence of topological indices belonging to these classes is from now open.

Next, from the SI and CI classes the so-called ***super-topological indices***, **SuTIs** are introduced with the following properties (Putz et al., 2008b):

- A super-topological index is constructed on an abstract space of degree defined by the number of the topological indices considered. In the present case, **deg(SuTI)**=4;
- A **SuT** index is introduced through a determinant of a symmetrical matrix, which has on its main diagonal the ordinary topological indices considered, and in rest the various combinations of 0 and 1, this way counting of all symmetrical possibilities of combined them;
- The individual topological indices in a **SuTI** are arranged, on principal diagonal, in the sub-matrices of their considered class, i.e., depending on the way in which they were calculated, *SI* or *CI*. Since they cannot interchange between different such classes, for example, the *simple based class* or the *connectivity based class*, the number of possible SuTis are somehow reduced;

- Finally, a **SuT** index is rooted to the degree of the topological space considered, here at **deg(SuTI)**=4, thus assuring that the result will be of topological index nature;
- As a general rule for constructing a **SuTI**, all interchanges between topological indices belonging to a specific topological class, here *SI* and *CI*, has to be assisted by proper combinations of figures 0 and 1 such that the resulted determinant be insensitive to these internal topological indices' permutations;
- With all these, the general analytic form of a **SuTI** looks like:

$$\mathbf{SuTI}^{(\mathbf{deg})} \equiv \left(\left|\begin{array}{ccc} \begin{pmatrix} SI_1 & 0\vee 1 & \cdots \\ 0\vee 1 & SI_2 & \cdots \\ \vdots & \vdots & \ddots \end{pmatrix} & 0\vee 1 & \cdots \\ 0\vee 1 & \begin{pmatrix} CI_1 & 0\vee 1 & \cdots \\ 0\vee 1 & CI_2 & \cdots \\ \vdots & \vdots & \ddots \end{pmatrix} & \cdots \\ & & (\ddots) \\ \vdots & \vdots & \vdots \end{array}\right|\right)^{1/\mathbf{deg}(\mathbf{SuTI})} \tag{3.189}$$

Actually, with the present (SI_1, SI_2) and (CI_1, CI_2) groups we may form the next super-topological indices (Putz et al., 2008b):

$$\mathbf{SuTI}_1^{(IV)} \equiv \left(\begin{vmatrix} SI_1 & 0 & 0 & 0 \\ 0 & SI_2 & 0 & 0 \\ 0 & 0 & CI_1 & 0 \\ 0 & 0 & 0 & CI_2 \end{vmatrix}\right)^{1/4}, \quad \mathbf{SuTI}_2^{(IV)} \equiv \left(\begin{vmatrix} SI_1 & 1 & 1 & 1 \\ 1 & SI_2 & 1 & 1 \\ 1 & 1 & CI_1 & 1 \\ 1 & 1 & 1 & CI_2 \end{vmatrix}\right)^{1/4} \tag{3.190}$$

$$\mathbf{SuTI}_3^{(IV)} \equiv \left(\begin{vmatrix} SI_1 & 1 & 0 & 0 \\ 1 & SI_2 & 0 & 0 \\ 0 & 0 & CI_1 & 1 \\ 0 & 0 & 1 & CI_2 \end{vmatrix}\right)^{1/4}, \quad \mathbf{SuTI}_4^{(IV)} \equiv \left(\begin{vmatrix} SI_1 & 0 & 1 & 1 \\ 0 & SI_2 & 1 & 1 \\ 1 & 1 & CI_1 & 0 \\ 1 & 1 & 0 & CI_2 \end{vmatrix}\right)^{1/4} \tag{3.191}$$

that does not depend on the way in which the SI and CI blocks are considered firstly.

Instead, when the diagonal order of blocks SI and CI is to be taken into account the next **SuT** indices may be constructed:

$$\mathbf{SuTI}_5^{(IV)} \equiv \left(\begin{Vmatrix} SI_1 & 1 & 0 & 0 \\ 1 & SI_2 & 0 & 0 \\ 0 & 0 & CI_1 & 0 \\ 0 & 0 & 0 & CI_2 \end{Vmatrix} \right)^{1/4}, \ \mathbf{SuTI}_6^{(IV)} \equiv \left(\begin{Vmatrix} SI_1 & 0 & 1 & 1 \\ 0 & SI_2 & 1 & 1 \\ 1 & 1 & CI_1 & 1 \\ 1 & 1 & 1 & CI_2 \end{Vmatrix} \right)^{1/4} \quad (3.192)$$

$$\mathbf{SuTI}_7^{(IV)} \equiv \left(\begin{Vmatrix} SI_1 & 0 & 0 & 0 \\ 0 & SI_2 & 0 & 0 \\ 0 & 0 & CI_1 & 1 \\ 0 & 0 & 1 & CI_2 \end{Vmatrix} \right)^{1/4}, \ \mathbf{SuTI}_8^{(IV)} \equiv \left(\begin{Vmatrix} SI_1 & 1 & 1 & 1 \\ 1 & SI_2 & 1 & 1 \\ 1 & 1 & CI_1 & 0 \\ 1 & 1 & 0 & CI_2 \end{Vmatrix} \right)^{1/4} \quad (3.193)$$

The other combinations reduce to that already presented through relations (3.190)–(3.193) or are unacceptable from the super-topological index above properties. These super-topological indices were so afar applied with limited performances so waiting for further successful consecration (Putz et al., 2008b).

However, further involvement of the distance-based information conveyed by $T(N)$ topological invariants given by Eqs. (3.180), (3.183), and (3.185) may be unfolded based on the natural partition of the atoms determined by the compact finite differences (CFD) formulations (3.172) and (3.173) of electronegativity and chemical hardness involved in energetic combinations and of their correlation hierarchy, as in next presented.

3.2.10.4 Topo-Reactivity by Timişoara-Parma Rule

One establishes the hierarchy of the above CFD methods (see Table 3.5) through assessing the best correlation model between the two forms of energies, namely of the *π-parabolic energy* combining the electronegativity and chemical hardness with the pi-electrons N_π (in molecular frontier orbitals); the present discussion follows (Putz, 2011c)

$$E_{\pi}(\chi,\eta) \cong -\chi\ N_{\pi} + \eta\ N_{\pi}^{2} \tag{3.194}$$

with the *π-reactive energies*

$$E_{\pi}(molecule) \cong E_{Bind}(molecule) + E_{Heat}(molecule) - E_{Total}(molecule) \tag{3.195}$$

The correlation results provide the hierarchy that ordinate the CFD models according with their richness in electronic frontier orbitals' information.

The founding CFD hierarchy should be further respected when implemented to "colour" the various parts of a molecules with electronegativity and chemical hardness information, according with the so-called *reactivity coloring (Timisoara-Parma) rule; the present discussion also follows* (Putz et al., 2013b-c):

The chemical descriptor (χ or η) values are distributed over all nodes of a molecule, grouped on successive reticules starting from the "central" most populated ones with bonding and nodes (frontier) electrons, while considering the equivalent/equidistant reticules till the exhausting the molecular bonding space, with paralleling decreasing CFD values of a descriptors (one common for all nodes of a reticule) until the exhausting the CFD models, being the last one considered for all remaining equivalent reticules of molecular space, if any.

Figure 3.12 gives such an example for a working PAH molecule, pentacene, with a working CFD hierarchy (Putz et al., 2013b-c) established through the identified CFD hierarchies of the pi-withreactive energies for a series of PAH molecules (Benzene, Naphthalene, Anthracene, Phenanthrene, Tetracene, Pyrene, Pentacene, Picene, Perylene, Benzo-a-pyrene, Quaterrylene, Coronene, Hexabenzocoronene, Dicoronylene) as reported in the Table 3.7.

Although (arbitrary) graph coloring has been previously employed in the literature for the fast-determination of the number of peaks and relative intensities of fullerenes in ^{13}C-NMR spectra (Ori & D'Mello, 1992, 1993; Ori et al., 2011), the proposed graph coloring method exhibits an enhanced chemical significance by assigning to each graph vertex v_i two new colors representing a portion of the overall molecular electronegativity (χ_i)

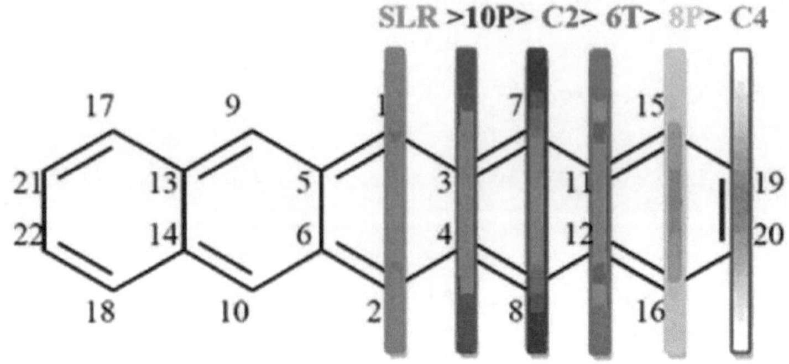

FIGURE 3.12 The chemical graph, with N=22 vertices, associated to hydrogen-depleted pentacene molecule. The coloring hierarchy of the CFD models (of Table 3.5) is indicated by rectangles colored according to distance of the vertices from the central atoms v_1, v_2 characterized by SLR data. Rectangles respect the molecular plane of symmetry, for example, v_5, v_6 and v_3, v_4 pairs respectively feature 10P and C2 data and so on (Putz et al., 2013b).

and chemical hardness (η_i). The distributions of the χ_i and η_i values with respect to the CFD hierarchy is based on the topological distance of a vertex v_i from the *central* axis of the molecule, see Figure 3.12. Based on the Timisoara-Parma method for enriching the topological descriptors with reactivity (frontier orbital) information, *the conventional or newly proposed topological index is "translated" into a topo-reactive index by considering the adjacency or the topological matrix of that index and filling it with molecular electronegativity values for nodes distributed in a "spectrum" according to the derived CFD prescription (see Table 3.5) that are employed to determine the "coloring" name. Then, one takes the square root of any coupling of the nodes within the molecule that form bonds and successive increasing order of roots for the coupling paths of the nodes in the molecule.*

From now on, the conventional or newly proposed topological index is "translated" into a *topo-reactive* one by considering the adjacency or the topological matrix of the given index and fulfill it with molecular electronegativity values for nodes according with (Timisoara-Parma) rule when the topological index (*T*) is basically employed (Putz et al., 2013b-c)

TABLE 3.7 The Hierarchies of the Correlation Factors (R) for the Compact Finite Difference (CFD) Models (of Table 3.5) for Electronegativity and Chemical Hardness from the Regression of the π-Reactive and Parabolic Energies, Eqs. (9.134) and (9.135), for a Set of PAH Molecules*

E_{PI}REACT	E_PC2	E_PC4	E_PC6	E_PSP	E_P6T	E_P8T	E_P8P	E_P10P	E_PSLR
R	0.97353	0.97208	0.96926	0.97149	0.97257	0.97064	0.97255	0.97576	0.97969

*PAH molecules (Benzene, Naphthalene, Anthracene, Phenanthrene, Tetracene, Pyrene, Pentacene, Picene, Perylene, Benzo-a-pyrene, Quaterrylene, Coronene, Hexabenzocoronene, Dicoronylene) (Putz et al., 2013b-c), computed using the semi-empirical PM3 method (Hypercube, 2002).

$$T(\chi) = \det{}^{1/N_\pi}\left\|\hat{T}_\chi\right\| \tag{3.196}$$

then take square roots for any couplings of whatever nodes within the molecule forming bonds, and successive increasing order of roots for any coupling paths of nodes from molecule; the same is doing for chemical hardness

$$T(\eta) = \det{}^{1/N_\pi}\left\|\hat{T}_\eta\right\| \tag{3.197}$$

to jointly form the topological-reactivity energy, say

$$E^T_{(\chi,\eta)} = -T(\chi)N_\pi + T(\eta)N_\pi^2 \tag{3.198}$$

as a topo-reactivity colored counterpart for the above pi-parabolic-energy, Eq. (3.194).

However, the key to our method consists of computing the colored or *reactive* versions of $\hat{T}_\chi$ and $\hat{T}_\eta$ for the distance matrix $\hat{D}$, as required by the above computational scheme, see Table 3.8. These operators, also-called *topo-electronegativity* and *topo-chemical hardness* matrices, properly carry the chemical information leading to the *reactive forms of the topological index* for electronegativity $T(\chi)$ and chemical hardness $T(\eta)$. The extraction of both the $T(\chi)$ and $T(\eta)$ reactive indices from the newly defined operators $\hat{T}_\chi$ and $\hat{T}_\eta$ may follow several mathematical routes. One is described in what follows (Putz et al., 2013b).

One starts from the fact that both matrices $\hat{T}_\chi$ and $\hat{T}_\eta$ are symmetric preserving the symmetry and the null-main diagonal of the $\hat{D}$ template since rooting in an invariant topological index. The vertex coloring allows for the calculation of both colored versions of the topo-reactive descriptors $T(\chi)$ and $T(\eta)$, given in terms of the respective matrices $\hat{T}_\chi$ and $\hat{T}_\eta$ by Eqs. (3.196) and (3.197), respectively. First, the $t(\chi)_{ij}$ generic element of $\hat{T}_\chi$ considers the shortest path between v_i and v_j (e.g., $v_i \rightarrow v_\alpha \rightarrow v_\beta \ldots v_\gamma \rightarrow v_\delta \rightarrow v_j$) with the length of this path is the chemical distance $d_{i,j}$ that appears in the topological index definition. The bond $v_i \rightarrow v_\alpha$ is colored by the *products* $\chi_i\chi_\alpha$ and $\eta_i\eta_\alpha$, which indicates that there is a certain level of interaction present between the two atoms. When the path reaches the v_β node, the color becomes $\chi_i\chi_\alpha\chi_\beta$ until the $d_{i,j}$ edges are all colored by the products

TABLE 3.8 Top: Connectivity Lists of the Hydrogen-Depleted Pentacene Molecule of Figure 3.12, With the Associated Symmetric Matrix $\hat{D}$ of the Chemical Distances in Bottom (Putz et al., 2013b)*

Site	Connected Sites			Site	Connected Sites		
v_1	3	5	–	v_{12}	8	11	16
v_2	4	6	–	v_{13}	9	14	17
v_3	1	7	4	v_{14}	10	13	18
v_4	2	3	8	v_{15}	11	19	–
v_5	6	9	1	v_{16}	20	12	–
v_6	2	5	10	v_{17}	21	13	–
v_7	3	11	–	v_{18}	14	22	–
v_8	4	12	–	v_{19}	15	20	–
v_9	5	13	–	v_{20}	19	16	–
v_{10}	6	14	–	v_{21}	17	22	–
v_{11}	7	15	12	v_{22}	21	18	–

$\hat{D}$	v_1	v_2	v_3	v_4	v_5	v_6	v_7	v_8	v_9	v_{10}	v_{11}	v_{12}	v_{13}	v_{14}	v_{15}	v_{16}	v_{17}	v_{18}	v_{19}	v_{20}	v_{21}	v_{22}
v_1	0	3	1	2	1	2	2	3	2	3	3	4	3	4	4	5	4	5	5	6	5	6
v_2	3	0	2	1	2	1	3	2	3	2	4	3	4	3	5	4	5	4	6	5	6	5
v_3	1	2	0	1	2	3	1	2	3	4	2	3	4	5	3	4	5	6	4	5	6	7
v_4	2	1	1	0	3	2	2	1	4	3	3	2	5	4	4	3	6	5	5	4	7	6
v_5	1	2	2	3	0	1	3	4	1	2	4	5	2	3	5	6	3	4	6	7	4	5
v_6	2	1	3	2	1	0	4	3	2	1	5	4	3	2	6	5	4	3	7	6	5	4

TABLE 3.8 Continued

$\hat{D}$	v_1	v_2	v_3	v_4	v_5	v_6	v_7	v_8	v_9	v_{10}	v_{11}	v_{12}	v_{13}	v_{14}	v_{15}	v_{16}	v_{17}	v_{18}	v_{19}	v_{20}	v_{21}	v_{22}
v_7	2	3	1	2	3	4	0	3	4	5	1	2	5	6	2	3	6	7	3	4	7	8
v_8	3	2	2	1	4	3	3	0	5	4	2	1	6	5	3	2	7	6	4	3	8	7
v_9	2	3	3	4	1	2	4	5	0	3	5	6	1	2	6	7	2	3	7	8	3	4
v_{10}	3	2	4	3	2	1	5	4	3	0	6	5	2	1	7	6	3	2	8	7	4	3
v_{11}	3	4	2	3	4	5	1	2	5	6	0	1	6	7	1	2	7	8	2	3	8	9
v_{12}	4	3	3	2	5	4	2	1	6	5	1	0	7	6	2	1	8	7	3	2	9	8
v_{13}	3	4	4	5	2	3	5	6	1	2	6	7	0	1	7	8	1	2	8	9	2	3
v_{14}	4	3	5	4	3	2	6	5	2	1	7	6	1	0	8	7	2	1	9	8	3	2
v_{15}	4	5	3	4	5	6	2	3	6	7	1	2	7	8	0	3	8	9	1	2	9	10
v_{16}	5	4	4	3	6	5	3	2	7	6	2	1	8	7	3	0	9	8	2	1	10	9
v_{17}	4	5	5	6	3	4	6	7	2	3	7	8	1	2	8	9	0	3	9	10	1	2
v_{18}	5	4	6	5	4	3	7	6	3	2	8	7	2	1	9	8	3	0	10	9	2	1
v_{19}	5	6	4	5	6	7	3	4	7	8	2	3	8	9	1	2	9	10	0	1	10	11
v_{20}	6	5	5	4	7	6	4	3	8	7	3	2	9	8	2	1	10	9	1	0	11	10
v_{21}	5	6	6	7	4	5	7	8	3	4	8	9	2	3	9	10	1	2	10	11	0	1
v_{22}	6	5	7	6	5	4	8	7	4	3	9	8	3	2	10	9	2	1	11	10	1	0

$\chi_i\chi_\alpha\chi_\beta\cdots\chi_\delta\chi_\gamma\chi_j$ and $\eta_i\eta_\alpha\eta_\beta\cdots\ \eta_\delta\eta_\gamma\eta_j$, respectively; the $t(\eta)_{ij}$ entries are analogously derived for $\hat{T}_\eta$.

Due to obvious dimensional reasons, the final formulae for the elements of the *topo-reactive matrices* (also-called reactive members or colored members or weights) consider the geometric average of the $(1+d_{ij})$ colors of the atoms present in the path $v_i \rightarrow v_j$ (Putz et al., 2013b):

$$t(\chi)_{ij} = \left[\hat{T}_\chi\right]_{ij} = \left(\prod_{v_i \xrightarrow{\alpha} v_j} \chi_\alpha\right)^{1/(1+d_{ij})} \tag{3.199}$$

$$t(\eta)_{ij} = \left[\hat{T}_\eta\right]_{ij} = \left(\prod_{v_i \xrightarrow{\alpha} v_j} \eta_\alpha\right)^{1/(1+d_{ij})} \tag{3.200}$$

In the above expressions, index α covers the $(1+d_{ij})$ atoms in the path; the forms of the reactive members in Eqs. (3.199) and (3.200) illustrate the

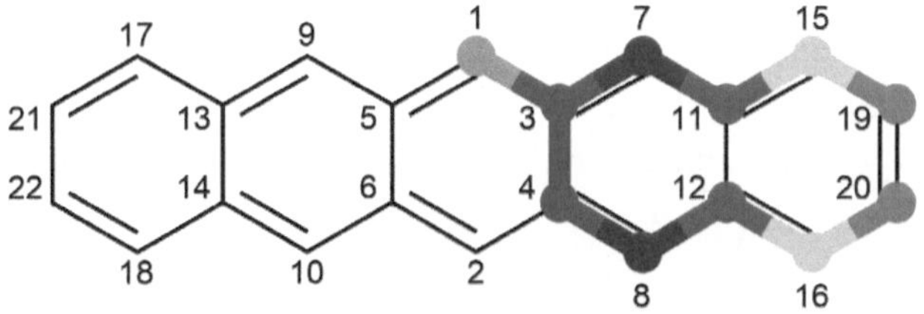

SLR>10P>C2>6T>8P>C4

$$t(\chi)_{1,3} = \sqrt{\chi_1 \times \chi_3} = 2.59999$$

$$t(\chi)_{1,7} = \sqrt[3]{\chi_1 \times \chi_3 \times \chi_7} = 3.14901$$

$$t(\chi)_{1,4} = \sqrt[3]{\chi_1 \times \chi_3^{\,2}} = 2.55798$$

$$t(\chi)_{1,8} = \sqrt[4]{t(\chi)_{1,4}^3 \times \chi_8} = 2.96530$$

$$t(\chi)_{1,12} = \sqrt[5]{t(\chi)_{1,8}^4 \times \chi_{12}} = 2.94976$$

$$t(\chi)_{1,20} = \sqrt[7]{\chi_{20} \times \prod_\lambda \chi_\lambda} = 3.06979 \quad \lambda = 1,3,4,8,12,16$$

FIGURE 3.13 The colors of the molecular paths $v_i \rightarrow v_j$ are determined by the CFD hierarchy within the Wiener indexing framework. Examples of $t(\chi)_{1j}$ reactive members are computed here for pentacene. Entry $t(\chi)_{13}$ and the SRL and 10P parameters are considered for v_1 and v_3, respectively. By symmetry, the $t(\chi)_{15}$, $t(\chi)_{24}$, and $t(\chi)_{26}$ elements are equal. The $t(\chi)_{17}$ colored member includes v_1, v_3, v_7 data, whereas $t(\chi)_{14}$ uses v_1, v_3, $v_{4=}v_3$. $t(\chi)_{1,20}$ includes v_1, v_3, v_4, v_8, v_{12}, v_{16}, v_{20} data, and reactive members $t(\eta)_{1j}$ are computed in similar ways (Putz et al., 2013b).

essential role played by *molecular topological information* stored in the minimal path $v_i \rightarrow v_j$ for predicting chemical-physical properties of the molecule. Figure 3.13 shows some computational sequences of reactive members $t(\chi)_{ij}$ for pentacene, using the hierarchy of colored-reactivity as provided by Table 3.9 for electronegativity; the same will be for $t(\eta)_{ij}$ topo-reactive weights (Putz et al., 2013b-c).

Clearly, both topo-reactive invariants $T(\chi)$ and $T(\eta)$ given in Eqs. (3.196) and (3.197) are the generic representations of a wide class of topological invariants obtainable from the matrices $\hat{T}_\chi$ and $\hat{T}_\eta$ using more elaborate mathematical operations.

Note that the obtained working matrices do not need to be further normalized with dimensional factors to become dimensionless, which is due to the generated matrix containing energies that will behave in a manner similar to the energy matrices in quantum theory. The associate diagonalization followed by diagonal summation will lead to a topo-electronic-like energy and take into account the primary information contained therein, i.e., connectivity (by nodes and paths) and reactivity (by electronegativity and chemical hardness and their CFD schemes). Therefore, one will have, for instance, Wiener-electronegativity and Wiener chemical hardness indices for comparing the Wiener (N_π) shapes for the given molecules to their observed/experimental/measured biological activity (toxicity, carcinogenicity, etc.).

3.2.11 LOGISTIC QSAR

Unlike the time-independent QSARs, the temporal analysis for the chemical-biological species interaction currently requires *further conceptual* tools for modeling. The present *logistic-QSAR* approach parallels the substrate-enzyme reaction for the consecrated Michaelis-Menten mechanism of the Chapter 1 of this volume, see also the highlights of the first 100 years of enzyme kinetics by (Brown, 1892, 1902; Henri, 1901, 1902; Michaelis & Menten, 1913; Haldane, 1930; Pauling, 1946; Voet & Voet, 1995; Cornish-Bowden, 1999; Copeland, 2000; Schnell & Maini, 2003), to provide the ligand-receptor (L-R) logistic kinetics in general and the ligand progress curve $L(t)$ in particular, depending on the 50%-effective concentration (EC_{50}) dose for the recorded activity (A) for a given species:

TABLE 3.9 Electronegativity Values (in electron-Volts, eV) for the Pentacene Using Eq. (3.172) and the Numerical Schemes in Table 3.5, Aiming the Topo-Reactive Coloring As in Figure 3.12, and Further Used to Compute the Coloring Waits in the Figure 3.13*

	χC2	**χC4**	**χC6**	**χSP**	**χ6T**	**χ8T**	**χ8P**	**χ10P**	**χSLR**
Pentacene	4.619324	4.61212	4.611329	1.510958	2.888452	2.576307	2.494174	2.476	2.730185

*Putz et al. (2013b-c).

$$A = \ln\left(\frac{1}{EC_{50}}\right) \tag{3.201}$$

Because $L(t)$ usually also depends on kinetic parameters such as the initial chemical concentration L_0 and the maximum biological uptake β_{max}, eventually under the natural exponential form, i.e., the nuclear radioactivity equation – here considered as; the present discussion follows (Putz & Putz, 2013b)

$$L(t) \cong L_0 \exp\left(-\frac{\beta_{max} t}{EC_{50}}\right) \tag{3.202}$$

Its complete unfolding requires *further computational* ligand-receptor activity modeling. This may be performed by using QSAR methodology (Topliss & Costello, 1972; Topliss & Edwards, 1979; Topliss, 1983), which can be achieved by correlating the observed activity with the above reactivity indices within multi-linear regressions yielding computed or *predicted activity* (A^*). However, for reasons that will be immediately revealed, only linear equations will be considered:

$$A^* = A^*(\chi \vee \eta \vee \omega \vee \pi) \tag{3.203}$$

Following Eq. (3.201), one can immediately make the correspondence between the initial ligand concentration and computed activity

$$L_0 = \exp\left(-\left\|A^*\right\|\right) \tag{3.204}$$

with $\|\bullet\|$ accounting for the algebraic (Banach) norm in the chemical space (with the dimension equal to the cardinal of the set of chemicals considered in the QSAR), see also Eq. (3.62), while for the EC_{50} parameter, a similar relationship holds at the level of recorded activities, namely

$$EC_{50} = \exp\left(-\left\|A\right\|\right) \tag{3.205}$$

For maximum biological uptake (via interaction with a ligand), one can employ the working definition

$$\beta = -\frac{d}{dt}L(t) \rightarrow -\frac{L_t - L_0}{\Delta t} \tag{3.206}$$

With complete consumption of the ligand ($L_t \rightarrow 0$), a maximum value is achieved.

$$\beta_{\max} = \frac{L_0}{\Delta t_\infty} \tag{3.207}$$

The "infinite" time interval may be shaped by considering the re-scaling first, reversing Eq. (1.56)

$$t = e^{\frac{1}{1-\tau}} - e \rightarrow \begin{cases} 0 \ldots \tau \rightarrow 0 \\ \infty \ldots \tau \rightarrow 1 \end{cases} \tag{3.208}$$

followed by differentiation

$$dt = \frac{d\tau}{(1-\tau)^2}\exp\left(\frac{1}{1-\tau}\right) \tag{3.209}$$

For practical considerations, the interval (0, ∞) projected into (0, 1) can safely use the setting $\tau = 1/2$ as a sufficient condition for maximal biological uptake, so that the associated working time interval (the so-called "*receptor time*") is $\Delta t_\infty = 4e^2$, using Eqs. (3.204) and (3.207) along the working parameters one yields (Putz & Putz, 2013b):

$$\beta_{\max} = \frac{1}{4}\exp\left(-\left\|A^*\right\| - 2\right) = \frac{L_0}{4e^2} \tag{3.210}$$

The working ligand progress curve (3.202) directly depends on recorded and computed activity, which in turn depend on the chemical reactivity indices considered, employing what can be called *quantitative reactivity-activity relationships* (QRAR). Yet, computationally, for each density functional theory (DFT) framework, see the Volume I of the present five-volume series (Putz, 2016a) and each reactivity index (see Section 3.2.9), a different progress curve for a given ligand-receptor interaction is obtained. The immediate inference can be made that faster consumption of $L(t)$ is involved for the more reactive index, and therefore, a more preeminent reactivity principle is associated. This allows for a given case study (Putz & Putz, 2013b):

- to formulate chemical-biological interactions with the help of conceptual-computational reactivity indices of electronegativity, chemical hardness, electrophilicity, and chemical power;
- to obtain a hierarchy of the allied chemical principles (electronegativity, chemical hardness, electrophilicity, and chemical power) for a given pool of molecules of certain species and to check whether they are maintained across many species.

This algorithm is schematically presented in Figure 3.14.

However, nowadays, in the post-genomic era, the development of kinetic models that allow simulation of complicated metabolic pathways and protein interactions is becoming increasingly important (Noble, 2002; Crampin et al., 2004). Yet, unfortunately, the difference between an *in vivo* biological system and homogeneous *in vitro* conditions is large, as shown by Schnell & Turner (2004). Accordingly, mathematical treatments of biochemical kinetics have been developed from the law of mass action *in vitro,* but the modifications required to bring them in line with stochastic *in vivo* situations are still under development (Savageau, 1969, 1999; Turner et al., 2004).

In this line of research, generally, the mechanism of the biological activity produced by a substance usually involves the combination of interactions between the molecules of that substance, called the effector or ligand (L), with a receptor (R), a protein, a biologically macromolecule, or a complex of macromolecules within the cell. The intensity of the biological action is illustrated in an ordinary way as the logarithm of the inverse of the concentration C to produce a specific biological answer, see Eq. (3.201). Because the C_{50} concentration (or, more often called as EC_{50} the molar *effective* concentration that produces 50% of the maximum biological activity) is used in many situations, it can be shown that the biological activity A is proportional to the affinity of the molecule or ligand L_i for the receptor R, which is at the basis of the explicated biological action. However, the most rational hypothesis concerning the mode of action of bioactive substances presumes that biological activity produced by a ligand L is proportional to the complexation degree of the receptor R with L. In this situation, the presumed biological activity of α%

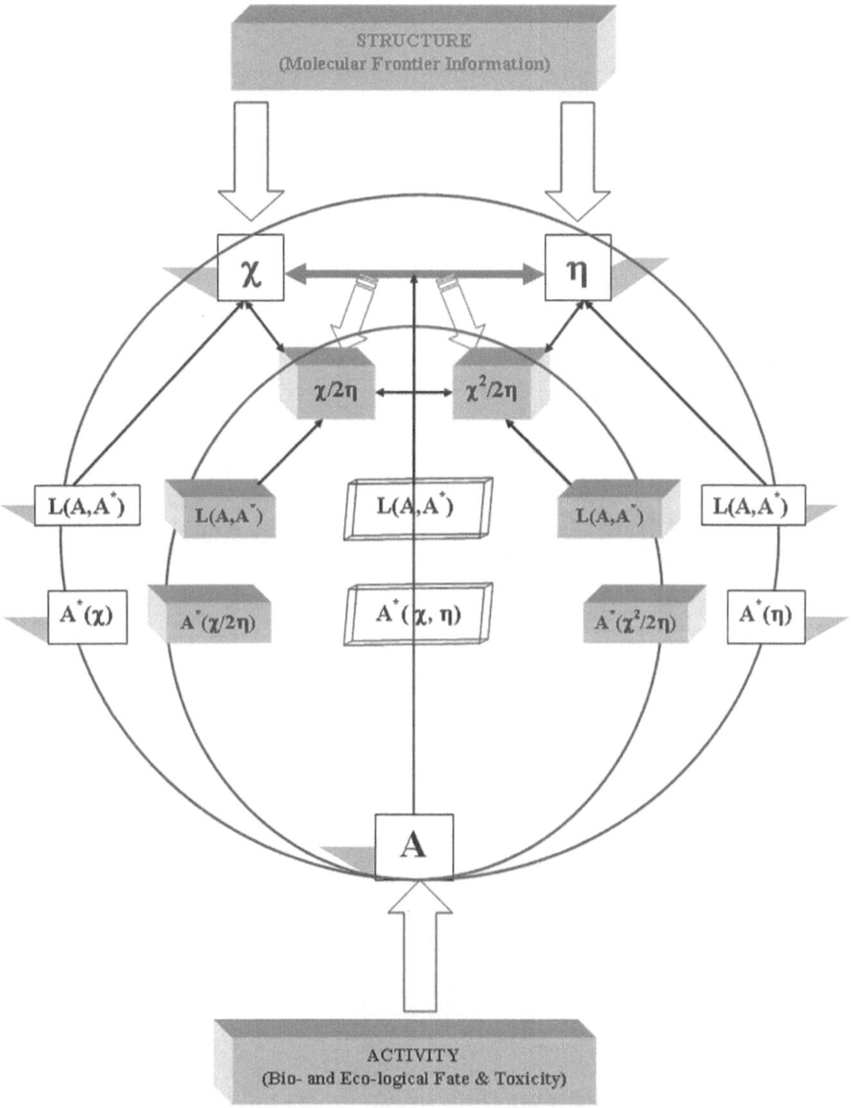

FIGURE 3.14 The workflow of Structure–Activity relationships involving conceptual and computational density functional theory through molecular frontier information (e.g., HOMO & LUMO), which are primarily employed as electronegativity (χ) and chemical hardness (η) indices and are then combined to form chemical power ($\chi/2\eta$) and electrophilicity ($\chi^2/2\eta$) reactivity measures. These values are correlated with observed biological and ecological activity (A) to provide the QSAR models (A^*) to finally produce ligand progress curves (L) that provide a hierarchy of chemical reactivity principles involved in biological activity within a given DFT computational framework and species of interest (Putz & Putz, 2013b).

TABLE 3.10 The Face-To-Face Ligand-Receptor (L-R) and Substrate-Enzyme (S-E) Kinetics*

Property	***L – R* kinetics**	***S – E* kinetics**
Reaction	$R + L \overset{K}{\leftrightarrow} C^{*}$	$E + S_i \underset{k_{-1}}{\overset{k_1}{\leftrightarrow}} ES \overset{k_2}{\rightarrow} E + P$
Species	R	E
Chemicals	L	S
Equation	$\frac{\alpha}{100} = \frac{[L]}{[L] + K}$	$\frac{v}{V_{\max}} = \frac{[S]}{[S] + K_M}$
Constant	$K = \frac{[R][L]}{[C^{*}]}$	$K_M = \frac{k_{-1} + k_2}{k_1}$
Graph	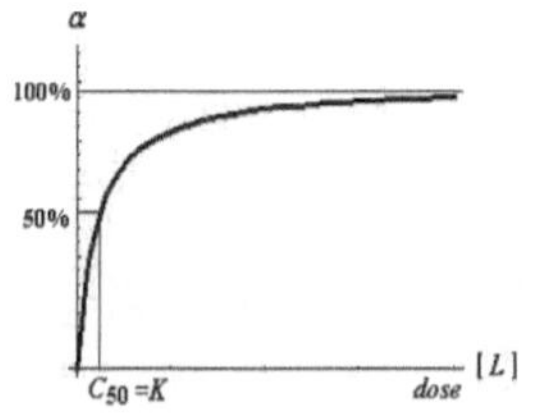	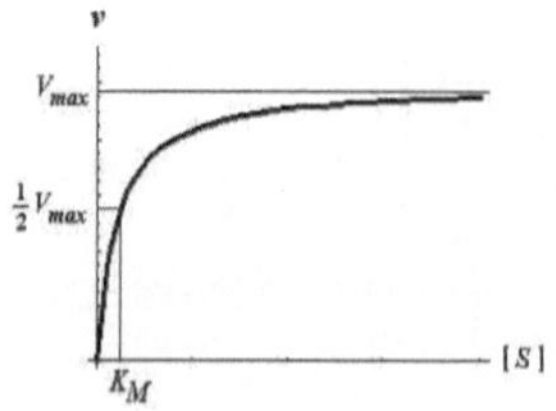

*Lacrămă et al. (2008); Putz (2012a-b); Putz & Putz (2013b).

(compared to a maximum of 100%) is produced by the concentration $[L]$ of ligand, which closely agrees with substrate-enzyme kinetics modeling, see Table 3.10.

This way, we may safely generalize the chemical-biological-kinetics (of Section 1.2) in terms of the general rate of biological uptake β while respecting the chemical concentration $[L]$

$$\beta = \frac{\beta_{\max}[L]}{[L] + EC_{50}} \tag{3.211}$$

Recognizing that the temporal link in Eq. (3.211) between the biological activity and chemical concentration is represented as

$$\beta = -\frac{d}{dt}[L](t) \tag{3.212}$$

the full temporal version can be widely formulated as (Voet & Voet, 1995; Cornish-Bowden, 1999; Copeland, 2000; Schnell & Maini, 2003), paralleling the Michaelis-Menten formulation from Eq. (1.16)

$$-\frac{d}{dt}[L](t)=\frac{\beta_{\max}[L](t)}{[L](t)+EC_{50}} \tag{3.213}$$

However, by following the considerations of the Section 1.3 one may further generalizes Eq. (3.213) such that the range of reaction rates is expanded and provides the logistic kinetic equation (1.37), reloaded here as the logistic expression

$$-\frac{1}{\beta_{\max}}\frac{d}{dt}[L](t)=1-e^{-\frac{[L](t)}{EC_{50}}} \tag{3.214}$$

Also here, worth noted that under initial conditions, the logistic equation (3.214) gives an initial velocity of reaction (β_0^*) that is uniformly higher than that calculated by Michaelis-Menten (3.213) at all initial concentrations of the ligand, except for the case where $[L_0] \to 0$, when both are zero (see Figure 1.1 with correspondences of Table 3.10).

Actually, by working out the same procedure as exposed in Section 1.3.3, one arrives at the logistic progress curve for ligand consumption in an analytically elementary form (Putz & Putz, 2013b):

$$[L]_L(t)=EC_{50}\ln\left(1+e^{-\frac{\beta_{\max}t}{EC_{50}}}\left(e^{\frac{[L_0]}{EC_{50}}}-1\right)\right) \tag{3.215}$$

This time-dependent solution (3.215) substitutes an elementary logarithmic dependency for the *W*-Lambert function. It is nevertheless remarkable that the solution of a generalized logistic kinetic version of the Michaelis-Menten instantaneous equation provides an analytically exact solution. It clearly reduces to the above equation (3.213) in the first order expansion of the chemical concentration time evolution with respect to the 50-effect concentration (EC_{50}) observed.

The original chemical-biological-kinetics (3.213) gave the analytical solution for the actual working kinetics that provides the logistical solution (3.215) whose reliability was previously tested on various enzyme kinetics mechanisms. This testing produced remarkable results that constituted a trusted background for employing it in the currently envisaged ecotoxicological studies.

Remarkably, rearranging the logistic solution of the chemical-biological interaction (3.215) under the equivalent form (Putz & Putz, 2013b)

$$e^{\frac{[L]_L(t)}{EC_{50}}} - 1 = e^{-\frac{\beta_{max} t}{EC_{50}}}\left(e^{\frac{[L_0]}{EC_{50}}} - 1\right) \tag{3.216}$$

provides for a relatively higher concentration EC_{50} >> (specific to environmental toxicological fate studies) for the working equation employed in Eq. (3.202) as the basis for advancing quantitative reactivity-activity relationships.

As such, for a *logistic-spectral analysis*, based on the molecular *M*-data for *N*-chemical species and toxicity activities of Table 3.1, the next steps are considered in producing the chemical-biological progress curves according to which chemical species "dissolve" in biological/environmental receptors, for each envisaged molecular-set or model of structural parameters or computational framework:

- For the activities of Table 3.1, the Spectral-SAR (Putz & Lacrămă, 2007; Putz & Putz, 2011; Putz, 2012a-b) relationships are formulated in the optimum form abstracted from (3.3), namely

$$\left|A^*\right\rangle^{ENDPOINT} = b_0\left|X_0\right\rangle + b_1\left|X_1\right\rangle + ... + b_M\left|X_M\right\rangle \tag{3.217}$$

- the predicted spectral norm is computed in the *N*-dimensional chemical space:

$$\left\|A^*\right\rangle\| = \sqrt{\left\langle A^* \middle| A^*\right\rangle} = \sqrt{\sum_{k=0}^{M} b_k^2 \left\langle X_k \middle| X_k\right\rangle} = \sqrt{\sum_{k=0}^{M} b_k^2 \left(\sum_{j=1}^{N} x_{kj}^2\right)} = \sqrt{\sum_{j=1}^{N} \left(A_j^*\right)^2} \tag{3.218}$$

- the initial chemical concentration within the logistical chemical-biological progress picture is related to the predicted S-SAR activity norms (3.218), according to (3.205);
- in the same mechanistic line of the chemical-biological interaction framework, the maximum biological effect is seen as the decrease of the initial chemical concentration in the effector time, see Eq. (3.210);
- along the EC_{50} parameter computed following the generalization of (3.201) towards the algebraic version (3.205), one has all the "ingredients" for progress curves that represent the logistical consumption/

metabolization/fate of the chemical species following interaction with the biological species/ecological environment; these are written in the working form, first with the aim of (3.210) and (3.215), as (Putz & Putz, 2013b):

$$[L]_L(t) = EC_{50} \ln\left(1 + e^{-\frac{1}{4e^2}\frac{[L_0]}{EC_{50}}t}\left(e^{\frac{[L_0]}{EC_{50}}} - 1\right)\right) \tag{3.219}$$

Finally, by replacing the (3.204) and (3.205) relationships, the (observed) activity–(predicted) activity form of the ligand progress curve

$$[L]_L(t) = e^{-\||A\rangle\|} \ln\left(1 + e^{-\frac{\exp(\||A\rangle\| - \||A^*\rangle\|)}{4e^2}t}\left(e^{\exp(\||A\rangle\| - \||A^*\rangle\|)} - 1\right)\right) \tag{3.220}$$

or toxicity-activity expressionl

$$[L]_L(t) = EC_{50} \ln\left(1 + e^{-\frac{\exp(-\||A^*\rangle\|)}{4e^2 EC_{50}}t}\left(e^{\frac{\exp(-\||A^*\rangle\|)}{EC_{50}}} - 1\right)\right) \tag{3.221}$$

are created for each structural parameter (reactivity indices considered) computed for each DFT framework considered and are correlated with each set of recorded species activities. Through this comparison, one may draw conclusions based upon the chemical reactivity principles governing the specific chemical-biological interaction under study.

Having this logistic-QSAR approach formulation one actually "closed the circle" of conceptual QSAR formulations on chemical orthogonal spaces since containug the influence of molecular structural causes on recorded biological activity in various relationships forms and variants til lteh temporal dependence through appealing to the logistic substrate-enzyme kinetics (of Chapter 1 of the present volume) here adapted/translated to the ligand-receptor paradigm, so that consistently circumscribing the present structure-activity relationships models to the linear, non-linear and logistic models of chemical-biological bondings. For most of the conceptually presented SAR models concrete QSAR realization will be presented in the next sections combing various chemical and biological species towards modern assessing of the chemical-biological bonding models.

3.3 QU-SAR CASE STUDIES ON CHEMICAL ORTHOGONAL SPACE

The following QSAR case studies have multi-fold characteristics:

- All combines the quantum structural information with recorded or otherwise determined bio/eco/carcino/toxico/pharmacology activities, thus justifying also by this feature the Quantum-SAR (Qu-SAR) general approach;
- Introduces, whenever possible, in the broad field of research and interest the applied Qu-SAR is addresses, i.e., at the bio/eco/carcino/toxico/pharmacology level;
- They widely correspond and are based on the Qu-SAR methods conceptually exposed in Section 3.2, here applied on paradigmatic series of interests
- The chemical-biological binding model is searched for being unveil, through establishing the structural causes hierarchies (either by spectral-SAR approach and/or with the aid of frontier orbitals' diagrammatic electronic dynamics) towards producing the observed/measured activity as the manifested effect of chemical-biological interaction, such that fulfilling the QSAR-OECD-5th demand in revealing of the mechanism of action to be further used and considered for the given specific quantum interaction aiming drug design or more complex modeling.

3.3.1 SPECTRAL-SAR ECOTOXICOLOGY (ON T. pyriformis)

Quite often, despite the tendency to submit a large class of descriptors to a QSAR analysis, this is not the best strategy (Klopman et al., 1995), at least in ecotoxicology, and whenever a specific mode of action or the elucidation of the causal mechanistically scheme is envisaged; the present discussion follows (Putz & Lacrămă, 2007).

More focused studies in ecotoxicology, and especially regarding *T. pyriformis*, have found that hydrophobicity (*LogP*) and electrophilic phenomena play particular places in explaining the ecotoxicology of the species. While hydrophobicity describes the penetration power of xenobiotics though biological membranes, other descriptors may reflect the electronic and specific interaction between the ligand and target site of receptor (Putz & Lacrămă, 2007). Moreover, it was convincingly argued that the classical Hammett constant can be successfully rationalized by a pure structural index as the energy of the lowest unoccupied molecular orbital (*LUMO*) stands for (Cronin et al., 2002). These facts open the attractive perspective of considering the ecotoxicological studies through employing the Hansch-type structure-activity expansion (3.63) thus also providing enough information from transport, EA and specific interaction at the molecular level, respectively.

However, in the present study, besides considering *LogP* as compulsory descriptor, the molecular polarizability (*POL*) will be considered for modelling the EA for its inherent definition that implies the radius of the electrostatic sphere of electrostatic interaction. This way, the first stage of binding, through the radius of interaction, is accounted; see the Volume II of the present five-volume work (Putz, 2016b). Then, the steric descriptor is chosen here, for generality, as the total molecular energy (E_{TOT}) in its ground state, for the reason that it is calculated at the optimum molecular geometry where the stereo-specificity is included (Putz & Lacrămă, 2007). Under these circumstances the ecotoxic activity to *Tetrahymena pyriformis*, determined in a population growth impairment assay with a 40 h static design and population density measured spectrophotometrically as the endpoint A=Log (1/IGC$_{50}$) (Cronin et al., 2002, 2004; Schultz, 1997, 1999; Schultz et al., 2002, 2003), from a series of xenobiotics of which majority are of phenol type, is in Table 3.11 listed (Putz & Lacrămă, 2007).

It is worth mentioning that the number of compounds is in relevant ratio with the number of descriptors used, according with the so-called Topliss-Costello rule (i.e., $N/M \geq 5$), and that both chemical variability and congenericity are fulfilled since most of them reflect the phenolic toxicity.

The standard QSAR analysis of data of Table 3.11 for all possible models of actions reveals the multivariate equations displayed in Table 3.12, together with their associate statistics (see Sections 2.2 and 2.4.3)

TABLE 3.11 The Series of the Xenobiotics of those Toxic Activities **A**= Log(1/IGC_{50}) were Considered (Cronin et al., 2002, 2004) along Structural Parameters LogP, POL (Å^3), and E_{TOT} (kcal/mol) as Accounting for the Hydrophobicity, Electronic (Polarizability) and Steric (Total Energy at Optimized 3D Geometry) Effects, Respectively*

No.	Compound		*A*	$\|1\rangle$	*Log P*	*POL*	E_{TOT}
	Name	**Formulae**	$\|Y\rangle$	$\|X_0\rangle$	$\|X_1\rangle$	$\|X_2\rangle$	$\|X_3\rangle$
1	methanol	CH_3OH	–2.67	1	–0.27	3.25	–11622.9
2	ethanol	C_2H_5OH	–1.99	1	0.08	5.08	–15215.4
3	butan-1-ol	C_4H_9OH	–1.43	1	0.94	8.75	–22402.8
4	butanone	C_4H_8O	–1.75	1	1.01	8.2	–21751.8
5	pentan-3-one	$C_5H_{10}O$	–1.46	1	1.64	10.04	–25344.6
6	phenol	C_6H_5OH	–0.21	1	1.76	11.07	–27003.1
7	aniline	$C_6H_5NH_2$	–0.23	1	1.26	11.79	–24705.9
8	3-cresol	CH_3-C_6H_4-OH	–0.06	1	2.23	12.91	–30597.6
9	4-methoxiphenol	OH-C_6H_4-O-CH_3	–0.14	1	1.51	13.54	–37976.3
10	2-hydroxyaniline	OH-C_6H_4-NH_2	0.94	1	0.98	12.42	–32095.4
11	Benzaldehyde	C_6H_5-CHO	–0.2	1	1.72	12.36	–29946.9
12	2-cresol	CH_3-C_6H_4-OH	–0.27	1	2.23	12.91	–30597.2
13	3,4-dimeyhylphenol	$C_6H_3(CH_3)_2OH$	0.12	1	2.7	14.74	–34190.8
14	3-nitrotoluene	CH_3-C_6H_4-NO_2	0.05	1	0.94	13.98	–42365.1
15	4-chlorophenol	C_6H_5-O-Cl	0.55	1	2.28	13	–35307.6
16	2,4-dinitroaniline	$C_6H_3(NO_2)NH_2$	0.53	1	–1.75	15.22	–63030.2
17	2-methyl-1–4-naphtoquinone	$C_{11}H_8O_2$	1.54	1	2.39	20.99	–49768.3

TABLE 3.11 Continued

No.	Compound		*A*	$\|1\rangle$	*Log P*	*POL*	E_{TOT}
	Name	**Formulae**	$\|Y\rangle$	$\|X_0\rangle$	$\|X_1\rangle$	$\|X_2\rangle$	$\|X_3\rangle$
18	1,2-dichlorobenzene	$C_6H_4Cl_2$	0.53	1	3.08	14.29	–36217.2
19	2,4-dinitrophenol	$C_6H_3(NO_2)OH$	1.08	1	1.67	14.5	–65318
20	1,4-dinitrobenzene	$C_6H_4N_2O_4$	1.3	1	1.95	13.86	–57926.7
21	2,4-dinitrotoluene	$C_7H_6(NO_2)_2$	0.87	1	2.42	15.7	–61520.7
22	2,6-ditertbutil 4-methyl phenol	$C_{15}H_{23}OH$	1.8	1	5.48	27.59	–59316.5
23	2,3,5,6-tetrachloroaniline	$C_6H_3NCl_4$	1.76	1	3.34	19.5	–57920.2
24	penthaclorophenol	C_6Cl_5OH	2.05	1	–0.54	20.71	–68512.4
25	phenylazophenol	$C_{12}H_{10}N_2O$	1.66	1	4.06	22.79	–55488.9
26	pentabromophenol	C_6Br_5OH	2.66	1	5.72	24.2	–66151.5

*Putz & Lacrămă (2007).

TABLE 3.12 QSAR Equations Through Standard Multi-Linear Routine and the Associate Statistics for All Possible Correlation Models Considered from Data of Table 3.11*

Model	Variables	QSAR Equation	r	s	F
Ia	*logP*	$A^{Ia} = -0.547836 + 0.435669 logP$	0.539	1.15	9.834
Ib	*POL*	$A^{Ib} = -2.84021 + 0.2166 POL$	0.908	0.574	112.15
Ic	E_{TOT}	$A^{Ic} = -2.50233 - 0.00007\, E_{TOT}$	0.882	0.644	84.015
IIa	*logP, POL*	$A^{IIa} = -2.91377 - 0.08109 logP + 0.23233 POL$	0.911	0.58	55.930
IIb	*logP*, E_{TOT}	$A^{IIb} = -2.64602 + 0.22991 logP - 0.00006\, E_{TOT}$	0.922	0.54	65.339
IIc	*POL*, E_{TOT}	$A^{IIc} = -2.98407 + 0.13427 POL - 0.00003\, E_{TOT}$	0.939	0.478	86.503
III	*logP, POL*, E_{TOT}	$A^{III} = -2.94395 + 0.06335 logP + 0.11206 POL - 0.00004 E_{TOT}$	0.941	0.48	56.598

*Putz & Lacrămă (2007).

$$r = \sqrt{1 - \frac{SR}{SQ}} \tag{3.222}$$

$$s = \sqrt{\frac{SR}{N - M - 1}} \tag{3.223}$$

$$F_{M, N-M-1} = \frac{N - M - 1}{M}\left(\frac{SQ}{SR} - 1\right) \tag{3.224}$$

as correlation factor, standard error of estimate and Fisher index, respectively, in terms of the total number of residues, measuring the spreading of the input activities with respect to their estimated counterparts

$$SR = \sum_{i=1}^{N}\left(A_i - A_i^{PREDICTED}\right)^2 \tag{3.225}$$

and the total sum of squares

$$SQ = \sum_{i=1}^{N}\left(A_i - \overline{A}\right)^2 \tag{3.226}$$

measuring the dispersion of the measured activities around their average

$$\overline{A} = \frac{1}{N}\sum_{i=1}^{N} A_i \tag{3.227}$$

while the number of compounds and descriptors were fixed to $N = 26$ and $M = 3$, in each endpoint case, respectively.

Before attempting a mechanistic analysis of the results, let us apply the SPECTRAL-SAR technique to the same data of Table 3.11 by using the spectral determinant (5.12) with the completed orthogonal and spectral coefficients of Eqs. (5.8) and (5.10), in each considered model of eco-toxic action, respectively. More explicitly, in equations (3.228)–(3.234), the spectral equations are presented with their determinant forms that once expanded produce the spectral multi-linear dependencies of Table 3.13 (Putz & Lacrămă, 2007):

(Ia):

$$\begin{vmatrix} |Y\rangle^{Ia} & 0.270385 & 0.435669 \\ |X_0\rangle & 1 & 0 \\ |X_1\rangle & 1.87808 & 1 \end{vmatrix} = 0, \quad \begin{vmatrix} |Y\rangle^{Ia} & 0.268751 & -0.547836 \\ |X_1\rangle & 1 & 0 \\ |X_0\rangle & 0.304687 & 1 \end{vmatrix} = 0 \tag{3.228}$$

(Ib):

TABLE 3.13 Spectral Structure Activity Relationships (SPECTRAL-SAR) Through Determinants of Eqs. (3.228)–(3.234) for All Possible Correlation Models Considered from the Data of Table 3.11*

Models	Vectors	SPECTRAL-SAR Equation
Ia	$\|X_0\rangle, \|X_1\rangle$	$\|Y\rangle^{Ia} = -0.547836\|X_0\rangle + 0.435669\|X_1\rangle$
Ib	$\|X_0\rangle, \|X_2\rangle$	$\|Y\rangle^{Ib} = -2.84021\|X_0\rangle + 0.216598\|X_2\rangle$
Ic	$\|X_0\rangle, \|X_3\rangle$	$\|Y\rangle^{Ic} = -2.50233\|X_0\rangle - 0.000067863\|X_3\rangle$
IIa	$\|X_0\rangle, \|X_1\rangle, \|X_2\rangle$	$\|Y\rangle^{IIa} = -2.91377\|X_0\rangle - 0.0810929\|X_1\rangle + 0.232325\|X_2\rangle$
IIb	$\|X_0\rangle, \|X_1\rangle, \|X_3\rangle$	$\|Y\rangle^{IIb} = -2.64602\|X_0\rangle + 0.229913\|X_1\rangle - 0.0000608117\|X_3\rangle$
IIc	$\|X_0\rangle, \|X_2\rangle, \|X_3\rangle$	$\|Y\rangle^{IIc} = -2.98407\|X_0\rangle + 0.134274\|X_2\rangle - 0.0000324573\|X_3\rangle$
III	$\|X_0\rangle, \|X_1\rangle, \|X_2\rangle, \|X_3\rangle$	$\|Y\rangle^{III} = -2.94395\|X_0\rangle + 0.0633549\|X_1\rangle - 0.112056\|X_2\rangle - 0.0000363728\|X_3\rangle$

*Putz & Lacrămă (2007).

$$\begin{vmatrix} |Y\rangle^{Ib} & 0.270385 & 0.216598 \\ |X_0\rangle & 1 & 0 \\ |X_2\rangle & 14.3612 & 1 \end{vmatrix} = 0, \begin{vmatrix} |Y\rangle^{Ib} & 0.0441181 & -2.84021 \\ |X_2\rangle & 1 & 0 \\ |X_0\rangle & 0.0607278 & 1 \end{vmatrix} = 0 \quad (3.229)$$

(Ic):

$$\begin{vmatrix} |Y\rangle^{Ic} & 0.270385 & -0.000067863 \\ |X_0\rangle & 1 & 0 \\ |X_3\rangle & -40857.5 & 1 \end{vmatrix} = 0, \begin{vmatrix} |Y\rangle^{Ic} & -0.0000157064 & -2.50233 \\ |X_3\rangle & 1 & 0 \\ |X_0\rangle & -0.0000208433 & 1 \end{vmatrix} = 0 \quad (3.230)$$

(IIa):

$$\begin{vmatrix} |Y\rangle^{IIa} & 0.270385 & 0.435669 & 0.232325 \\ |X_0\rangle & 1 & 0 & 0 \\ |X_1\rangle & 1.87808 & 1 & 0 \\ |X_2\rangle & 14.3612 & 2.22431 & 1 \end{vmatrix} = 0 \quad (3.231)$$

(IIb):

$$\begin{vmatrix} |Y\rangle^{IIb} & 0.270385 & 0.435669 & -0.0000608117 \\ |X_0\rangle & 1 & 0 & 0 \\ |X_1\rangle & 1.87808 & 1 & 0 \\ |X_3\rangle & -40857.5 & -3383.5 & 1 \end{vmatrix} = 0 \quad (3.232)$$

(IIc):

$$\begin{vmatrix} |Y\rangle^{IIc} & 0.270385 & 0.216598 & -0.0000324573 \\ |X_0\rangle & 1 & 0 & 0 \\ |X_2\rangle & 14.3612 & 1 & 0 \\ |X_3\rangle & -40857.5 & -2536.37 & 1 \end{vmatrix} = 0 \quad (3.233)$$

(III):

$$\begin{vmatrix} |Y\rangle^{III} & 0.270385 & 0.435669 & 0.232325 & -0.0000363728 \\ |X_0\rangle & 1 & 0 & 0 & 0 \\ |X_1\rangle & 1.87808 & 1 & 0 & 0 \\ |X_2\rangle & 14.3612 & 2.22431 & 1 & 0 \\ |X_3\rangle & -40857.5 & -3383.5 & -3306.57 & 1 \end{vmatrix} = 0 \tag{3.234}$$

Remarkably, one may easily note the striking similitude of the equations in Tables 3.12 and 3.13, respectively. Moreover, in equations (3.228)–(3.233) the spectral determinant was written in few possible ways of orthogonalization, nevertheless leading to the same results in Table 3.13. That is the computational proof that SPECTRAL-SAR indeed provides a viable alternative to standard QSAR at each level of modeling, being independent of number of descriptors, compounds, or order of orthogonalization. We advocate on the computational advantage of SPECTRAL-SAR though lesser steps of computation and by the full analyticity of the delivered structure-activity determinant related equation. However, conceptually, SPECTRAL-SAR achieves additional degree of novelty with respect to normal QSAR since the spectral equation is given in terms of vectors rather than variables. Such feature marks a fundamental advancement since this way we can deal at once with whole available data (of activity and descriptors) within a generalized vectorial space. Consequently, we may also use the spectral norm of the activity (3.62) as the general tool by which various models can be compared no matter of dimensionality and of multi-linear degree since they all reduce to a single number. This could help fulfilling the QSAR's old dream in providing a conceptual basis for the comparison of various models and endpoints. Even more, while also accurately reproducing the statistics of the standard QSAR, the actual SPECTRAL-SAR permits the introduction of an alternative way of computing correlation factors by using the above spectral norm concept (Putz & Lacrămă, 2007).

As such the so-called algebraic SPECTRAL-SAR correlation factor is defined as the ratio of the spectral norm of the predicted activity versus that of the measured one, see Eqs. (3.55) and (3.86). For the present case of the measured spectral norm of *T. pyriformis* activity, $\left\| |Y\rangle^{MEASURED} \right\| = 6.83243$, with the algebraic SPECTRAL-SAR correlation factors for the actual predicted models are given in Table 3.14 along the associated spectral norms

TABLE 3.14 The Predicted Spectral Norm, the Statistic and the Algebraic Correlation Factors of the SPECTRAL-SAR Models of Table 3.13, Computed upon the General Eqs. (3.62), (3.222), and (3.86) Since the Entry Data of Table 3.11 Are Employed, Respectively*

	Ia	*Ib*	*Ic*	*IIa*	*IIb*	*IIc*	*III*
$\left\| \|Y\rangle^{PREDICTED} \right\|$	3.86176	6.22803	6.0607	6.24858	6.32297	6.43641	6.44557
$r_{S-SAR}^{STATISTIC}$	0.53905	0.90759	0.88193	0.91074	0.92214	0.9395	0.9409
$r_{S-SAR}^{ALGEBRAIC}$	0.56521	0.91154	0.88705	0.91455	0.92543	0.94204	0.94338

*Putz & Lacrămă (2007).

for activity and the standard statistical correlation factor values (Putz & Lacrămă, 2007).

The findings in Table 3.14 are twice relevant: first, because it is clear that the spectral norm parallels the statistic correlation factor; second, because, despite the introduced algebraic correlation factor does the same job in general, it poses slightly higher values on a systematic basis. In other words, one can say that in an algebraic sense the SPECTRAL-SAR

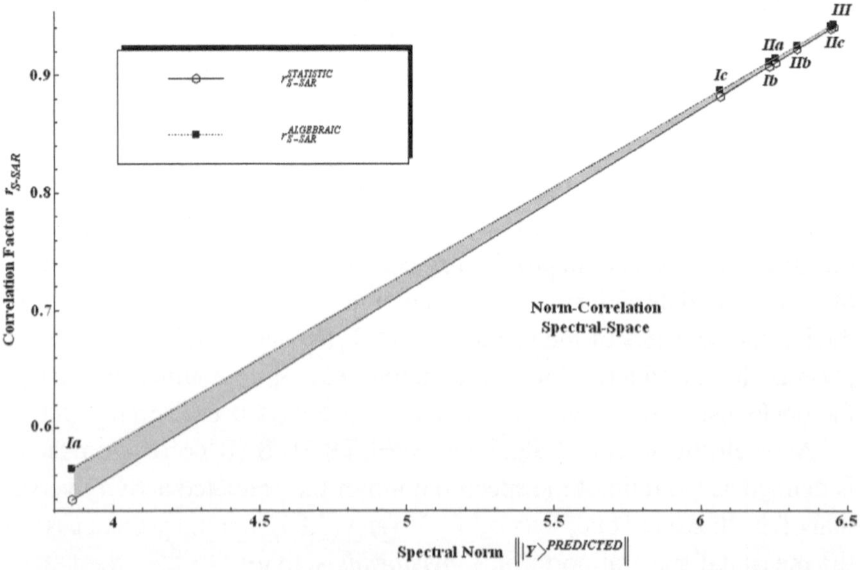

FIGURE 3.15 Norm correlation spectral space of the statistical and algebraic correlation factors against the spectral norm of the predicted SPECTRAL-SAR models of Table 3.14, respectively (Putz & Lacrămă, 2007).

furnishes systematically higher correlation factors than the standard QSAR does. This feature is also depicted in Figure 3.15 from where it is also noted that both correlation factors tend to approach each other near the ideal correlation factor, i.e., in the proximity of $r = 1.00$.

Nevertheless, we should note at this point that while a certain model does not satisfy the correlation factor criteria for being validated, i.e., $r > 0.84$, as is the case of the model (*Ia*) when only hydrophobicity is taken into account, this does not mean that the descriptor or chemical domain is less relevant; it is merely an indication that this descriptor may be further considered in a multivariate combination with others until produce better model. Indeed, both within standard QSAR and SPECTRAL-SAR approaches all models except (*Ia*) are characterized by relevant statistics (Putz & Lacrămă, 2007).

Next, aiming to see whether the obtained models can provide us a mechanistic model of chemical-biological interaction of tested xenobiotics on *T. pyriformis* species, the introduced spectral norm is employed in conjunction with algebraic or statistic correlation factors to compute the *spectral paths* between these models. Such an endeavor may lead to an intra-species analysis of models and form the first step for designing of integrated test batteries (or an expert system) at the inter-species level of ecotoxicology.

TABLE 3.15 Synopsis of the Statistic and Algebraic Values of Paths Connecting the SPECTRAL-SAR Models of Table 3.13 in the Norm-Correlation Spectral-Space of Figure 3.15*

Path	Value	
	Statistic	Algebraic
Ia-IIa-III	2.61485	2.61132
Ia-IIb-III	2.61485	2.61132
Ia-IIc-III	2.61485	2.61132
Ib-IIa-III	**0.220072**	**0.219855**
Ib-IIb-III	**0.220072**	**0.219855**
Ib-IIc-III	**0.220072**	**0.219855**
Ic-IIa-III	0.389359	0.388969
Ic-IIb-III	0.389359	0.388969
Ic-IIc-III	0.389359	0.388969

*Putz & Lacrămă (2007).

In this respect, Table 3.15 presents the computed spectral distance between the models of the measured Log(1/IGC_{50}) endpoints of Table 3.11 though considering all paths' combinations that contain a single model for each class, with one and two descriptors, towards the closest model with respect to the ideal one, i.e., (*III*). It follows that the paths are grouped according to the intermediary passing model while extreme models (initial and final) are kept fixed. Such ordered paths can be rationalized once a selection criterion is further introduced. Since paths are involved, one may learn from the well-established principle of nature according to which the events are linked by closest paths (in all classical and quantum spaces).

Therefore, we may formulate the *SPECTRAL-SAR least path principle* as follows: the hierarchy of models is driven by the minimum distance between endpoints (predicted norm of activities) of different classes of descriptors and of their combinations; whenever multiple minimum paths

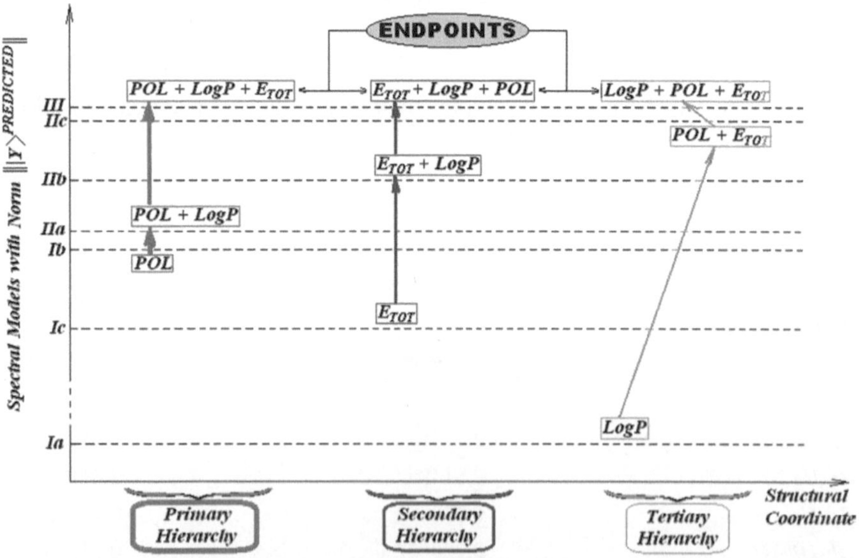

FIGURE 3.16 Spectral-structural models, designed through the principle of minimal SPECTRAL-SAR paths of Table 3.15, emphasizing the primary, secondary and tertiary hierarchies forward the endpoints of the Tetrahymena pyriformis eco-toxicological activity according with data of Table 3.11, SPECTRAL-SAR equations of Table 3.13, and of the associated spectral norms computed upon Eq. (3.62) (Putz & Lacrămă, 2007).

are possible, the principle applies iteratively downwards between individual intermediate models of paths, starting with that one with minimal spectral norm (Putz & Lacrămă, 2007).

In our case, according to the enounced minimum spectral path principle, the diagram of Figure 3.16 is constructed. It emphasizes different mechanistic hierarchies of the *T. pyriformis* toxicophores. It comes out that, for instance, while three minimum paths result from Table 3.15, namely *Ib-IIa-III*, *Ib-IIb-III*, and *Ib-IIc-III*, only one is selected as giving the primary hierarchy, that is *Ib-IIa-III*, based on the fact that the spectral norm of *IIa* is the closest one to *Ib*. This is a purely mechanistic result since the correlation order in Table 3.14 would require that *IIc* be the next model chosen when starting from model *Ib*. At this point, we see that what is ordered from a statistical point of view may be degenerate in path length between the spectral norms.

Therefore, it appears that statistics might not be the most adequate criterion for SAR validity, since models with different correlations factors may be equally inter-related through spectral norms (Putz & Lacrămă, 2007).

Used exclusively, the statistic criteria will give little information about the subsidiary inter-species correlations in a unitary picture. On the contrary, the spectral path principle is able to generate a scheme of connected paths between the models employing the natural principle of minimal action: *minimal length between spectral norms of different categories of endpoints is more favorable and comes firstly into a process driven by the succession of activities*.

Thus, once the path *Ib-IIa-III* is naturally selected as the primary hierarchy of the ecotoxicity mechanism of *T. pyriformis*, one can expect that, with actual interpretation of the minimum spectral paths, the envisioned sequence of actions towards the measured one can be causally modeled as the action of polarizability followed by that of hydrophobicity and finally by that of total energy, through the optimization of molecular geometry during the chemical-biological interactions involved.

This picture tells that the covalent interaction is the most dominant one, in this case, and drives the approach between the xenobiotics and the cells of organism; then enters into action the transfer through cellular membrane and finally the stabilization is assured by the stereo-specificity of the compounds linked to the receptor site. This way, a molecular mechanism may be coherently formulated in terms of norms of actions and of their inter-distances.

Whenever the primary route is inhibited or consumed, the second hierarchy of action follows by excluding the models previously involved and based on the same least principle of action. The second initial model will be chose that which is nearest to the first one on the spectral norm scale. Then, from all equivalent paths the next step is made toward the closes neighbor in the spectral norm sense.

The second hierarchy results along the endpoints path *Ic-IIb-III*, see Figure 3.16. This tells us that, by some subsidiary, slower action, the stereo-specificity selection is the first stage of the chemical-biological interaction analyzed, followed by membrane transport and only then by the stabilization of chemical bonds through polarizability.

If even the secondary route is somehow repressed, the third way of ecotoxicological action of *T. pyriformis* is also revealed as in Figure 3.16 illustrated by the path *Ia-IIc-III*, being constructed on the minimal activity action grounds as well.

It is not surprising that the application of minimal action principle on the spectral activity norms furnished many, however ordered, ways in which chemical-biological interaction is present in nature. This is in accordance with the evolutionary truth that the Nature reserves the privilege to develop many paths to achieve an action (Putz & Lacrămă, 2007; Putz, 2012b).

There results that the present SPECTRAL-SAR approach gives these new possibilities of hierarchically modeling of activities, over-passing the statistical analysis limited to single choices. Nevertheless, further work has to be performed by employing SPECTRAL-SAR method and of its minimal spectral path principle on many species and class of compounds in order to better validate the present results and algorithm.

3.3.2 ALGEBRAIC VS. STATISTIC CORRELATION (FOR ALIPHATIC AMINES' TOXICITY)

Amines and their derivatives are compounds, which are extensively used as drugs, cosmetics, dyes, in synthesis of pesticides and as synthetic intermediates. Aliphatic amines are considered to be strong organic bases

(Nelson, 1985); however, a broad classification scheme was developed by Verhaar and co-workers in 1992 in which compounds are assigned to one of four general classes of toxic mode of action corresponding to short-term exposures with lethality as endpoint: inert chemicals, less inert chemicals, reactive chemicals and specific-acting chemicals (Ramos et al., 1997).

Primary aliphatic amines are slightly more toxic than baseline toxicity and were classified as less inert compounds acting by a "polar narcosis" mechanism. In general, narcosis in aquatic organisms is described as a non-specific reversible functional disturbance of biological membranes due to the accumulation of chemicals in hydrophobic phases within the organism. Instead, secondary and tertiary aliphatic amines were classified as inert compounds (narcotics or baseline toxicity), because of their non-specific mode of action and a general coherency between lipophylicity and toxicity was suggested (van Wezel & Opperhuizen, 1995).

Given the multiple uses of amines in a wide variety of environmental and industrial fields, for example, for pharmaceuticals, agrochemicals, paints and coatings, or specialty chemicals (plastic additives, monomer synthesis, water treatment, etc., the present interest in QSAR modeling of their toxicity among either primary, secondary and tertiary structures; the present discussion follows (Putz et al., 2009a).

As such, the rat oral 50% lethal dose (LD_{50}) toxicity is here considered for the amines of Table 3.16. The envisaged structure-activity analysis will be performed at the Hansch level description, see Eq. (3.63), through correlating the biotoxicity $A=Log_{10}(1/LD_{50})$ with molecular parameters as hydrophobicity LogP, electronic polarizability POL and steric total energy E_{tot} computed at the optimum geometric molecular configuration computed within the HyperChem environment (Hypercube, 2002).

Since the usual statistic analysis demands the trial and test steps in validation the molecules of Table 3.16 were classified accordingly being included in the trial set those that best fulfill the normal distribution of input data (LD50), as evidenced from Figure 3.17. Worth noted that in each aliphatic amine's category as primary, secondary, and tertiary there gets out either trial or test toxicants assuring therefore the required spread of input data for statistical analysis.

TABLE 3.16 The Series of Primary, Secondary and Tertiary Aliphatic Amides Together with Associated 50% Lethal Toxicity Dose LD_{50} (mg/kg) (When Original Data (Greim et al., 1998) Was Given As An Interval the Average Range Was Here Considered) Among the Employed Genuine (Non-Biased) Activity $A=Log_{10}(1/LD_{50})-6$ and Structural Parameters As Hydrophobicity LogP, Polarizability POL (Å^3) and Total Energy At Optimal Molecular Configuration E_{tot} (kcal/mol) Were Computed Within Hyperchem Environment (Hypercube, 2002; Putz et al. 2009a)*

Index	*CAS No. CAS Name*	**Structural Formula**	LD_{50}	*A*	*LogP*	*POL*	E_{tot}
▼ **PRIMARY ALIPHATIC AMINES** ▼							
1■	75–04–7 Ethylamine	$H_3C-CH_2-NH_2$	400	–2.602	–0.27	5.80	–12912.76
2 ■	107–10–8 Propylamine	$H_3C-CH_2-CH_2-NH_2$	470	–2.672	0.20	7.63	–16506.47
3 ■	109–73–9 Butylamine	$H_3C-CH_2-CH_2-CH_2-NH_2$	528	–2.722	0.59	9.47	–20100.16
4 ■	75–31–0 Isopropylamine	$H_3C - CH(CH_3) - NH_2$	550	–2.740	0.14	7.63	–16505.39
5 ■	13952–84–6 Sec-Butylamine	$H_3C - CH(NH_2) - CH_2 - CH_3$	545	–2.736	0.61	9.47	–20097.96
6 ▲	108–91–8 Cyclohexylamine	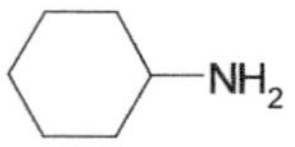	385	–2.585	0.97	12.36	–26650.52
7 ■	2869–34–3 1-Tridecanamine	$C_{13}H_{27}-NH_2$	820	–2.913	4.16	25.98	–52443.68
8 ▲	111–86–4 1-Octanamine	$CH_3-(CH_2)_7-NH_2$	350	–2.544	2.18	16.80	–34475.05
9 ▲	2016–57–1 1-Decanamine	$CH_3-(CH_2)_9-NH_2$	280	–2.447	2.97	20.47	–41662.51
10 ■	124–22–1 Dodecylamine	$CH_3-(CH_2)_{11}-NH_2$	1020	–3.008	3.77	24.14	–48849.95

TABLE 3.16 Continued

Index	*CAS No. CAS Name*	**Structural Formula**	LD_{50}	*A*	*LogP*	*POL*	E_{tot}
11 ■	2016–42–4 tetradecylamine	CH_3-$(CH_2)_{13}$-NH_2	1100	–3.041	4.56	27.81	–56037.41
12 ■	124–30–1 Octadecylamine	CH_3-$(CH_2)_{17}$-NH_2	2000	–3.301	6.14	35.15	–70412.31
13 ■	112–90–3 Oleyl amine	CH_3 - $(CH_2)_7$ - CH = CH - $(CH_2)_7$ - CH_2 - NH_2	1100	–3.041	5.88	34.96	–69753.33
▼SECONDARY ALIPHATIC AMINES▼							
14 ▲	109–89–7 Dietylamine	CH_3-CH_2-NH-CH_2-CH_3	540	–2.732	0.48	9.47	–20090.93
15 ▲	142–84–7 Dipropylamine	CH_3-CH_2-CH_2-NH-CH_2-CH_2-CH_3	695	–2.841	1.42	13.14	–27278.29
16 ▲	13360–63–9 Butylamine-N-ethyl	CH_3-CH_2-CH_2-CH_2-NH-CH_2-CH_3	308.5	–2.489	1.35	13.14	–27278.29
17 ▲	111–92–2 Dibutylamine	CH_3-CH_2-CH_2-CH_2-NH-CH_2-CH_2-CH_2-CH_3	369.5	–2.567	2.21	16.80	–34465.66
18 ▲	101–83–7 Dicyclohexylamine	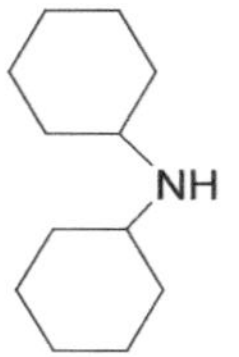	286.5	–2.457	2.97	22.60	–47562.52
19 ■	106–20–7 Dihexylamine-2,2-diethyl	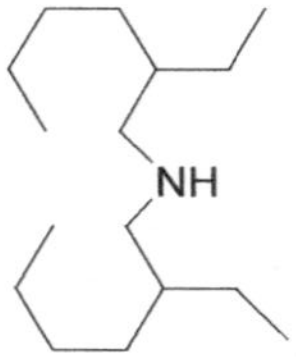	1015	–3.006	5.39	31.48	–63205.53

TABLE 3.16 Continued

Index	*CAS No. CAS Name*	**Structural Formula**	LD_{50}	*A*	*LogP*	*POL*	E_{tot}
20 ▲	110–89–4 Piperidine	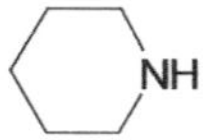	485	–2.686	0.52	10.53	–23051.82
21 ▲	110–91–8 Morpholine		1475	–3.169	–0.55	9.33	–26841.67
22 ▲	6485–55–8 cis-2,6-dimetyl- morpholine	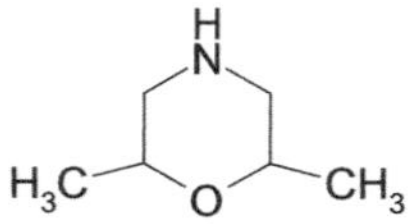	2380	–3.377	0.28	13.00	–34022.86
▼**TERTIARY ALIPHATIC AMINES**▼							
23 ■	75–50–3 Trimethylamine	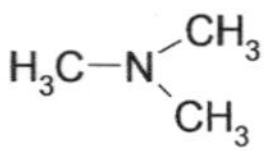	770	–2.886	0.16	7.63	–16487.84
24 ▲	593–81–7 Trimethylamine-hydrochloride	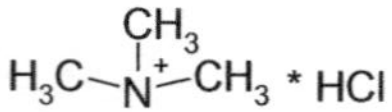	3090	–3.49	0.07	9.37	–25441.78
25 ■	598–56–1 N, N-dimethylethyl–amine	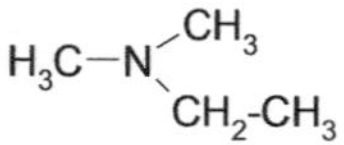	606	–2.782	0.50	9.47	–20079.71
26 ■	121–44–8 Triethylamine	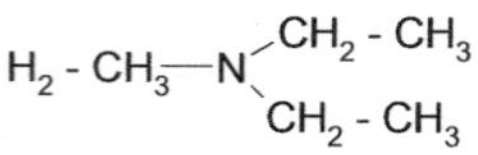	595	–2.775	1.18	13.14	–27262.98

TABLE 3.16 Continued

Index	*CAS No. CAS Name*	**Structural Formula**	LD_{50}	*A*	*LogP*	*POL*	E_{tot}
27 ■	98–94–2 Cyclohexylamine-N, N-dimethyl	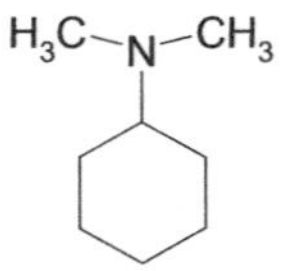	499	–2.698	1.74	16.03	–33812.66
28 ▲	102–82–9 Tributylamine	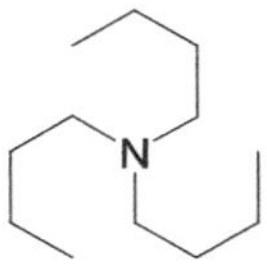	654	–2.816	3.78	24.14	–48824.4
29 ▲	112–18–5 Dodecylamine-N, N-dimethylamine	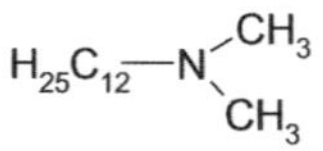	1315	–3.119	4.53	27.81	–56016.79
30 ▲	4088–22–6 1-Octadecanamine-N-methyl-N-octadecyl	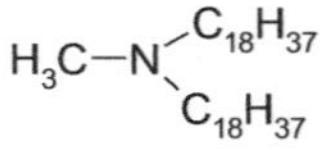	2000	–3.301	13.67	70.02	–138667.22

*The molecules included in the "trial" set have their index marked with a filled square while those considered in the "test" set were marked with the filled grey triangle, see also the Figure 3.17 (Putz & Lacrămă, 2007).

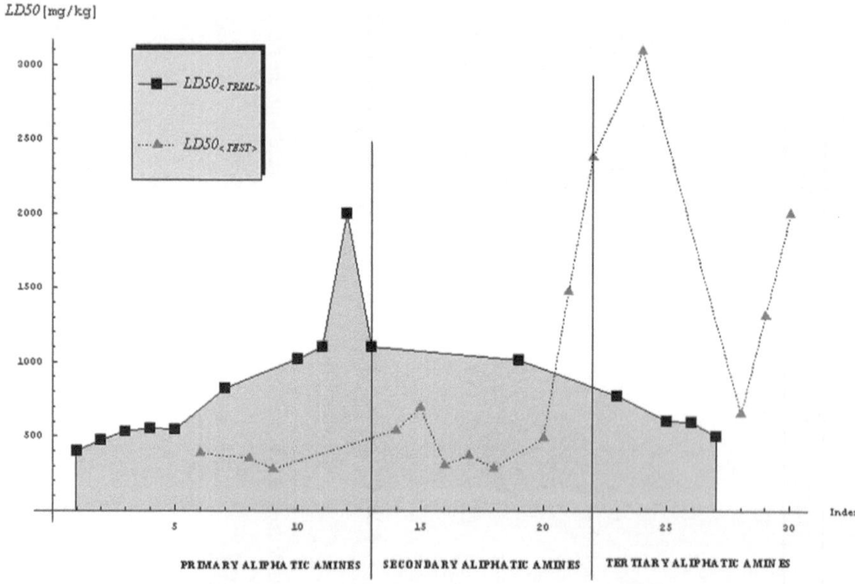

FIGURE 3.17 The plot of the primary, secondary and tertiary amines' toxicity LD50 of Table 3.16 grouped within trial and test sets, after (Putz et al., 2009a).

The considered endpoints are containing uni-, bi-, and three- parameters models with regression equation together with statistic and algebraic results presented in the Table 3.17.

The basic statistical equations and tests employed are built on the foreground observed, predicted and residues (errors) squared sums of differences:

$$SO = \frac{1}{N-1}\sum_{i=1}^{N}\left(A_i^{obs} - \overline{A}\right)^2 \tag{3.235}$$

$$SP = \frac{1}{M}\sum_{i=1}^{N}\left(A_i^{pred} - \overline{A}\right)^2 \tag{3.236}$$

$$SE = \frac{1}{N-M-1}\sum_{i=1}^{N}\left(A_i^{obs} - A_i^{pred}\right)^2 \tag{3.237}$$

with $\overline{A} = N^{-1}\sum_{i=1}^{N} A_i^{obs}$; they are traditionally evaluated and validated as:

Standard error of estimates test

$$SEE = \sqrt{SE} \tag{3.238}$$

TABLE 3.17 Structure Activity Relationships for all Possible Correlation Models Considered from the Data in Table 3.16 Together with the Statistical (Simple Correlation Factor, Standard Error of Estimation SEE, Explained Variance EV, Student *t*-test, Fischer F-test) and Algebraic (Correlation Factor r^{ALG} and Norm-Length $\|\bullet\|$) for each Considered Endpoint*

Mode	*End point*	*Structure–activity Relationships*	*Statistical descriptors*								*Algebraic descriptors*	
			$R^{♠}$	SEE^{*}	ρ_{SEE}^{*}	$EV^{¥}$	$t_{comp}^{\S}$	$\rho_{t\text{-}comp}^{\S}$	$F_{comp}^{\#}$	$\rho_{F\text{-}comp}^{\#}$	$\lVert Y_{MODE}^{predicted}\rVert^{♦}$	$r^{ALG♣}$
Ia	LogP	–26981 –0.0705451LogP	0.887662	0.09004	7.764	0.772	**6.3676**	**2.08432**	**48.305**	**5.3258**	11.1003228	0.999573
Ib	POL	–2.59298 –0.0151562POL	0.880475	0.09269	7.541	0.758	6.1299	2.00651	44.839	4.94366	11.1000384	0.999547
Ic	E_{tot}	–2.58076 $+7.73519\cdot10^{-6}E_{tot}$	0.878731	0.09332	7.490	0.755	6.0752	1.98861	44.06	4.85777	11.0999697	0.999541
IIa	LogP, POL	–3.16218 –0.37495LogP +0.0659889POL	0.901464	0.08809	7.936	0.781	6.2708	2.01893	26.023	3.75512	11.1008756	0.999622
IIb	LogP, E_{tot}	–3.19725 –0.362757LogP $-3.24015\cdot10^{-5}E_{tot}$	0.904036	**0.08699**	**8.036**	**0.787**	6.3719	2.05148	26.837	3.87258	11.1009795	0.999632
IIc	POL, E_{tot}	–2.79473 –0.25427POL $-1.22285\cdot10^{-4}E_{tot}$	0.893712	0.0913	7.656	0.765	5.9869	1.92753	23.809	3.43564	11.1005641	0.999594
III	LogP, POL, E_{tot}	–3.17409 –0.304737LogP –0.0896505POL $+7.18245\cdot10^{-5}E_{tot}$	**0.905222**	0.09032	7.739	0.771	5.789	1.82676	16.6391	2.6751	11.1010275	0.999636

♠ see Eq. (3.240) and the following note: (*) see Eq. (3.238); ¥ see Eq. (3.240); § see Eqs. (3.241) and (3.242); # see Eqs. (3.243) and (3.244); ♦ see Eq. (3.62) with $\lVert Y^{observed}\rVert=11.1051$; ♣ see Eq. (3.86).

*Putz et al. (2009a).

while its relevance to the goodness of the fit is tested by comparing it with the max-min of activity range throughout the index ratio

$$\rho_{SEE} = \left|A_{\max}^{obs} - A_{\min}^{obs}\right| / \sqrt{SE} > 5 \tag{3.239}$$

Explained variance:

$$EV = R_{adjusted}^2 = 1 - \frac{SE}{SO} \tag{3.240}$$

from where the simple statistical correlation factor R follows skipping the degree of freedom dependence in Eqs. (3.235)–(3.237) for the involved sums;

The Student-t test:

$$\underset{(N-M-2)}{t_{computed}} = \sqrt{N-1-(M+1)}\sqrt{\frac{EV}{1-EV}} > \underset{\substack{(1-\alpha=0.99,\\ N-M-2)}}{t_{tabulated}} = \begin{cases} 3.169, & N=15, M=3 \\ 3.106, & N=15, M=2 \\ 3.055, & N=15, M=1 \end{cases} \tag{3.241}$$

while the degree of fulfilling of this criteria is indicated by the ration index

$$\rho_{t-comp} = \underset{(N-M-2)}{t_{computed}} / \underset{\substack{(1-\alpha=0.99,\\ N-M-2)}}{t_{tabulated}} > 2 \tag{3.242}$$

The Fisher test:

$$\underset{(M,N-M-1)}{F_{computed}} = \frac{SP}{SE} > \underset{\substack{(1-\alpha=0.99;\\ M,N-M-1)}}{F_{tabulated}} = \begin{cases} 6.22, & N=15, M=3 \\ 6.93, & N=15, M=2 \\ 9.07, & N=15, M=1 \end{cases} \tag{3.243}$$

while the degree of fulfilling of this criteria is indicated by the ration index

$$\rho_{F-comp} = \underset{(M,N-M-1)}{F_{computed}} / \underset{\substack{(1-\alpha=0.99;\\ M,N-M-1)}}{F_{tabulated}} > 4 \tag{3.244}$$

From the Table 3.17 one gets, indeed, a somehow confusing situation since, according with individual criteria of statistical validation, there resulted that each group of statistical indices provide different model as the most robust one among all trials.

Actually, based only on direct comparison of simple statistical correlation factors the model III is predicted as the most reliable one from the result of Table 3.17. Instead, when analyzed throughout the minimum of SEE, and maximum of its associate observed activity range normalized index, as well by the maximum of EV across all investigated models of Table 3.17 the LogP-E_{tot} model (IIc) is selected as the recommended regression.

Moreover, when the Student-t and Fisher tests respecting the degree with which their computed values exceed the tabulate ones, see Eqs. (3.241), (3.242) and (3.243), (3.244), the model Ia it is yielded as the optimum one out from the same set of models investigated. We may equally called this situation as "statistical paradox" in which having a collection of models depending on the statistical tool used different model will be assessed as the most robust one.

Fortunately, having introduced the idea of paths across models such paradox is somehow solved out since the different "successful" models are arranged in a linked manner along a "statistical path", here under the form

$$\text{Ia} \rightarrow \text{IIb} \rightarrow \text{III} \tag{3.245}$$

This would be the end of the statistical analysis of a QSAR study, since from such path also the mechanistic interpretation of the bio(eco)logical-chemical-action may be formulated as originating in hydrophobic cause (LogP from model Ia) followed by entering "in action" of the stericity (through E_{tot} from model IIb over the LogP presence already included in the model Ia) and being finally stabilized by the ionic or electrostatic character of the interaction (by POL of model III in addition to LogP and E_{tot} from the previous models).

Going now to the algebraic information of models, the specific norms and algebraic correlation factors are presented in Table 3.17 for each of studied models; as a note, in all cases the algebraic factor is clearly dominating its statistical counterpart being however so high that many more digits have to be considered for making them different; this is nothing than the reflection of the very specific (or local) character of the modeled interaction. In other words there is suggested that, among all similar models

TABLE 3.18 Synopsis of the Algebraic Paths Connecting the S-SAR models of Table 3.17 in the Norm-Correlation Spectral-Space; the Statistical Path Emerging from Analysis of Statistical Descriptors of Table 3.17 is Also Marked

Algebraic Paths	Value	
Ia-IIa-III	0.00070756	
Ia-[IIb]-III	*0.00070756*	*Statistical-path*
Ia-[IIc]-III	**0.00070756**	*Algebraic-α path*
Ib-IIa-III	0.000993177	*Algebraic-β path*
Ib-IIb-III	0.000993177	
Ib-IIc-III	0.000993177	
Ic-IIa-III	0.00106214	
Ic-IIb-III	0.00106214	*Algebraic-γ path*
Ic-IIc-III	0.00106214	

produced on the same base and algorithm from a given data, the difference in their reliability and prognosis is as sharp as the degree of interaction characterized.

Making a step forward, with the algebraic information of Table 3.17 all the possible (ergodic) paths are then constructed and their lengths computed according with the recipe of Eq. (3.89) with results displayed in the Table 3.18 (Putz et al., 2009a).

The algebraic QSAR analysis is completed with the search of the minimum path out from those listed in Table 3.18 by means of the least action principle of Eq. (3.88). In present case there is noted that all paths originating in each uni-parameter models produce the same length toward the observed action (in fact towards the computed endpoint III since all path from III to observed endpoint are trivially equal); in this situation the discrimination among them is made on above described global-to-local minimum principle. Practically one choose the global minimum path among all existing path lengths: it results that there are three such paths all starting from Ia model; then the local differentiation is made by looking to the closest algebraic norm among IIa, IIb and IIc models respecting the norm of Ia endpoint: it results that the model IIc displays in Table 3.17 such feature so that it is the most susceptible to be considered as the "first move" out of model Ia. This way the "algebraic path"

$$\text{Ia}\rightarrow\text{IIc}\rightarrow\text{III} \tag{3.246}$$

was selected by algebraic methodology as the primary or alpha path in linking the causes relaying on considered structural parameters towards the observed effects as measured eco-toxicity. Note that in the same manner the other two paths are determined from the remaining paths, based on the same systematic principle, while avoiding in inclusion of previous selected models in the rest of the paths (i.e., if the model IIb was already included in the alpha path it wil be excluded from the search of beta path and so on).

Finally we need to discuss about the two above selected paths, namely the statistical and algebraic ones, Eqs. (3.245) and (3.246), respectively.

Firstly, about their similitude, they both belong to the same algebraic path length, as clearly visualized from Table 3.18; this nevertheless, emphasizes on the power of least action principle in selecting as alpha or the global minimum path both the statistical and algebraic hierarchy; note that in the previous "algebraic" language the same length means the same "intensity of interaction" that is equally provided by statistical and algebraic approach. Observe that without the least action principle and of its present particularizations in Table 3.18, the statistical results obtained from Table 3.17 would remain without a consistent rationale; now, not only the "statistical path" was seen as one of the possible paths in connecting the possible endpoints for a given set of structural parameters but also was recovered among the shortest paths by the least action principle therefore validating the results of Table 3.17 within the present algebraic frame of QSAR analysis.

Yet, it remains to distinguish between the statistical and algebraic paths in order to establish which of them is more reliable in describing the molecular mechanisms of action in producing the observed bioeffects. For deciding upon this issue the test set of data of Table 3.16 is now employed to all involved models in paths of Eqs. (3.245) and (3.246) to produce the predicted-test activities to be compared with those observed in Table 3.16. The results are depicted in Figure 3.18; at the first sight they are of poor quality but this is due to the indeed non-normal distribution of selected test-data as clearly evidenced from Figure 3.17. However, also because

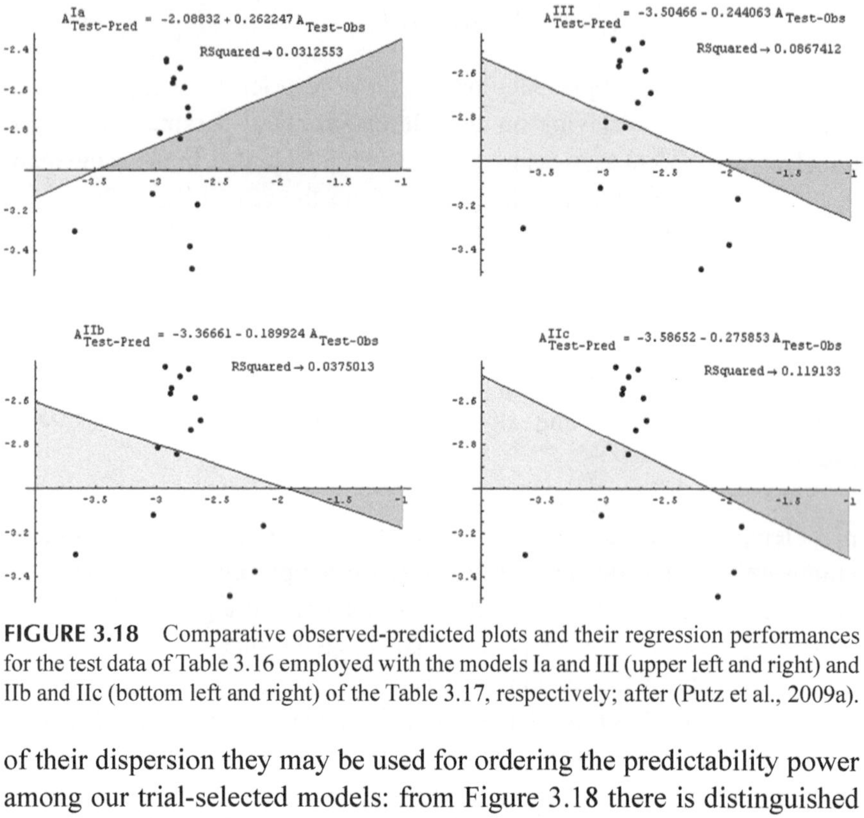

FIGURE 3.18 Comparative observed-predicted plots and their regression performances for the test data of Table 3.16 employed with the models Ia and III (upper left and right) and IIb and IIc (bottom left and right) of the Table 3.17, respectively; after (Putz et al., 2009a).

of their dispersion they may be used for ordering the predictability power among our trial-selected models: from Figure 3.18 there is distinguished that the model IIc has not only smooth supremacy over its "concurrent" model IIb but display better predictability power over all other models involved, Ia and III. Therefore, we may conclude that the alpha algebraic path of Eq. (3.246) it the optimum one both by minimum length among the endpoint parameters involved over the trial molecules but also over the test compounds of Table 3.16.

The resulted optimum in trial-tested path of endpoints among the strcuturla parameters LogP, POL and E_{tot} computed for sets of aliphatic molecules of Table 3.16 in modeling their LD50 toxicity on rats stands therefore the path of Eq. (3.245); it prescribes that the superposition of the structural (quantum) molecular states characterized by $|Log\ P\rangle$ (model Ia based) and $|POL,\ E_{tot}\rangle$ (model IIc based) determine the observed effects in toxicity. Such as, the selected path gives information not only on the order of structural parameters' causes but also on their mode of action.

By all these there was practically proved the reliability of the algebraic approach to explain the eventually "statistically paradoxes" as well to overcome them in a systematically analytical manner. Still, a more general fundament for the algebraic QSAR approach containing statistical descriptors, validity and predictability effects should be studied at the pure mathematical level regardless of the chemical or biological systems involved.

3.3.3 PROJECTIVE QSAR (FOR IONIC LIQUIDS ON Daphnia Magna)

Since their emergence a decade ago, ionic liquids have had a constantly growing influence on organic, bio- and green chemistry, due to the unique physico-chemical properties manifested by their typical salt structure: a heterocyclic nitrogen-containing organic cation (in general) and an inorganic or organic anion (Pernak & Chwala, 2003), with melting points below 100°C and no vapor pressure (Pernak et al., 2003). The latter property leads to the practical replacement of conventional volatile organic compounds (VOCs) from the point of view of atmospheric emissions, though they do present the serious drawback that a small amount of IL could enter the environment through groundwater (Sheldon, 2005). This risk makes it necessary to perform further eco-toxicological studies of IL on various species, in order to improve the "design rules" for synthesized IL with minimal toxicity to environment integrated organisms; the present discussion follows (Lacrămă et al., 2007; Putz & Putz, 2013a).

Ionic liquids display variable stability in terms of moisture and solubility in water, polar and nonpolar organic solvents (Docherty & Kulpa, 2005). Various values of ionic liquid hydrophobicity and polarity may be tailored (Sheldon, 2005) with the help of nucleoside chemistry (Freemantle, 2007) according to the main principles of green chemistry (Jastorff et al., 2005; Anastas & Warner, 1988): the new chemicals must be designed to preserve effectiveness of function while reducing toxicity, and not persisting in the environment at the end of their usage, but breaking down into inoffensive degradation products.

In this respect, the costs of all approaches for sustainable product design can be reduced using SAR and QSAR methods (Putz & Lacrămă, 2007;

Lacrămă et al., 2007; Putz et al., 2007; Chicu & Putz, 2009; Putz, 2012b). It has already been proved that the anti-microbial activity of quaternary ammonium chlorides is lipophilicity-dependent (Jastorff et al., 2003). While the 1-octanol-water partition coefficient could be seen only as the first approximation for compound lipophylicity, bioaccumulation and toxicity in fish, as well as sorption to soil and sediments assumes that lipophylicity is the main factor of anti-microbial activity (Pernak et al., 2003; Jastorff et al., 2003). Nevertheless, aiming at a deeper understanding of the specific mechanistic description of IL eco-toxicity, it is worth considering that the ionic liquid properties are more comprehensively quantified through lipophylicity, polarizability and total energy as a unitarily complex of factors in developing appropriate structure-activity relationship (SAR) studies (Lacrămă et al., 2007; Putz & Putz, 2013a).

Long- and short-term tests are used to explore the toxicity of the ionic liquid.

The results from using the Ames test for mutagenity with the *Salmonella typhyimurium* species indicated that none of the imidazolium, pyridinium and quaternary ammonium ionic liquids caused mutations. However, the Ames test is a short-term test, and it cannot be possible to fully predict the carcinogenicity in animals (Docherty et al., 2006). The antimicrobial activity (against strains of Gram-positive and –negative bacteria and fungi) increase with increasing alkyl chain lengths in pyridinium, imidazolium and quaternary ammonium salts (Pernak & Chwala, 2003). However, because bacteria have a short generation time, they are an ideal starting point for investigating the SAR in ionic liquids and serve as a basis for further toxicity tests to higher organisms and more complex systems; a few examples are listed below (Docherty & Kulpa, 2005; NTP & NIEHS, 2004):

- [BMIM][Br] (1-n-butyl-3-methylimidayolium bromide) was observed to be considerably less toxic than [BMPy][Br] (1-n-butyl-3-methylpyridinium bromide);
- Cation toxicity increases with the increase in the hydrophobicity of the molecules, from Bpy (1-n-butyl-3-pyridinium) to BMPy (1-n-butyl-3-methylpyridinium) to BdMPy (1-n-butyl-3-dimethylpyridinium);
- Butyl substituted ionic liquids have considerably lower logP values and are more water soluble;

- In comparison with some commonly used industrial solvents (acetone, methanol, ethyl acetate, and dichloromethane), the octyl and hexyl substituted ionic liquids are more toxic.
- [BMIM][Cl] (1-n-butyl-3-methylimidayolium bis (trifluoromethylsulfonyl) imide chloride) and [BPy][Cl] (1-n-butyl-3-pyridinium chloride) may have similar activity to the pesticide parquet.

At the organism levels, which are *Vibrio fisheri* and *Daphnia magna* in this review, the toxicity generally increases with the cation type as ammonium < pyridinium < imidazolium < triazolium < tetrazolium. The choline ionic liquids (with a negatively charged oxygen atom) and the quaternary ammonium ionic liquids with short chain lengths were relatively nontoxic to *Vibrio fisheri*. Furthermore, the methylation of the aromatic ring of the cation reduces the toxicity to *Vibrio fisheri* and *Daphnia magna* (Couling et al., 2006). The toxicity increases with the number of nitrogen atoms, whereas the anions play a secondary role in the toxicity (even if the presence of positively charged atoms on the anion is predicted to slightly increase the toxicity) (Couling et al., 2006).

Note that *Daphnia* is an important link between the microbial and higher trophic levels (Couling et al., 2006; McQueen et al., 1986). By acting on *Daphnia magna*, ionic liquids with longer alkyl chain substituents have toxicities comparable to phenol, whereas those with shorter substituents (for example, [BMIM][Br]) were more toxic to *Daphnia magna* than benzene and methanol (Bernot et al., 2005; Couling et al., 2006).

The lethal concentration to *D. magna* (also-called the acute toxicity) was observed to be considerably lower for ILs that employed imidazolium as cation than salts with Na^+; therefore, the toxicity was related to the imidazolium cation and not to the anions. With respect to the life history of the freshwater crustacean, salts with a sodium cation ($NaPF_6$ sodium hexafluorophosphate and $NaBF_4$, sodium tetrafluoroborate) affect the reproduction of *D. magna* at high concentrations. Sub-lethal effects on the life-history traits occurred when the concentrations of the ionic liquids were an order of magnitude lower than those for acute effects. The clonally variation was a possible reason why the average brood size in controls with those for ionic liquids. This was the first study on aquatic eukaryotes. The EC50 values on the toxicity to a single species are necessary for further studies at the community and ecosystems levels (Bernot et al., 2005).

Overall, qualitatively, the toxic effects were not observed to be notably different for the imidazolium and pyridinium ionic liquids, although the toxicity was greater with an increase in the length of the side chain. The K_{ow} values for the IL were lower than those of chemicals that bioaccumulate in the tissue of organisms. Quantitative analysis by the presented QSAR/Spectral-SAR algorithm follows.

The main problem in assessing the viable QSAR studies to predict ionic liquid toxicities concerns the *anionic-cationic interaction* superimposed on the anionic and cationic subsystems containing ionic liquids. There are basically two complementary ways of attaining this goal. One may address the search of special rules for assessing the anionic-cationic structural separately from the individual anionic and cationic ones, and then generating the QSAR models. Yet, because the cationic and anionic effects on liquid toxicity are merely separately studied at the moment, the appropriate strategy would be to firstly derive the anionic and cationic QSARs and only then to move on to a QSAR of the ionic liquid viewed as an anionic-cationic interaction (Lacrămă et al., 2007; Putz & Putz, 2013a).

As previously exposed, when the ionic liquids activity is evaluated two different additive models for modeling anionic-cationic interaction can be examined (see Section 3.2.5).

When turning to the study of the action of ionic liquids on the *Daphnia* species, the pool of molecules in Table 3.17a are employed for assessing the SAR-Ionic Liquid Ecotoxicological |0+> Model for IL-*Daphnia* chemical-biological interaction. This fact is confirmed by employing Eqs. (3.69), (3.79) and (3.81) for the cationic and anionic data of Table 3.17a and later computing the S-SAR determinants (3.9) associated with all models and combinations presented in Table 3.18a, which leads to the internal angle computations (on |1+> IL states) and the results in Table 3.19 and assures the application of the model |0+> is in accordance with the prescription given by Eq. (3.83).

Within the Spectral-SAR algorithm and allied eco-toxicological principles, the results allow the specific conclusions (Putz et al., 2007; Putz & Putz, 2013a):

- From the toxicological actions of Table 3.18a, it can be observed that both anionic and cationic fragments have important contrib-utions to

TABLE 3.17A The Actions of the Studied Ionic Liquids on the *Daphnia magna* Species with the Toxic Activities Aexp= Log(EC50) (Couling et al., 2006), while the Marked Values Were Taken from Bernot et al. (2005) Along With the Structural Parameters LogP, POL, and ETOT to Account for the Hydrophobicity, Electronic (Polarizability) and Steric (Total Energy At Optimized 3D Geometry) Effects, Computed With the HyperChem Program (Hypercube, 2002), for Each Cation and Anion Fragment, and for the Anionic-Cationic |0+> Composed State By Means of Eqs. (3.69), (3.79) and (3.81), Respectively (Putz et al., 2007; Lacrămă et al., 2007; Putz & Putz, 2013a)

Ionic Liquid Compound		A_{exp}	LogP			POL [Å³]			E_{TOT} [kcal/mol]		
Structure	***Name***	$\|Y_{exp}>$	$\|X_{1C}>$	$\|X_{1A}>$	$\|X_{1AC}>$	$\|X_{2C}>$	$\|X_{2A}>$	$\|X_{2AC}>$	$\|X_{3C}>$	$\|X_{3A}>$	$\|X_{3AC}>$
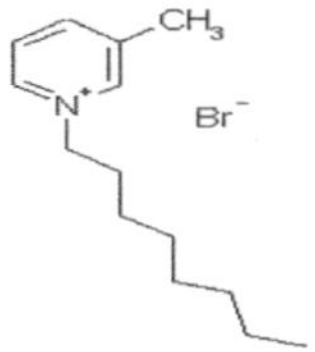	1-n-octyl-3-methylpyridinium bromide	–2.60	4.90	0.94	4.92	26.69	3.01	87.08	–371060.81	–1596918.25	–1967840
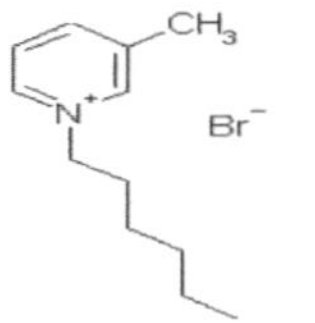	1-n-hexyl-3-methylpyridinium bromide	–2.41	4.11	0.94	4.15	23.02	3.01	78.87	–322641.81	–1596918.25	–1919410
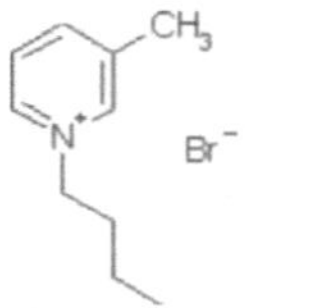	1-n-butyl-3-methylpyridinium bromide	–1.24	3.32	0.94	3.41	19.35	3.01	70.37	–274222.62	–1596918.25	–1870990

TABLE 13.17A Continued

Ionic Liquid Compound		A_{exp}	LogP			POL [Å^3]			E_{TOT} [kcal/mol]		
Structure	*Name*	$\|Y_{exp}>$	$\|X_{1C}>$	$\|X_{1A}>$	$\|X_{1AC}>$	$\|X_{2C}>$	$\|X_{2A}>$	$\|X_{2AC}>$	$\|X_{3C}>$	$\|X_{3A}>$	$\|X_{3AC}>$
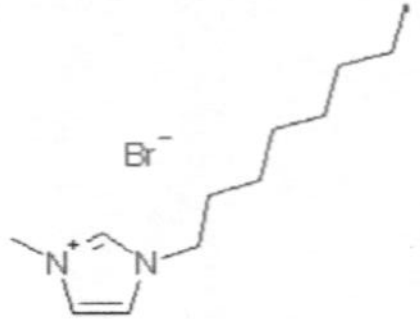	1-n-octyl-3-methylimidazolium bromide	–4.33	2.26	0.94	2.5	24.56	3.01	82.35	–357484.59	–1596918.25	–1954260
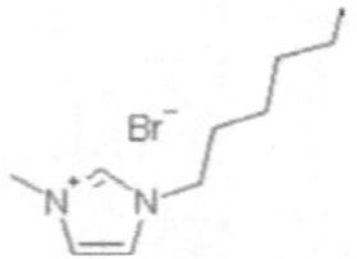	1-n-hexyl-3-methylimidazolium bromide	–2.22	1.47	0.94	1.93	20.89	3.01	73.98	–309065.84	–1596918.25	–1905830
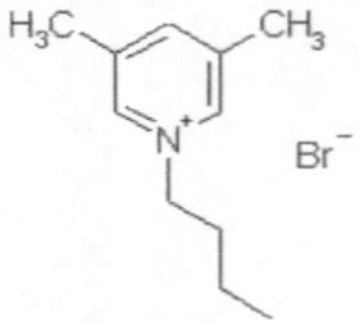	1-n-butyl-3,5-dimethylpyridinium bromide	–1.01	3.78	0.94	3.84	21.18	3.01	74.65	–298437.03	–1596918.25	–1895210
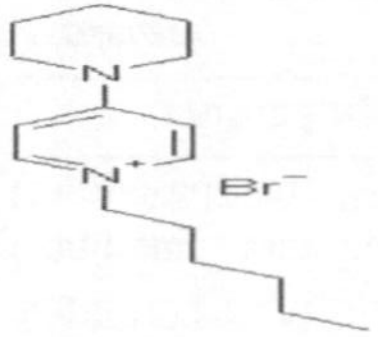	1-n-hexyl-4-piperidino pyridinium bromide	–3.66	4.63	0.94	4.65	30.93	3.01	96.25	–452857.03	–1596918.25	–2049640

TABLE 13.17A Continued

Ionic Liquid Compound		A_{exp}	LogP			POL [Å^3]			E_{TOT} [kcal/mol]		
Structure	***Name***	$\|Y_{exp}>$	$\|X_{1C}>$	$\|X_{1A}>$	$\|X_{1AC}>$	$\|X_{2C}>$	$\|X_{2A}>$	$\|X_{2AC}>$	$\|X_{3C}>$	$\|X_{3A}>$	$\|X_{3AC}>$
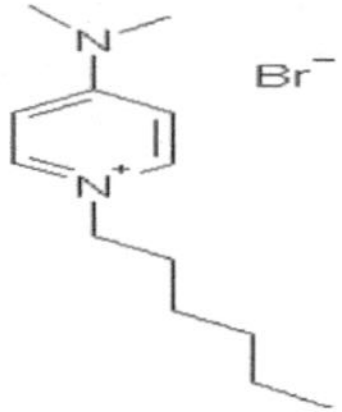	1-n-hexyl-4-dimethylamino pyridinium bromide	–3.28	3.91	0.94	3.96	26.2	3.01	86.00	–380945.12	–1596918.25	–1977720
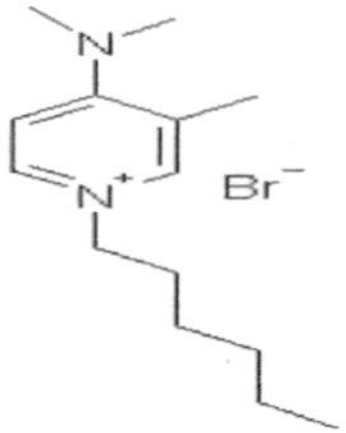	1-n-hexyl-3-methyl-4-dimethylamino pyridinium bromide	–2.79	4.37	0.94	4.40	28.04	3.01	90.03	–405145.97	–1596918.25	–2001920
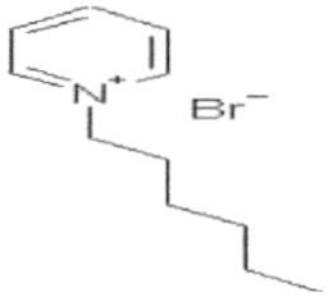	1-n-hexylpyridinium bromide	–1.93	3.64	0.94	3.71	21.18	3.01	74.65	–298427.37	–1596918.25	–1895200

TABLE 13.17A Continued

Ionic Liquid Compound		A_{exp}	LogP			POL [Å^3]			E_{TOT} [kcal/mol]		
Structure	***Name***	$\|Y_{exp}>$	$\|X_{1C}>$	$\|X_{1A}>$	$\|X_{1AC}>$	$\|X_{2C}>$	$\|X_{2A}>$	$\|X_{2AC}>$	$\|X_{3C}>$	$\|X_{3A}>$	$\|X_{3AC}>$
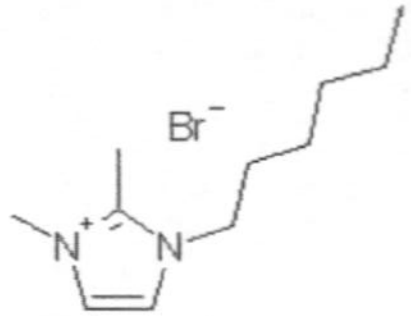	1-n-hexyl-2,3-dimethylimidazolium bromide	–2.19	1.67	0.94	2.06	22.72	3.01	78.19	–333284.94	–1596918.25	–1930060
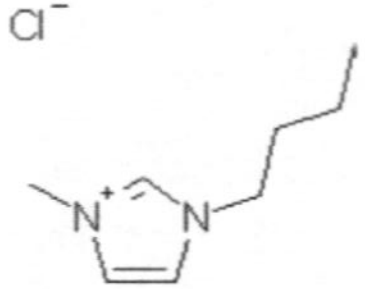	1-n-butyl-3-methylimidazolium chloride	–1.07*	0.68	0.63	1.34	17.22	2.32	59.60	–260646.64	–285190.78	–545677
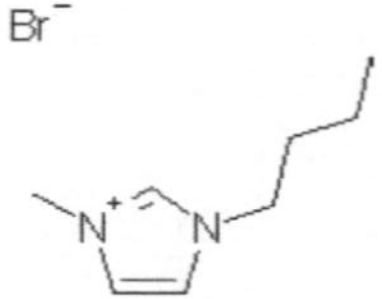	1-n-butyl-3-methylimidazolium bromide	–1.43*	0.68	0.94	1.51	17.22	3.01	65.26	–260646.64	–1596918.25	–1857410
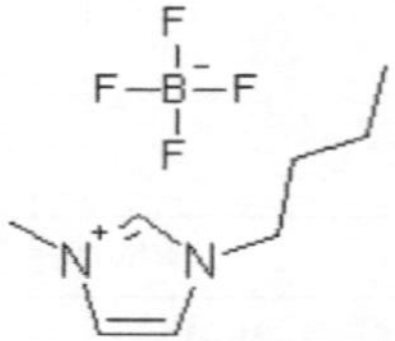	1-n-butyl-3-methylimidazolium tetrafluoroborate	–1.32*	0.68	1.37	1.78	17.22	2.46	60.80	–260646.64	–261310.59	–521798

TABLE 13.17A Continued

Ionic Liquid Compound		A_{exp}	LogP			POL [Å³]			E_{TOT} [kcal/mol]		
Structure	***Name***	$\|Y_{exp}>$	$\|X_{1C}>$	$\|X_{1A}>$	$\|X_{1AC}>$	$\|X_{2C}>$	$\|X_{2A}>$	$\|X_{2AC}>$	$\|X_{3C}>$	$\|X_{3A}>$	$\|X_{3AC}>$
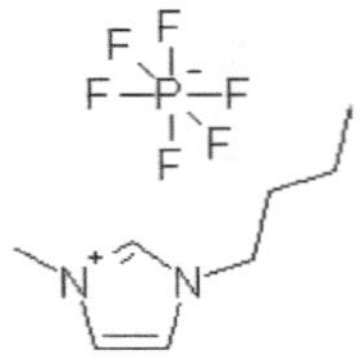	1-n-butyl-3-methylimidazolium hexafluorophosphate	–1.15*	0.68	2.06	2.28	17.22	1.78	54.62	–260646.64	–580264.94	–840746
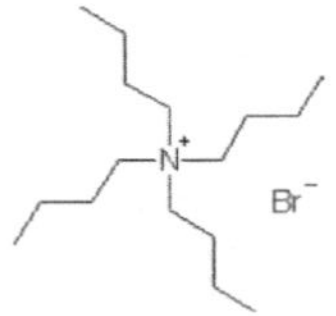	Tetrabutyl ammonium bromide	–1.53	4.51	0.94	4.54	30.91	3.01	96.21	–422421.97	–1596918.25	–2019200
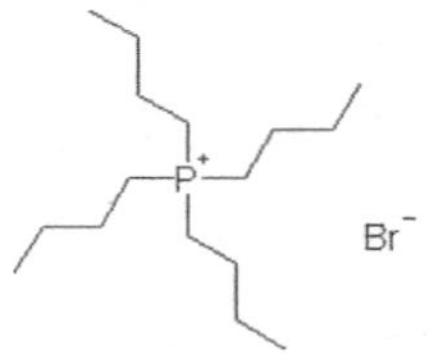	Tetrabutyl phosphonium bromide	–2.05	2.89	0.94	3.02	30.91	3.01	96.21	–600149.625	–1596918.25	–2196930

TABLE 3.18A Spectral Structure Activity Relationships (SPECTRAL-SAR) of the Ionic Liquids Toxicity of Table 3.17 Against the *Daphnia magna* Species, and the Associated Computed Spectral Norms, Computed Upon Eq. (3.62), With the Observed/Recorded Result $\| |YEXP> \| = 9.59481$, Statistic and Algebraic Correlation Factors, Computed Upon Eqs. (3.87) & (3.86) (Putz & Lacrămă, 2007; Putz, 2012b), Throughout the Possible Correlation Models Considered From the Anionic, Cationic, and Ionic Liquid |0+> and |1+> States, see Eqs. (3.84) and (3.85), Respectively (Lacrămă et al., 2007; Putz et al., 2007; Putz & Putz, 2013a)

Mode	Vectors' Predicted	$\| \|Y\rangle^{Mode} \|$	$r_{S-SAR}^{STATISTIC}$	$r_{S-SAR}^{ALGEBRAIC}$
Ia	$\|Y_{A-Ia}>= f(\|X_0>, \|X_{1A}>)$	8.83127	0.266552	0.920421
	$\|Y_{C-Ia}>=f(\|X_0>, \|X_{1C}>)$	8.92169	0.420761	0.929845
	$\|Y_{AC-Ia}>^{0+}=f(\|X_0>, \|X_{1AC}>)$	8.89048	0.374616	0.926593
	$\|Y_{AC-Ia}>^{1+}=\|Y_{A-Ia}>+\|Y_{C-Ia}>$	17.6883	2.21964 *i*	1.84353
Ib	$\|Y_{A-Ib}>=f(\|X_0>, \|X_{2A}>)$	8.94784	0.455964	0.932572
	$\|Y_{C-Ib}>=f(\|X_0>, \|X_{2C}>)$	9.06691	0.59121	0.944981
	$\|Y_{AC-Ib}>^{0+}=f(\|X_0>, \|X_{2AC}>)$	*9.08979*	*0.613973*	*0.947366*
	$\|Y_{AC-Ib}>^{1+}=\|Y_{A-Ib}>+\|Y_{C-Ib}>$	17.9079	2.19638 *i*	1.86641
Ic	$\|Y_{A-Ic}>= f(\|X_0>, \|X_{3A}>)$	8.96309	0.475327	0.934161
	$\|Y_{C-Ic}>=f(\|X_0>, \|X_{3C}>)$	8.95817	0.469161	0.933648
	$\|Y_{AC-Ic}>^{0+}=f(\|X_0>, \|X_{3AC}>)$	*8.99267*	*0.510889*	*0.937244*
	$\|Y_{AC-Ic}>^{1+}=\|Y_{A-Ic}>+\|Y_{C-Ic}>$	17.8233	2.20167 *i*	1.8576
IIa	$\|Y_{A-IIa}>=f(\|X_0>, \|X_{1A}>, \|X_{2A}>)$	8.96021	0.47173	0.933861
	$\|Y_{C-IIa}>=f(\|X_0>, \|X_{1C}>, \|X_{2C}>)$	9.06885	0.59317	0.945183
	$\|Y_{AC-IIa}>^{0+}=f(\|X_0>, \|X_{1AC}>, \|X_{2AC}>)$	*9.1014*	*0.62522*	*0.948575*
	$\|Y_{AC-IIa}>^{1+}=\|Y_{A-IIa}>+\|Y_{C-IIa}>$	17.9161	2.1931 *i*	1.86727
IIb	$\|Y_{A-IIb}>= f(\|X_0>, \|X_{1A}>, \|X_{3A}>)$	8.96426	0.476781	0.934283
	$\|Y_{C-IIb}>=f(\|X_0>, \|X_{1C}>, \|X_{3C}>)$	8.99112	0.50908	0.937082
	$\|Y_{AC-IIb}>^{0+}=f(\|X_0>, \|X_{1AC}>, \|X_{3AC}>)$	*8.99774*	*0.51675*	*0.937772*
	$\|Y_{AC-IIb}>^{1+}=\|Y_{A-IIb}>+\|Y_{C-IIb}>$	17.8793	2.21324 *i*	1.86343
IIc	$\|Y_{A-IIc}>=f(\|X_0>, \|X_{2A}>, \|X_{3A}>)$	8.96808	0.481497	0.93468
	$\|Y_{C-IIc}>=f(\|X_0>, \|X_{2C}>, \|X_{3C}>)$	9.09155	0.615686	0.947549
	$\|Y_{AC-IIc}>^{0+}=f(\|X_0>, \|X_{2AC}>, \|X_{3AC}>)$	9.09116	0.615307	0.947508
	$\|Y_{AC-IIc}>^{1+}=\|Y_{A-IIc}>+\|Y_{C-IIc}>$	17.9586	2.19937 *i*	1.8717
III	$\|Y_{A-III}>=f(\|X_0>, \|X_{1A}>, \|X_{2A}>, \|X_{3A}>)$	8.96926	0.482946	0.934803
	$\|Y_{C-III}>=f(\|X_0>, \|X_{1C}>, \|X_{2C}>, \|X_{3C}>)$	9.12145	0.644232	0.950666
	$\|Y_{AC-III}>^{0+}=f(\|X_0>, \|X_{1AC}>, \|X_{2AC}>, \|X_{3AC}>)$	9.10319	0.62694	0.948762
	$\|Y_{AC-III}>^{1+}=\|Y_{A-III}>+\|Y_{C-III}>$	17.9531	2.17933 *i*	1.87113

TABLE 3.19 The Values of the Cosines of the Anion-Cationic Vectorial Angles, Computed Upon Eq. (3.83) (Lacrămă et al., 2007; Putz, 2012b) for All Considered Modes of Action of Table 3.18 and for the |1+> States of the Considered Ionic Liquids (Putz et al., 2007; Putz & Putz, 2013a)

Mode	***Ia***	***Ib***	***Ic***	***IIa***	***IIb***	***IIc***	***III***
$\cos\theta_{AC}$	0.985468	0.976338	0.978196	0.975018	0.983081	0.97768	0.969683

the "length" and "intensity" of the ionic liquids ecotoxicity through the computed spectral norms and algebraic correlation factors, respectively, which are close to the experimental one, i.e., to 9.59481 value;

- In all cases, the mode of action where all three Hansch factors were considered (mode *III* with $LogP+POL+E_{TOT}$) records the best norm and correlations that are the closest description of the ionic liquids-*Daphnia magna* chemical-biological interaction;
- The cationic influence is observed with the dominant contribution over the anionic effects in ecotoxicity, in all considered Hansch modes of action;
- The statistical correlation factors always yield smaller values than the corresponding algebraically ones, see Table 3.18;
- There are recorded imaginary statistical correlations of the computed $|Y_{AC\text{-}Mode}>^{1+}$ endpoints that indicate certain limitations of its use for activity modeling in ecotoxicology; for these cases, the algebraically outputs provide almost the sum of the anionic and cationic length and intensity endpoint activity. This result can be phenomenologically explained by the so-called "resonance effect" when the angles between the anionic and cationic endpoint vectors are almost zero, as clearly evidenced by the cosine values of Table 3.19;
- Within the |0+> model, all of the lengths and intensities of the endpoints $|Y_{AC\text{-}Mode}>^{0+}$ behave as an average of the anionic and cationic ecotoxicological effects with a smooth increase over the individual cationic effects for the modes *Ib* (*POL*), *Ic* (E_{TOT}), *IIa* (*LogP+POL*), and *IIb* ($LogP+E_{TOT}$); however, further selection for the binding mechanism is performed by identifying the minimum analysis of the Spectral-paths, Table 3.20.

The Spectral path analysis is unfolded for the |1+> models by "ergodic" selection of the models per paths, with the ecotoxicological results in

TABLE 3.20 Synopsis of the Statistical and Algebraic Values of the Paths Connecting the SPECTRAL-SAR Models of Table 3.18, in the Norm-Correlation Spectral-Space, for *Daphnia magna* Species Against the Ionic Liquids Toxicity of Table 3.17; the Primary, Secondary and Tertiary – the So-Called Alpha (α), Beta (β) and Gamma (γ) Paths, Are Indicated According to the "Selection" and "Validation" Principles in Norm-Correlation Spectral Space When the Statistic and Algebraic Variants of the Correlation Factors Are Respectively Used (Putz et al., 2007; Putz & Putz, 2013a)

Path	**Cationic**		**Anionic**		**Ionic Liquid**			
					State |0+>		**State |1+>**	
	Statistic	***Algebraic***	***Statistic***	***Algebraic***	***Statistic***	***Algebraic***	***Statistic***	***Algebraic***
Ia-IIa-III	0.299988	0.200851	**0.256742** γ	0.13874	0.330033	**0.213862** γ	0.260535	0.266181
Ia-IIb-III	**0.300103** γ	0.200851	0.2567	0.13874	**0.330581** γ	0.213862	**0.25639** γ	**0.266181** γ
Ia-IIc-III	0.299895	**0.200851** γ	0.25666	**0.13874** γ	0.33011	0.213862	0.269477+ $R^* i$	0.277223
Ib-IIa-III	**0.07607** α	**0.0548409** α	0.034447	**0.0215298** β	0.0186427	0.0134673	**0.0418672** α	**0.0454552** α
Ib-IIb-III	0.299514	0.207241	0.0344468	0.0215298	0.286398	0.198562	0.0886683	0.102966
Ib-IIc-III	0.0760732	0.0548409	**0.0344464** β	0.0215298	**0.0186427** α	**0.0134673** α	0.0506137+ $R^* i$	0.0564973

TABLE 3.20 Continued

Path	Cationic		Anionic		Ionic Liquid			
					State |0+>		State |1+>	
	Statistic	*Algebraic*	*Statistic*	*Algebraic*	*Statistic*	*Algebraic*	*Statistic*	*Algebraic*
Ic-IIa-III	0.23953	0.16417	0.0190146	0.0119873	**0.160257** $\boxed{\beta}$	0.111113	0.126723	0.130484
Ic-IIb-III	0.23952	**0.16417** $\boxed{\beta}$	**0.00980164** $\boxed{\alpha}$	**0.00619966** $\boxed{\alpha}$	0.160264	**0.111113** $\boxed{\beta}$	**0.120323** $\boxed{\beta}$	**0.130484** $\boxed{\beta}$
Ic-IIc-III	**0.239484** $\boxed{\beta}$	0.16417	0.00980164	0.00619966	0.16027	0.111113	0.135252+ R^{*} ***i***	0.141526

Table 3.20 and correspondingly interpreted as follows (Putz et al., 2007; Putz & Putz, 2013a):

- the additive parametric and endpoint models, |0+> and |1+>, provide the same hierarchies of the paths for the chemical-biological actions;
- the statistical imaginary correlation values for the ionic liquids |1+> are avoided from the mechanistic principle and do not belong to any selected path in Table 3.20;
- the dominant cationic effects can also be noted here at the least paths level because the nature of the cationic mechanism is preserved to the ionic liquids nature according with the spectral path equations:

$$\alpha_C + \beta_A = \alpha_{IL} \tag{3.247}$$

$$\beta_C + \alpha_A = \beta_{IL} \tag{3.248}$$

$$\gamma_C + \gamma_A = \gamma_{IL} \tag{3.249}$$

- the results of all SPECTRAL-SAR ecotoxicological principles applied to ionic liquids-*Daphnia magna* case of chemical-eco-biological interaction can be unitarily presented in the Figure 3.19, where the spectral hypersurface was generated by the 3D interpolation of all lengths (norms) for all the endpoint modes of Table 3.18, for all cationic, anionic, |0+> and |1+> states of ionic liquids of Table 3.17. The alpha dominant paths are easily identified according to Table 3.20, as originating in the *Ib*, i.e., on *POL*arizability or van der Waals molecular mode of action, while the beta and gamma ones starts with the steric (*Ic*: E_{TOT}) and hydrophobic (*Ia*: *Log*P) specific chemical-biological binding, respectively.

Overall, one may assess the sets of Eqs. (3.247)–(3.249) as specific for ionic liquids action over biological marine species within the additive and parametric models |1+> and |0+>, respectively, that should be further confirmed or extended by future studies with other structural parameters (beyond Hansch descriptors, i.e., by topological and quantum molecular factors) and/or with other species on similar congeneric ILs and working algebraic chemical-biological interaction models.

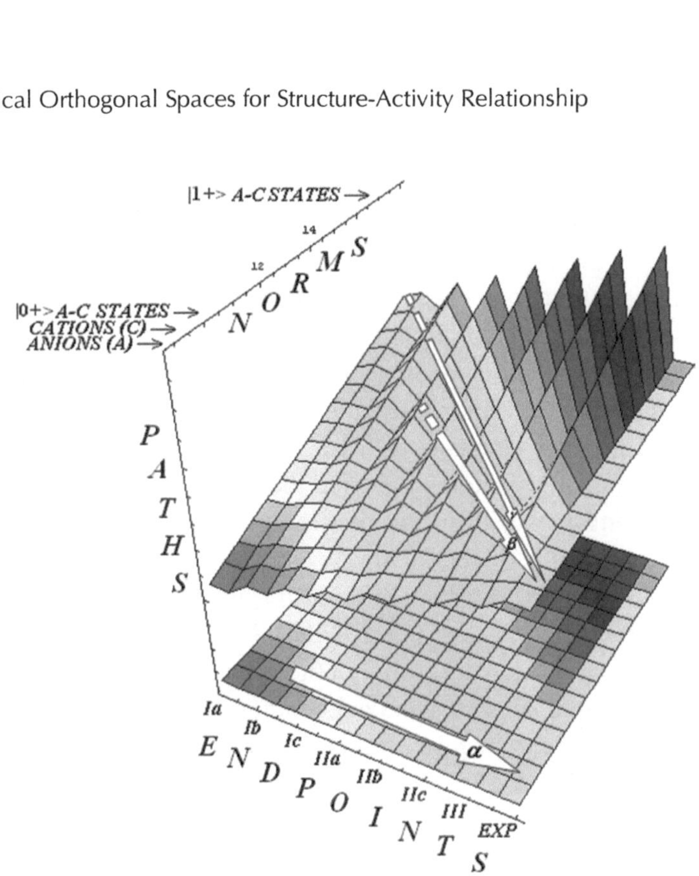

FIGURE 3.19 The spectral hypersurface of the structural hierarchical paths toward the recorded (*EXP*) ecotoxicological activity (in the extreme right hypersurface region) of the ionic liquids of Table 3.35 on *Daphnia magna* species: the alpha path (α) initiates on the polarizability (*Ib*) anionic-cationic interaction (in the left-bottom hypersurface region), being followed by the beta path (β), which originates on the steric (*Ic*) anionic-cationic interaction (in the left-top hypersurface region hypersurface region), and successively by the gamma path (γ) based on the hydrophobic (*Ia*) anionic-cationic interaction (in the extreme left-top hypersurface region) of the norm-correlation spectral space of Table 8 with the decaying order of the thickness of the connecting arrows, respectively (Putz et al., 2007; Putz & Putz, 2013a).

3.3.4 CATASTROPHE QSAR (FOR ANTI-HIV PYRIDINONES)

Thom's theory of catastrophes has acquired much popularity for its simple yet valuable description of the system-environment interaction that includes phenomena such as steady state equilibrium and life cycles (Thom, 1973). In particular, biological systems come first under catastrophe modeling because they display a causal action-reaction response to various natural or imposed constraining limits. As an example, the reactions of organisms to vital toxicological threats were developed into the survival attractor concept by employing butterfly bifurcation phenomenology, which is closely related to the cusp catastrophe, thus revealing the close connection with the turning points around singularity points of the fundamental central field laws of attraction (Viret, 1994). The cusp catastrophe was further implemented in the physiological processes of predation and generation, thus giving mathematical support to Heidegger's philosophical concept of *entity* and having the major consequence of translating the ontological entities into computer language (Lacorre, 1997). Following this line of applications, Jungian psychology entered the topological approach phase through modeling personal unconscious and conscious states using the swallowtail catastrophe (Viret, 2010). As a consequence, neuro-self-organization was advanced by reduction to cusp synergetics as an archetypal precursor of epileptic seizures (Cerf, 2006). Nevertheless, in chemistry the catastrophe approach enters through the need to unitarily characterize elementary processes such as chemical bonding, leading to the so-called bonding evolution theory and reformulation of the electronic localization functions (Silvi & Savin, 1994), see also the Volume II of the present five-volume work (Putz, 2016b). In the last decade, catastrophe theory was successfully grounded on Hilbert space modeling with the density matrix and nonlinear evolution as specific tools for the non-commutative (quantum) systems (Aerts et al., 2003). At this point, the interesting connection with the linear superposition of quantum states may be generalized in a non-linear manner with direct correspondence for widespread quantitative structure-activity relationship (QSAR) treatments of the "birth and death of an organism"; the present discussion follows (Putz et al., 2011b).

In this context, the present contribution provides in silico assistance to clinical efforts in current antiretroviral therapy by contributing to the development of a given class of actual anti-HIV-1 compounds and identifying their viral inhibitory mechanisms and influential structural factors. Continuous efforts both in theory and in clinical practice are made to provide new and valid data for HIV infection management. Note that acquired immunodeficiency deficiency syndrome (AIDS) was first recognized in 1981. Only 25 compounds have been approved for use in HIV infected patients, and they are distributed among several classes of antiretroviral drug type (De Clercq, 2009a, b): nucleoside reverse transcriptase inhibitors (NRTIs); nucleotide reverse transcriptase inhibitors (NtRTIs); non-nucleoside reverse transcriptase inhibitors (NNRTIs); protease inhibitors

(PIs); cell entry (or fusion) inhibitors (FIs); co-receptor inhibitors (CRIs); and integrase inhibitors (INIs). Among these, it is well known that most NNRTIs have a low genetic barrier to resistance, i.e., high viral resistance may be induced by a single mutation at the NNRTI binding site (El Safadi et al., 2007). It is this particular feature that makes NNRTIs so well adapted for a comprehensive catastrophe theory application. Although NNRTIs are an open battlefield for research, being highly active in naïve and drug-resistant HIV infected patients (Ivetac & McCammon, 2009), QSAR methods are cost-effective approaches to developing new and potent molecules with increased anti-HIV activity (Gupta, 2002; Prabhakar et al., 2006; Prajapati et al., 2009; Zhan et al., 2011; Chen et al., 2011; Rebehmed et al., 2008; Afantitis et al., 2006; Marino et al., 2006; Mandal & Roy, 2009; Bak & Polanski, 2006; Duda-Seiman et al., 2006). As a viable alternative to the available 3D-QSARs, the present endeavor makes the first steps toward generalizing multi-linear QSAR to non-linear catastrophe QSAR analysis and toward providing a conceptual-computational framework in which both the interactions occurring between the pyridinone derivatives and the NNRTI binding site and the structural domains influential for HIV-1 RT inhibitory activity are accounted for (Medina-Franco et al., 2004).

As a working molecular series, the interesting series of pyridinone derivatives in Table 3.21 is herein employed (Medina-Franco et al., 2004) because of their potential for improving and complementing the currently available four NNRTIs that have been approved by the U.S. FDA for HIV/AIDS treatment (Nevirapine-Viramune®, Delavirdine-Rescriptor®, Efavirenz-Sustiva®, Etravirine-Intelence®), all of which bind to the hydrophobic pocket of HIV-1 reverse transcriptase (Lu & Chen, 2010). The pyridinone derivatives were divided into a training set of 23 compounds and a test set of 9 compounds according to the methods of normal/Gaussian (G) and non-normal/non-Gaussian (NG) fitted activity (Figure 3.20); the present discussion follows (Putz et al., 2011b).

The catastrophe-QSAR algorithm of Section 3.1.6 was applied to the molecules of Table 3.21, and the trial results are presented in the Tables 3.22–3.26 (Putz et al., 2011b).

For the trial set of molecules from Figure 3.20 and Table 3.21, the results in Tables 3.22 and 3.23 can be interpreted as follows (Putz et al., 2011b):

TABLE 3.21 Actual Working Reverse Transcriptase Pyridinone Inhibitors Grouped in Gaussian (G) and non-Gaussian (NG) Molecular Congeneric Sets (Putz et al., 2011b) With Their Structural Information (Hydrophobicity, LogP; Molecular Polarizability POL [Å^3] and Total Optimized Energy of Formation H[kcal/mol]) Computed Upon the Semi-Empirical PM3 Method (Hypercube, 2002), Along With Their Observed Activity A=Log(1/IC50) (Medina-Franco et al., 2004)

No.	Type	Working Molecules		A^{Obs}	QSAR parameters		
		Structure	Name	log(1/IC$_{50}$)	LogP	POL (Å^3)	H (kcal/mol)
1.	G1	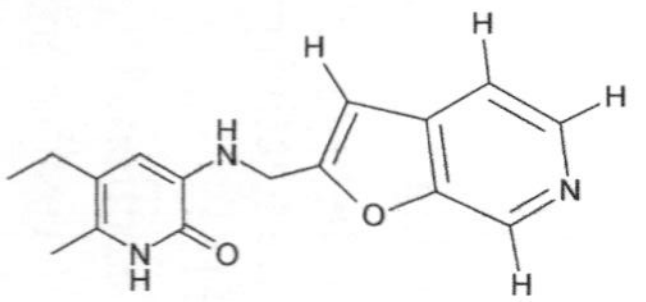	3-{[(6′-azabenzofuran-2′-yl) methyl]amino}-5-ethyl-6-methylpyridin-2(1H)-one	3.98	–0.54	31.21	–14.67
2.	G2	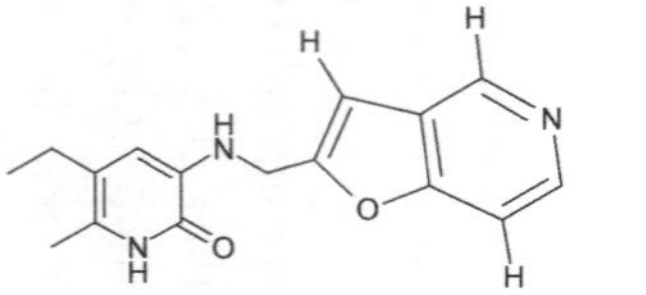	3-{[(5′-azabenzofuran-2′-yl) methyl]amino}-5-ethyl-6-methylpyridin-2(1H)-one	4.49	–0.54	31.21	–16.195
3.	G3	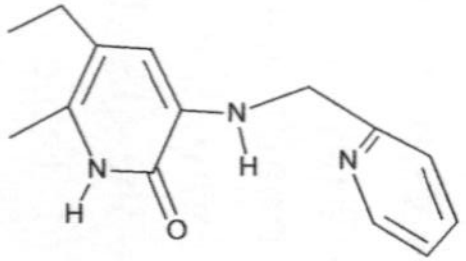	3-{[(pyridine-2′-yl) methyl]amino}-5-ethyl-6-methylpyridin-2(1H)-one	4.82	0.21	27.87	–5.854
4.	G4	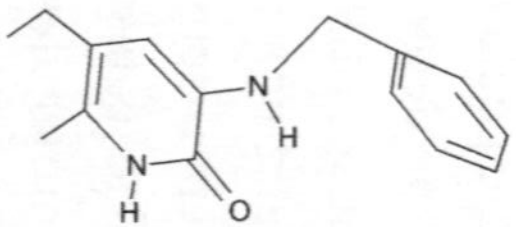	3-benzylamino-5-ethyl-6-methylpyridin-2(1H)-one	5.27	0.67	28.58	–11.659

TABLE 3.21 Continued

No.	Type	Working Molecules		A^{0bs}	QSAR parameters		
		Structure	Name	$log(1/IC_{50})$	LogP	POL (Å³)	H (kcal/mol)
5.	G5	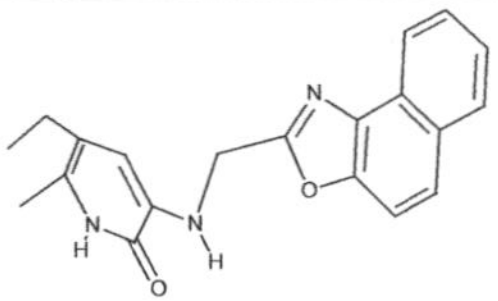	3-{[(1, '3′-naftoxazol-2′-yl) methyl]amino}-5-ethyl-6-methylpyridin-2(1H)-one	5.57	1.20	38.48	–1.878
6.	G6	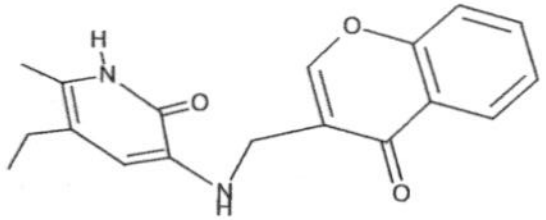	3-{[(1′-benzopyran-4′-one-3′-yl) methyl]amino}-5-ethyl-6-methylpyridin-2(1H)-one	5.96	–0.71	33.84	–61.455
7.	G7	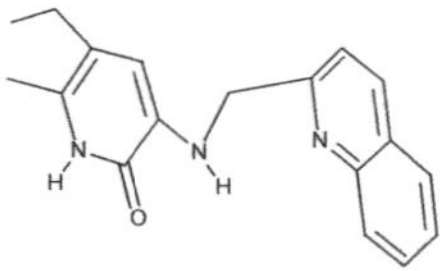	3-{[(benzopyridine-2′-yl) methyl]amino}-5-ethyl-6-methylpyridin-2(1H)-one	6.28	1.16	35.14	11.246
8.	G8	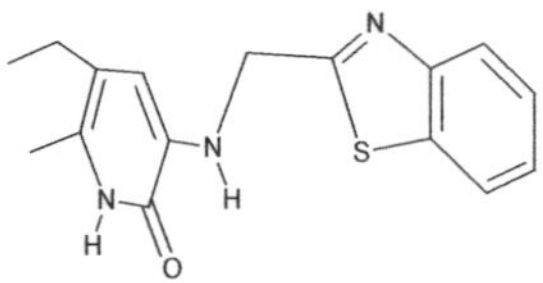	3-{[(1, '3′-benzothiazole-2′-yl) methyl]amino}-5-ethyl-6-methylpyridin-2(1H)-one	6.46	0.54	33.57	17.808
9.	G9	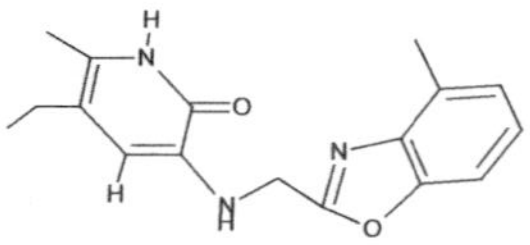	3-{[(4′-methyl-benzoxazole-2′-yl) methyl]amino}-5-ethyl-6-methylpyridin-2(1H)-one	6.92	0.67	33.05	–27.613

TABLE 3.21 Continued

No.	*Type*	*Working Molecules*		A^{0bs}	*QSAR parameters*		
		Structure	***Name***	$log(1/IC_{50})$	***LogP***	***POL*** (Å^3)	***H (kcal/mol)***
10.	G10	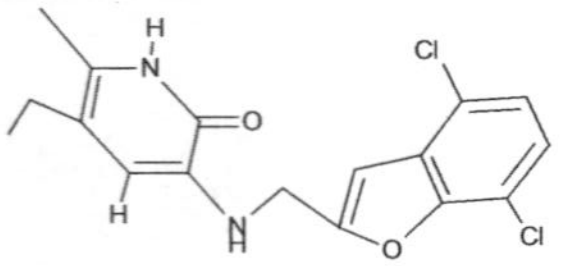	3-{[(4, ’7′-dichloro-benzofuran-2′-yl) methyl] amino}-5-ethyl-6-methylpyridin-2(1H)-one	7.24	0.88	35.78	–33.749
11.	G11	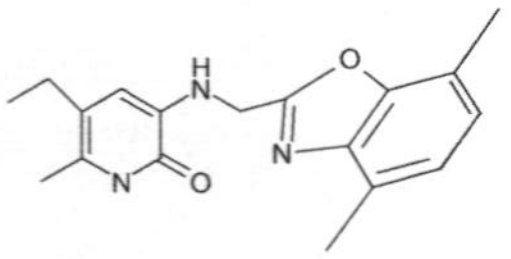	3-{[(4, ’7′-dimethyl-benzoxazol-2′-yl) methyl] amino}-5-ethyl-6-methylpyridin-2(1H)-one	7.7	1.13	34.88	–38.048
12.	G12	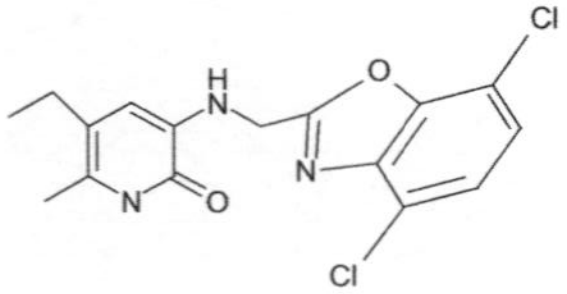	3-{[(4, ’7′-dichloro-benzoxazol-2′-yl) methyl] amino}-5-ethyl-6-methylpyridin-2(1H)-one	7.72	1.24	35.07	–30.071
13.	G13	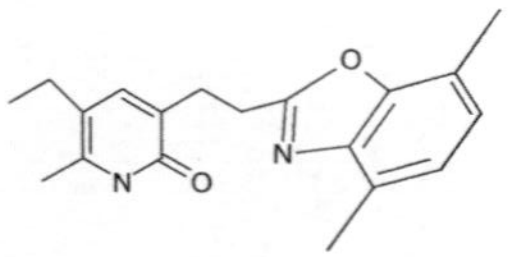	3-[(4, ’7′-dimethyl-benzoxazol-2′-yl) ethyl]-5-ethyl-6-methylpyridin-2(1H)-one	7.55	2.62	35.37	–47.701
14.	G14	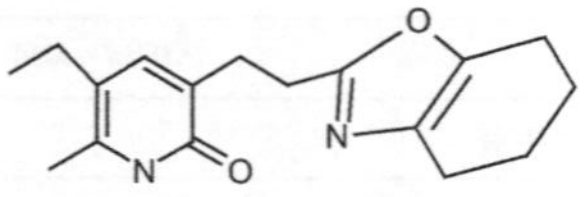	3-[(4, ’5, ’6, ’7′-tetrahydro-benzoxazole-2′-yl) ethyl]-5-ethyl-6-methylpyridin-2(1H)-one	7.24	–0.02	32.08	–63.299

TABLE 3.21 Continued

No.	Type	Working Molecules		A^{0bs}	QSAR parameters		
		Structure	*Name*	$log(1/IC_{50})$	*LogP*	*POL* ($Å^3$)	*H (kcal/mol)*
15.	G15	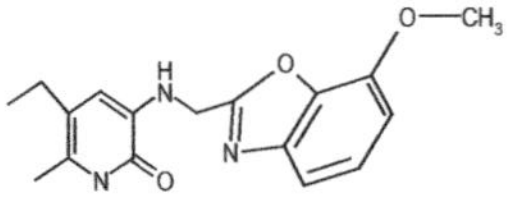	3-{[(4′-methoxy-benzoxazole-2′-yl) methyl]amino}-5-ethyl-6-methylpyridin-2(1H)-one	6.74	–0.05	33.68	–54.452
16.	G16	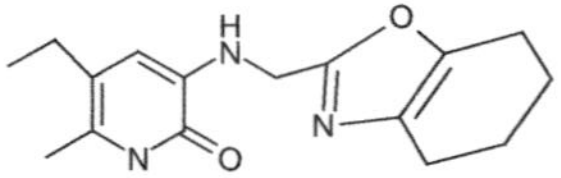	3-[(4, ’5, ’6, ’7′-tetrahydro-benzoxazole-2′-yl) methyl]amino}-5-ethyl-6-methylpyridin-2(1H)-one	6.55	–1.50	31.59	–50.643
17.	G17	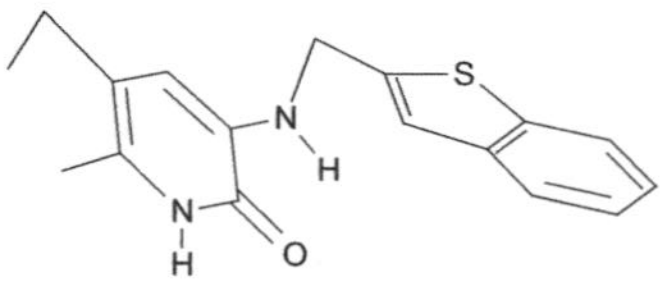	3-{[(benzothiophene-2′-yl) methyl] amino}-5-ethyl-6-methylpyridin-2(1H)-one	6.30	0.19	34.28	11.703
18.	G18	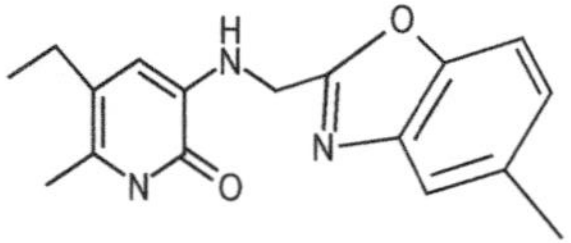	3-{[(5′-methylbenzoxazole-2′-yl) methyl]amino}-5-ethyl-6-methylpyridin-2(1H)-one	5.90	0.67	33.05	–27.741
19.	G19	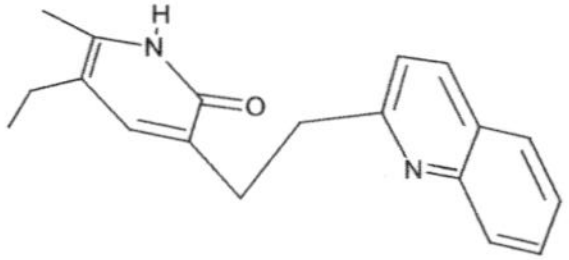	3-[(benzopyridine-2′-yl) ethyl]5-ethyl-6-methylpyridin-2(1H)-one	5.61	2.71	35.62	3.331

TABLE 3.21 Continued

No.	Type	Working Molecules		A^{0bs}	QSAR parameters		
		Structure	Name	log(1/IC$_{50}$)	LogP	POL (Å^3)	H (kcal/mol)
20.	G20	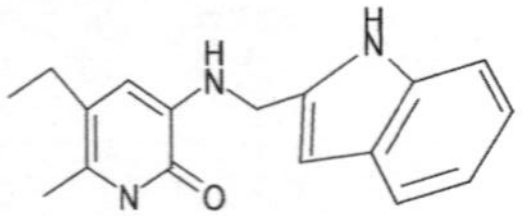	3-{[(indol-2′-yl) methyl] amino}-5-ethyl-6-methylpyridin-2(1H)-one	5.36	–0.34	32.63	4.727
21.	G21	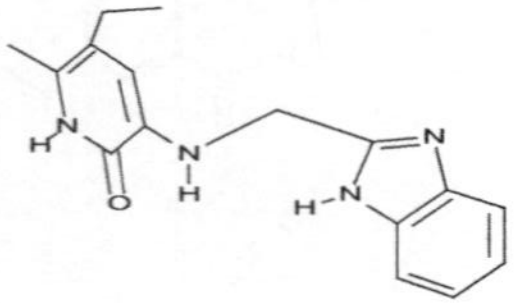	3-{[(quinazolin-2′-yl) methyl]amino}-5-ethyl-6-methylpyridin-2(1H)-one	5.12	0.02	31.92	8.171
22.	G22	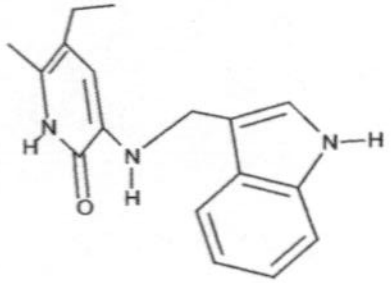	3-{[(indol-3′-yl)methyl] amino}-5-ethyl-6-methylpyridin-2(1H)-one	4.65	–0.43	32.63	2.957
23.	G23	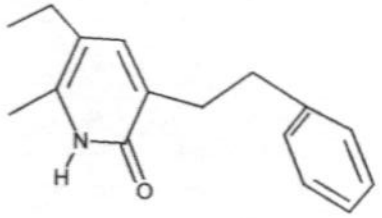	3-(β-phenilethyl)-5-ethyl-6-methylpyridin-2(1H)-one	4.30	2.36	29.06	–23.245
24.	NG1	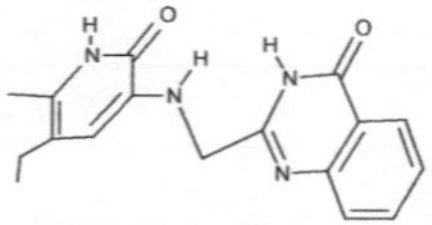	3-{[(4′-quinozolone-2′-yl) methyl]amino}-5-ethyl-6-methylpyridin-2(1H)-one	5.60	–0.47	33.85	–36.959

TABLE 3.21 Continued

No.	Type	Working Molecules		A^{Obs}	QSAR parameters		
		Structure	*Name*	*log(1/IC$_{50}$)*	*LogP*	*POL (Å^3)*	*H (kcal/mol)*
25.	NG2	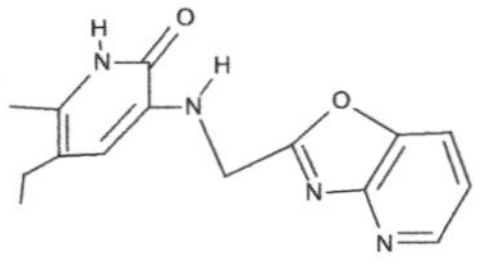	3-{[(3, '4′-diazobenzofuran-2′-yl) methyl]amino}-5-ethyl-6-methylpyridin-2(1H)-one	5.72	0.05	30.50	–8.120
26.	NG3	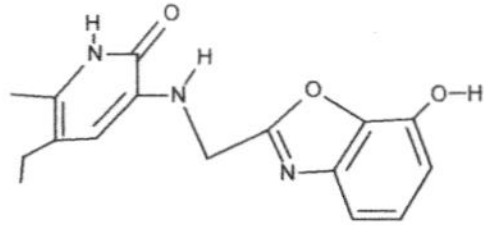	3-{[(7′-hydroxy-benzoxazole-2′-yl) methyl]amino}-5-ethyl-6-methylpyridin-2(1H)-one	6.36	–0.08	31.85	–62.189
27.	NG4	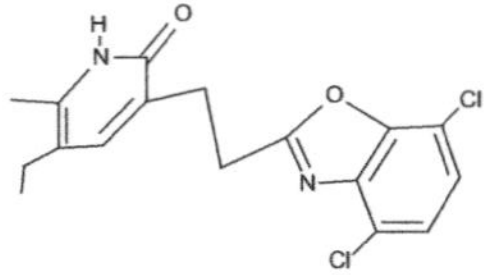	3-[(4, '7′-dichloro-benzoxazole-2′-yl) ethyl]-5-ethyl-6-methylpyridin-2(1H)-one	7.85	2.72	35.55	–39.459
28.	NG5	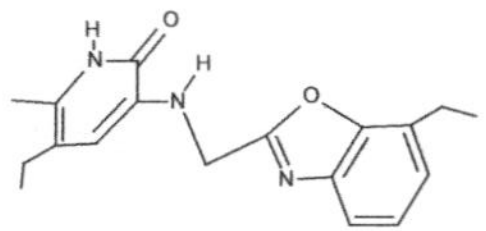	3-{[(7′-ethyl-benzoxazole-2′-yl) methyl]amino}-5-ethyl-6-methylpyridin-2(1H)-one	6.59	1.06	34.88	–34.478
29.	NG6	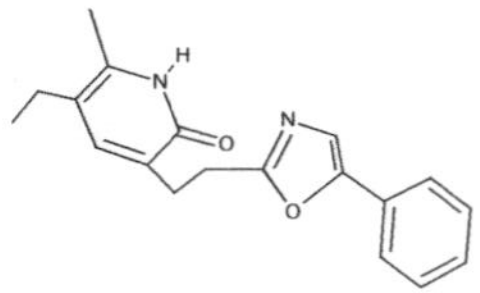	3-[(5′-phenyl-oxazole-2′-yl) ethyl]-5-ethyl-6-methylpyridin-2(1H)-one	6.41	0.96	35.17	–21.361

TABLE 3.21 Continued

No.	*Type*	*Working Molecules*		A^{0bs}	*QSAR parameters*		
		Structure	*Name*	*log(1/IC$_{50}$)*	*LogP*	*POL (Å^3)*	*H (kcal/mol)*
30.	NG7	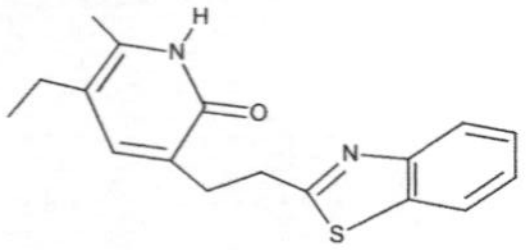	3-[(benzothiazole-2′-yl) ethyl]-5-ethyl-6-methylpyridin-2(1H)-one	6.43	2.02	34.06	8.873
31.	NG8	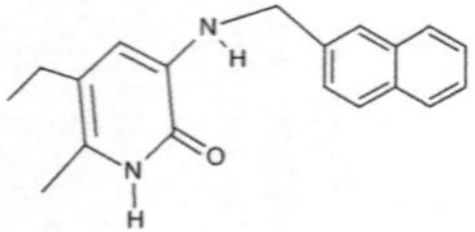	3-{[(2′naphtyl) methyl] amino}-5-ethyl-6-methylpyridin-2(1H)-one	6.34	1.67	35.85	5.495
32.	NG9	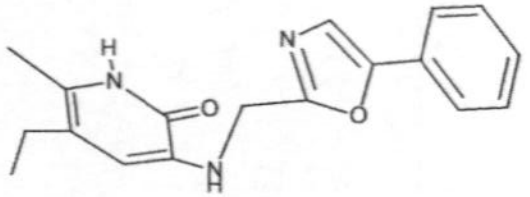	3-{[(5′-phenyl-oxazole-2′-yl) methyl]amino}-5-ethyl-6-methylpyridin-2(1H)-one	5.63	–0.53	34.69	–10.850

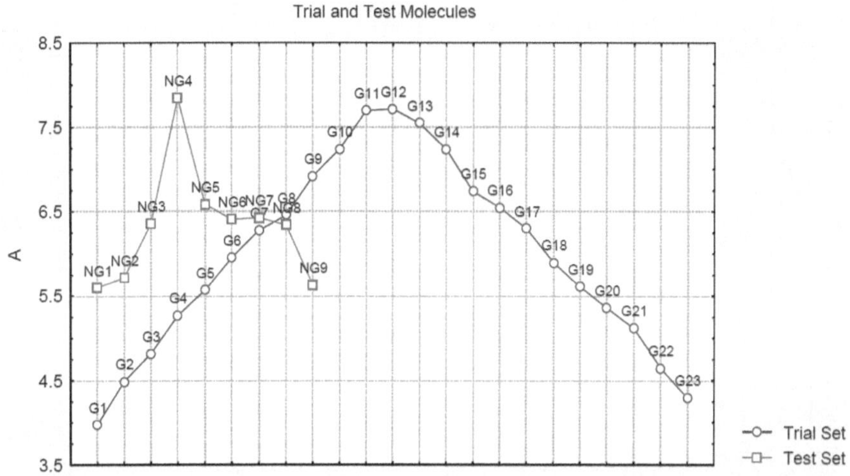

FIGURE 3.20 Gaussian (G) and non-Gaussian (NG) screening of the observed activities of the working molecules in Table 3.21 grouped into trial and test congener series (Putz et al., 2011b).

- First, it is clear that consideration of the catastrophe (polynomial) correlations is an improvement over the fashioned multi-linear QSAR statistics; see also Eq. (3.109).
- The hydrophobicity indicator gives generally low correlations with any polynomial (linear, multilinear or catastrophe) approach, being a quite irrelevant linear QSAR descriptor (Table 3.22) but improving up to twice its influence within the swallow tail and butterfly phenomenologies once its fifth and sixth power involvement are considered. Nevertheless, this provides a sign of the value of catastrophe-QSAR for achieving a deeper understanding of the molecular mechanics of specific interactions when the normal multi-linear QSAR does not assign transport descriptors with much predictive power.
- The relative statistical power, as defined by Eq. (3.112), does not always parallel the Pearson coefficient or the relative correlation factors, as is evident from Tables 3.22 and 3.23. However, because it includes more statistical information, we consider a model as relevant when it has greater individual output of this newly introduced statistical index. In particular, neither the linear nor the multilinear QSAR framework provides a good fit between the statistical correlation and the relative statistical power using the structural

TABLE 3.22 Correlation Equations for the Group-I Models of Table 3.4 and the Molecular Structures and Data of Table 3.21*

Catastrophe	QSAR Model	$R_{Pearson}$ [a]	R_{ALG} [b]	r [c]	*t-Stud.*	t [d]	*Fisher*	f [e]	Π [f]
QSAR (I)	$\lvert Y_I^{LogP}\rangle = 5.861\lvert 1\rangle + 0.240\lvert LogP\rangle$	0.228	0.984	4.317	22.344	7.854	1.150	0.143	8.963
	$\lvert Y_I^{POL}\rangle = -2.257\lvert 1\rangle + 0.249\lvert POL\rangle$	0.554	0.989	1.784	–0.832	–0.292	9.284	1.158	2.147
	$\lvert Y_I^{H}\rangle = 5.57\lvert 1\rangle - 0.021\lvert H\rangle$	0.476	0.987	2.074	20.597	7.24	6.156	0.768	7.57
Fold (F)	$\lvert Y_F^{LogP}\rangle = 5.854\lvert 1\rangle + 0.738\lvert LogP\rangle - 0.106\lvert LogP^3\rangle$	0.382	0.986	2.581	22.936	8.062	1.705	0.213	8.468
	$\lvert Y_F^{POL}\rangle = -24.206\lvert 1\rangle + 1.26\lvert POL\rangle - 3\cdot 10^{-4}\lvert POL^3\rangle$	0.601	0.989	1.646	–1.422	–0.45	5.650	0.704	1.859
	$\lvert Y_F^{H}\rangle = 5.58\lvert 1\rangle - 0.016\lvert H\rangle - 2\cdot 10^{-6}\lvert H^3\rangle$	0.481	0.987	2.053	20.095	7.063	3.01	0.375	7.365
Cusp (C)	$\lvert Y_C^{LogP}\rangle = 5.707\lvert 1\rangle + 0.426\lvert LogP\rangle + 0.372\lvert LogP^2\rangle - 0.071\lvert LogP^4\rangle$	0.348	0.985	2.832	16.120	5.666	0.872	0.109	6.335
	$\lvert Y_C^{POL}\rangle = 431.26\lvert 1\rangle - 35.694\lvert POL\rangle + 0.833\lvert POL^2\rangle - 10^{-4}\lvert POL^4\rangle$	0.713	0.992	1.391	2.240	0.787	6.558	0.818	1.796
	$\lvert Y_C^{H}\rangle = 5.006\lvert 1\rangle + 0.042\lvert H\rangle + 0.003\lvert H^2\rangle - 10^{-6}\lvert H^4\rangle$	0.764	0.993	1.300	19.802	6.960	8.864	1.105	7.166

TABLE 3.22 Continued

Catastrophe	QSAR Model	$R_{Pearson}^{(a)}$	$R_{ALG}^{(b)}$	$r^{(c)}$	*t-Stud.*	$t^{(d)}$	*Fisher*	$f^{(e)}$	$\Pi^{(f)}$
Swallow tail (ST)	$\left\|Y_{ST}^{LogP}\right\rangle = 5.649\left\|1\right\rangle + 1.608\left\|LogP\right\rangle + 0.326\left\|LogP^2\right\rangle - 0.978\left\|LogP^3\right\rangle + 0.093\left\|LogP^5\right\rangle$	0.575	0.989	1.720	18.665	6.561	2.222	0.277	6.788
	$\left\|Y_{ST}^{POL}\right\rangle = 1476.244\left\|1\right\rangle - 156.079\left\|POL\right\rangle + 5.791\left\|POL^2\right\rangle - 0.079\left\|POL^3\right\rangle + 5.5\cdot10^{-6}\left\|POL^5\right\rangle$	0.715	0.992	1.387	0.45	0.158	4.708	0.587	1.515
	$\left\|Y_{ST}^{H}\right\rangle = 4.884\left\|1\right\rangle + 0.031\left\|H\right\rangle + 0.004\left\|H^2\right\rangle + 5.2\cdot10^{-5}\left\|H^3\right\rangle + 4\cdot10^{-10}\left\|H^5\right\rangle$	0.763	0.993	1.302	15.608	5.486	6.263	0.781	5.692
Butterfly (B)	$\left\|Y_{B}^{LogP}\right\rangle = 5.646\left\|1\right\rangle + 1.464\left\|LogP\right\rangle + 0.303\left\|LogP^2\right\rangle - 0.688\left\|LogP^3\right\rangle - 0.041\left\|LogP^4\right\rangle + 0.027\left\|LogP^6\right\rangle$	0.578	0.989	1.711	15.169	5.332	1.704	0.212	5.604
	$\left\|Y_{B}^{POL}\right\rangle = -16485.827\left\|1\right\rangle + 2491.049\left\|POL\right\rangle - 146.094\left\|POL^2\right\rangle + 4.037\left\|POL^3\right\rangle - 0.047\left\|POL^4\right\rangle + 2.9\cdot10^{-6}\left\|POL^6\right\rangle$	0.718	0.992	1.382	–0.355	–0.125	3.619	0.451	1.459
	$\left\|Y_{B}^{H}\right\rangle = 4.876\left\|1\right\rangle + 0.110\left\|H\right\rangle + 0.004\left\|H^2\right\rangle - 2.3\cdot10^{-4}\left\|H^3\right\rangle - 7.67\cdot10^{-6}\left\|H^4\right\rangle + 6.3\cdot10^{-10}\left\|H^6\right\rangle$	0.856	0.996	1.163	19.088	6.709	9.349	1.166	6.908

[a] the statistical Pearson correlation factor (3.102); [b] computed from Eq. (3.111); [c] computed from Eq. (3.113);
[d] computed from Eq. (3.114) with $t_{Tabulated\,(0.99;20)} = 2.845$; [e] computed from Eq. (3.115) with $F_{Tabulated\,(0.99;1,21)} = 8.02$; [f] computed from Eq. (3.112).
*Putz et al. (2011b).

TABLE 3.23 Correlation Equations for the Group-II Models of Table 3.4 and the Molecular Structures and Data of Table 3.21*

Catastrophe	QSAR Model	$R_{Pearson}$ [(a)]	R_{ALG} [(b)]	r [(c)]	*t-Stud.*	t [(d)]	*Fisher*	f [(e)]	Π [(f)]
QSAR (II)	$\lvert Y_{II}^{LogP,POL}\rangle = -2.044\lvert 1\rangle + 0.051\lvert LogP\rangle + 0.242\lvert POL\rangle$	0.556	0.989	1.778	–0.702	–0.245	4.464	0.763	1.9504
	$\lvert Y_{II}^{LogP,H}\rangle = 5.379\lvert 1\rangle + 0.304\lvert LogP\rangle - 0.023\lvert H\rangle$	0.556	0.989	1.778	18.564	6.489	4.468	0.764	6.771
	$\lvert Y_{II}^{POL,H}\rangle = -2.637\lvert 1\rangle + 0.248\lvert POL\rangle - 0.021\lvert H\rangle$	0.728	0.992	1.363	–1.151	–0.402	11.302	1.932	2.398
Hyperbolic umbilic (HU)	$\lvert Y_{HU}^{LogP,POL}\rangle = -39.499\lvert 1\rangle - 2.463\lvert LogP\rangle + 2.043\lvert POL\rangle + 0.104\lvert (LogP)(POL)\rangle - 0.145\lvert LogP^3\rangle - 6\cdot 10^{-4}\lvert POL^3\rangle$	0.715	0.992	1.387	–2.215	–0.774	3.561	0.609	1.701
	$\lvert Y_{HU}^{LogP,H}\rangle = 5.319\lvert 1\rangle + 1.083\lvert LogP\rangle - 0.002\lvert H\rangle - 0.003\lvert (LogP)(H)\rangle - 0.161\lvert LogP^3\rangle - 9\cdot 10^{-6}\lvert H^3\rangle$	0.736	0.992	1.3485	19.328	6.756	4.019	0.687	6.923
	$\lvert Y_{HU}^{POL,H}\rangle = -13.192\lvert 1\rangle + 0.766\lvert POL\rangle + 0.122\lvert H\rangle - 0.004\lvert (POL)(H)\rangle - 2\cdot 10^{-4}\lvert POL^3\rangle - 5.1\cdot 10^{-7}\lvert H^3\rangle$	0.755	0.993	1.315	–0.79	–0.276	4.503	0.770	1.549
Elliptic umbilic (EU)	$\lvert Y_{EU}^{LogP,POL}\rangle_A = -69.262\lvert 1\rangle - 0.556\lvert LogP\rangle + 4.531\lvert POL\rangle + 0.443\lvert LogP^2\rangle - 0.068\lvert POL^2\rangle + 0.002\lvert (LogP)(POL^2)\rangle - 0.322\lvert LogP^3\rangle$	0.757	0.993	1.312	–2.548	–0.891	3.582	0.612	1.670
	$\lvert Y_{EU}^{LogP,POL}\rangle_B = 644.623\lvert 1\rangle + 0.022\lvert LogP\rangle - 59.934\lvert POL\rangle + 0.467\lvert LogP^2\rangle + 1.855\lvert POL^2\rangle - 0.015\lvert (POL)(LogP^2)\rangle - 0.019\lvert POL^3\rangle$	0.722	0.992	1.374	1.866	0.652	2.908	0.497	1.600

TABLE 3.23 Continued

Catastrophe	QSAR Model	$R_{Pearson}$ [a]	R_{ALG} [b]	r [c]	*t-Stud.*	t [d]	*Fisher*	f [e]	Π [f]
	$\left\|Y_{EU}^{LogP,H}\right\rangle_A = 5.022\|1\rangle + 0.974\|LogP\rangle + 0.025\|H\rangle$ $+0.530\|LogP^2\rangle + 0.001\|H^2\rangle$ $+2.87\cdot10^{-4}\|(LogP)(H^2)\rangle - 0.359\|LogP^3\rangle$	0.843	0.995	1.181	20.638	7.214	6.542	1.118	7.395
	$\left\|Y_{EU}^{LogP,H}\right\rangle_B = 4.779\|1\rangle + 0.643\|LogP\rangle + 0.029\|H\rangle$ $-0.211\|LogP^2\rangle + 0.004\|H^2\rangle$ $+0.001\|(H)(LogP^2)\rangle + 5\cdot10^{-5}\|H^3\rangle$	0.851	0.995	1.170	17.047	5.958	7.015	1.199	6.189
	$\left\|Y_{EU}^{POL,H}\right\rangle_A = 807.822\|1\rangle - 74.631\|POL\rangle - 0.02\|H\rangle$ $+2.291\|POL^2\rangle + 0.005\|H^2\rangle$ $-2\cdot10^{-4}\|(POL)(H^2)\rangle - 0.023\|POL^3\rangle$	0.857	0.996	1.162	3.124	1.092	7.346	1.256	2.029
	$\left\|Y_{EU}^{POL,H}\right\rangle_B = 11.888\|1\rangle - 0.562\|POL\rangle + 0.068\|H\rangle$ $+0.011\|POL^2\rangle + 0.004\|H^2\rangle$ $-4\cdot10^{-5}\|(H)(POL^2)\rangle + 4\cdot10^{-5}\|H^3\rangle$	0.853	0.996	1.167	0.532	0.186	7.120	1.217	1.696
Parabolic umbilic (PU)	$\left\|Y_{PU}^{LogP,POL}\right\rangle_A = 474.915\|1\rangle + 0.021\|LogP\rangle - 39.256\|POL\rangle$ $+0.454\|LogP^2\rangle + 0.914\|POL^2\rangle$ $-0.015\|(LogP^2)(POL)\rangle - 10^{-4}\|POL^4\rangle$	0.722	0.992	1.374	1.817	0.635	2.905	0.497	1.593
	$\left\|Y_{PU}^{LogP,POL}\right\rangle_B = -67.522\|1\rangle - 1.539\|LogP\rangle + 4.444\|POL\rangle$ $+0.573\|LogP^2\rangle - 0.067\|POL^2\rangle$ $+0.002\|(POL^2)(LogP)\rangle - 0.115\|LogP^4\rangle$	0.703	0.992	1.411	–2.219	–0.776	2.611	0.446	1.671

TABLE 3.23 Continued

Catastrophe	QSAR Model	$R_{Pearson}$ [(a)]	R_{ALG} [(b)]	r [(c)]	*t-Stud.*	t [(d)]	*Fisher*	f [(e)]	Π [(f)]
	$\left\|Y_{PU}^{LogP,H}\right\rangle_A = 4.852\|1\rangle + 0.700\|LogP\rangle + 0.041\|H\rangle$ $-0.240\|LogP^2\rangle + 0.004\|H^2\rangle$ $+0.002\|(LogP^2)(H)\rangle - 10^{-6}\|H^4\rangle$	0.874	0.996	1.140	20.243	7.075	8.645	1.478	7.317
	$\left\|Y_{PU}^{LogP,H}\right\rangle_B = 5.10\|1\rangle + 0.552\|LogP\rangle + 0.020\|H\rangle$ $+0.460\|LogP^2\rangle + 9.57\cdot10^{-4}\|H^2\rangle$ $+1.93\cdot10^{-4}\|(H^2)(LogP)\rangle - 0.099\|LogP^4\rangle$	0.767	0.993	1.295	16.828	5.882	3.815	0.652	6.058
	$\left\|Y_{PU}^{POL,H}\right\rangle_A = 8.876\|1\rangle - 0.366\|POL\rangle + 0.069\|H\rangle$ $+0.008\|POL^2\rangle + 0.003\|H^2\rangle$ $-3.7\cdot10^{-5}\|(POL^2)(H)\rangle - 4.5\cdot10^{-7}\|H^4\rangle$	0.841	0.995	1.183	0.386	0.135	6.447	1.102	1.623
	$\left\|Y_{PU}^{POL,H}\right\rangle_B = 595.212\|1\rangle - 48.906\|POL\rangle - 0.019\|H\rangle$ $+1.129\|POL^2\rangle + 5\cdot10^{-3}\|H^2\rangle$ $-1.49\cdot10^{-4}\|(H^2)(POL)\rangle - 1.73\cdot10^{-4}\|POL^4\rangle$	0.856	0.996	1.163	3.074	1.074	7.292	1.246	2.015

[(a)] the statistical Pearson correlation factor (3.102); [(b)] computed from Eq. (3.111); [(c)] computed from Eq. (3.113);
[(d)] computed from Eq. (3.114) with $t_{\substack{Tabulated\\(0.99;19)}} = 2.861$; [(e)] computed from Eq. (3.115) with $F_{\substack{Tabulated\\(0.99;2,20)}} = 5.85$; [(f)] computed from Eq. (3.112).

TABLE 3.24 Single-Structure Matrices of the Euclidean Distances $\Delta\Pi_l$ of the QSAR and Catastrophe Models' Relative Statistics of Table 3.22 Employing Eq. (3.116)*

(a)

LogP	***F***	***C***	***ST***	***B***
QSAR	1.750	2.645	2.905	3.627
F		2.411	1.732	2.865
C			1.437	1.174
ST				1.231

(b)

POL	*F*	*C*	*ST*	*B*
QSAR	0.517	1.198	0.828	0.830
F		1.317	0.717	0.524
C			0.670	0.983
ST				0.314

(c)

H	***F***	***C***	***ST***	***B***
QSAR	0.431	0.89	1.916	1.127
F		1.054	1.793	1.242
C			1.509	0.292
ST				1.29

*Putz et al. (2011b).

parameter combinations considered. Instead, parabolic catastrophe correlations, the *cusp and butterfly models*, are revealed to be quite relevant, in particular regarding the formation energy (H) for which they show the highest Pearson correlation and relative statistical power values in comparison with the other descriptors plugged into these models. Unfortunately, for the two-variable descriptor models of Table 3.23, no consistency was found between the highest Pearson value and the relative statistical power apart from a few degenerate cases of descriptors for the parabolic models where the highest relative statistical power value corresponds with the highest Pearson correlation. Note that for the degenerate cases of Table 3.23, when two mixed descriptors can be combined in two distinct ways, the working model is considered to have maximum relative statistical power.

TABLE 3.25 Differences $\Delta^2\Pi_I$ Between the Single-Structure Matrices of the Euclidean Distances in Table 3.24*

(a)

|LogP–POL|	***F***	***C***	***ST***	***B***
QSAR	1.233	1.446	2.076	2.797
F		1.094	1.015	2.341
C			0.767	0.191
ST				0.917

(b)

|LogP–H|	*F*	*C*	*ST*	*B*
QSAR	1.32	1.755	0.988	2.501
F		1.358	0.062	1.624
C			0.072	0.882
ST				0.059

(c)

|POL-H|	***F***	***C***	***ST***	***B***
QSAR	0.086	0.309	1.088	0.297
F		0.264	1.076	0.717
C			0.839	0.691
ST				0.976

*Putz et al. (2011b).

However, because the two-fold aim of the present research is to find the best predictive model and the molecular mechanism of action for the given set of molecules, the statistical indices of Tables 3.22 and 3.23 are employed to compute the first- and second-order differences (or distances) in relative statistical power as described by Eqs. (3.116)–(3.119) of Section 3.1.6. They correspond to the inter-descriptor/inter-modeling paths of molecular actions, whose minimum values are identified according to the prescription of Eq. (3.120).

Through this minimal relative statistical power path recipe, once the models and descriptors predicted to be on the forefront of the structure-action interaction are selected, they are then further filtered with the testing

TABLE 3.26 Single-Structure Matrices of the Euclidean Distances $\Delta\Pi_{II}$ of the QSAR and Catastrophe Models' Relative Statistics of Table 3.23 Employing Eq. (3.116); Note that for the Degenerate Models of Table 3.23 That One is Employed that Displays Higher Relative Statistical Power (Π)*

(a)

LogP^POL	***HU***	***EU***	***PU***
QSAR	0.675	0.810	1.005
HU		0.139	1.414
EU			1.531

(b)

POL	***F***	***C***	***ST***
QSAR	0.512	0.917	1.123
HU		0.964	0.878
EU			1.152

(c)

H	***F***	***C***	***ST***
QSAR	1.170	1.652	1.640
HU		1.46	1.440
EU			0.02

*Putz et al. (2011b).

set to finally identify the best predictive model and reveal the mechanism of action by means of the structural descriptors considered.

In the present case of the HIV inhibitors in Table 3.21, the data computed from Tables 3.22 and 3.23 provide the results for Tables 3.24–3.26, to be discussed herein (Putz et al., 2011b):

- Table 3.24: At the individual descriptor level, the cusp and butterfly models are very close to each other for LogP and the forming energy H, which is even more relevant for the hydrophobicity, because for the forming energy it transpires from Table 3.22 that the butterfly model practically reduces to the cusp model because the sixth contribution virtually vanishes. However, for the structural influence on polarizability (POL) the butterfly and swallow-tail are the closest models. When one considers the hierarchy of the individual descriptors according to their QSAR-I models in Table 3.22 in terms of the reduction in relative statistical power

$$LogP \rightarrow H \rightarrow POL \tag{3.250}$$

through combining it with the catastrophes involved in Table 3.24, one correspondingly obtains the evolution cycle of the models:

$$(\ldots \rightarrow)[Butterfly] \rightarrow [Cusp] \rightarrow [Butterfly] \rightarrow [SwallowTail](\rightarrow \ldots) \tag{3.251}$$

- Table 3.25: When the second order distance difference is considered between the individual inter-modeling paths of Table 3.24 can nevertheless be considered through the further variations of paths of Table 3.24. Also, the QSAR-I and the fold (F) catastrophe model intervene in changing the influence on specific interactions from POL to H. Therefore, by counting the minimum hierarchy of these paths, the distance ordering is obtained as follows:

$$(LogP \div H) \rightarrow (H \div POL) \rightarrow (POL \div LogP) \tag{3.252}$$

which, remarkably, confirms the descriptors' cycles of influence in accordance with the first order prescription of Eq. (3.250). However, a more detailed succession is recorded for the inter-model evolution:

$$(\ldots \rightarrow)[Butterfly] \cong [SwallowTail] \rightarrow [QSAR - I] \cong [Fold] \rightarrow [Cusp] \cong [Butterfly](\ldots \rightarrow) \tag{3.253}$$

When comparing cycles (3.253) with (3.251), it seems that the QSAR-I and Fold models appear in (3.253) at the second cycle after the first one is performed on the prescription of (3.251). For this reason also, the direct second order inter-descriptor-inter-models analysis is undertaken, and the results are reported in Table 3.26, to be discussed hereafter.

- Table 3.26: Interestingly, in terms of the two structural descriptors, the QSAR model is present even though its individual statistics are not the highest in Table 3.23; however, judging by the ordering of minimum paths recorded, the coupling descriptors hierarchy is established as:

$$(H \& POL) \rightarrow (POL \& LogP) \rightarrow (LogP \& H) \tag{3.254}$$

which is associated with the models' evolution

$$(\ldots \rightarrow)[PU] \rightarrow [EU] \rightarrow [HU] \rightarrow [QSAR](\rightarrow [PU] \ldots \rightarrow) \tag{3.255}$$

One should make "contact" between the descriptor hierarchies [(3.250), (3.252), (3.254)] and the models' cycles [(3.251), (3.253) and (3.255)] by means of the predictivity powers of the models along the minimum paths identified in Tables 3.24 and 3.26 with the single and double descriptors, respectively, for the non-Gaussian (NG) molecules of Table 3.21 and Figure 3.20. The results are systematically presented in Tables 3.27 and 3.28 (Putz et al., 2011b).

The results of correlation tests in Table 3.27 indicate the structure index–model activity hierarchy (Putz et al., 2011b):

$$\left|Y_B^{LogP}\right\rangle > \left|Y_B^{POL}\right\rangle > \left|Y_C^{LogP}\right\rangle > \left|Y_{ST}^{POL}\right\rangle > \left|Y_C^{H}\right\rangle > \left|Y_B^{H}\right\rangle \tag{3.256}$$

TABLE 3.27 Predicted Activity As Computed for the Non-Gaussian Molecules of Table 3.21 With the Models of Table 3.22 Founded Along the Minimum Paths of Table 3.24; for Each Predicted Model, Its Correlation With the Observed Activity is Indicated at the Bottom of the Table (Putz et al., 2011b)

Model / **Molecule**	$\left\|Y_C^{LogP}\right\rangle$	$\left\|Y_C^{H}\right\rangle$	$\left\|Y_{ST}^{POL}\right\rangle$	$\left\|Y_B^{LogP}\right\rangle$	$\left\|Y_B^{POL}\right\rangle$	$\left\|Y_B^{H}\right\rangle$
NG1	5.586	6.179	5.294	5.094	–20.595	5.687
NG2	5.729	4.885	4.294	5.719	–9.764	4.360
NG3	5.676	0.415	4.708	5.531	–13.457	–7.932
NG4	5.729	6.156	5.149	6.657	–29.709	5.259
NG5	6.487	6.141	5.309	6.705	–25.700	5.923
NG6	6.399	5.438	5.258	6.708	–27.365	5.219
NG7	6.903	5.631	5.319	5.311	–21.540	5.984
NG8	6.904	5.334	5.027	5.995	–31.693	5.566
NG9	5.580	4.9357	5.328	5.054	–24.666	4.383
R-Pearson	***0.195***	***0.129***	***0.174***	***0.701***	***0.488***	***0.026***

Somehow the influences of POL and H are reversed relative to the prescription by trial succession of Eq. (3.250), revealing hydrophobicity as the main influential factor. However, due to the fact that the predicted activities of POL in Table 3.27 are all in the "opposite evolution direction" with respect to the activities recorded in Table 3.21, i.e., they are all negative, the uni-parametric tests and their associated hierarchy (3.256) are discarded, and one looks toward the second class of QSAR and catastrophe algorithms.

Instead, the test correlations of Table 3.28 provide the structure-activity ordering for the bi-parameter-models

$$\left|Y_{II}^{LogP,H}\right\rangle > \left|Y_{HU}^{LogP,POL}\right\rangle > \left|Y_{HU}^{LogP,H}\right\rangle > \left|Y_{PU}^{POL,H}\right\rangle_B > \left|Y_{EU}^{LogP,POL}\right\rangle_A > \left|Y_{EU}^{POL,H}\right\rangle_A \tag{3.257}$$

Remarkably, the hierarchy (3.257) starts with the QSAR model, which is revealed to be at the top of the validated catastrophe models with statistical performance even higher than through the predicted equation of Table 3.23 and the trial set of Table 3.21. Moreover, the QSAR-II model involves parameters (LogP & H) that are followed by the hyperbolic umbilic (HU) model in terms of (LogP & POL) parameters, in this way

TABLE 3.28 Predicted Activity As Computed for the Non-Gaussian Molecules of Table 3.21 With the Models of Table 3.23 Founded Along the Minimum Paths of Table 3.26; for Each Predicted Model, Its Correlation With the Observed Activity is Indicated At the Bottom of the Table (Putz et al. 2011b)

Model / **Molecule**	$\left\|Y_{II}^{LogP,H}\right\rangle$	$\left\|Y_{HU}^{LogP,POL}\right\rangle$	$\left\|Y_{HU}^{LogP,H}\right\rangle$	$\left\|Y_{EU}^{LogP,POL}\right\rangle_A$	$\left\|Y_{EU}^{POL,H}\right\rangle_A$	$\left\|Y_{PU}^{POL,H}\right\rangle_B$
NG1	6.0865	5.918	5.308	5.387	5.351	7.210
NG2	5.581	5.839	5.399	5.448	4.816	4.578
NG3	6.785	6.132	7.526	5.686	1.423	7.234
NG4	7.115	6.642	6.037	6.289	5.480	7.765
NG5	6.495	7.382	6.853	7.277	6.033	7.629
NG6	6.163	7.291	6.426	7.104	7.338	7.647
NG7	5.790	7.388	6.087	7.615	6.879	6.547
NG8	5.761	7.560	6.330	7.640	7.895	7.447
NG9	5.467	5.755	4.786	5.177	7.586	7.303
R-Pearson	***0.778***	***0.468***	***0.454***	***0.431***	***0.057***	***0.451***

recovering the original mono-structural influences as anticipated by Eqs. (3.250) and (3.252). Thus, the series of models in Eq. (3.257) is validated, and it will be further employed to establish the models' successions and the molecular structural pattern of inhibiting anti-HIV-1 drug resistance.

To this end, apart from the first and last models of Eq. (3.257), which are associated with the maximum (0.778) and minimum (0.057) test performance, the middle catastrophe models provide closely related performance in the range (0.431, 0.468). Their graphical 3D-representation of the parametric domains LogP: (–1.50, 2.72), POL: (27.87, 38.48) and H: (–63.299, 17.808) of all (trial and test) structures in Table 3.21 are displayed in Figure 3.21. Next, it is apparent that they can be coupled according to the same spanned domains, thus forming the activity models' differences $\left|Y_{II}^{LogP,H}\right\rangle - \left|Y_{HU}^{LogP,H}\right\rangle$, $\left|Y_{HU}^{LogP,POL}\right\rangle - \left|Y_{EU}^{LogP,POL}\right\rangle_{A'}$, $\left|Y_{EU}^{POL,H}\right\rangle_{A} - \left|Y_{PU}^{POL,H}\right\rangle_{B'}$, plotted in the top of Figure 3.22. Through registering the parameters and the models' successions:

$$[QSAR] \xrightarrow{LogP,H} [HU] \xrightarrow{LogP,POL} [EU] \xrightarrow{POL,H} [PU] \qquad (3.258)$$

one may reach the following important conceptual-computational conclusions (Putz et al., 2011b):

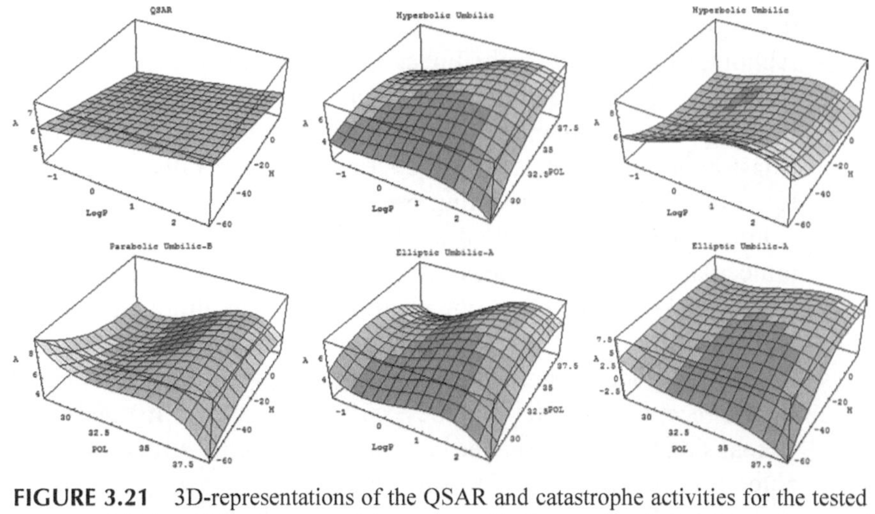

FIGURE 3.21 3D-representations of the QSAR and catastrophe activities for the tested models of Table 3.28 in the range of the structural indicators (LogP, Pol, H) as abstracted from Table 3.21 (Putz et al., 2011b).

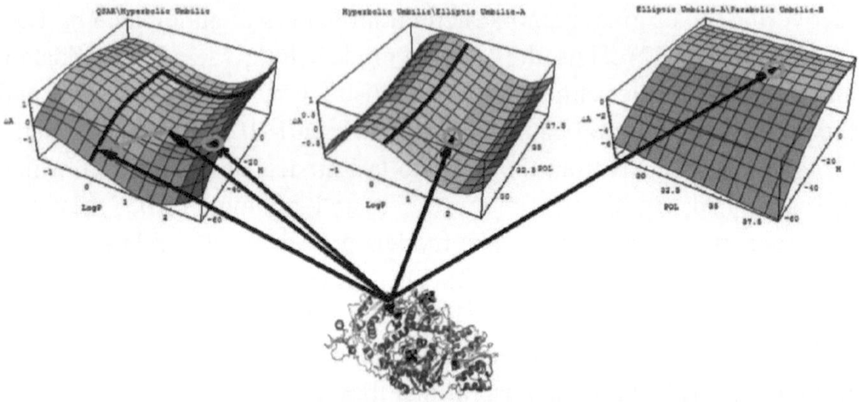

FIGURE 3.22 Determination of the structural domains of pyridinone-derivative type non-nucleoside reverse transcriptase inhibitors in the same range of structural descriptors by employing the principles of hydrophobicity, minimum polarizability, binding energy, and the minimum difference between the polynomial activity models of Figure 3.21 (Putz et al., 2011b); the hydrophobic pocket was identified in the p66 subunit of HIV-1-rt of specific transferase R221239 (Himmel et al., 2005; EBI, 2011).

- The HIV-1 inhibitory activity is triggered by a hydrophobic interaction followed by energetic stabilization of the ligand/substrate (pyrididone derivative/viral protein) interaction here modeled by the heat of molecular formation and eventually completed by the ionic field influence herein represented by the polarizability descriptor.
- Although the QSAR multi-linear model should not be excluded from the molecular modeling of complex bio-chemical interactions, it should be complemented with other polynomial correlational catastrophe-type models that produce significant results comparable to those of other 3D-modeling procedures such as docking-based comparative molecular field analysis (CoMFA) and comparative molecular similarity indices analysis (CoMSIA) (Medina-Franco et al., 2004). At this point one should stress that although CoMFA and CoMSIA results are superior to the catastrophe-QSARs, due to the grid-docking procedure that highly favors the modeling of ligand-receptor interaction, they suffer of lacking in analytical "expansion" of the modeled activity in terms of structural variables (i.e., on the behavior space); instead, Catastrophe-QSARs systematically improve over ordinary multi-linear regressions, allow the interpretation of the structure variable behavior, provide

thoroughly the design of the applicability domain, while identifying the most potent active molecules, as CoMFA/CoMSIA predicts (see below).

However, the issue remains of establishing the molecular structure most suitable for HIV-1 inhibitory activity among the considered pool of pyridinone derivatives in Table 3.21. To this end, the representations in Figure 3.22 are synergistically employed to identify the molecular structural domains that optimally promote binding of the pyridine derivative to the hydrophobic pocket in the p66 subunit of HIV-1 through searching for joint fulfillment of the following structural parameters and inter-model evolutionary generic principles (Putz et al., 2011b):

- LogP: For positive values, the compound behaves hydrophobically and requires dissolution in an organic solvent; by contrast, for negative values the compound is hydrophilic and can be dissolved directly in an aqueous buffer. For LogP equal to 0, the compound partitions at a 1:1 organic-to-aqueous phase ratio, meaning that it is likely soluble in both organic and aqueous solvents and in cellular environments; thus, values of LogP equal to or greater than zero are selected to achieve hydrophobicity and suitability for the cellular environment (Leo et al., 1971; Cronin & Mark, 2006), while characterizing the stacking bonding of aromatic rings (Selassie, 2003);
- H: Because the formation of a compound from its elements usually is an exothermic process, most heats of formation are negative, and this is also a characteristic of the dynamic equilibrium of ligand-substrate interactions (Masterton et al., 1983); note that the advantage of using heat of formation as QSAR descriptor resides in the following: it thermodynamically relates with the free energy $\Delta G = -RT \ln K_{eq}$ by the equilibrium constant K_{eq} which parallels the recorded activity at thermodynamic level (Medina-Franco et al., 2004); it nevertheless expands the Gibbs free energy from the hydrogen to covalent bonding strength (Selassie, 2003);
- PO: It is expected that “the natural direction of evolution of any system is towards a state of minimum polarizability” (Chattaraj & Sengupta, 1996), while accounting for the dipolar interaction (Selassie, 2003);
- Activity Models: Represent the same chemical-biological process providing their differences with respect to structural domains are minimized to zero.

These principles are applied to the activity models' differences at the top of Figure 3.22, and they lead to the identification of the structural domain (and even points) characteristic of the pyridinone derivative most well-adapted to inhibiting the HIV-1 life cycle. The graphical results in Figure 3.22 suggest that the ordering of the structural indicators is (Putz et al., 2011b):

$\left|Y_{II}^{LogP,H}\right\rangle - \left|Y_{HU}^{LogP,H}\right\rangle$:

$$\{LogP:(0, 1.5)\&H:(-55, -40)\ kcal/mol\} \cup \{LogP \cong 2.5\&H \cong -40\ kcal/mol\} \quad (3.259)$$

$\left|Y_{HU}^{LogP,POL}\right\rangle - \left|Y_{EU}^{LogP,POL}\right\rangle_A$:

$$LogP \cong 1\ \&\ POL \cong 32\ Å^3 \quad (3.260)$$

$\left|Y_{EU}^{POL,H}\right\rangle_A - \left|Y_{PU}^{POL,H}\right\rangle_B$:

$$POL \cong 34.5\ Å^3;\ H \cong -10\ kcal/mol \quad (3.261)$$

The "solution" of system (3.259)–(3.261) gives the actual molecules in Table 3.21 predicted to be the most potent binding inhibitors, namely compounds **27** (LogP≅2.72, H≅ −39.459kcal/mol, POL≅ 35.55Å^3), **28** (LogP≅1.06, H≅ −34.478kcal/mol, POL≅ 34.88Å^3), and **29** (LogP≅0.96, H≅−21.361kcal/mol, POL≅ 35.17Å^3). Most impressively, these molecules were also predicted by the much more sophisticated methods of CoMFA and CoMSIA as having increased binding affinity between the aromatic ring (or wing 2 of the pyridinone derivative) and amino acid Tyr181 of the first molecule and Tyr188 of the last two. These two amino acids are very important in the inhibition of RT by NNRTIs because the most common mutations are Tyr181Cys and Tyr188Cys, and they are responsible for the emergence of viruses resistant to pyridinone derivatives. Therefore, designing pyridinone compounds that allow aromatic ring stacking interactions with Tyr181 and Tyr 188 may prevent these mutations and increase the activity of these anti-HIV drugs.

Worth noting that all the modeling stages required by the OECD – QSAR principles (OECD, 2007; Putz et al., 2011a) are implemented here in a synergistic manner, namely (Putz et al., 2011b):

(i) *A defined endpoint*: The hydrophobic binding of the inhibitor in the pocket of the p66 subunit of reverse-transcriptase was confirmed herein through the identification of hydrophobicity as the major influence among all the mono-nonlinear catastrophes employed; see Eqs. (3.250) & (3.251).

(ii) *An unambiguous algorithm*: The Spectral-SAR minimum path principle (Putz & Lacrămă, 2007; Putz et al., 2009a; Putz & Putz, 2011; Putz, 2012b) is here generalized to include relevant combination of statistical information (e.g., the correlation factor R, Student's t-test, Fischer's F-test) to provide an equal footing multi-dimensional Euler distance [see Eqs. (3.112)–(3.120)], thus avoiding the previously identified discrepancy in judging the mid-range performance in terms of correlation or other statistical factors (Putz et al., 2009a).

(iii) *A defined domain of applicability*: By performing linear vs. nonlinear QSARs, the present strategy allows for the identification of recommended applicable structural domains through setting their difference to zero via inter-model activity minimization, which is equivalent to assuring the "smoothness" of the inhibitor-protein binding evolution towards the final steric inhibition output.

(iv) *Appropriate measures of goodness-of-fit, robustness and predictivity*: The trial results were evaluated by external validation employing a testing set, which was selected by means of Gaussian vs. non-Gaussian distributions of the compounds' activities, an improvement over the earlier arbitrariness of sampling the compounds only within a certain activity range. For instance, for linear QSAR the predicted correlation was superior to the tested correlation, thus confirming the reliability of this validation technique.

(v) *A mechanistic interpretation*: The selected succession of catastrophe-QSARs indicates that the inhibitor-HIV protein binding mutations that are involved in "birth and death" processes are associated with "waves" of induced activity in certain structural domain variants (see Figure 3.21). Moreover, the flat QSAR hypersurface should be complemented with catastrophe analysis to determine the specific structural domains for optimum interactions (see Figure 3.22) and for the associated molecular structure design of NNRT inhibitors.

Overall, the QSAR presented here combined with catastrophe polynomial structure activity relationships provides a reliable conceptual and computational tool for identifying the mechanisms underlying ligand-subtract interactions and the structural domains best able to promote them. Consequently, this method should be further integrated into automated data processing and tested on other complex open systems with bio- or eco- toxicological relevance, especially where evolutionary life-cycles are present.

3.3.5 QUANTUM AMPLITUDE (ON BREAST ANTI-CANCER BIOACTIVITY)

Although in general considered beneficial with the protective role in many age-related diseases – flavonoids (see Figure 3.23 – with the general scheme in no.0) should be more carefully studied since their not entirely elucidated pharmacokinetics (Middleton et al., 2000; Sargent et al., 2001; Havsteen, 2002; Zhang et al., 2004a-b, 2005).

For instance, recently, there was inferred that for certain flavonoids such as Chrysin, Biochanin A and Apigenin a very low micromolar concentration is able in producing 50% (EC_{50}) of the maximum increase in mitoxantrone (MX) inhibitor substrate accumulation (interaction) with breast cancer resistance protein (BCRP) helping in reversing the multidrug resistance (MDR) mechanism of overexpressing MCF-7 MX100 cancer cells (Sargent et al., 2001; Zhang et al., 2004a-b, 2005).

Therefore, in order to assess the molecular role and structural- related mechanisms for potential lead compounds in the drug design for anticancer treatment, a series of representative classes of flavonoids have been in next employed, see Figure 3.23, with their recorded biological activities (A) among the computed transport (hydrophobicity-LogP), the electrostatic (polarizability POL), and steric (total energy at optimized 3D-configuration E_{TOT}) Hansch correlation variables (Hansch, 1969), see Table 3.29, to successively provide the QSAR, S-SAR and finally to unfold the Qua-SAR analysis; this discussion follows (Putz et al., 2009b).

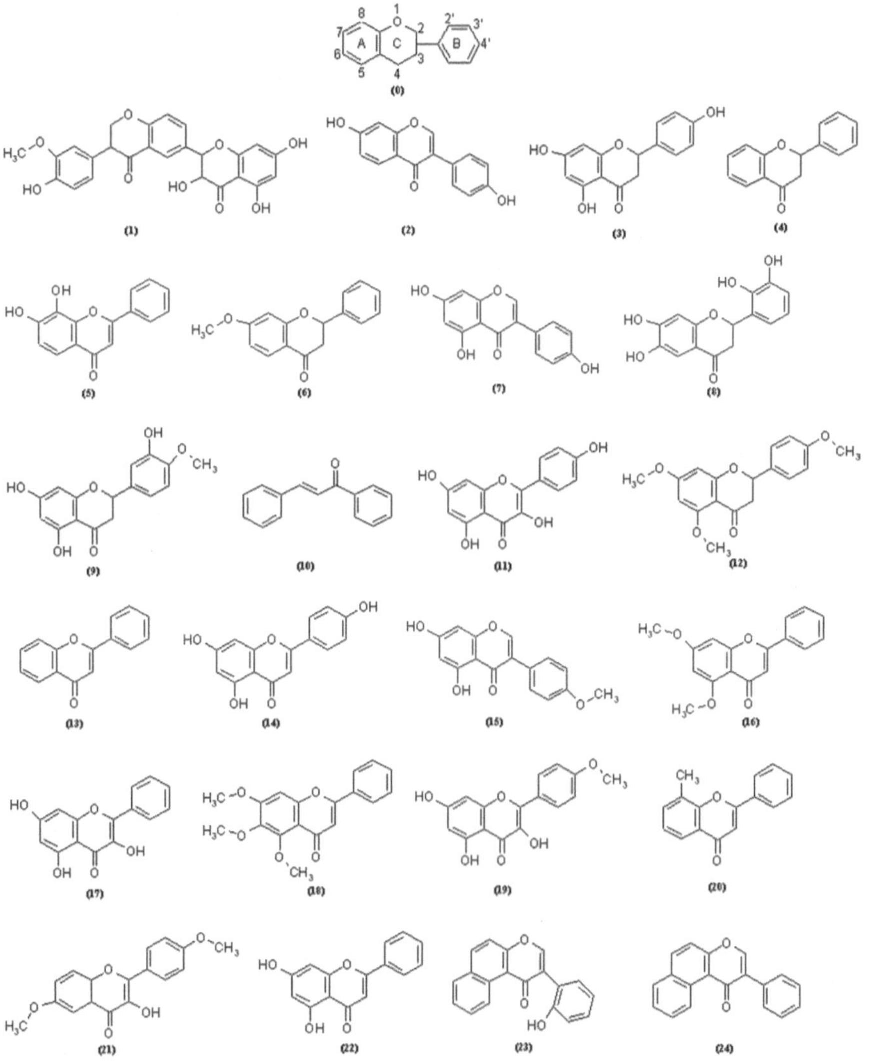

FIGURE 3.23 The studied flavonoids (Putz et al., 2009b) (with basic structure of as no.0 while the others are in the Table 3.29 characterized by associate QSAR data), covering the flavones, isoflavones, chalcones, flavonols and flavanones, as they assist the increase of mitoxantrone (MX) accumulation in BCRP-overexpressing MCF-7 MX100 breast caner cells (Zhang et al., 2005)

Note that in Table 3.29 the molecules were displayed in ascendant order of their recorded activities, from no. 1 to no. 24, for having present which is superior to which in each time of their reciprocal quotation. Such arrangement allows the construction of activity differences chart, see Table 3.30,

TABLE 3.29 The Flavonoids of Figure 3.23 Arranged Upon Their Ascending Observed Activities, Defined as A= $-\log_{10}(EC_{50}[\mu M])$ (Zhang et al., 2005), Along the Associate Computed Structural Parameters (Putz et al., 2009b) Like the Hydrophobicity (LogP), Electronic Cloud Polarizability (POL) and the Ground State Configurationally Optimized Total Energy (E_{TOT}) (Hypercube, 2002)

No.	*Molecular Name*	*Activity A*	*Structural parameters*		
			LogP	*POL(Å³)*	E_{TOT}*(kcal/mol)*
(1)	Silybin	3.74	2.03	45.68	–146625.1875
(2)	Daidzein	4.24	1.78	26.63	–76984.7109
(3)	Naringenin	4.49	1.99	27.46	–85032.9218
(4)	Flavanone	4.6	2.84	25.55	–62849.3125
(5)	7,8-Dihydroxyflavone	4.7	1.75	26.63	–76982.1328
(6)	7–Methoxyflavanone	4.79	2.59	28.02	–73823.8046
(7)	Genistein	4.83	1.50	27.27	–84380.7578
(8)	6,2′, 3′ -7-Hydroxyflavanone	4.85	1.70	28.10	–92422.6640
(9)	Hesperetin	4.91	1.73	29.93	–96003.9921
(10)	Chalcone	4.93	3.68	25.49	–55450.1093
(11)	Kaempferol	5.22	0.56	27.90	–91770.5859
(12)	4′-5,7-Trimethoxyflavanone	5.25	2.08	32.96	–95768.9062
(13)	Flavone	5.4	2.32	25.36	–62196.3437
(14)	Apigenin	5.78	1.46	27.27	–84379.8593
(15)	Biochanin A	5.79	1.53	29.10	–87961.2812
(16)	5,7-Dimethoxyflavone	5.85	1.81	30.30	–84139.4687
(17)	Galangin	5.92	0.85	27.27	–84376.8359
(18)	5,6,7–Trimethoxyflavone	5.96	1.56	32.77	–94976.1875
(19)	Kaempferide	5.99	0.60	29.74	–95351.3984
(20)	8-Methylflavone	6.21	2.79	27.19	–65789.9218
(21)	6,4′-Dimethoxy-3-hydroxy-flavone	6.35	0.41	31.13	–92162.7187
(22)	Chrysin	6.41	1.75	26.63	–76986.1171
(23)	2′-Hydroxy-α-naphtoflavone	7.03	3.07	33.26	–82027.8359
(24)	7,8-Benzoflavone	7.14	3.35	32.63	–74634.5234

with great utility in establishing *the inter-endpoint molecular activity differences* of Eq. (3.128) entering *quantum-SAR factor* of Eq. (3.240).

Next, for computing the other influential activity difference in Qua-SAR, namely the *inter-endpoint norm difference* of Eq. (3.127), the

TABLE 3.30 The Anti-Symmetric Matrix of the Inter-Molecular Activity Differences for the Working Flavonoids of Table 3.29 (Putz et al., 2009b)

1	**2**	**3**	**4**	**5**	**6**	**7**	**8**	**9**	**10**	**11**	**12**	**13**	**14**	**15**	**16**	**17**	**18**	**19**	**20**	**21**	**22**	**23**	**24**	
0	0.5	0.75	0.86	0.96	1.05	1.09	1.11	1.17	1.19	1.48	1.51	1.66	2.04	2.05	2.11	2.18	2.22	2.25	2.47	2.61	2.67	3.29	3.4	**1**
	0	0.25	0.36	0.46	0.55	0.59	0.61	0.67	0.69	0.98	1.01	1.16	1.54	1.55	1.61	1.68	1.72	1.75	1.97	2.11	2.17	2.79	2.9	**2**
		0	0.11	0.21	0.3	0.34	0.36	0.42	0.44	0.73	0.76	0.91	1.29	1.3	1.36	1.43	1.47	1.5	1.72	1.86	1.92	2.54	2.65	**3**
			0	0.1	0.19	0.23	0.25	0.31	0.33	0.62	0.65	0.8	1.18	1.19	1.25	1.32	1.36	1.39	1.61	1.75	1.81	2.43	2.54	**4**
				0	0.09	0.13	0.15	0.21	0.23	0.52	0.55	0.7	1.08	1.09	1.15	1.22	1.26	1.29	1.51	1.65	1.71	2.33	2.44	**5**
					0	0.04	0.06	0.12	0.14	0.43	0.46	0.61	0.99	1	1.06	1.13	1.17	1.2	1.42	1.56	1.62	2.24	2.35	**6**
						0	0.02	0.08	0.1	0.39	0.42	0.57	0.95	0.96	1.02	1.09	1.13	1.16	1.38	1.52	1.58	2.2	2.31	**7**
							0	0.06	0.08	0.37	0.4	0.55	0.93	0.94	1	1.07	1.11	1.14	1.36	1.5	1.56	2.18	2.29	**8**
								0	0.02	0.31	0.34	0.49	0.87	0.88	0.94	1.01	1.05	1.08	1.3	1.44	1.5	2.12	2.23	**9**
									0	0.29	0.32	0.47	0.85	0.86	0.92	0.99	1.03	1.06	1.28	1.42	1.48	2.1	2.21	**10**
										0	0.03	0.18	0.56	0.57	0.63	0.7	0.74	0.77	0.99	1.13	1.19	1.81	1.92	**11**
											0	0.15	0.53	0.54	0.6	0.67	0.71	0.74	0.96	1.1	1.16	1.78	1.89	**12**
												0	0.38	0.39	0.45	0.52	0.56	0.59	0.81	0.95	1.01	1.63	1.74	**13**
													0	0.01	0.07	0.14	0.18	0.21	0.43	0.57	0.63	1.25	1.36	**14**
														0	0.06	0.13	0.17	0.2	0.42	0.56	0.62	1.24	1.35	**15**
															0	0.07	0.11	0.14	0.36	0.5	0.56	1.18	1.29	**16**
																0	0.04	0.07	0.29	0.43	0.49	1.11	1.22	**17**
																	0	0.03	0.25	0.39	0.45	1.07	1.18	**18**
																		0	0.22	0.36	0.42	1.04	1.15	**19**

TABLE 3.30 Continued

1	2	3	4	5	6	7	8	9	10	11	12	13	14	15	16	17	18	19	20	21	22	23	24	
																			0	0.14	0.2	0.82	0.93	**20**
																				0	0.06	0.68	0.79	**21**
																					0	0.62	0.73	**22**
																						0	0.11	**23**
																							0	**24**

C=10 possible endpoint models with data of Table 3.29 are in Table 3.31 presented. However, worth remarking that the traditional hydrophobicity factor LogP seems to have quite little or even no-influence from traditional statistical correlation (model Ia).

The first conclusion is that flavonoids have practically no exclusive or primarily role in drug transporting to BCRP site; still, the electrostatic influence through POL is practically missing as well (model Ib), while the stericity through E_{TOT} unfolds some statistically sensitive role in ligand (MX)-receptor (BCRP) binding (model Ic). The last assertion may be sustained also by going to the two-correlated parameters

TABLE 3.31 QSAR Equations Through Spectral-SAR Multi-Linear Procedure, See Section 3.2.2 and (Putz & Lacrămă, 2007), for All Possible Correlation Models Considered From Data of Table 3.29 (Putz et al., 2009b); Here $|X_0\rangle$ is the Unitary Vector $|11\ldots1_{24}\rangle$, While the Structural Variables Are Set As $|X_1\rangle = LogP$, $|X_2\rangle = POL$, and $|X_3\rangle = E_{TOT}$; the Predicted Activities' Norms Where Calculated with Eq. (3.123), While the Algebraic Correlation Factor of Eq. (3.124) Uses the Measured Activity of $\| |A\rangle \| = 26.9357$ Computed Upon Eq. (3.123) With Data of Table 3.29; $R_{Statistic}$ is the Traditional Pearson correlation Factor, See Chapter 2 of the Present Volume and Topliss (1983)

Model	Variables	(Q/S-)SAR Equation	$\| \|Y\rangle \|^{PREDICTED}$	$R_{Algebraic}$	$R_{Statistic}$
Ia	$\|X_0>, \|X_1>$	$\|Y>^{Ia} = 5.39837\|X_0> + 0.0179106\|X_1>$	26.6138	0.988049	0.0175601
Ib	$\|X_0>, \|X_2>$	$\|Y>^{Ib} = 5.67735\|X_0> - 0.00834411\|X_2>$	26.61425	0.988065	0.0409922
Ic	$\|X_0>, \|X_3>$	$\|Y>^{Ic} = 6.48303\|X_0> + 0.0000124625\|X_3>$	26.6344	0.988812	0.252513
IIa	$\|X_0>, \|X_1\rangle, \|X_2>$	$\|Y>^{IIa} = 5.64318\|X_0> + 0.0178242\|X_1\rangle - 0.00833676\|X_2>$	26.614349	0.988069	0.0445618
IIb	$\|X_0>, \|X_1>, \|X_3>$	$\|Y>^{IIb} = 6.93331\|X_0> - 0.120924\|X_1> + 0.0000150708\|X_3>$	26.638	0.988947	0.273909
IIc	$\|X_0>, \|X_2>, \|X_3>$	$\|Y>^{IIc} = 4.99884\|X_0> + 0.122989\|X_2> + 0.0000376701\|X_3>$	26.6681	0.990063	0.409837
III	$\|X_0>, \|X_1>, \|X_2>, \|X_3>$	$\|Y>^{III} = 5.59424\|X_0> - 1.05993\|X_1> + 0.400704\|X_2> + 0.000117452\|X_3>$	26.7758	0.994064	0.708509

endpoint models, when one can see the confirmation of the stericity role through E_{TOT} correlation variable: while combination LogP∧POL does not improve the statistical correlation of model IIa significantly over single-parameter LogP∨POL correlations, the total energy presence provides better and better correlation behavior as it is combined with LogP (the model IIb) and with POL (the model IIc), respectively. Instead, when all the Hansch structural variables are taken into account the model III is generated with appreciable statistical correlation respecting the other computed combinations.

Overall, there can not be inferred that LogP and POL does have no influence on correlation only because when alone they do not correlate at all with flavonoids' bioactivity, because their cumulative presence in model III highly improves the single E_{TOT} correlation of model Ic as well as mixed correlations of bi-variable models IIb and IIc. Therefore, the mechanistic "alchemy" of structural features on molecular activity seems enough complex when all hydrophobicity, electrostatic and stericity influences combine as they are reciprocally activating one each other with a superior resultant in modeling ligand-receptor binding.

Yet, special discussion deserves the algebraic correlation factors in Table 3.31 (Putz et al., 2009b): there is clear that as they are not measuring the dispersive character of the local computed (molecular) points against the average recorded activity as statistical metrics do, their values are all close to unity and close among them as well; however, they are modeling another reality of computation, being closer to path integral approach than to differential analysis, through indexing the global behavior or the total length of the computed vector to the recorded one. Still, while between the algebraic and statistical correlations only an indirect connection exists (Putz et al., 2008a), the one-to-one hierarchical ordering of models is always recorded thus supporting the usefulness of using algebraically scale when the shrink of correlation factors is more favorable. For instance, in the present case, as above revealed, according to the statistical analysis, there seems that LogP (Ia) and POL (Ib) have no influence in correlation, while when combined with E_{TOT} in model III they considerably enrich the single E_{TOT} correlation power of model Ic. Such behavior shows that orthogonal, i.e., independent, descriptors may provide better results when are combined than when considered apart due to the increase of the (inter) correlation space.

Having the QSAR analysis performed, the specific Spectral-SAR stage can be unfolded by means of the (*K*=9, *M*=3) ergodic paths with the spectral Euclidian lengths given by Eq. (3.125) in both statistical and algebraic frameworks, as shown in Table 3.32. Next, the least *M*=3 paths with the dominant *M*-factors influence are selected by applying the above exposed recursive rule of *least path principle* resumed by Eq. (3.126). Remarkably, there follows that the resulted alpha (α), beta (β), and gamma (γ) most influential paths are identically shaped no mater whether statistical or algebraically schemes are undertaken. This result, although not necessarily viewed as a general rule, shows that in this specific case the algebraically analysis leaves with systematically the same mechanistically results as those obtained with statistical tools. However, once more, we stress on that algebraically measure may give more realistic inside in the Q(Spectral)-SAR phenomenology since its inner vectorial and norm-based algorithm accounting for each individual molecular contribution to the whole activity "basin" rather than respecting the average activity (Putz et al., 2009b).

Going now to the individual molecular level analysis, Table 3.33 lists the residues activities between computed and observed activities for each

TABLE 3.32 Synopsis of Paths Connecting the Endpoints of Table 3.21 in the Norm-Correlation Spectral-Space (Putz et al., 2009b)

Path	Value	
	Algebraic	Statistic
Ia-IIa-III	0.162142	0.710311
Ia-IIb-III	0.162142	0.713422
Ia-IIc-III	γ 0.162142	γ 0.713533
Ib-IIa-III	β 0.161697	β 0.686875
Ib-IIb-III	0.161697	0.690271
Ib-IIc-III	0.161697	0.690059
Ic-IIa-III	0.181617	0.892215
Ic-IIb-III	α 0.141579	α 0.477638
Ic-IIc-III	0.141579	0.478416

TABLE 3.33 Residual Activities $A_i - Y_i^{Model}$ of the Compounds of Table 3.29 for the Spectral-SAR Models of Table 3.31 Ordered According With the Alpha, Beta and Gamma Paths of Table 3.32; That Residue Which is Closes to Zero in Each Considered Endpoint is Marked By a Line Border (Putz et al., 2009b)

No.	Models						
	α		*β*		*γ*		
	Ic	*IIb*	*Ib*	*IIa*	*Ia*	*IIc*	*III*
1	–0.915706	–0.738065	–1.5562	–1.55854	–1.69473	–1.35359	–0.785284
2	–1.2836	–1.31784	–1.21515	–1.2129	–1.19025	–1.13401	–1.09626
3	–0.933302	–0.921149	–0.958225	–0.959719	–0.944015	–0.682916	–0.0110057
4	–1.09977	–1.04269	–0.864162	–0.880793	–0.849239	–1.17367	–0.840236
5	–0.823636	–0.861503	–0.755151	–0.752361	–0.729716	–0.674109	–0.668387
6	–0.772996	–0.717525	–0.653552	–0.665745	–0.654761	–0.874038	–0.615978
7	–0.60143	–0.650231	–0.619811	–0.612569	–0.595239	–0.344115	–0.190835
8	–0.481207	–0.484848	–0.592885	–0.589214	–0.578821	–0.123257	0.653106
9	–0.376575	–0.367246	–0.517615	–0.514493	–0.519358	–0.153417	0.432251
10	–0.86198	–0.722625	–0.534663	–0.566265	–0.534284	–1.11502	–0.464907
11	–0.119334	–0.262529	–0.224554	–0.200562	–0.188403	0.246777	–0.181659
12	–0.0395047	0.0115343	–0.152333	–0.155471	–0.185627	–0.194929	–0.098518
13	–0.307904	–0.31541	–0.0657478	–0.0731081	–0.0399254	–0.374895	–0.591958
14	0.348559	0.294919	0.330189	0.338144	0.355478	0.605851	0.716663
15	0.403192	0.367358	0.355459	0.362152	0.364224	0.525694	0.488215
16	0.415563	0.403619	0.425472	0.427166	0.419209	0.294139	–0.0847295
17	0.488521	0.361109	0.470189	0.489017	0.506403	0.745737	0.209751

TABLE 3.33 Continued

No.	Models						
	α		β		γ		
	Ic	*IIb*	*Ib*	*IIa*	*Ia*	*IIc*	*III*
18	0.660616	0.646707	0.556082	0.562214	0.533687	0.508577	0.0433466
19	0.695292	0.566274	0.560799	0.584065	0.580881	0.925367	0.314017
20	0.546881	0.605582	0.759522	0.743771	0.761657	0.345406	0.404994
21	1.01555	0.855242	0.932398	0.959039	0.944284	0.994295	–0.458865
22	0.886414	0.848557	0.954849	0.957639	0.980284	1.03604	1.04208
23	1.56925	1.70416	1.63017	1.60938	1.57664	1.03055	0.996681
24	1.58711	1.7366	1.73491	1.70914	1.68163	0.939523	0.787543

of considered models, distributed along the already identified least paths. At this instance, the most fitted molecule is outlined out of each endpoint; most impressive, the actual research selected the same molecule as the best-fitted one along the both α and β paths, namely the molecule no. 12 (4′-5,7-Trimethoxyflavanone) and the molecule no. 13 (Flavone), respectively. Moreover, these molecules are not among the most potent one respecting the observed activity of Table 3.29, being situated at the middle to second-half panel of the 24 molecules considered.

Such result tells us that the maximum-recorded activity is not necessary that one induced by *specific* chosen structural variables (here as LogP, POL, and E_{TOT}). This is the case of the most fitted molecule on the most correlated endpoint (III) appeared to be no.3 (Naringenin), with low activity on the observed range compared with the no. 25 (7,8 – Benzoflavone) in Table 3.29. Consequently, one can said that the first half of the observed activities in Table 3.29 may be attributed to certain physico-chemical indicators with clear mechanistically role, while the rest of observed activities may be due to other unidentified specific structural descriptors or even to non-specific ones (rooting in the sub-quantum nature of the particular observer-observed system). Nevertheless, this lower activity prescribed by the computational results is in accordance with the so-called "homeopathic principle" prescribing cure by moderate-to-low active drugs while better monitoring their effects through controlled physico-chemical descriptors (Putz et al., 2009b).

For the shake of comparison, the actual Spectral(Qua)SAR results are to be compared with the consecrated Principal Component Analysis (PCA) (Miller & Miller, 2000). This way, Figure 3.24 illustrates the graphical 3D correlations among the descriptors LogP, POL and E_{TOT} used in this study; it offers the visual way for assessing the almost no-correlation of LogP with other concerned variables, POL and E_{TOT}, respectively. This lead with conclusion that LogP is almost orthogonal (independent) on (respecting) the other two Hansch variables. Instead, when further performing the factor analysis, the Table 3.35 is obtained while clearly revealing the scarce correlation carried by considering LogP variable alone. This is in close agreement with the Spectral-SAR results, see above.

In any case, the hydrophobicity description and its descriptor cannot be rejected only by factor analysis since it drives (firstly or latter)

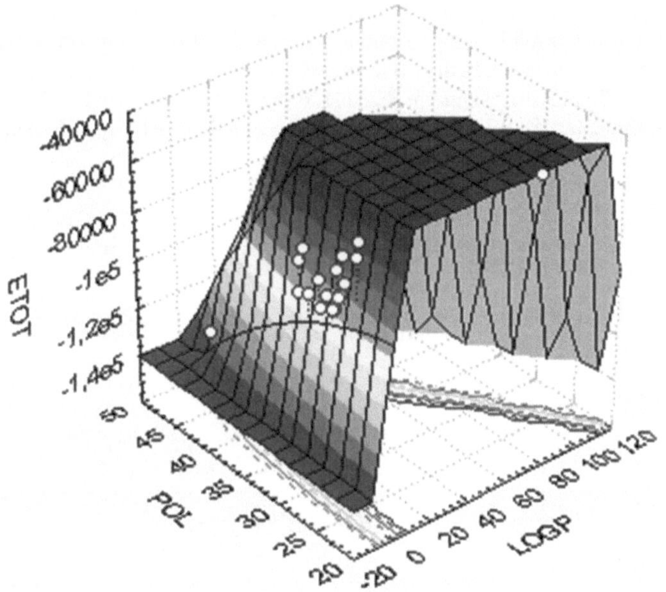

FIGURE 3.24 Quadratic 3D representation of LogP *vs.* POL *vs.* E_{TOT} variables' fit (Putz et al., 2009b) employing the data of Table 3.29 (Statistica, 1995).

the inter-membrane interaction that is essential for drug-cell binding. Spectral- and Qua-SAR highly proved the important role hydrophobicity plays in combination with electrostatic (POL) and steric (E_{TOT}) interactions. Moreover, while PCA shows the POL factor influence equals that of E_{TOT}, whereas their role in correlation is sensible different in Spectral-SAR analysis (compare model Ib-last column of Table 3.31 with POL-last column of Table 3.35). However, again, this discrepancy is in the favor of S-SAR since the PCA results are due to the sensitive degree of POL-E_{TOT} correlation (see Figure 3.24), from where the PCA yield that POL and E_{TOT} display similar correlation power, while S-SAR includes also the orthogonalization of POL and E_{TOT} variables prior correlation takes effect and better discriminates among their influence in bonding (Putz et al., 2009b).

Nevertheless, going ahead with the Spectral-SAR results the Qua-SAR factors may be immediately recover by employing the molecular activity differences from Table 3.30 for the fittest molecules of Table 3.33 along the models of the most influential paths in molecular mechanism towards

TABLE 3.34 Determination of the Quantum-SAR, see Eq. (3.131) with Eqs. (3.127) and (3.128), Associate With Certain Couple of Molecules Involved in Activating Specific Structural Quantum Indices (Or Their Combinations) Driving Spectral Paths of Table 3.32, By Employing Minimum Residue Recipe Throughout Table 3.33 for Each Considered Endpoint, As Well As the Associate Recorded Bioactivity Differences of Table 3.30, Respectively (Putz et al., 2009b)

Path	$\Delta Y_{I\|I'}^{PATH}$ (IEND)#	$\Delta A_{i\|j}^{I\|I'}$ (IEMAD)♣	$q_{i\|j}^{I\|I';PATH}$ *	q^{PATH}
α	$\Delta Y_{Ic\|IIb}^{\alpha}=0.00364573$	$A_{12\|12}^{Ic\|IIb}=0$	$q_{12\|12}^{Ic\|IIb;\alpha}=0.991641$	$q^{\alpha}=0.125464$
	$\Delta Y_{IIb\|III}^{\alpha}=0.137836$	$A_{12\|3}^{IIb\|III}=-0.76$	$q_{12\|3}^{IIb\|III;\alpha}=0.126521$	
β	$\Delta Y_{Ib\|IIa}^{\beta}=0.0000989324$	$A_{13\|13}^{Ib\|IIa}=0$	$q_{13\|13}^{Ib\|IIa;\beta}=0.999772$	$q^{\beta}=0.0848036$
	$\Delta Y_{IIa\|III}^{\beta}=0.161487$	$A_{13\|3}^{IIa\|III}=-0.91$	$q_{13\|3}^{IIa\|III;\beta}=0.0848229$	
γ	$\Delta Y_{Ia\|IIc}^{\gamma}=0.0542592$	$A_{13\|8}^{Ia\|IIc}=-0.55$	$q_{13\|8}^{Ia\|IIc;\gamma}=0.248737$	$q^{\gamma}=0.0847168$
	$\Delta Y_{IIc\|III}^{\gamma}=0.107771$	$A_{8\|3}^{IIc\|III}=-0.36$	$q_{8\|3}^{IIc\|III;\gamma}=0.340588$	

#Inter-Endpoint Norm Difference, Eq. (3.127); ♣Inter-Endpoint Molecular Activity Difference, Eq.(3.128); *Note that here the basic relation of Eq. (3.131) was considered in decimal base since originally, the associated activities in Table 3.29 were as such defined.

TABLE 3.35 Principal Component Analysis For the Data of Table 3.29 Within Unrotated (Unnormalized) Factor Score Coefficients, by Statistica (1995); see (Putz et al., 2009b)

	PC1	**PC2**	**PC3**	**Multiple PC1-PC3 factors' R^2**
Eigenvalue:	1.958158	0.892127	0.149715	
% total variance:	65.27195	29.73757	4.99049	
Variable		**Factors' coefficients**		
LogP	0.232179	–0.997780	0.20467	0.083712
POL	–0.472177	–0.302902	–1.79349	0.716820
E_{TOT}	0.483556	0.183309	–1.84956	0.728872

MX-BCRP binding. The resulted *IEND* and *IEMAD* of Table 3.34 are combined to produce the quantum-SAR factors of Eq. (3.131) type for each two-molecules-two-models on specific paths, while the "metabolization power" *per* path is finally obtained by their couplings, according with multiplicative quantum rule of Eq. (3.137). Worth noting that the overall quantum-SAR factors of paths are in total agreement with the previous spectral-SAR selected path hierarchy, i.e., the α path is associated with the highest Qu-SAR factor, being followed by that of β path and by that of γ one in last column of Table 3.34.

This result may be quite important if such behavior may be proved to hold in general since it would allow the effective quantification of paths according with their metabolization power. However, such endeavor exceeds the present communication purpose and will remain as a future challenge in Qua-SAR studies (Putz et al., 2009b).

Finally, all QSAR, Spectral-SAR and Qua-SAR computational results may be collected and resumed by associate "spectral" scheme for evolution of the fittest molecular structures along the endpoint models for the (M=)3 selected mechanistic paths of actions, see Figure 3.25. Note that algebraic correlation environment was chose as the "vertical" indicator

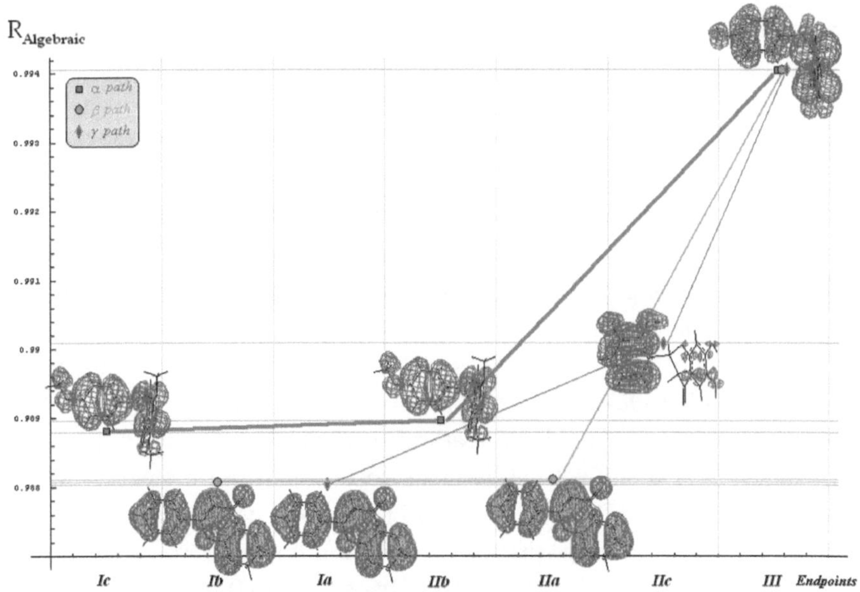

FIGURE 3.25 Spectral representation of the endpoints employed in designing the bioactivity mechanism for the molecules of Table 3.29, according with the algebraic correlation factors of Eq. (3.124) in Table 3.31, across the shortest (three) paths identified from Table 3.32, while marking the fittest molecules' orbital 3D-distribution for each considered model, i.e., molecule no. 12 (4′-5,7-Trimethoxyflavanone) for the models Ic and IIb, molecule no. 13 (Flavone) for models Ib, Ia, and IIa, molecule no. 8 (6,2′,3′-7-Hydroxyflavanone) for model IIc, and molecule no. 3 (Naringenin) for model III, respectively (Putz et al., 2009b).

for the degree with which a certain model reaches the observed activity in the vectorial norm sense [equally, the norm themselves could be used for ordinate axis (Putz & Lacrămă, 2007)].

Going now to comment upon the "metabolization power" as indicated by the quantum-SAR factors on Figure 3.25, one can firstly observe that for the α path the "first movement" from the Ic (E_{TOT}) to IIb (LogP∧E_{TOT}) corresponds to quantum free motion so that the null IEMAD for molecule no. 12 (4′-5,7-Trimethoxyflavanone) is carried; here, the quantum metabolization factor $q_{12|12}^{Ic|IIb;\alpha}$ is consumed only for strongly activating the membrane transporter feature (LogP) of the same molecule. Instead, on the last passage of the α path the factor $q_{12|3}^{IIb|III;\alpha}$ is responsible for converting the electrostatic (POL) influence of the flavonoids no. 12 towards no.3 (Naringenin) activity as well as for reverse-O-methylation (methoxylation) of oxigens in positions 5 and 7 (on ring A) and 4′ (on ring B) respecting the molecular pattern no.0 in Figure 3.23, respectively. Such result is in fully accordance with the reverse quantum influence that is at the foreground of quantum-SAR factor conversion prescribed by Eq. (3.130), i.e., quantifying the power of back transformation of molecular EC50s respecting the "arrows" of IEND and IEMAND in Eq. (3.131). However, the fact that such transformation is the first one acting at molecular level is sustained also by optimized 3D configurations of involved molecules no. 12 and 3, being both with rings A and B spatially bent in Figure 3.25 respecting the ring C of the planar pattern no.0 of Figure 3.23.

Somehow different situation is met for β path in Figure 3.25; in its first part a higher Qu-SAR factor $\left(q_{13|13}^{Ib|IIa;\beta} > q_{12|12}^{Ic|IIb;\alpha}\right)$ is need for activating the transporter hydrophobicity feature in model IIa (LogP∧POL) starting from model Ib (POL), while in its second part the molecule no. 13 (Flavone) is shown to be metabolized in molecule no. 3 (Naringenin) by a direct hidroxilation in positions 5, 7 (on ring A) and 4′ (on ring B), the same as before, respecting the molecular pattern no.0 in Figure 3.23, by a smaller Qu-SAR factor, $q_{13|3}^{IIa|III;\beta} << q_{12|3}^{IIb|III;\alpha}$, compared with that involved in the previous alpha path. Despite these, the overall quantum factor of beta path is lower than that of alpha, meaning a decrease capacity of metabolization since direct addition is involved, contrarily to the ordinary "inverse" Quantum-SAR transformation of Eq. (3.130), while stericity (here founded as the most influential QSAR variable) is triggered by more steric energy difference consumed between the planar optimized configuration of molecule no. 13 on that spatially bended of molecule no. 3 in Figure 3.25.

Even more metabolization "operations" take place along the γ path of Figure 3.25: there is started on the same planar configuration of molecule no.13 (Flavone); then, the Qu-SAR factor $q_{13|8}^{Ia|IIc;\gamma}$ turn it into molecule no. 8 (6,2′,3′-7-Hydroxyflavanone) by hydroxylation on the indicated positions (6 and 7 for ring A and 2′ and 3′ for ring B respecting the pattern molecule no.0 of Figure 3.23) while activating electrostatic and steric factors in model IIc (POL∧-E_{TOT}) from independent hydrophobicity factor of model Ia (LogP) – a complex movement that explain why this molecular path comes at the final, with less probability and potency; nevertheless, this path has on its last passage no less complex transformation, i.e., turning the molecule no. 8 into no. 3 one by combined reverse hydroxylation in positions 2′ and 3′ with direct hydroxylation of position 4′ on B ring and with movement from ortho (6)-to-para (5) of hydroxyl group on ring A respecting pattern molecule no. 0 of Figure 3.23, respectively; the transformation efficiency $q_{8|3}^{IIc|III;\gamma}$ is a bit higher than on the first part of the path since it require less steric energy consumption to bent the ring C respecting the A-B ones while accounting for electronic delocalization density (orbitals) over them until the configuration of molecule no. 3 is reached in Figure 3.25 (Putz et al., 2009b).

Overall, there is clear that the Qua-SAR scheme offers a quantification recipe along the most effective spectral paths combined with most fitted molecules for a trial basin of analogues compounds and structural variables. In the present case there was revealed that the energetic steric factor E_{TOT} seems to mainly drive the mechanistic molecular transformation in MX-BCRP binding phenomenology, while the molecule no. 3 (Naringenin) appears as the fittest molecules belonging to the most relevant endpoint, in clear disjunction with the roughly molecular selection upon initial input observed activity data. That is, the Naringenin (no. 3) is showed to be the best adapted molecule for the actual LogP∧POLÙE_{TOT} structural (independent) factors being metabolized from molecules as 6,2′,3′-7-Hydroxyflavanone (no. 8), 4′-5,7-Trimethoxyflavanone (no. 12), and Flavone (no.13) by specific molecular mechanistically paths. However, there appears that these molecules are not linked even through the paths with the most active compounds of Table 3.29; statistically, this can be explained by the so-called "regression towards the mean" effects, in the sense that the best correlations translated to the compounds found in the middle of the mentioned sorted Table 3.29; from the structural point of view such behavior may attributed to the specific parameters used for correlations that best describe molecules with specific groups, most favorable for the descriptor's nature.

On the other side, the present study affirms the position 7 of ring A and position 4′ of ring B respecting the pattern molecule no. 0 of Figure 3.23 as the most suitable ones for producing an increase in BCRP inhibition activity, given that these positions belong to the α and β paths and being common to the rest of spectral paths as well. Instead, the position that does not appear at all in any of the α, β, or γ paths, namely position 8 on ring B may present adverse drug interactions (Putz et al., 2009b).

Further Qua-SAR studies are necessary and will be developed for exploring other bio- and eco- active compounds for their interactions with organs and organisms; they may hopefully lead with a coherent analytical picture of chemical-biological bonding focused on selecting the most adapted molecules and of the most privileged molecular positions for delivering controlled structural based chemical reactivity and biological activity.

3.3.6 RESIDUAL VS. STRUCTURAL ALERTS' QSAR (FOR RATS' TOCICOLOGY BY HIGH DIVERSITY MOLECULES)

It is widely recognized that cancer and carcinogenesis are the main challenges facing 21st Century medicinal chemistry (Croce, 2008; Dingli & Nowak, 2006), particularly in the area of preventative toxicology (Danaei et al., 2005; Merlo et al., 2006; Ward et al., 2006; Pagano et al., 2006) as it assumes an idealized toxicity against organisms and acts through a subtle, undiscovered molecular mechanism. The basic mechanism in cancer cell proliferation is through a variety of compounds, making it difficult to assess specific ligand-receptor interaction patterns (Roukos, 2009; Knudson, 2001).

There is a reasonable basis for cancer apoptosis in the *electrophilic theory* of Miller & Miller (1977, 1981), which assumes a positively charged or polarized nature of the ligand (carcinogenic alkylating agents, originally). Currently, there is a more integrated and general view of *genotoxic carcinogenicity* (Arcos & Argus, 1995) that is closely related to mutagenic phenomena through a covalent binding to DNA, followed by direct damage by means of a unified (or by reactive intermediates) electrophilic mechanism of action. In contrast, *epigenetic carcinogenesis* (Woo, 2003) activates through a variety of specific and different mechanisms that do not involve covalent binding to DNA but to more congeneric (or similar) molecules, with a specific (or local) mechanism of action for each particular set of compounds.

Even though epigenetic carcinogenesis has typically been treated with the quantitative structure-activity relationship (QSAR) *principle of congenericity* (Benigni et al., 2007), the present case study will focus on genotoxic carcinogenesis because of its chemical bonding at the DNA level. In addition, the statistical physicochemical combination analysis for a variety of toxicants produces a molecular mechanistic model of action with a comprehensive physicochemical interpretation.

With the ever-increasing costs of traditional animal testing and the large number of industrial chemicals that need toxicological evaluation, international programs like Europe's REACH (Registration, Evaluation and Authorization of Chemicals) expressly endorse in silico (computational) ecotoxicological studies as alternative approaches to reduce experimental hazard, especially when "testing does not appear necessary" (Worth et al., 2007). This strategy is particularly useful in the first phases of validation for a new compound, before entering the industrial mainstream. This process primarily consists of preliminary screening based on models of literature and their extrapolations (Phase I), followed by the read-across, grouping and construction of new models employing the available commercial or non-commercial models, such as OncoLogic (Woo & Lai, 2005), HazardExpert (Lewis et al., 2002), Derek (Marchant, 1996), ToxTree (Benigni et al., 2009), Multicase (Matthews & Contrera, 1998), and CAESAR (Price, 2008; Benfenati, 2010) (Phase II), and eventually concluding with in vitro or in vivo assays (Phase III).

Phases I and II are theoretical-computational and, when approached through statistical or multivariate methods, the OECD (Organization for Economic Cooperation and Development) principles for a QSAR study must include the following information (OECD, 2007; Putz et al., 2011a):

"(i) a defined endpoint, (ii) an unambiguous algorithm, (iii) a defined domain of applicability, (iv) appropriate measures of goodness-of-fit, robustness and predictivity, and (v) a mechanistic interpretation."

In this context, the goal of the present work was to advance a general QSAR modeling approach employing the residues of direct correlation with definite physico-chemical descriptors to a second (or looping) correlation with the residual QSAR method. This was then applied to a non-congeneric series of rat toxicants to discover a general mechanism for genotoxic carcinogenesis in accordance with OECD-QSAR principles; the present discussion follows (Putz, 2011a; Putz et al., 2011a, c).

(i) The actual *defined endpoint:* targeted the carcinogenic activity in rats (*Rattus norvegicus*), as measured by TD_{50} values (in *mg/kg body wt/ day*) derived from the Carcinogenic Potency Database (CPD) (Fjodorova et al., 2010). Activity is expressed here as a function of the TD50 values, $A=Log(1/TD_{50})$. The working series of molecules, were chosen to have a *high diversity molecular structure* and fulfilling the Topliss and Costello (1972) rule according which their cardinal should be at least 5-times larger the number of structural descriptors used. They are separately shown in Tables 3.36 and 3.7, as calibration/trial/trainin sets using Gaussian screening and as test set using quasi-Gaussian distribution screening (Figure 3.26), respectively. The parameters recommended by (Hansch et al., 2001) (hydrophobicity, polarizability and total energy) and special reactivity indices (electronegativity and chemical hardness), all computed using the semiempirical PM3 method, were used for both full molecules and structural alerts for the molecules found in Tables 3.36 and 3.7.

However, as noted by Hansch, "*there is no substitute for extensive experience…in physical organic chemistry and QSAR*" (Hansch et al., 2001). Highly diverse molecular groups were employed in assessing the observed genotoxic carcinogenesis/mutagenicity. Several particular choices or "degrees of freedom" can be considered in to bring the analysis in line with the traditional QSAR dogma of "congeneric molecules."

(ii) *The unambiguous algorithm* is addressed by the following stages (Putz, 2011a; Putz et al., 2011c):

- *Physicochemical parameters*: meaningful physicochemical parameters as hydrophobicity, polarizability, total energy, electronegativity,

TABLE 3.36 Molecules From the Gaussian Training Set (Figure 3.26) and Corresponding Rat TD50 Toxicities (in mg/kg body wt/day) (Fjodorova et al., 2010) and Activities A=Log(1/TD50) Using Semi-Empirical PM3-Computed [Hyperchem (Hypercube, 2002)] Structural Parameters: Hydrophobicity (LogP), Polarizability (POL) [Å3], Total Optimized Energy (Etot) [kcal/mol], Along Computed Frontier Orbilal Related Electronegativity (3.160) [eV] and Chemical Hardness (3.163) [eV] (Putz et al., 2011c)

No.	Full Molecule: Chemical Structure	Full Molecule: Name Formula (CASRN)	Toxicity: TD_{50} Activity: *A=Log[1/TD_{50}]*	Structural Alert (SA)	LogP: Full Molecules *Structural Alert*	POL [Å³]: Full Molecules *Structural Alert*	Etot [kcal/mol]: Full Molecules *Structural Alert*	χ [eV]: Full Molecules *Structural Alert*	η [eV]: Full Molecules *Structural Alert*
1.	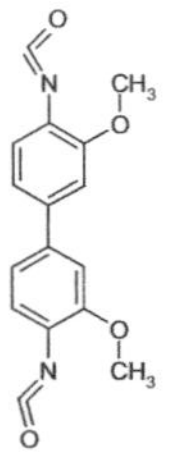	3,3'-Dimethoxy-4,4'-biphenylene diisocyanate (3,3'-Dimethoxy-benzidine-4,4'-diisocyanate) $C_{16}H_{12}N_2O_4$ (91–93–0)	1630 *2.79*		2.07 *−0.46*	30.03 *3.27*	−82478.58594 *−13584.68848*	4.74077805 *5.0497393*	3.85074395 *5.54064075*
2.	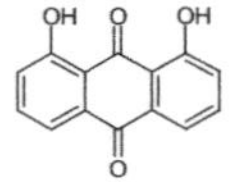	Chrysazin (Danthron) $C_{14}H_8O_4$ (117–10–2)	245 *3.61*	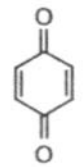	1.87 *1.52*	24.44 *10.8*	−68162.28125 *−31325.97266*	5.4079765 *6.3137325*	3.9369375 *4.6074975*
3.	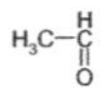	Acetaldehyde C_2H_4O (75–07–0)	153 *3.82*	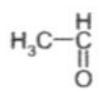	−0.58 *+0.58*	4.53 *4.53*	−13662.00781 *−13662.00781*	4.94880425 *4.94880425*	5.75505575 *5.75505575*
4.	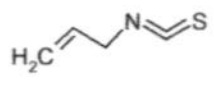	Allyl isothiocyanate C_4H_5NS (57–06–7)	96 *4.02*	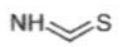	1.17 *0.19*	11.74 *6.43*	−20700.27344 *−11094.80273*	4.9388987 *5.032356*	4.2117593 *4.346452*

TABLE 3.36 Continued

No.	Full Molecule: Chemical Structure	Full Molecule: Name Formula (CASRN)	Toxicity: TD_{50} Activity: $A=Log[1/TD_{50}]$	Structural Alert (SA)	LogP: Full Molecules *Structural Alert*	POL [$Å^3$]: Full Molecules *Structural Alert*	Etot [kcal/mol]: Full Molecules *Structural Alert*	χ [eV]: Full Molecules *Structural Alert*	η [eV]: Full Molecules *Structural Alert*
5.	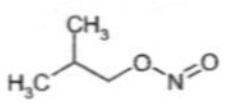	Isobutyl nitrite $C_4H_9NO_2$ (542–56–3)	54.1 *4.27*	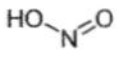	1.63 *0.38*	9.96 *2.62*	−31363 *−17580.39258*	5.294418075 *5.4457523*	5.263031925 *5.3349777*
6.	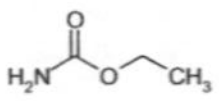	Urethane $C_3H_7NO_2$ (51–79–6)	41.3 *4.38*	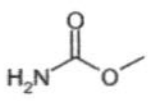	−0.06 *−0.44*	8.35 *4.68*	−27989.58203 *−21103.80273*	4.573154 *4.474373*	5.741656 *5.576267*
7.		Ethylene oxide C_2H_4O (75–21–8)	21.3 *4.67*		−0.16 *−0.16*	4.31 *4.31*	−13626.54297 *−13626.54297*	4.4747555 *4.4747555*	6.8617045 *6.8617045*
8.	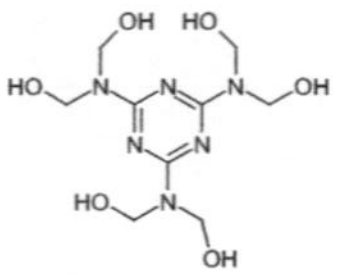	Hexa(hydroxymethyl) melamine $C_9H_{18}N_6O_6$ (531–18–0)	10.2 *4.99*	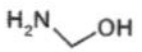	1.96 *0.08*	27.19 *5.08*	−108827.09 *−14382.44336*	4.05956015 *3.782247*	4.50969485 *7.116263*
9.	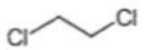	1,2-Dichloroethane $C_2H_4Cl_2$ (107–06–2)	8.04 *5.09*	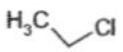	1.59 *1.22*	8.3 *6.37*	−21506.41406 *−14559.67578*	5.0714835 *4.586889*	5.6050665 *5.823301*
10.	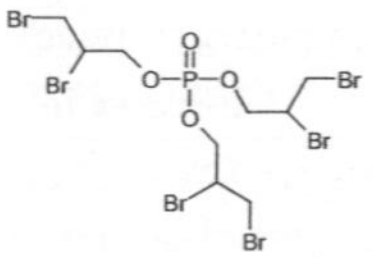	Tris(2,3-dibromopropyl) phosphate $C_9H_{15}Br_6O_4P$ (126–72–7)	3.83 *5.42*	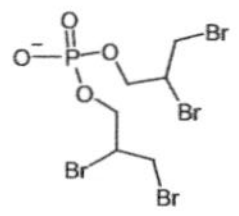	5.37 *3.73*	35.91 *25.15*	−108827.09 *−82903.73*	5.6512295 *5.3925243*	4.5231705 *4.63098575*

TABLE 3.36 Continued

No.	Full Molecule: Chemical Structure	Full Molecule: Name Formula (CASRN)	Toxicity: TD_{50} Activity: *A=Log[1/TD50]*	Structural Alert (SA)	LogP: Full Molecules *Structural Alert*	POL [Å³]: Full Molecules *Structural Alert*	Etot [kcal/mol]: Full Molecules *Structural Alert*	χ [eV]: Full Molecules *Structural Alert*	η [eV]: Full Molecules *Structural Alert*
11.		Beta-Propiolactone $C_3H_4O_2$ (57–57–8)	1.46 *5.84*		−0.25 *−0.25*	6.23 *6.23*	−23148.73047 *−23148.73047*	5.2018966 *5.2018966*	6.0842834 *6.0842834*
12.	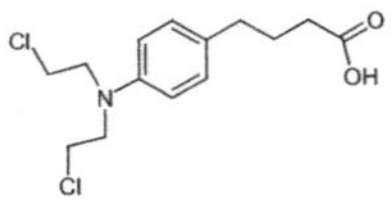	Chlorambucil $C_{14}H_{19}Cl_2NO_2$ (305–03–3)	0.896 *6.048*	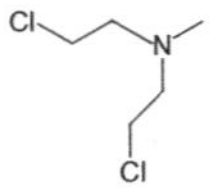	4.14 *1.2*	31.04 *13.32*	−76933.42969 *−32495.47656*	4.350258535 *4.4258064*	4.405313465 *5.31470165*
13.	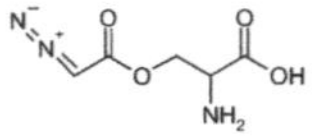	Azaserine $C_5H_7N_3O_4$ (115–02–6)	0.793 *6.10*	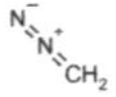	−1.03 *−0.04*	14.25 *4.11*	−54439.625 *−10877.61426*	5.2656847 *4.431137*	4.7215543 *4.794258*
14.	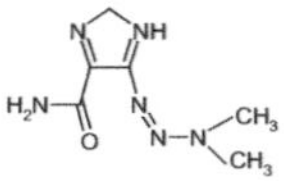	Dacarbazine $C_6H_{10}N_6O$ (4342–03–4)	0.71 *6.15*	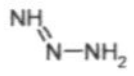	−0.92 *0.48*	17.95 *4.18*	−49126.58594 *−12249.66113*	4.9880568 *4.2572947*	4.1820822 *5.06465235*
15.	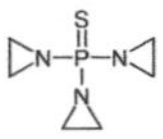	Thiotepa (Tris(aziridinyl)-phosphine sulfide) $C_6H_{12}N_3PS$ (52–24–4)	0.164 *6.789*		0.54 *−0.38*	17.63 *5.02*	−38905.46484 *−10956.04395*	5.2831755 *3.5910075*	3.8071835 *6.3290665*

TABLE 3.36 Continued

No.	Full Molecule: Chemical Structure	Full Molecule: Name Formula (CASRN)	Toxicity: TD_{50} Activity: *A=Log[1/ TD$_{50}$]*	Structural Alert (SA)	LogP: Full Molecules *Structural Alert*	POL [Å^3]: Full Molecules *Structural Alert*	Etot [kcal/ mol]: Full Molecules *Structural Alert*	χ [eV]: Full Molecules *Structural Alert*	η [eV]: Full Molecules *Structural Alert*
16.	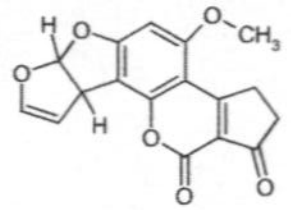	Aflatoxin-B1 $C_{17}H_{12}O_6$ (1162–65–8)	0.0032 *8.49*	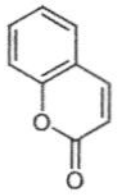	0.99 *1.82*	29.86 *15.7*	−91307.82331 *−40247.55469*	5.3273625 *5.2410253*	3.9567405 *4.2472247*
17.	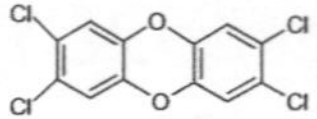	2,3,7,8-Tetrachl-orodibenzo-p-dioxin $C_{12}H_4Cl_4O_2$ (1746–01–6)	0.0000457 *10.34*	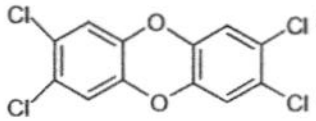	4.93 *4.93*	28.31 *28.31*	−76933.75 *−76933.75*	4.7914412 *4.7914412*	4.0075488 *4.0075488*
18.	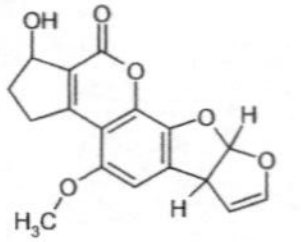	Aflatoxicol $C_{17}H_{14}O_6$ (29611–03–8)	0.00247 *8.61*	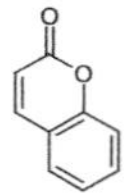	0.46 *1.82*	30.41 *15.7*	−91979.58594 *−40247.55469*	5.140259 *5.2410253*	3.945276 *4.2472247*
19.	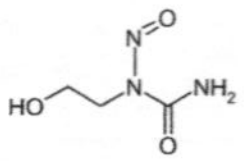	1-(2-Hydroxyethyl)-1-nitrosourea $C_3H_7N_3O_3$ (13743–07–2)	0.244 *6.61*		−0.95 *0.37*	10.92 *2.55*	−42184.19141 *−14202.18945*	5.42904375 *5.8565512*	5.08170625 *6.2971588*

TABLE 3.36 Continued

No.	Full Molecule: Chemical Structure	Full Molecule: Name Formula (CASRN)	Toxicity: TD_{50} Activity: *A=Log[1/ TD_{50}]*	Structural Alert (SA)	LogP: Full Molecules *Structural Alert*	POL [Å³]: Full Molecules *Structural Alert*	Etot [kcal/ mol]: Full Molecules *Structural Alert*	χ [eV]: Full Molecules *Structural Alert*	η [eV]: Full Molecules *Structural Alert*
20.	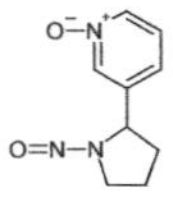	N'-Nitrosonornicotine-1-N-oxide $C_9H_{11}N_3O_2$ (78246–24–9)	0.876 *6.06*	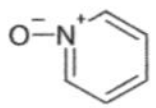	0.25 *0.12*	19.48 *10.35*	−53174.95313 *−25900.39453*	5.04527 *4.9295405*	4.273811 *4.3386305*
21.	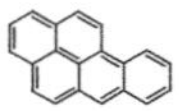	Benzo(a)pyrene $C_{20}H_{12}$ (50–32–8)	0.956 6.02	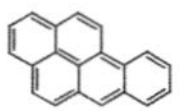	5.37 *5.37*	36.04 *36.04*	−58881.02734 *−58881.02734*	4.631374 *4.631374*	3.410258 *3.410258*
22.	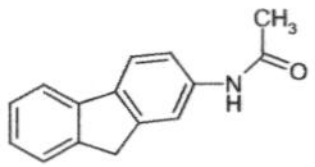	2-Acetylamino-fluorene $C_{15}H_{13}NO$ (53–96–3)	1.22 *5.91*	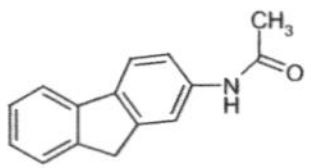	2.61 *2.61*	26.26 *26.26*	−56110.60547 *−56110.60547*	4.38615285 *4.38615285*	4.02819215 *4.02819215*
23.	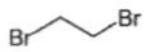	1,2-Dibromoethane $C_2H_4Br_2$ (106–93–4)	1.52 *5.82*	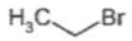	1.71 *1.29*	9.7 *7.07*	−28203.0625 *−15407.94336*	6.1527065 *5.5320367*	5.0695035 *5.37857335*
24.	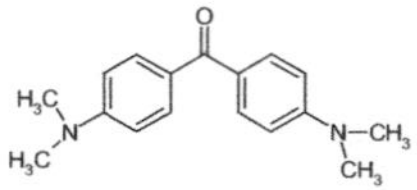	Michler's ketone $C_{17}H_2ON_2O$ (90–94–8)	5.64 *5.25*	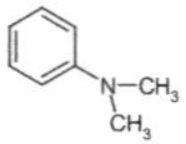	3.8 *2.31*	19.85 *15.46*	−67801.28125 *−29500.11719*	4.3453716 *3.6634669*	4.1924714 *4.2943161*
25.	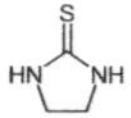	Ethylene thiourea (ETU) $C_3H_6N_2S$ (96–45–7)	8.13 *5.09*	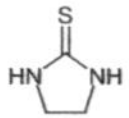	0.33 *0.33*	11.45 *11.45*	−22095.42578 *−22095.42578*	4.40057075 *4.40057075*	4.20081425 *4.20081425*

TABLE 3.36 Continued

No.	Full Molecule: Chemical Structure	Full Molecule: Name Formula (CASRN)	Toxicity: TD_{50} Activity: *A=Log[1/TD₅₀]*	Structural Alert (SA)	LogP: Full Molecules *Structural Alert*	POL [Å³]: Full Molecules *Structural Alert*	Etot [kcal/mol]: Full Molecules *Structural Alert*	χ [eV]: Full Molecules *Structural Alert*	η [eV]: Full Molecules *Structural Alert*
26.	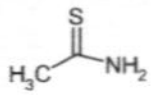	Thioacetamide C_2H_5NS (62–55–5)	11.5 *4.94*	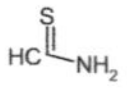	−0.21 *−0.42*	9.04 *7.21*	−15263.96289 *−11813.05762*	4.72959049 *4.7550568*	3.99513951 *4.0219202*
27.	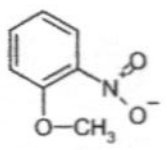	o-Nitroanisole $C_7H_7NO_3$ (91–23–6)	15.6 *4.81*	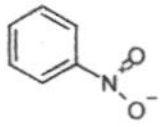	−0.18 *0.07*	14.75 *12.28*	−45613.03906 *−35381.23828*	5.5631575 *5.8686355*	4.3657605 *4.7339745*
28.	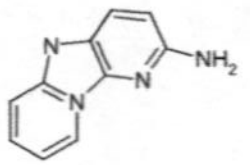	2-Aminodi-pyrido[1,2-a:3’,2’-d] imidazole (Glu-P-2) $C_{10}H_8N_4$ (67730–10–3)	42.3 *4.37*	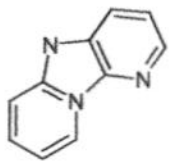	2.35 *2.9*	20.73 *19.38*	−45103.06641 *−40998.30859*	4.5267029 *4.7452532*	3.7506371 *3.87575785*
29.	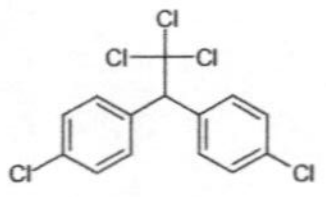	Dichlorodiphenyl-trichloroethane (DDT) $C_{14}H_9Cl_5$ (50–29–3)	84.7 *4.07*	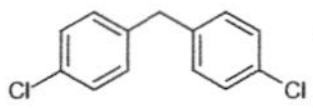	6.39 *4.92*	33.4 *25.23*	−77956.60156 *−52871.28516*	4.95182645 *5.0230205*	4.50488155 *3.2895935*
30.	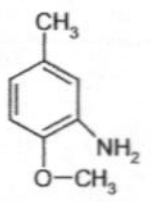	p-Cresidine $C_8H_{11}NO$ (120–71–8)	98 *4.01*	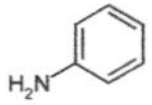	1.48 *1.26*	16.09 *11.79*	−36280.75391 *−22612.99212*	3.9300665 *3.7259962*	4.3473585 *4.3413348*

TABLE 3.36 Continued

No.	Full Molecule: Chemical Structure	Full Molecule: Name Formula (CASRN)	Toxicity: TD_{50} Activity: *A=Log[1/ TD_{50}]*	Structural Alert (SA)	LogP: Full Molecules *Structural Alert*	POL [Å³]: Full Molecules *Structural Alert*	Etot [kcal/mol]: Full Molecules *Structural Alert*	χ [eV]: Full Molecules *Structural Alert*	η [eV]: Full Molecules *Structural Alert*
31.	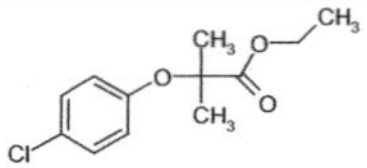	Ethyl 2-(4-chlorophenoxy)-2-methylpropionate (Clofibrate) $C_{12}H_{15}ClO_3$ (637–07–0)	169 *3.77*	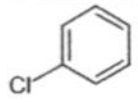	2.97 *2.56*	24.73 *12.36*	−65740.6875 *−25464.87109*	4.49111609 *4.6624658*	4.53578491 *4.72527225*
32.	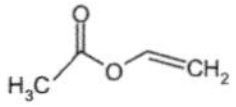	Vinyl acetate $C_4H_6O_2$ (108–05–4)	341 *3.47*	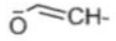	−0.01 *1.28*	8.65 *3.98*	−26598.12305 *−12920.42871*	4.6849081 *4.5472153*	5.2657279 *4.91445575*
33.	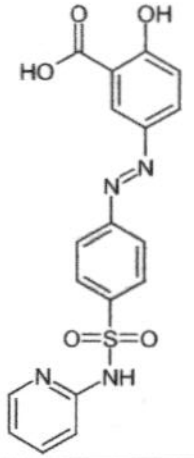	Salicylazo-sulfapyridine $C_{18}H_{14}N_4O_5S$ (599–79–1)	1590 *2.799*	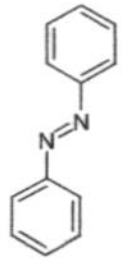	4.54 *4.35*	36.79 *22.15*	−107222.1719 *−43772.44922*	5.209331 *5.0515378*	3.898064 *4.2222632*

TABLE 3.37 Molecules From the Quasi-Gaussian Test Set (Figure 3.26), With the Activities and Structural Parameters As in Table 3.36 (Putz et al., 2011c)

No.	Full Molecule: Chemical Structure	Full Molecule: Name Formula (CASRN)	Toxicity: TD_{50} Activity: *A=Log[1/ TD_{50}]*	Structural Alert (SA)	LogP: Full Molecules *Structural Alert*	POL [Å³]: Full Molecules *Structural Alert*	Etot [kcal/ mol]: Full Molecules *Structural Alert*	χ [eV]: Full Molecules *Structural Alert*	η [eV]: Full Molecules *Structural Alert*
34	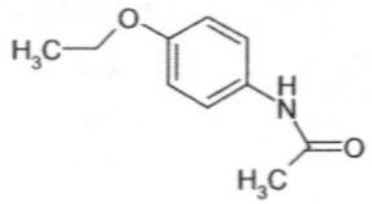	Phenacetin $C_{10}H_{13}NO_2$ (62–44–2)	1250 *2.90*	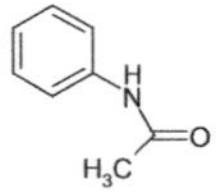	0.99 *−0.03*	19.85 *15.73*	−49230.08203 *−36279.96484*	4.063315 *4.1985829*	4.307675 *4.3648181*
35.	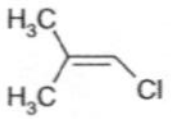	Dimethylvinyl chloride (DMVC) C_4H_7Cl (513–37–1)	31.8 *4.498*	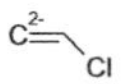	1.51 *0.45*	9.85 *4.05*	−20725.60325 *−13014.37793*	4.32596855 *5.6212095*	4.98083445 *4.2918295*
36.	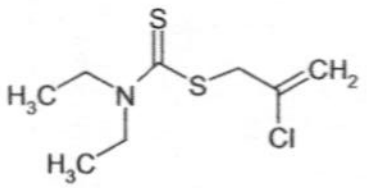	Sulfallate $C_8H_{14}ClNS_2$ (95–06–7)	26.1 *4.58*	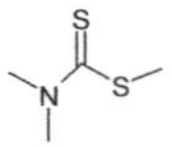	2.73 *0.62*	24.79 *10.21*	−46435.69922 *−16106.21777*	4.8447835 *5.093712*	3.8753115 *3.905288*
37.	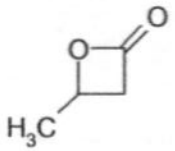	beta-Butyrolactone $C_4H_6O_2$ (3068–88–0)	13.8 *4.86*	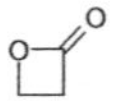	0.17 *−0.25*	8.06 *6.23*	−26599.55273 *−23148.73047*	5.1344426 *5.2020294*	6.0826774 *6.0841706*
38.	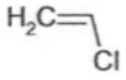	Vinyl Chloride C_2H_3Cl (75–01–4)	6.11 *5.21*	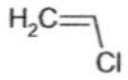	1.01 *1.01*	6.18 6.18	−13820.70898 *−13820.70898*	4.56666095 *4.56666095*	5.27117005 *5.27117005*
39.	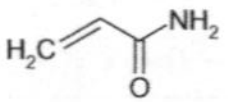	Acrylamide C_3H_5NO (79–06–1)	3.75 *5.43*	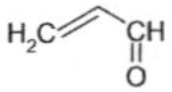	−0.28 *0.17*	7.52 *6.17*	−20478.92578 *−16372.65625*	4.77457395 *5.4404992*	4.91861805 *5.25305085*

TABLE 3.37 Continued

No.	Full Molecule: Chemical Structure	Full Molecule: Name Formula (CASRN)	Toxicity: TD_{50} *Activity: $A=Log[1/TD_{50}]$*	Structural Alert (SA)	LogP: Full Molecules *Structural Alert*	POL [Å³]: Full Molecules *Structural Alert*	Etot [kcal/mol]: Full Molecules *Structural Alert*	χ [eV]: Full Molecules *Structural Alert*	η [eV]: Full Molecules *Structural Alert*
40.	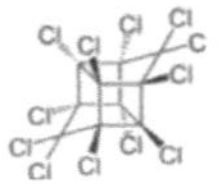	Mirex $C_{10}Cl_{12}$ (2385–85–5)	1.77 *5.75*		6.41 *6.41*	38.39 *38.39*	−114919.4688 *−114919.4688*	5.27780275 *5.2778028*	5.22349725 *5.22349725*
41.	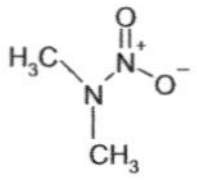	Dimethylnitramine $C_2H_6N_2O_2$ (4164–28–7)	0.547 *6.26*	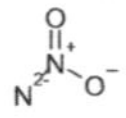	0.97 *1.32*	7.64 *3.18*	−28551.91406 *−20856.80078*	5.288693895 *7.5671675*	5.374516105 *4.9321725*
42.	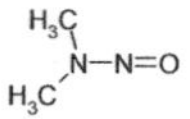	N-Nitrosodi-methylamine $C_2H_6N_2O$ (62–75–9)	0.0959 *7.02*	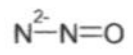	0.01 *0.37*	7.01 *2.55*	−21802.08203 *−14202.18945*	4.6046239 *5.8565512*	5.1639551 *6.2971588*
43.	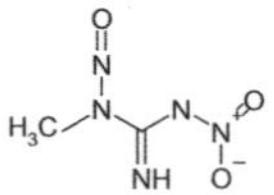	N-Methyl-N'-nitro-N-nitrosoguanidine (1-Methyl-3-nitro-1-nitroso-guanidine) $C_2H_5N_5O_3$ (70–25–7)	0.803 *6.1*	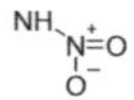	1.5 *0.84*	11.13 *3.97*	−46112.81641 *−21661.2832*	5.475207 *5.8694368*	4.654173 *5.785463245*
44.	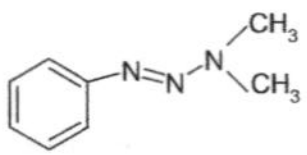	1-Phenyl-3,3-dimethyltriazene $C_8H_{11}N_3$ (7227–91–0)	2.31 *5.64*	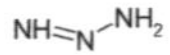	2.53 0.48	17.51 *4.18*	−36944.65625 *−12249.66113*	4.65555575 *4.2572947*	4.28693125 *5.06465235*

TABLE 3.37 Continued

No.	Full Molecule: Chemical Structure	Full Molecule: Name Formula (CASRN)	Toxicity: TD_{50} Activity: $A=Log[1/TD_{50}]$	Structural Alert (SA)	LogP: Full Molecules *Structural Alert*	POL [Å³]: Full Molecules *Structural Alert*	Etot [kcal/mol]: Full Molecules *Structural Alert*	χ [eV]: Full Molecules *Structural Alert*	η [eV]: Full Molecules *Structural Alert*
45.	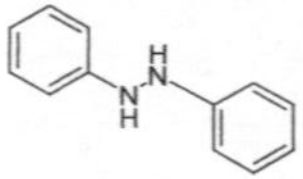	Hydrazobenzene $C_{12}H_{12}N_2$ (122–66–7)	5.59 *5.25*	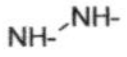	3.4 *−0.65*	22.8 *2.83*	−44481.07422 *−8164.909668*	3.65518885 *4.5593574*	3.99645815 *5.0568536*
46.	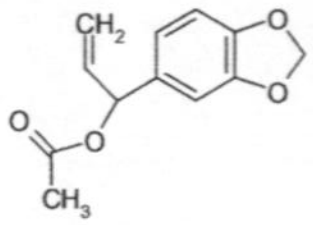	1'-Acetoxysafrole $C_{12}H_{12}O_4$ (34627–78–6)	25 *4.6*	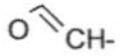	−0.11 *1.28*	22.47 *3.98*	−64108.48047 *−12920.42871*	4.516422835 *4.5472153*	4.517086165 *4.91445575*
47.	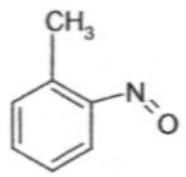	o-Nitrosotoluene C_7H_7NO (611–23–4)	50.7 *4.29*	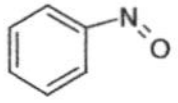	2.29 *1.82*	13.48 *11.65*	−32074.53516 *−28624.52734*	5.20234765 *5.2781928*	4.40152935 *4.43740625*
48.	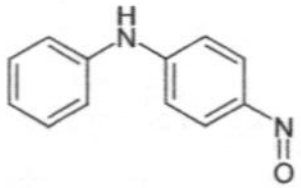	p-Nitrosodi-phenylamine $C_{12}H_{10}N_2O$ (156–10–5)	201 *3.7*	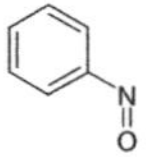	3.07 *1.82*	22.66 *11.65*	−50526.36328 *−28624.52734*	4.57337225 *5.2781928*	3.74357475 *4.43740625*
49.	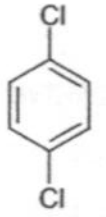	1,4-Dichlorobenzene (p-dichlorobenzene) $C_6H_4Cl_2$ (106–46–7)	644 *3.19*		3.08 *3.08*	14.29 *14.29*	−32415.54297 *−32415.54297*	4.73892295 *4.73892295*	4.49613405 *4.49613405*

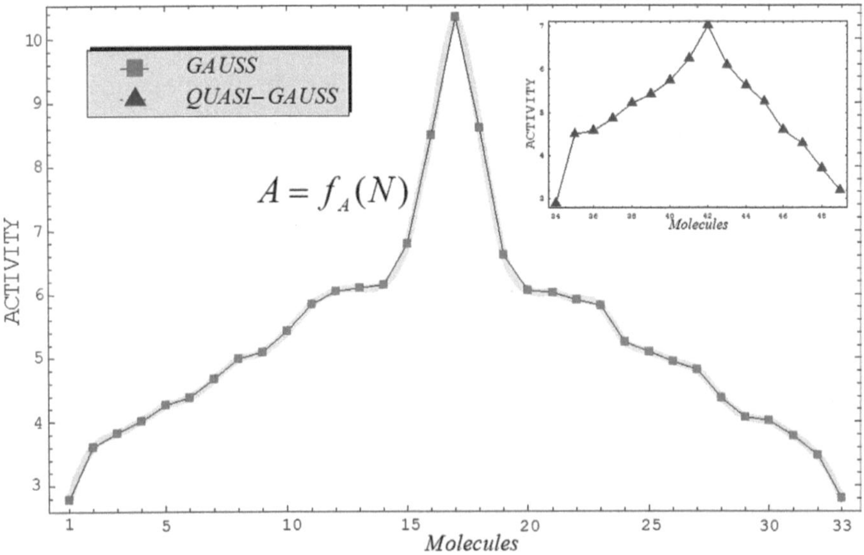

FIGURE 3.26 Graphical representation of the working activities for the molecules in Tables 3.36 and 3.7, classified to build up the "Gaussian" and "quasi-Gaussian" series that are specific to the training and testing QSAR purposes, respectively. The interpolating function, $A = f_A(N)$ to be used in Eq. (3.148) is also shown as the contour of the Gaussian set of trial molecules (Putz, 2011a; Putz et al., 2011c).

and chemical hardness should be considered in order to better interpret the derived models in terms of molecular mechanisms.

- *Universal hydrophobicity*: when a full molecule is identical to its structural alert, the sign of the structural alert may be flipped relative to that of the molecule for the action-reaction solubility characteristics. For instance, this can be applied to the LogP of the reagent because it is the logarithm (base 10) of the partition coefficient (P), the ratio of the compound's organic (oil)-to-aqueous phase concentrations.. Therefore, the opposite and equal values of the molecule and its identical structural alert induce a kind of "universality" in the solvation ability of the concerned toxicant. This approach may be applied to molecules with a recognized high toxicological or carcinogenic potential, and should not be overestimated in the molecular series employed. For the present trial series (Table 3.36), this approach was employed for molecule no. **3**, acetaldehyde (ethanal, C_2H_4O). Such an approach is justified, because this compound's average global production is about 10^6 tons/year (Eckert et al., 2006); it is a common electrophile in organic synthesis (Sowin et al., 2004) (in agreement with Miller's electrophilic theory (Miller & Miller, 1977, 1981) of genotoxic carcinogenesis: "there is *sufficient* evidence for the carcinogenicity of acetaldehyde (the major metabolite of ethanol) in experimental animals") (WHO, 1998); it is a probable carcinogen in humans (EPA, 1994a), but occurs naturally in coffee, bread and ripe fruit, and is produced by plants as part of their normal metabolism; and it can be spread through air, water, land or groundwater pathways and can be absorbed through inhalation, smoking or consumption (EPA, 1994b).

- *Equal steric properties*: molecules with similar carcinogenic properties may be considered to have equal optimized stericities, i.e., total energies, when their true values are in the same domains. Thus, non-carcinogenic molecules may be considered to be similar in some of their physicochemical properties, including stericity (in this case, associated with total energy). For instance, in the trial series of compounds (Table 3.36) molecules **8** and **10**, have energies of −94,064.03906 [eV] and −108,827.09 [eV], respectively (as calculated with PM3 and geometry optimization). These can be considered to have the same stericities in intra-cellular binding, due to their similar energies, similar activities in rats (as given by the CPD) (Gold, 2011), close positions in the Gaussian graph (Figure 3.26) and their identical carcinogenic characteristics (Huijbregts et al., 2005) such as damage factors, disease-specific factors, and the same uncertainty factor for the combined damage and effect factors. Consequently, the common value was set from the more carcinogenic molecule (**10**). However, as is the case with the above "universal hydrophobicity" adjustment, the equal stericity principle should be applied with caution (as a rule, it should be applied to less than 10% of the molecules in a series) and only to mark non-congeneric series of molecules with similarity physicochemical properties.

(iii) *The defined domain of applicability*, although conceptually included in one of the above stages of the unambiguous algorithm framework, is customarily specified separately for clarity. However, because the present application focused on modeling genotoxic carcinogenesis, this principle is redundant because of its implicit non-congeneric approach features. As such, the molecules in Tables 3.36 and 3.7 span many organic classes and derivatives, including amides, amines, aromatic systems, lactones, nitrites, quinines, cyanides, urethanes, ketones, and cycloalkanes. The QSAR analysis and mechanistic model was, therefore, expected to have non-local character (i.e., not depending on the series of toxicants involved) susceptible of general behavior.

(ii-bis) *The unambiguous algorithm* continues with *the analytical stage of the QSAR model* yielded the regression equations and their correlation factors and allied statistical descriptors. Table 3.38 gives the direct and residual QSAR models for all descriptor combinations considered for the trial molecules of Table 3.36 according to Eqs. (3.139) and (3.140), respectively. As anticipated, while the direct QSAR provided very low correlations, the residual-QSAR was characterized by the limiting case of unity factors of residuals, which raised the residual correlation factor as much as the complementary direct QSAR was lowered. The direct and residual QSAR complementary nature was, in this way, advanced. In particular, the lowest direct correlation, the LogP mechanism, corresponded to the highest residual QSAR. At the same time, when LogP was further synergistically combined with other structural influences like POL and Etot, the direct potency increased by a factor of one hundred, whereas the residual QSAR correlations decreased by only a few units. This proves the utility of the direct QSAR principle in assessing a statistical model that could be supplemented with further considerations, as with residual QSAR and other validity measures, to provide the best understanding of the analyzed phenomenon (Putz, 2011a; Putz et al., 2011c).

Table 3.39 compares the detailed self-consistent principle with the factor and averaged versions of the residual QSAR modeling of Eq. (3.141). If Eq. (3.141) is amended with the residual correlation factor or its complement to yield the observed-to-QSAR activity proportionality or if the averaged activity in Eq. (3.146) is replaced with expressions of Eqs. (3.147) ($\overline{A} = 5.285636$) and (3.148) ($\widetilde{A} = 5.20711$), then the results are systematically the same or very close to those reported in Table 3.38. In other words, whenever the model resembles the direct molecular variables' dependency, the direct QSAR statistical efficiency will be systematically reached.

Actually, in this case study, the two fundamental reactivity principles of electronegativity (Section 3.2.9.1) and chemical hardness (Section 3.2.9.2) involve *intra-electrophilic (intramolecular)* electron transfer from the HOMO and LUMO of the ligand molecule (or SA), such that, after donating

TABLE 3.38 The Parameters and Statistical Correlation Coefficients for the Residual-QSAR Algorithm of Eqs. (3.139) and (3.140), As Applied To the Molecules of Table 3.36 in All Possible Combinations of Variables (Putz, 2011a)

Structural variables	a_0		b_{i0}		R_0	a_1	b_1	R_1
LogP	5.297587		–0.007280		0.0091	5.285636	1	0.9999
POL	4.712835		0.029613		0.1832	5.285636	1	0.9831
Etot	4.676954		–0.000011		0.2033	5.285636	1	0.9791
LogP, POL	4.339331	–0.279746		0.072662	0.2925	5.285636	1	0.9563
LogP, Etot	4.578059	–0.162902		–0.000018	0.2608	5.285636	1	0.9654
POL, Etot	4.679442	–0.000978		–0.000012	0.2033	5.285636	1	0.9791
LogP, POL, Etot	4.341697	–0.273668	0.06646	–0.000002	0.2929	5.285636	1	0.9562

TABLE 3.39 Residual-QSAR Self-Consistent (SC), Factor (F1), Averaged (AV, with $\overline{A} = 5.285636$) Models of Eqs. (3.141), (5.145), and (3.146) for the Hansch Parameters of Table 3.38, With the Modeling and Predictive Powers for the "Gaussian" and "Quasi-Gaussian" Molecules of Tables 3.36 and 3.7 Represented By Their Associated Correlation Factors, Respectively (Putz et al., 2011c)

Structural Variables	Activity Model			
	Type	*Equation*	R_{Gauss}	$R_{Q\text{-}Gauss}$
Ia: LogP	SC	$A - 0.011951 + 0.00728[LogP]$	0.99996	0.99994
	F1	$-119.51 + 72.8[LogP]$	0.0091	0.1240
	AV	$\overline{A} - 0.011951 + 0.00728[LogP]$	0.0091	0.1240
Ib: POL	SC	$A + 0.572801 - 0.029613[POL]$	0.98307	0.97713
	F1	$33.8936 - 1.75225[POL]$	0.1832	0.23179
	AV	$\overline{A} + 0.572801 - 0.029613[POL]$	0.1832	0.23179
Ic: Etot	SC	$A + 0.608682 + 1.1 \times 10^{-5}[Etot]$	0.98362	0.97238
	F1	$29.1235 + 5.26316 \times 10^{-4}[Etot]$	0.2033	0.04250
	AV	$\overline{A} + 0.608682 + 1.1 \times 10^{-5}[Etot]$	0.2033	0.04250
IIa:LogP, POL	SC	$A + 0.946305 + 0.279746\,[LogP] - 0.072662[POL]$	0.95626	0.94916
	F1	$21.6546 + 6.40151\,[LogP] - 1.66275[POL]$	0.2925	0.21906
	AV	$\overline{A} + 0.946305 + 0.279746[LogP] - 0.072662[POL]$	0.2925	0.21906
IIb: ogP, Etot	SC	$A + 0.707577 + 0.162902\,[LogP] + 1.8 \times 10^{-5}[Etot]$	0.96686	0.96164
	F1	$20.4502 + 4.70815\,[LogP] + 5.20231 \times 10^{-4}[Etot]$	0.2608	0.0524
	AV	$\overline{A} + 0.707577 + 0.162902[LogP] + 1.8 \times 10^{-5}[Etot]$	0.2608	0.0524

TABLE 3.39 Continued

Structural Variables	Activity Model			
	Type	*Equation*	R_{Gauss}	$R_{Q\text{-}Gauss}$
IIc: POL,	SC	$A + 0.606194 + 0.000978\,[POL] + 1.2 \times 10^{-5}[Etot]$	0.97838	0.97017
Etot	F1	$29.0045 + 0.046793\,[POL] + 5.74163 \times 10^{-4}[Etot]$	0.2033	0.03654
	AV	$\overline{A} + 0.606194 + 0.000978[POL] + 1.2\times10^{-5}[Etot]$	0.2033	0.03654
III: LogP,	SC	$A + 0.943939 + 0.273668\,[LogP] - 0.06646\,[POL] + 2. \times 10^{-6}[Etot]$	0.95628	0.94927
POL,	F1	$21.5511 + 6.24813\,[LogP] - 1.51735\,[POL] + 4.56621 \times 10^{-5}[Etot]$	0.2929	0.19871
Etot	AV	$\overline{A} + 0.943939 + 0.273668[LogP] - 0.06646[POL] + 2.\times10^{-6}[Etot]$	0.2929	0.19871

one HOMO electron to the molecular or SA LUMO, *exo-electrophilic (intermolecular)* electron transfer between the new molecule or SA HOMO* and receptor LUMO occurs. This leaves a larger SA-HOMO–LUMO gap through ligand HOMO* relaxation (which is formally removed, so the gap between LUMO* and the second order ligand's HOMO is increased). This produces an overall *electrophilic docking* effect (Figure 3.27).

Note that other electrophilic mechanisms involving molecular or SA HOMO and LUMO frontier transformations and relaxations may be possible, but the two stages of EE and chemical hardness maximization should be equally satisfied. Together, the electronegativity and chemical hardness indices unambiguously describe a ligand-receptor docking mechanism via intra- and exo-electrophilic stages, generalizing the Millers' theory of direct electronic transfer between the molecular or SA-HOMO and receptor LUMO (Miller & Miller, 1981).

Structural alert QSARs for the trial compounds of Table 3.36 gave the structural alert activities A^{SA} shown in Table 3.40. Full molecular QSARs for the trial molecules of Table 3.36 are reported in Table 3.41 as "M" computed/predicted models. Combined QSAR predictions based on the molecular descriptors from Table 3.36 and the structural alert activities of Table 3.40 are reported in Table 3.41 as "*M∧SA*"; these results showcase

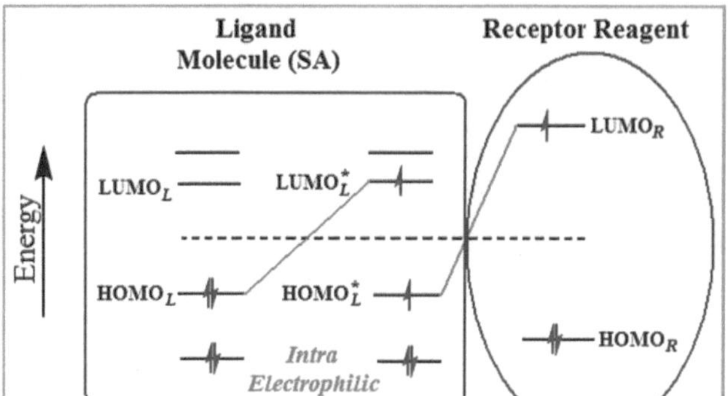

FIGURE 3.27 The electrophilic docking structure-reactivity algorithm correlating electronegativity and chemical hardness with chemical carcinogenesis. The algorithm starts with electronegativity docking (equalization) between the ligand and the receptor. Next, intra-molecular (in connection with specific structural alerts) maximization of the HOMO-LUMO gap (i.e., of chemical hardness) is accomplished by exo-electrophilic transfer of an electron from ligand to receptor (Putz et al., 2011c).

TABLE 3.40 QSAR Models for the Activities of the Trial Molecules With the Physicochemical Parameters of the SAs in Table 3.36 (Putz et al., 2011c)

Variable	QSAR model	Pred. Activity	R
χ^{SA}	$5.62 - 0.07\chi^{SA}$	A_{χ}^{SA}	0.026
η^{SA}	$6.016 - 0.149\eta^{SA}$	A_{η}^{SA}	0.086
POL^{SA}	$4.677 + 0.05POL^{SA}$	A_{POL}^{SA}	0.27
$LogP^{SA}$	$5.066 + 0.159LogP^{SA}$	A_{LogP}^{SA}	0.16
E_{tot}^{SA}	$4.402 - 0.00031\,E_{tot}^{SA}$	A_{Etot}^{SA}	0.35
χ^{SA}, η^{SA}	$6.51 - 0.09\chi^{SA} - 0.156\eta^{SA}$	$A_{\chi,\eta}^{SA}$	0.093
χ^{SA}, POL^{SA}	$5.1 - 0.08\eta^{SA} + 0.5POL^{SA}$	$A_{\chi,POL}^{SA}$	0.27
$\chi^{SA}, LogP^{SA}$	$5.54 - 0.1\chi^{SA} + 0.16LogP^{SA}$	$A_{\chi,LogP}^{SA}$	0.16
χ^{SA}, E_{tot}^{SA}	$5.58 - 0.255\,\chi^{SA} - 0.000032\,E_{tot}^{SA}$	$A_{\chi,Etot}^{SA}$	0.371
η^{SA}, POL^{SA}	$2.43 + 0.38\eta^{SA} + 0.08POL^{SA}$	$A_{\eta,POL}^{SA}$	0.316
$\eta^{SA}, LogP^{SA}$	$4.75 + 0.05\eta^{SA} + 0.179LogP^{SA}$	$A_{\eta,LogP}^{SA}$	0.167
η^{SA}, E_{tot}^{SA}	$2.6 + 0.314\eta^{SA} - 0.00004\,E_{tot}^{SA}$	$A_{\eta,Etot}^{SA}$	0.38
$POL^{SA}, LogP^{SA}$	$4.36 + 0.12POL^{SA} - 0.44LogP^{SA}$	$A_{POL,LogP}^{SA}$	0.337
POL^{SA}, E_{tot}^{SA}	$4.38 - 0.0625\,POL^{SA} - 0.00005\,E_{tot}^{SA}$	$A_{POL,Etot}^{SA}$	0.382

TABLE 3.40 Continued

Variable	QSAR model	Pred. Activity	R
$E_{tot}^{SA}, LogP^{SA}$	$4.09 - 0.412\,LogP^{SA} - 0.00006\,E_{tot}^{SA}$	$A_{Etot,LogP}^{SA}$	0.4305
$\chi^{SA}, \eta^{SA}, POL^{SA}$	$2.62 - 0.035\chi^{SA} + 0.382\eta^{SA} + 0.08POL^{SA}$	$A_{\chi,\eta,POL}^{SA}$	0.31
$\chi^{SA}, \eta^{SA}, LogP^{SA}$	$5.25 - 0.095\chi^{SA} + 0.049\eta^{SA} + 0.179LogP^{SA}$	$A_{\chi,\eta,LogP}^{SA}$	0.17
$\chi^{SA}, \eta^{SA}, E_{tot}^{SA}$	$3.78 - 0.254\,\chi^{SA} + 0.3141\,\eta^{SA} - 0.00004\,E_{tot}^{SA}$	$A_{\chi,\eta,Etot}^{SA}$	0.399
$\chi^{SA}, POL^{SA}, LogP^{SA}$	$4.52 - 0.0325\chi^{SA} + 0.12POL^{SA} - 0.439LogP^{SA}$	$A_{\chi,POL,LogP}^{SA}$	0.337
$\chi^{SA}, POL^{SA}, E_{tot}^{SA}$	$6.46 - 0.45\,\chi^{SA} - 0.092\,POL^{SA} - 0.0007\,E_{tot}^{SA}$	$A_{\chi,POL,Etot}^{SA}$	0.4141
$\chi^{SA}, LOGP^{SA}, E_{tot}^{SA}$	$5.75 - 0.36\,\chi^{SA} - 0.45\,LogP^{SA} - 0.00006\,E_{tot}^{SA}$	$A_{\chi,LogP,Etot}^{SA}$	0.451
$\eta^{SA}, POL^{SA}, LogP^{SA}$	$2.05 + 0.39\eta^{SA} + 0.16POL^{SA} - 0.449LogP^{SA}$	$A_{\eta,POL,LogP}^{SA}$	0.373
$\eta^{SA}, POL^{SA}, E_{tot}^{SA}$	$3.09 + 0.22\eta^{SA} - 0.034\,POL^{SA} - 0.00005\,E_{tot}^{SA}$	$A_{\eta,POL,Etot}^{SA}$	0.392
$\eta^{SA}, LogP^{SA}, E_{tot}^{SA}$	$3.23 + 0.156\eta^{SA} - 0.368\,LogP^{SA} - 0.0006\,E_{tot}^{SA}$	$A_{\eta,LogP,Etot}^{SA}$	0.436
$POL^{SA}, LogP^{SA}, E_{tot}^{SA}$	$4.07 + 0.015\,POL^{SA} - 0.45\,LogP^{SA} - 0.000057\,E_{tot}^{SA}$	$A_{POL,LogP,Etot}^{SA}$	0.4312

TABLE 3.41 Multi-Linear QSAR Models for the Trial Molecular Activities With the Full Molecular (M) Physicochemical Parameters of Table 3.36 and the Corresponding Activities of the Structural Alerts (ASA) From Table 3.40 (Putz et al., 2011c)

Variable	QSAR model	Pred. Activity	R
χ^M	$2.18+0.63\chi^M$	A_χ^M	0.187
χ^M, A_χ^{SA}	$-65.66+12.1A_\chi^{SA}+1.41\chi^M$	$A_\chi^{M\wedge ASA}$	0.295
η^M	$7.13-0.405\eta^M$	A_η^M	0.192
η^M, A_η^{SA}	$11.38-0731A_\eta^{SA}-0.49\eta^M$	$A_\eta^{M\wedge ASA}$	0.198
POL^M	$4.71+0.029POL^M$	A_{POL}^M	0.18
POL^M, A_{POL}^{SA}	$-0.438+1.1A_{POL}^{SA}-0.006POL^M$	$A_{POL}^{M\wedge ASA}$	0.275
$LogP^M$	$5.29-0.0007LogP^M$	A_{POL}^M	0.009
$LogP^M, A_{LogP}^{SA}$	$-13.64+3.71A_{LogP}^{SA}-0.421LogP^M$	$A_{POL}^{M\wedge ASA}$	0.325
E_{tot}^M	$4.67-0.00001E_{TOT}^M$	A_{Etot}^M	0.203
E_{tot}^M, A_{Etot}^{SA}	$-0.05+1.01A_{Etot}^{SA}+0.000001E_{tot}^M$	$A_{Etot}^{M\wedge ASA}$	0.358
χ^M, η^M	$3.92+0.68\chi^M-0.43\eta^M$	$A_{\chi,\eta}^M$	0.278
$\chi^M, \eta^M, A_{\chi,\eta}^{SA}$	$0.015+0.73\chi^M-0.35\eta^M+0.62A_{\chi,\eta}^{SA}$	$A_{\chi,\eta}^{M\wedge ASA}$	0.28
χ^M, POL^M	$1.36+0.67\chi^M+0.031POL^M$	$A_{\chi,POL}^M$	0.27
$\chi^M, POL^M, A_{\chi,POL}^{SA}$	$-6.009+0.86\chi^M-0.012POL^M+1.38A_{\chi,POL}^{SA}$	$A_{\chi,POL}^{M\wedge ASA}$	0.373

TABLE 3.41 Continued

Variable	QSAR model	Pred. Activity	R
$\chi^M, LogP^M$	$2.12+0.639\chi^M+0.013LogP^M$	$A^M_{\chi,LogP}$	0.188
$\chi^M, LogP^M, A^{SA}_{\chi,LogP}$	$-19.46+0.81\chi^M-0.44LogP^M+4.05A^{SA}_{\chi,LogP}$	$A^{M\wedge ASA}_{\chi,LogP}$	0.409
χ^M, E^M_{tot}	$1.64+0.62\chi^M-0.00001E^M_{tot}$	$A^M_{\chi,Etot}$	0.27
$\chi^M, E^M_{tot}, A^{SA}_{\chi,Etot}$	$-4.81+0.83\chi^M+0.00003E^M_{tot}+1.16A^{SA}_{\chi,Etot}$	$A^{M\wedge ASA}_{\chi,Etot}$	0.443
η^M, POL^M	$6.18-0.264\eta^M+0.015POL^M$	$A^M_{\eta,POL}$	0.204
$\eta^M, POL^M, A^{SA}_{\eta,POL}$	$1.47-0.64\eta^M-0.05POL^M+1.48A^{SA}_{\eta,POL}$	$A^{M\wedge ASA}_{\eta,POL}$	0.378
$\eta^M, LogP^M$	$7.66-0.49\eta^M-0.08LogP^M$	$A^M_{\eta,LogP}$	0.213
$\eta^M, LogP^M, A^{SA}_{\eta,LogP}$	$-12.17-0.52\eta^M-0.531LogP^M+3.919A^{SA}_{\eta,LogP}$	$A^{M\wedge ASA}_{\eta,LogP}$	0.403
η^M, E^M_{tot}	$5.993-0.24\eta^M-0.000008E^M_{tot}$	$A^M_{\eta,Etot}$	0.226
$\eta^M, E^M_{tot}, A^{SA}_{\eta,Etot}$	$1.585-0.442\eta^M+0.00001E^M_{tot}+1.182A^{SA}_{\eta,Etot}$	$A^{M\wedge ASA}_{\eta,Etot}$	0.426
$POL^M, LogP^M$	$4.339+0.72POL^M-0.279LogP^M$	$A^M_{POL,LogP}$	0.29
$POL^M, LogP^M, A^{SA}_{POL,LogP}$	$-0.81+0.04POL^M-0.31LogP^M+1.08A^{SA}_{POL,LogP}$	$A^{M\wedge ASA}_{POL,LogP}$	0.422
POL^M, E^M_{tot}	$4.67-0.00097POL^M-0.00001E^M_{tot}$	$A^M_{POL,Etot}$	0.203
$POL^M, E^M_{tot}, A^{SA}_{POL,Etot}$	$-0.299-0.0304POL^M-0.00008E^M_{tot}+1.091A^{SA}_{POL,Etot}$	$A^{M\wedge ASA}_{POL,Etot}$	0.3907

TABLE 3.41 Continued

Variable	QSAR model	Pred. Activity	R
$E_{tot}^{M}, LogP^{M}$	$4.578 - 0.162\,LogP^{M} - 0.000018\,E_{tot}^{M}$	$A_{Etot,LogP}^{M}$	0.26
$E_{tot}^{M}, LogP^{M}, A_{Etot,LogP}^{SA}$	$-0.603 - 0.2337\,LogP^{M} - 0.00009\,E_{tot}^{M} + 1.093\,A_{Etot,LogP}^{SA}$	$A_{Etot,LogP}^{M\wedge ASA}$	0.488
$\chi^{M}, \eta^{M}, POL^{M}$	$2.89 + 0.686\,\chi^{M} - 0.28\eta^{M} + 0.016\,POL^{M}$	$A_{\chi,\eta,POL}^{M}$	0.288
$\chi^{M}, \eta^{M}, POL^{M}, A_{\chi,\eta,POL}^{SA}$	$-3.19 + 0.86\,\chi^{M} - 0.696\eta^{M} - 0.06\,POL^{M} + 1.62\,A_{\chi,\eta,POL}^{SA}$	$A_{\chi,\eta,POL}^{M\wedge ASA}$	0.452
$\chi^{M}, \eta^{M}, LogP^{M}$	$4.458 + 0.65\,\chi^{M} - 0.5\eta^{M} - 0.06\,LogP^{M}$	$A_{\chi,\eta,LogP}^{M}$	0.287
$\chi^{M}, \eta^{M}, LogP^{M}, A_{\chi,\eta,LogP}^{SA}$	$-16.899 + 0.794\,\chi^{M} - 0.49\eta^{M} - 0.52\,LogP^{M} + 4.054\,A_{\chi,\eta,LogP}^{SA}$	$A_{\chi,\eta,LogP}^{M\wedge ASA}$	0.462
$\chi^{M}, \eta^{M}, E_{tot}^{M}$	$3.016 + 0.65\,\chi^{M} - 0.293\eta^{M} - 0.00007\,E_{tot}^{M}$	$A_{\chi,\eta,Etot}^{M}$	0.298
$\chi^{M}, \eta^{M}, E_{tot}^{M}, A_{\chi,\eta,Etot}^{SA}$	$-2.689 + 0.841\,\chi^{M} - 0.473\eta^{M} + 0.00001\,E_{tot}^{M} + 1.256\,A_{\chi,\eta,Etot}^{SA}$	$A_{\chi,\eta,Etot}^{M\wedge ASA}$	0.497
$\chi^{M}, POL^{M}, LogP^{M}$	$1.481 + 0.58\,\chi^{M} + 0.07\,POL^{M} - 0.2512\,LogP^{M}$	$A_{\chi,POL,LogP}^{M}$	0.338
$\chi^{M}, POL^{M}, LogP^{M}, A_{\chi,POL,LogP}^{SA}$	$-5.593 + 0.81\,\chi^{M} + 0.037\,POL^{M} - 0.281\,LogP^{M} + 1.254\,A_{\chi,POL,LogP}^{SA}$	$A_{\chi,POL,LogP}^{M\wedge ASA}$	0.483
$\chi^{M}, POL^{M}, E_{tot}^{M}$	$1.503 + 0.64\,\chi^{M} + 0.012\,POL^{M} - 0.000007\,E_{tot}^{M}$	$A_{\chi,POL,Etot}^{M}$	0.276
$\chi^{M}, POL^{M}, E_{tot}^{M}, A_{\chi,POL,Etot}^{SA}$	$-4.009 + 0.725\,\chi^{M} - 0.008\,POL^{M} + 0.00001\,E_{tot}^{M} + 1.123\,A_{\chi,POL}^{SA}$	$A_{\chi,POL,Etot}^{M\wedge ASA}$	0.469
$\chi^{M}, LogP^{M}, E_{tot}^{M}$	$1.961 + 0.538\,\chi^{M} - 0.134\,LogP^{M} - 0.000017\,E_{tot}^{M}$	$A_{\chi,LogP,Etot}^{M}$	0.304
$\chi^{M}, LogP^{M}, E_{tot}^{M}, A_{\chi,LogP,Etot}^{SA}$	$-4.251 + 0.68\,\chi^{M} - 0.219\,LogP^{M} - 0.00007\,E_{tot}^{M} + 1.166\,A_{\chi,LogP,Etot}^{SA}$	$A_{\chi,LogP,Etot}^{M\wedge ASA}$	0.552

TABLE 3.41 Continued

Variable	QSAR model	Pred. Activity	R
$\eta^M, POL^M, LogP^M$	$4.87 - 0.093\eta^M + 0.066 POL^M - 0.269 LogP^M$	$A^M_{\eta,POL,LogP}$	0.294
$\eta^M, POL^M, LogP^M, A^{SA}_{\eta,POL,LogP}$	$-0.253 - 0.153\eta^M + 0.033 POL^M - 0.321 LogP^M + 1.156 A^{SA}_{\eta,POL,LogP}$	$A^{M \wedge ASA}_{\eta,POL,LogP}$	0.466
η^M, POL^M, E^M_{tot}	$6.84 - 0.39\eta^M - 0.368 POL^M - 0.00018 E^M_{tot}$	$A^M_{\eta,POL,Etot}$	0.239
$\eta^M, POL^M, E^M_{tot}, A^{SA}_{\eta,POL,Etot}$	$3.415 - 1.08\eta^M - 0.13 POL^M - 0.00002 E^M_{tot} + 1.598 A^{SA}_{\eta,POL,Etot}$	$A^{M \wedge ASA}_{\eta,POL,Etot}$	0.507
$\eta^M, LogP^M, E^M_{tot}$	$6.176 - 0.303\eta^M - 0.177 LogP^M - 0.000015 E^M_{tot}$	$A^M_{\eta,LogP,Etot}$	0.287
$\eta^M, LogP^M, E^M_{tot}, A^{SA}_{\eta,LogP,Etot}$	$1.433 - 0.449\eta^M - 0.261 LogP^M - 0.00003 E^M_{tot} + 1.172 A^{SA}_{\eta,LogP,Etot}$	$A^{M \wedge ASA}_{\eta,LogP,Etot}$	0.525
$POL^M, LogP^M, E^M_{tot}$	$4.341 + 0.06 POL^M - 0.273 LogP^M - 0.000002 E^M_{tot}$	$A^M_{POL,LogP,Etot}$	0.298
$POL^M, LogP^M, E^M_{tot}, A^{SA}_{POL,LogP,Etot}$	$-0.598 + 0.031 POL^M - 0.288 LogP^M - 0.000002 E^M_{tot} + 1.065 A^{SA}_{POL,LogP,Etot}$	$A^{M \wedge ASA}_{POL,LogP,Etot}$	0.495

how consideration of the structural alerts allows systematic improvement of the predictions over the molecular indicators. Note that the structural alert parameters may be combined with the molecular ones only at the level of full molecules; in this way, full molecular parameters are combined with predicted activity at the molecular level as provided by structural alerts modeling.

Residual QSARs for the structural alert models derived in *OECD-QSAR-step (i)* are computed, considering the predicted activities in Table 3.40; the results are presented in Table 3.42. Considerable correlation was found, indicating the indirect influence of structural alerts on mutagenicity and carcinogenesis.

In OECD-QSAR-step (iii) the structural residual alert QSARs were obtained by selecting models from *OECD-QSAR-step (ii)* that reproduce the structural alerts' parameter correlations (3.158) as given in Eq. (3.157). The residual-alert methodology may lead to new equations besides those presented in Table 3.40. The results are displayed in Table 3.43, with correlation performances reported for the trial molecules of Table 3.36 and the test compounds of Table 3.37.

(iv) *The validity and predictivity* principle is considered to be one of the most important stages of QSAR analysis. Although internal and external validation statistical procedures exist, the former is often overestimated. This has been confirmed in situations when the external validation sets were well predicted, even with poor cross-validated performance (Franke & Gruska, 2003). As a general rule, external validation tests are considered the true standard to assess prediction in QSAR modeling. Focusing on the special case of genotoxicity, one should consider all residual QSAR models obtained within previous QSAR principles (i.e., the self-consistent and factor/averaged residual QSAR models of Table 3.39, in particular) while remembering that the last ones resemble the direct QSAR statistical performances. The external validation set is presented in Table 3.37 and was identified through the quasi-Gaussian shape of the Figure 3.26 inset. The testing set and associated statistical performances are reported in the last column of Table 3.39. These need to be interpreted in light of the searched mechanistic model, or the predictive power lies only in the range of the residual QSARs, with no real information contained therein. This will be realized by applying the final principle of the OECD-QSAR framework.

TABLE 3.42 Residual-QSARs for the Structural Alert Models of Table 3.40 (Putz et al., 2011c)

Variable	QSAR model	Pred. Activity	R
RA_{χ}^{SA}	$5.2856 + RA_{\chi}^{SA}$	ARA_{χ}^{SA}	0.999
RA_{η}^{SA}	$5.2856 + RA_{\eta}^{SA}$	ARA_{η}^{SA}	0.996
RA_{POL}^{SA}	$5.2856 + RA_{POL}^{SA}$	ARA_{POL}^{SA}	0.961
RA_{LogP}^{SA}	$5.2856 + RA_{LogP}^{SA}$	ARA_{LogP}^{SA}	0.986
RA_{Etot}^{SA}	$5.2856 + RA_{Etot}^{SA}$	ARA_{Etot}^{SA}	0.933
$RA_{\chi}^{SA}, RA_{\eta}^{SA}$	$5.2856 + 0.886\,RA_{\chi}^{SA} + 0.114\,RA_{\eta}^{SA}$	$ARA_{\chi,\eta}^{SA}$	0.999
$RA_{\chi}^{SA}, RA_{POL}^{SA}$	$5.2856 + 0.987\,RA_{\chi}^{SA} + 0.013\,RA_{POL}^{SA}$	$ARA_{\chi,POL}^{SA}$	0.999
$RA_{\chi}^{SA}, RA_{LogP}^{SA}$	$5.2856 + 0.963\,RA_{\chi}^{SA} + 0.037\,RA_{LogP}^{SA}$	$ARA_{\chi,LogP}^{SA}$	0.999
$RA_{\chi}^{SA}, RA_{Etot}^{SA}$	$5.2856 + 0.98\,RA_{\chi}^{SA} + 0.022\,RA_{Etot}^{SA}$	$ARA_{\chi,Etot}^{SA}$	0.999
$RA_{\eta}^{SA}, RA_{POL}^{SA}$	$5.2856 + 1.194\,RA_{\eta}^{SA} - 0.206\,RA_{POL}^{SA}$	$ARA_{\eta,POL}^{SA}$	0.997
$RA_{\eta}^{SA}, RA_{LogP}^{SA}$	$5.2856 + 1.101\,RA_{\eta}^{SA} - 0.103\,RA_{LogP}^{SA}$	$ARA_{\eta,LogP}^{SA}$	0.996
$RA_{\eta}^{SA}, RA_{Etot}^{SA}$	$5.2856 + 1.105\,RA_{\eta}^{SA} - 0.118\,RA_{Etot}^{SA}$	$ARA_{\eta,Etot}^{SA}$	0.996
$RA_{POL}^{SA}, RA_{LogP}^{SA}$	$5.2856 - 0.709\,RA_{POL}^{SA} + 1.684\,RA_{LogP}^{SA}$	$ARA_{POL,LogP}^{SA}$	0.991
$RA_{POL}^{SA}, RA_{Etot}^{SA}$	$5.2856 + 1.642\,RA_{POL}^{SA} - 0.670\,RA_{Etot}^{SA}$	$ARA_{POL,Etot}^{SA}$	0.966
$RA_{LogP}^{SA}, RA_{Etot}^{SA}$	$5.2856 + 1.395\,RA_{LogP}^{SA} - 0.430\,RA_{Etot}^{SA}$	$ARA_{LogP,Etot}^{SA}$	0.991

TABLE 3.42 Continued

Variable	QSAR model	Pred. Activity	R
$RA_{\chi}^{SA}, RA_{\eta}^{SA}, RA_{POL}^{SA}$	$5.2856 + 0.852\,RA_{\chi}^{SA} + 0.176\,RA_{\eta}^{SA} - 0.030\,RA_{POL}^{SA}$	$ARA_{\chi,\eta,POL}^{SA}$	0.999
$RA_{\chi}^{SA}, RA_{\eta}^{SA}, RA_{LogP}^{SA}$	$5.2856 + 0.884\,RA_{\chi}^{SA} + 0.123\,RA_{\eta}^{SA} - 0.007\,RA_{LogP}^{SA}$	$ARA_{\chi,\eta,LogP}^{SA}$	0.999
$RA_{\chi}^{SA}, RA_{\eta}^{SA}, RA_{Etot}^{SA}$	$5.2856 + 0.892\,RA_{\chi}^{SA} + 0.104\,RA_{\eta}^{SA} + 0.004\,RA_{Etot}^{SA}$	$ARA_{\chi,\eta,Etot}^{SA}$	0.999
$RA_{\chi}^{SA}, RA_{POL}^{SA}, RA_{LogP}^{SA}$	$5.2856 + 0.941\,RA_{\chi}^{SA} - 0.045\,RA_{POL}^{SA} + 0.102\,RA_{LogP}^{SA}$	$ARA_{\chi,POL,LogP}^{SA}$	0.999
$RA_{\chi}^{SA}, RA_{POL}^{SA}, RA_{Etot}^{SA}$	$5.2856 + 1.005\,RA_{\chi}^{SA} - 0.081\,RA_{POL}^{SA} + 0.080\,RA_{Etot}^{SA}$	$ARA_{\chi,POL,Etot}^{SA}$	0.999
$RA_{\chi}^{SA}, RA_{LogP}^{SA}, RA_{Etot}^{SA}$	$5.2856 + 0.983\,RA_{\chi}^{SA} - 0.003\,RA_{LogP}^{SA} + 0.023\,RA_{Etot}^{SA}$	$ARA_{\chi,LogP,Etot}^{SA}$	0.999
$RA_{\eta}^{SA}, RA_{POL}^{SA}, RA_{LogP}^{SA}$	$5.2856 + 0.951\,RA_{\eta}^{SA} - 0.418\,RA_{POL}^{SA} + 0.450\,RA_{LogP}^{SA}$	$ARA_{\eta,POL,LogP}^{SA}$	0.997
$RA_{\eta}^{SA}, RA_{POL}^{SA}, RA_{Etot}^{SA}$	$5.2856 + 1.212\,RA_{\eta}^{SA} - 0.290\,RA_{POL}^{SA} + 0.067\,RA_{Etot}^{SA}$	$ARA_{\eta,POL,Etot}^{SA}$	0.997
$RA_{\eta}^{SA}, RA_{LogP}^{SA}, RA_{Etot}^{SA}$	$5.2856 + 0.955\,RA_{\eta}^{SA} + 0.213\,RA_{LogP}^{SA} - 0.185\,RA_{Etot}^{SA}$	$ARA_{\eta,LogP,Etot}^{SA}$	0.997
$RA_{POL}^{SA}, RA_{LogP}^{SA}, RA_{Etot}^{SA}$	$5.2856 - 0.423\,RA_{POL}^{SA} + 1.617\,RA_{LogP}^{SA} - 0.227\,RA_{Etot}^{SA}$	$ARA_{POL,LogP,Etot}^{SA}$	0.991
$RA_{\chi}^{SA}, RA_{\eta}^{SA}, RA_{POL}^{SA}, RA_{LogP}^{SA}$	$5.2856 + 0.815\,RA_{\chi}^{SA} + 0.170\,RA_{\eta}^{SA} - 0.082\,RA_{POL}^{SA} + 0.093\,RA_{LogP}^{SA}$	$ARA_{\chi,\eta,POL,LogP}^{SA}$	0.999
$RA_{\chi}^{SA}, RA_{POL}^{SA}, RA_{LogP}^{SA}, RA_{Etot}^{SA}$	$5.2856 + 0.966\,RA_{\chi}^{SA} - 0.121\,RA_{POL}^{SA} + 0.083\,RA_{LogP}^{SA} + 0.074\,RA_{Etot}^{SA}$	$ARA_{\chi,POL,LogP,Etot}^{SA}$	0.999
$RA_{\eta}^{SA}, RA_{POL}^{SA}, RA_{LogP}^{SA}, RA_{Etot}^{SA}$	$5.2856 + 0.966\,RA_{\eta}^{SA} - 0.461\,RA_{POL}^{SA} + 0.422\,RA_{LogP}^{SA} + 0.038\,RA_{Etot}^{SA}$	$ARA_{\eta,POL,LogP,Etot}^{SA}$	0.999
$RA_{\chi}^{SA}, RA_{\eta}^{SA}, RA_{LogP}^{SA}, RA_{Etot}^{SA}$	$5.2856 + 0.902\,RA_{\chi}^{SA} + 0.117\,RA_{\eta}^{SA} - 0.034\,RA_{LogP}^{SA} + 0.016\,RA_{Etot}^{SA}$	$ARA_{\chi,\eta,LogP,Etot}^{SA}$	0.999
$RA_{\chi}^{SA}, RA_{\eta}^{SA}, RA_{POL}^{SA}, RA_{Etot}^{SA}$	$5.2856 + 0.858\,RA_{\chi}^{SA} + 0.193\,RA_{\eta}^{SA} - 0.138\,RA_{POL}^{SA} + 0.088\,RA_{Etot}^{SA}$	$ARA_{\chi,\eta,POL,Etot}^{SA}$	0.999

TABLE 3.42 Continued

Variable	QSAR model	Pred. Activity	R
$RA_{\chi}^{SA}, RA_{\eta}^{SA}, RA_{POL}^{SA}, RA_{LogP}^{SA}, RA_{Etot}^{SA}$	$5.2856 + 0.830RA_{\chi}^{SA} + 0.188RA_{\eta}^{SA} - 0.171RA_{POL}^{SA} + 0.070RA_{LogP}^{SA} + 0.083RA_{Etot}^{SA}$	$ARA_{\chi,\eta,POL,LogP,Etot}^{SA}$	0.999
$RA_{(\chi,\eta)}^{SA}$	$5.2856 + RA_{(\chi,\eta)}^{SA}$	$ARA_{(\chi,\eta)}^{SA}$	0.99
$RA_{(\chi,POL)}^{SA}$	$5.2856 + RA_{(\chi,POL)}^{SA}$	$ARA_{(\chi,POL)}^{SA}$	0.96
$RA_{(\chi,LogP)}^{SA}$	$5.2856 + RA_{(\chi,LogP)}^{SA}$	$ARA_{(\chi,LogP)}^{SA}$	0.985
$RA_{(\chi,Etot)}^{SA}$	$5.2856 + RA_{(\chi,Etot)}^{SA}$	$ARA_{(\chi,Etot)}^{SA}$	0.928
$RA_{(\eta,POL)}^{SA}$	$5.2856 + RA_{(\eta,POL)}^{SA}$	$ARA_{(\eta,POL)}^{SA}$	0.985
$RA_{(\eta,LogP)}^{SA}$	$5.2856 + RA_{(\eta,LogP)}^{SA}$	$ARA_{(\eta,LogP)}^{SA}$	0.985
$RA_{(\eta,Etot)}^{SA}$	$5.2856 + RA_{(\eta,Etot)}^{SA}$	$ARA_{(\eta,Etot)}^{SA}$	0.92
$RA_{(POL,LogP)}^{SA}$	$5.2856 + RA_{(POL,LogP)}^{SA}$	$ARA_{(POL,LogP)}^{SA}$	0.93
$RA_{(POL,Etot)}^{SA}$	$5.2856 + RA_{(POL,Etot)}^{SA}$	$ARA_{(POL,Etot)}^{SA}$	0.923
$RA_{(Etot,LogP)}^{SA}$	$5.2856 + RA_{(Etot,LogP)}^{SA}$	$ARA_{(Etot,LogP)}^{SA}$	0.902
$RA_{(\chi,\eta),(POL,LogP)}^{SA}$	$5.2856 + 1.200\,RA_{(\chi,\eta)}^{SA} - 0.212\,RA_{(POL,LogP)}^{SA}$	$ARA_{(\chi,\eta),(POL,LogP)}^{SA}$	0.996
$RA_{(\chi,\eta),(POL,Etot)}^{SA}$	$5.2856 + 1.0003\,RA_{(\chi,\eta)}^{SA} - 0.0003\,RA_{(POL,Etot)}^{SA}$	$ARA_{(\chi,\eta),(POL,Etot)}^{SA}$	0.995
$RA_{(\chi,\eta),(Etot,LogP)}^{SA}$	$5.2856 + 0.9991\,RA_{\chi,\eta}^{SA} + 0.0009\,RA_{Etot,LogP}^{SA}$	$ARA_{(\chi,\eta),(Etot,LogP)}^{SA}$	0.995
$RA_{(\chi,POL),(\eta,LogP)}^{SA}$	$5.2856 - 0.490\,RA_{(\chi,POL)}^{SA} + 1.471\,RA_{(\eta,LogP)}^{SA}$	$ARA_{(\chi,POL),(\eta,LogP)}^{SA}$	0.988

TABLE 3.42 Continued

Variable	QSAR model	Pred. Activity	R
$RA^{SA}_{(\chi,POL),(\eta,Etot)}$	$5.2856 + 1.033\,RA^{SA}_{(\chi,POL)} - 0.036\,RA^{SA}_{(\eta,Etot)}$	$ARA^{SA}_{(\chi,POL),(\eta,Etot)}$	0.96
$RA^{SA}_{(\chi,POL),(Etot,LogP)}$	$5.2856 + 0.945\,RA^{SA}_{(\chi,POL)} + 0.062\,RA^{SA}_{(Etot,LogP)}$	$ARA^{SA}_{(\chi,POL),(Etot,LogP)}$	0.961
$RA^{SA}_{(\chi,LogP),(\eta,POL)}$	$5.2856 + 1.319\,RA^{SA}_{(\chi,LogP)} - 0.339\,RA^{SA}_{(\eta,POL)}$	$ARA^{SA}_{(\chi,LogP),(\eta,POL)}$	0.987
$RA^{SA}_{(\chi,LogP),(\eta,Etot)}$	$5.2856 + 1.148\,RA^{SA}_{(\chi,LogP)} - 0.167\,RA^{SA}_{(\eta,Etot)}$	$ARA^{SA}_{(\chi,LogP),(\eta,Etot)}$	0.986
$RA^{SA}_{(\chi,LogP),(POL,Etot)}$	$5.2856 + 1.086\,RA^{SA}_{(\chi,LogP)} - 0.097\,RA^{SA}_{(POL,Etot)}$	$ARA^{SA}_{(\chi,LogP),(POL,Etot)}$	0.985
$RA^{SA}_{(\chi,Etot),(\eta,POL)}$	$5.2856 - 0.084\,RA^{SA}_{(\chi,Etot)} + 1.081\,RA^{SA}_{(\eta,POL)}$	$ARA^{SA}_{(\chi,Etot),(\eta,POL)}$	0.948
$RA^{SA}_{(\chi,Etot),(\eta,LogP)}$	$5.2856 - 0.388\,RA^{SA}_{(\chi,Etot)} + 1.353\,RA^{SA}_{(\eta,LogP)}$	$ARA^{SA}_{(\chi,Etot),(\eta,LogP)}$	0.99
$RA^{SA}_{(\chi,Etot),(\eta,LogP)}$	$5.2856 + 0.308\,RA^{SA}_{(\chi,Etot)} + 0.706\,RA^{SA}_{(POL,LogP)}$	$ARA^{SA}_{(\chi,Etot),(\eta,LogP)}$	0.944
$RA^{SA}_{(\eta,POL),(Etot,LogP)}$	$5.2856 + 0.870\,RA^{SA}_{(\eta,POL)} + 0.145\,RA^{SA}_{(Etot,LogP)}$	$ARA^{SA}_{(\eta,POL),(Etot,LogP)}$	0.949
$RA^{SA}_{(\eta,LogP),(POL,Etot)}$	$5.2856 + 1.155\,RA^{SA}_{(\eta,LogP)} - 0.173\,RA^{SA}_{(POL,Etot)}$	$ARA^{SA}_{(\eta,LogP),(POL,Etot)}$	0.987
$RA^{SA}_{(\eta,Etot),(POL,LogP)}$	$5.2856 + 0.342\,RA^{SA}_{(\eta,Etot)} 0.685\,RA^{SA}_{(POL,LogP)}$	$ARA^{SA}_{(\eta,Etot),(POL,LogP)}$	0.947
$RA^{SA}_{(\chi,\eta,POL)}$	$5.2856 + RA^{SA}_{(\chi,\eta,POL)}$	$ARA^{SA}_{(\chi,\eta,POL)}$	0.948
$RA^{SA}_{(\chi,\eta,LogP)}$	$5.2856 + RA^{SA}_{(\chi,\eta,LogP)}$	$ARA^{SA}_{(\chi,\eta,LogP)}$	0.985
$RA^{SA}_{(\chi,\eta,Etot)}$	$5.2856 + RA^{SA}_{(\chi,\eta,Etot)}$	$ARA^{SA}_{(\chi,\eta,Etot)}$	0.916
$RA^{SA}_{(\chi,POL,LogP)}$	$5.2856 + RA^{SA}_{(\chi,POL,LogP)}$	$ARA^{SA}_{(\chi,POL,LogP)}$	0.941

TABLE 3.42 Continued

Variable	QSAR model	Pred. Activity	R
$RA^{SA}_{\{\chi,POL,Etot\}}$	$5.2856 + RA^{SA}_{\{\chi,POL,Etot\}}$	$ARA^{SA}_{\{\chi,POL,Etot\}}$	0.91
$RA^{SA}_{\{\chi,LogP,Etot\}}$	$5.2856 + RA^{SA}_{\{\chi,LogP,Etot\}}$	$ARA^{SA}_{\{\chi,LogP,Etot\}}$	0.89
$RA^{SA}_{\{\eta,POL,LogP\}}$	$5.2856 + RA^{SA}_{\{\eta,POL,LogP\}}$	$ARA^{SA}_{\{\eta,POL,LogP\}}$	0.927
$RA^{SA}_{\{\eta,POL,Etot\}}$	$5.2856 + RA^{SA}_{\{\eta,POL,Etot\}}$	$ARA^{SA}_{\{\eta,POL,Etot\}}$	0.919
$RA^{SA}_{\{\eta,LogP,Etot\}}$	$5.2856 + RA^{SA}_{\{\eta,LogP,Etot\}}$	$ARA^{SA}_{\{\eta,LogP,Etot\}}$	0.899
$RA^{SA}_{\{POL,LogP,Etot\}}$	$5.2856 + RA^{SA}_{\{POL,LogP,Etot\}}$	$ARA^{SA}_{\{POL,LogP,Etot\}}$	0.902

TABLE 3.43 Residual-Alert QSARs For the Models of Table 3.42 That Fulfill Eq. (3.157) With Highest Trial Correlation Factors*

No. Crt.	Variabile	QSAR Model	R_{trial}	R_{test}
I_a	χ^{SA}, E_{tot}^{SA}	$ARA^{SA} = 158.7 - 34\chi^{SA} - 0.003 E_{tot}^{SA}$	0.368	0.168
		$A^{SA} = 5.58 - 0.255\chi^{SA} - 0.000032 E_{tot}^{SA}$	0.371	0.127
I_b	η^{SA}, POL^{SA}	$ARA^{SA} = -77.866 + 14.75\eta^{SA} + 0.833 POL^{SA}$	0.078	0.505
		$A^{SA} = 2.43 + 0.38\eta^{SA} + 0.08 POL^{SA}$	0.316	0.043
I_c	$\eta^{SA}, LogP^{SA}$	$ARA^{SA} = -408.2 + 82\eta^{SA} + 8 LogP^{SA}$	0.063	0.725
		$A^{SA} = 4.75 + 0.05\eta^{SA} + 0.179 LogP^{SA}$	0.167	0.052
I_d	η^{SA}, E_{tot}^{SA}	$ARA^{SA} = -64.769 + 12.615\eta^{SA} - 0.0023 E_{tot}^{SA}$	0.384	0.087
		$A^{SA} = 2.6 + 0.314\eta^{SA} - 0.00004 E_{tot}^{SA}$	0.38	0.131
I_e	$POL^{SA}, LogP^{SA}$	$ARA^{SA} = 2.76 + 1.4 POL^{SA} - 10.68 LogP^{SA}$	0.040	0.222
		$A^{SA} = 4.36 + 0.12 POL^{SA} - 0.44 LogP^{SA}$	0.337	0.357
I_f	POL^{SA}, E_{tot}^{SA}	$ARA^{SA} = 19.821 - 2.928 POL^{SA} - 0.0071 E_{tot}^{SA}$	0.369	0.015
		$A^{SA} = 4.38 - 0.0625 POL^{SA} - 0.00005 E_{tot}^{SA}$	0.382	0.132
I_g	$LogP^{SA}, E_{tot}^{SA}$	$ARA^{SA} = 3.142 - 6.314 LogP^{SA} - 0.0028 E_{tot}^{SA}$	0.386	0.016
		$A^{SA} = 4.09 - 0.412 LogP^{SA} - 0.00006 E_{tot}^{SA}$	0.430	0.007
$II_a^1 = II_f^1$	$\chi^{SA}, POL^{SA}, E_{tot}^{SA}$	$ARA^{SA} = 84 - 17.5\chi^{SA} - POL^{SA} - 0.005 E_{tot}^{SA}$	0.373	0.056
		$A^{SA} = 6.46 - 0.45\chi^{SA} - 0.092 POL^{SA} - 0.0007 E_{tot}^{SA}$	0.414	0.018

TABLE 3.43 Continued

No. Crt.	Variabile	QSAR Model	R_{trial}	R_{test}
II_a^2	$\chi^{SA}, LogP^{SA}, E_{tot}^{SA}$	$ARA^{SA} = 108 - 22.66\chi^{SA} - 0.133\,LogP^{SA} - 0.0023\,E_{tot}^{SA}$	0.371	0.149
		$A^{SA} = 5.75 - 0.36\chi^{SA} - 0.45\,LogP^{SA} - 0.00006\,E_{tot}^{SA}$	0.451	0.136
II_b^1	$\chi^{SA}, \eta^{SA}, POL^{SA}$	$ARA^{SA} = -210 + 29.5\chi^{SA} + 13\eta^{SA} + 0.5POL^{SA}$	0.018	0.592
		$A^{SA} = 2.62 - 0.035\chi^{SA} + 0.382\eta^{SA} + 0.08POL^{SA}$	0.31	0.027
$II_b^2 = II_e^2$	$\eta^{SA}, POL^{SA}, LogP^{SA}$	$ARA^{SA} = -44.705 + 8.294\eta^{SA} + 1.176\,POL^{SA} - 4.467\,LogP^{SA}$	0.122	0.286
		$A^{SA} = 2.05 + 0.39\eta^{SA} + 0.16POL^{SA} - 0.449\,LogP^{SA}$	0.373	0.112
$II_b^3 = II_d^1 = II_f^2$	$\eta^{SA}, POL^{SA}, E_{tot}^{SA}$	$ARA^{SA} = -85.72 + 16.363\eta^{SA} + 1.272\,POL^{SA} + 0.0018\,E_{tot}^{SA}$	0.304	0.178
		$A^{SA} = 3.09 + 0.22\eta^{SA} - 0.034\,POL^{SA} - 0.00005\,E_{tot}^{SA}$	0.392	0.152
$II_c = II_d^2$	$\eta^{SA}, LogP^{SA}, E_{tot}^{SA}$	$ARA^{SA} = -42.58 + 8.352\eta^{SA} - 1.941\,LogP^{SA} - 0.0029\,E_{tot}^{SA}$	0.382	0.039
		$A^{SA} = 3.23 + 0.156\eta^{SA} - 0.368\,LogP^{SA} - 0.0006\,E_{tot}^{SA}$	0.436	0.012
II_e^1	$\chi^{SA}, POL^{SA}, LogP^{SA}$	$ARA^{SA} = -154 + 32.5\chi^{SA} + POL^{SA} - 8LogP^{SA}$	0.019	0.399
		$A^{SA} = 4.52 - 0.0325\chi^{SA} + 0.12POL^{SA} - 0.439\,LogP^{SA}$	0.337	0.370
$II_e^3 = II_f^3$	$POL^{SA}, LogP^{SA}, E_{tot}^{SA}$	$ARA^{SA} = 2.151 + 0.636\,POL^{SA} - 7.787\,LogP - 0.0021\,E_{tot}^{SA}$	0.394	0.038
		$A^{SA} = 4.07 + 0.015\,POL^{SA} - 0.45\,LogP^{SA} - 0.000057\,E_{tot}^{SA}$	0.431	0.034
III_1	$\chi^{SA}, \eta^{SA}, POL^{SA}, LogP^{SA}$	$ARA^{SA} = 42 - 2.052\chi^{SA} - 3.842\eta^{SA} + 12.894\,POL^{SA} - 13.842\,LogP^{SA}$	0.289	0.277
		$A^{SA} = 1.921 + 0.24\chi^{SA} + 0.4\eta^{SA} + 0.16POL^{SA} - 0.45\,LogP^{SA}$	0.373	0.007

TABLE 3.43 Continued

No. Crt.	Variabile	QSAR Model	R_{trial}	R_{test}
III_2	$\chi^{SA}, POL^{SA}, LogP^{SA}, E_{tot}^{SA}$	$ARA^{SA} = -34.928 - 5.571\chi^{SA} + 6POL^{SA} - 22.142LogP^{SA} - 0.0006E_{tot}^{SA}$	0.372	0.296
		$A^{SA} = 5.894 - 0.39\chi^{SA} - 0.016POL^{SA} - 0.41LogP^{SA} - 0.00007E_{tot}^{SA}$	0.452	0.106
III_3	$\eta^{SA}, POL^{SA}, LogP^{SA}, E_{tot}^{SA}$	$ARA^{SA} = -19.228 + 4.085\eta^{SA} + 0.657POL^{SA} - 1.914LogP^{SA} + 0.0002E_{tot}^{SA}$	0.21	0.184
		$A^{SA} = 2.72 + 0.23\eta^{SA} + 0.04POL^{SA} - 0.45LogP^{SA} - 0.00005E_{tot}^{SA}$	0.441	0.033
III_4	$\chi^{SA}, \eta^{SA}, LogP^{SA}, E_{tot}^{SA}$	$ARA^{SA} = 384 - 63\chi^{SA} - 17\eta^{SA} - 5LogP^{SA} - 0.004E_{tot}^{SA}$	0.348	0.300
		$A^{SA} = 4.94 - 0.35\chi^{SA} + 0.13\eta^{SA} - 0.41LogP^{SA} - 0.00006E_{tot}^{SA}$	0.455	0.066
III_5	$\chi^{SA}, \eta^{SA}, POL^{SA}, E_{tot}^{SA}$	$ARA^{SA} = 36.66 + 27.33\chi^{SA} + 3.666\eta^{SA} - 172POL^{SA} - 0.0003E_{tot}^{SA}$	0.274	0.225
		$A^{SA} = 5.587 - 0.41\chi^{SA} + 0.12\eta^{SA} - 0.07POL^{SA} - 0.00006E_{tot}^{SA}$	0.416	0.021

* These are compared with the respective direct structural alert models of Table 3.40 using their correlation performances for the trial and test molecules in Tables 3.36 and 3.37, respectively (Putz et al., 2011c).

(v) *The possibility of advancing a mechanistic interpretation* may be achieved by applying the statistical information from all trial and test sets and residual-QSAR modeling levels. If uniform criteria are implemented, one implements this principle by the *minimum* (*statistical*) *path principle*. Like all natural optimum principles, it assumes the shortest statistical path selected among all possible paths connecting the QSAR models. In all trial and test cases, it synergistically includes the primary path of action in terms of the physicochemical descriptors. Consequently, this principle also provides the second and third paths and the entire hierarchy of structural causes successively triggering the investigated endpoint effect with the observed actions. The minimum path principle ultimately reveals the structural causes and corresponding mechanistic picture, linking them to the observed action and providing the described biological effect. Depending on the QSAR model and statistical information to be processed, the statistical paths can be computed in various forms. For example, with the aid of Euclidean measure, similar studies recently presented the Spectral-SAR algebraic version of the consecrated QSAR applied to various eco-toxicological scenarios (Putz & Lacrămă, 2007; Putz et al., 2010; Chicu & Putz, 2009). Accordingly, the correlation factors of Table 3.39 were combined through all statistical path combinations (3.121) and (3.122) (Putz et al., 2011c). The numbers of paths built from connected, distinct models were indexed with k orders (dimension of correlation space or the number of structural variables included in a given model) from k=1 to k=M. Each path was then computed by the Euclidean formula (3.125) over the number of combinations of structural indicators potentially considered. Then follows the minimum principle (3.126) serarching over the endpoint residual-QSAR regression models computed with 1, 2, ..., M structural parameters, respectively.

The results are collected in Table 3.44, where the first (alpha), second (beta), and third (gamma) statistical paths are indicated. They were computed by the described optimal procedure with the amendment that, in the case of equal correlation paths, the minimum path was considered to originate on the QSAR model with the highest correlation factor. Once a path was selected, the next hierarchical path was chosen as the minimum among the remaining ones, such that all considered endpoints were involved only once (except for all variables containing endpoint – the

TABLE 3.44 Synopsis of the Statistical Paths Connecting the Correlation Factors For the Models of Table 3.39 (Putz et al., 2011c)

Statistical Path	Self-Consistent res-QSARs		Factor and Averaged res-QSARs	
	Gauss	*Q-Gauss*	*Gauss*	*Q-Gauss*
Ia-IIa-III	**0.04372**$^{\gamma}$	**0.05089**$^{\gamma}$	**0.2838**$^{\gamma}$	0.11541
Ia-IIb-III	0.04368	0.05067	0.2838	0.21791
Ia-IIc-III	0.04368	0.05067	0.2838	**0.24963**$^{\gamma}$
Ib-IIa-III	0.02683	0.02808	0.1097	**0.03308**$^{\alpha}$
Ib-IIb-III	0.02679	**0.02786**$^{\beta}$	**0.1097**$^{\beta}$	0.3257
Ib-IIc-III	**0.02679**$^{\alpha}$	0.02786	0.1097	0.35742
Ic-IIa-III	0.02738	0.02333	0.0896	0.19691
Ic-IIb-III	**0.02734**$^{\beta}$	0.02311	0.0896	**0.15621**$^{\beta}$
Ic-IIc-III	0.02734	**0.02311**$^{\alpha}$	**0.0896**$^{\alpha}$	0.16813

model III – that is a common horizon to all other combinations). With this method, the correlation information was combined and employed in the most general and natural manner, providing suitable structural paths to cause the observed activity. This also assured unity/specificity along the ergodicity of the paths' maps. Similar rules apply in deciding the overall models of Table 3.44, which is most representative to the alpha, beta and gamma paths (Putz et al., 2011c).

The path that is reached the most times throughout all the residual-QSARs was considered adjudicated for a given path type. In particular, the procedure started with the alpha path, which corresponds to the following chain of models (Table 3.44) (Putz et al., 2011c):

$$\alpha : Ic \rightarrow IIc(POL, Etot) \rightarrow III \tag{3.262}$$

It is then followed by the beta path identified by the models' sequence

$$\beta : Ib \rightarrow IIb(LogP, Etot) \rightarrow III \tag{3.263}$$

and, finally, by the gamma path's progression

$$\gamma : Ia \rightarrow IIa(LogP, POL) \rightarrow III \tag{3.264}$$

All these paths were selected more than once from all of the computed residual-QSARs in Table 3.44. In addition, part of the alpha path is identified first, and the rest should fulfill the ergodicity rule invoked above at this level (i.e., characterizing the models' sequence not previously consumed).

By analyzing the results of Eqs. (3.262)–(3.264) to understand the molecular mechanics from inter- to intracellular space, we can see that the intermediate residual-QSARs that approximate the interaction of structures with the environment can be retained. This method was inspired by the Husserl (1980) phenomenology method, which puts the core of the event in parenthesis and excludes the very incipient moments (i.e., the initial, transient stage does not decisively count in evolution) and those of the very final recordings (i.e., when all causes are mixed) to understand properly the evolutionary causes of some event. As a result, the molecular mechanism of genotoxic carcinogenesis may be a result of the succession of several linked structural causes,

$$\underbrace{(POL \to Etot)}_{\alpha} \Rightarrow \underbrace{(LogP \to Etot)}_{\beta} \Rightarrow \underbrace{(LogP \to POL)}_{\gamma} \tag{3.265}$$

beginning with the associated scenario (Figure 3.28).

A molecule is first polarized (POL) upon entering intercellular space due to the plasmatic environment's solvent effects. It then rotates to the optimal steric position (Etot) to realize cellular membrane transduction by activating its hydrophobicity (LogP). It may travel this way though the cellular space while binding to DNA elements via further steric interactions (Etot) and while remaining polarized. It may eventually break some parts of DNA residues and carry them in the extra-cellular space (LogP), where the enriched molecule will suffer further polarization (POL) from solvent interactions with the new molecular structure. The mechanism then enters a new ligand-DNA cycle, while the remaining DNA will enter mutagenesis. Remarkably, each considered structural (causal) indicator acted twice at the level of one interaction cycle in the obtained mechanism (3.265) in accordance with the self-consistent nature of the present residual-QSAR analysis (see Eq. (3.141)) (Putz, 2011a).

More detailed mechanisms of action may describe genotoxic carcinogenesis if additional physicochemical information is considered, but the steps of analysis would be the same. Additional, detailed intermediate

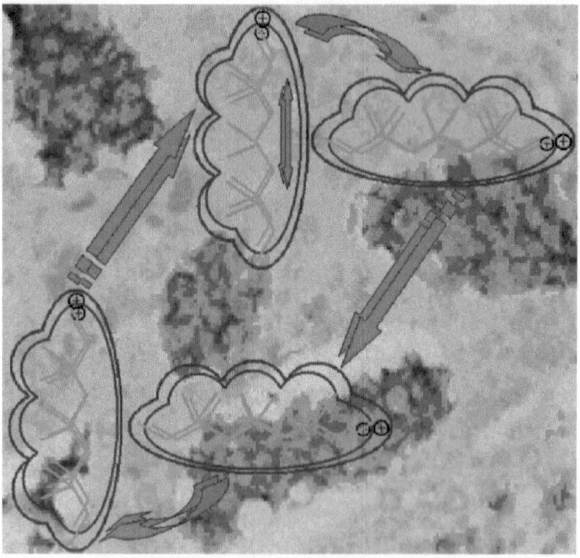

FIGURE 3.28 Illustration of the molecular mechanism for genotoxic carcinogenesis according to the present residual-QSAR correlation-path hierarchy superimposed over an immunohistochemcial analysis of paraffin-embedded sections of rat intestinal cancer (Putz, 2011a) using the Caspase-2 antibody (2011).

steps would need to be added, while preserving the mechanisms' self-consistency and cyclic character through the statistical paths. The electrophilic influence (through polarization) should also be included as a natural generalization of Millers' theory.

To this aim one may return to the structural-alert analysis and employs the *euclidean paths for residual-alert OECD-QSARs* of *step (ii-bis)* (Table 3.43) for the trial molecules of Table 3.36 and the test compounds of Table 3.37. The models were arranged so that each model emerges from the previous one on the basis of their common descriptors; the results are reported in Table 3.45 by employing the Euclidean path between two successive QSAR models (computed endpoints), Eq. (3.125). The *optimum paths for residual-alert QSARs* are derived by searching the minimum paths and the associated hierarchy according to the formal constraint of Eq. (3.126) (Putz & Lacrămă, 2007; Lacrămă et al., 2007; Putz et al., 2009b, 2010; Chicu & Putz, 2009) for the residual-QSAR regression (endpoint) models computed with 1, …, *k*, …, $m \leq M$ structural parameters (under the *OECD-QSAR step (v)*), as prescribed by

TABLE 3.45 Trial-Test Averages of the Correlations' Connected Paths Between the Endpoint Models of Table 3.43, Computed Using the Euler Eq. (3.125) see (Putz et al., 2011c)

Endpoint Paths	$\Delta R_{ARA^{SA}}$			$\Delta R_{ARA^{SA}}$		
	Trial	**Test**	**Average**	**Trial**	**Test**	**Average**
$I_a \rightarrow II_a^1 = II_f^1 \rightarrow III_2$	0.005099	0.264847	0.134973	0.057384	0.140089	0.098736
$I_a \rightarrow II_a^2 \rightarrow III_2$	0.003162	0.148222	**0.075692**[α]	0.080006	0.031320	**0.055663**[β]
$I_a \rightarrow II_a^2 \rightarrow III_4$	0.023194	0.152190	0.087692	0.080099	0.070576	0.075337
$I_b \rightarrow II_b^2 = II_e^2 \rightarrow III_3$	0.098386	0.241588	0.169987	0.088729	0.104890	0.09681
$I_b \rightarrow II_b^3 = II_d^1 = II_f^2 \rightarrow III_3$	0.244769	0.327055	0.285912	0.090426	0.161375	0.125901
$I_c \rightarrow II_b^2 = II_e^2 \rightarrow III_3$	0.105948	0.450693	0.278321	0.216933	0.099201	0.158067
$I_c \rightarrow II_b^2 = II_e^2 \rightarrow III_1$	0.177115	0.439092	0.308104	0.206000	0.120933	0.163467
$I_c \rightarrow II_c = II_d^2 \rightarrow III_3$	0.362415	0.701156	0.531786	0.269046	0.045177	0.157112
$I_c \rightarrow II_c = II_d^2 \rightarrow III_4$	0.320806	0.733973	0.52739	0.269670	0.067201	0.168436
$I_d \rightarrow II_b^3 = II_d^1 \rightarrow III_3$	0.123434	0.091197	0.107316	0.050447	0.120838	0.085643
$I_d \rightarrow II_b^3 = II_d^1 \rightarrow III_5$	0.085440	0.102420	**0.09393**[β]	0.026832	0.132672	**0.079752**[γ]
$I_d \rightarrow II_c = II_d^2 \rightarrow III_3$	0.172011	0.152738	0.162375	0.056222	0.120838	0.08853
$I_d \rightarrow II_c = II_d^2 \rightarrow III_4$	0.034058	0.265377	0.149718	0.059135	0.130678	0.094907
$I_e \rightarrow II_b^2 = II_e^2 \rightarrow III_3$	0.120282	0.120415	**0.120349**[γ]	0.077369	0.257421	0.167395

TABLE 3.45 Continued

Endpoint Paths	$\Delta R_{ARA^{SA}}$			$\Delta R_{ARA^{SA}}$		
	Trial	**Test**	**Average**	**Trial**	**Test**	**Average**
$I_e \rightarrow II_e^3 = II_f^3 \rightarrow III_3$	0.431226	0.187882	0.309554	0.094530	0.323154	0.208842
$I_f \rightarrow II_b^3 = II_d^1 = II_f^2 \rightarrow III_3$	0.114284	0.163110	0.138697	0.050009	0.120668	0.085339
$I_f \rightarrow II_e^3 = II_f^3 \rightarrow III_3$	0.185690	0.147800	0.166745	0.050009	0.098005	0.074007
$I_g \rightarrow II_a^2 = II_g^1 \rightarrow III_4$	0.027459	0.201221	0.11434	0.021377	0.146768	0.084073
$I_g \rightarrow II_c = II_d^2 \rightarrow III_3$	0.172046	0.146812	0.159429	0.007810	0.021587	**0.014699**[a]
$I_g \rightarrow II_c = II_d^2 \rightarrow III_4$	0.034234	0.262011	0.148123	0.019924	0.054230	0.037077
$I_g \rightarrow II_e^3 = II_f^3 \rightarrow III_3$	0.184173	0.147648	0.165911	0.010049	0.027018	0.018534

alert-QSAR model of Eqs. (3.153)–(3.158). These paths were computed for both trial and test compounds, and their average values (Table 3.45) for the residual-alert and direct-alert models (Table 3.43) are reported (Putz et al., 2011c).

The average column of Table 3.45 shows two sets of first (alpha), second (beta) and third (gamma) pathways in the ergodic pathways (Putz, 2012a, b), i.e., those uniquely contained QSAR models across all possible combinations, namely (Putz et al., 2011c):

those based on *residual-alert QSARs*:

$\boldsymbol{\alpha}: I_a \rightarrow II_a^2 \rightarrow III_2$

$$\boxed{\chi^{SA}, E_{tot}^{SA}} \ldots \boxed{\chi^{SA}, LogP^{SA}, E_{tot}^{SA}} \ldots \boxed{\chi^{SA}, POL^{SA}, LogP^{SA}, E_{tot}^{SA}} \quad (3.266)$$

$\boldsymbol{\beta}: I_d \rightarrow II_b^3 = II_d^1 \rightarrow III_5$

$$\boxed{\eta^{SA}, E_{tot}^{SA}} \ldots \boxed{\eta^{SA}, POL^{SA}, E_{tot}^{SA}} \ldots \boxed{\chi^{SA}, \eta^{SA}, POL^{SA}, E_{tot}^{SA}} \quad (3.267)$$

$\boldsymbol{\gamma}: I_e \rightarrow II_b^2 = II_e^2 \rightarrow III_3$

$$\boxed{POL^{SA}, LogP^{SA}} \ldots \boxed{\eta^{SA}, POL^{SA}, LogP^{SA}} \ldots \boxed{\eta^{SA}, POL^{SA}, LogP^{SA}, E_{tot}^{SA}} \quad (3.268)$$

and those based on *direct-alert QSARs:*

$\boldsymbol{\alpha}: I_g \rightarrow II_c = II_d^2 \rightarrow III_3$

$$\boxed{LogP^{SA}, E_{tot}^{SA}} \ldots \boxed{\eta^{SA}, LogP^{SA}, E_{tot}^{SA}} \ldots \boxed{\eta^{SA}, POL^{SA}, LogP^{SA}, E_{tot}^{SA}} \quad (3.269)$$

$\boldsymbol{\beta}: I_a \rightarrow II_a^2 \rightarrow III_2$

$$\boxed{\chi^{SA}, E_{tot}^{SA}} \ldots \boxed{\chi^{SA}, LogP^{SA}, E_{tot}^{SA}} \ldots \boxed{\chi^{SA}, POL^{SA}, LogP^{SA}, E_{tot}^{SA}} \quad (3.270)$$

$\boldsymbol{\gamma}: I_d \rightarrow II_b^3 = II_d^1 \rightarrow III_5$

$$\boxed{\eta^{SA}, E_{tot}^{SA}} \ldots \boxed{\eta^{SA}, POL^{SA}, E_{tot}^{SA}} \ldots \boxed{\chi^{SA}, \eta^{SA}, POL^{SA}, E_{tot}^{SA}} \quad (3.271)$$

The remaining issue is to decide among these two pathways, while noting, for instance, the α- and β- residual-alert-QSARs are reproduced as β- and γ-direct-alert-QSARs, respectively.

The intent of OECD QSAR Principle 5 (Section 3.1.5) is not to reject models that have no apparent mechanistic basis but to ensure that some consideration is given to the possibility of a mechanistic association between the descriptors used in a model and the endpoint being predicted and to ensure that this association is documented. Since the physico-chemical QSAR parameters were chosen in this study, a mechanistic interpretation of the models is possible. This nevertheless follows specific steps, some of them integrated in the previously discussed OECD-QSAR principles (Putz et al., 2011c):

- With the alert-QSAR models, in either residual or direct forms, (Table 3.43), Euclidean measures between all computed models that successively that fall along the pathways of Table 3.45 are constructed (see also the Step IV above).
- The first optimum paths are selected on the ergodic basis, by applying Eqs. (3.125) and (3.126) for the residual and direct alert-QSARs, respectively (the so-called Step V).
- The two classes of paths, see Eqs. (3.125) and (3.126), are compared on the basis of their electrophilic-docking (sub)-mechanisms identified within the unambiguous algorithm stage of the second OECD-QSAR principle. Comparison of the alpha-paths of the two alert-QSAR routes reveals that only residual-alert-QSAR correctly displays the involvement of the electronegativity in docking. As a consequence, the electrophilic-docking mechanistic interpretation of genotoxic carcinogenesis will be based only on the residual-alert-QSARs; this confirms the recent assessment of residual-QSAR as the in silico modeling technique best suited for treating chemical carcinogenesis. The present approach generalizes this in two ways: by detailing the mechanistic scenario with the electronegativity-to-chemical hardness reactivity-stability influence, and by considering the structural alert information in QSAR modeling rather than working with the entire molecular structural information.
- The explicit mechanistic scenario is based on the information contained within Eqs. (3.266)–(3.268), which gives rise to a natural

sequence that makes a *closed loop* over all three main interactions paths, given by (Putz et al., 2011c)

$$\begin{aligned}&\alpha\text{-}Steric\ movement\ \boxed{E_{tot}^{SA}} \rightarrow \alpha\text{-}Electronegativity\ Docking\ \boxed{\chi^{SA}}\\&\rightarrow \alpha\text{-}Cellular diffusion \boxed{LogP^{SA}} \rightarrow \alpha\text{-}Polarizability movement \boxed{POL^{SA}}\\&\rightarrow \beta\text{-}Electrophilic docking \boxed{\eta^{SA}} \rightarrow \beta\text{-}Polarizability movement \boxed{POL^{SA}}\\&\rightarrow \beta\text{-}Electronegativity Docking \boxed{\chi^{SA}} \rightarrow \gamma\text{-}Cellular\ diffusion \boxed{LogP^{SA}}\\&\rightarrow \gamma\text{-}Electrophilic\ docking \boxed{\eta^{SA}} \rightarrow \gamma\text{-}Steric\ movement \boxed{E_{tot}^{SA}}\end{aligned} \tag{3.272}$$

which is formally represented in Figure 3.29.

The cycle of Eq. (3.272) provides insight into the residual looping mechanism of the molecule or structural alert; receptor interaction,

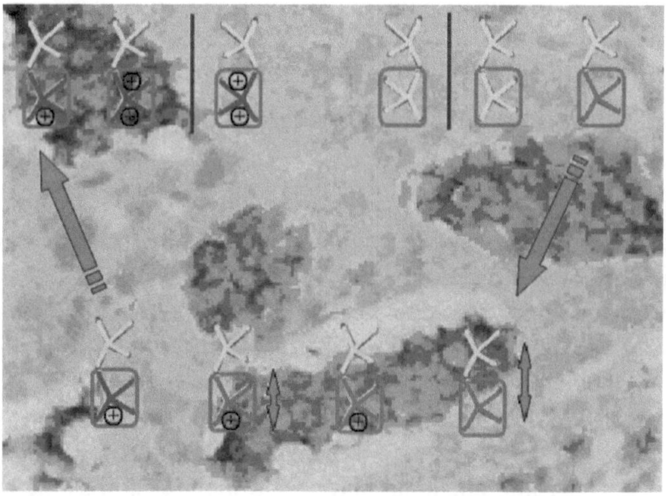

FIGURE 3.29 Illustration on a ligand-receptor cyclic interaction coordinate of the molecular mechanism of genotoxic carcinogenesis as given by the residual-alert-QSAR correlation-path hierarchy of Eqs. (3.266)–(3.268) then summarized in Eq. (3.272). The mechanism is superimposed over an immunohistochemcial analysis of paraffin-embedded sections of rat intestinal cancer using the (Caspase-2 antibody, 2011). In these evolving molecular graphs (the SA region is circumvented), steric movement is represented by mirroring, electronegativity docking by changing SA colors, diffusion by translation arrows; polarizability by vibration arrows, and electrophilic docking (the final stage including the maximum hardness principle) by positive charging (Putz et al., 2011c).

especially for electrophilic docking, here was related to electronegativity and chemical action, as compared with the previous global molecular studies which were limited to Hansch parameters only (Putz, 2011a). During one such interaction loop, the SA-molecule acquires a charge of +2, thus entering the next electrophile-nucleophile interaction loop with even more reactivity; this eventually leads to amplified biological activity manifested by exacerbated apoptosis due to breaking newly formed bonds in DNA. This is in close agreement with Millers' (1981) observation.

One can go further by choosing the first five instead of the first three interaction paths from the data of Table 3.45, because this number is the cardinal of the employed correlation parameters in actual residual-alert-QSARs. However, though electronegativity and chemical hardness are closely related to the total energy, see Eqs. (3.160) and (3.163), using only the first three interacting residual-alert pathways seems appropriate for the present purpose. For future studies, the extra index of electrophilicity (Parr et al., 1999) can be also explicitly incorporated to test its conceptual value in the electrophilic theory of chemical carcinogenesis (Parthasarathi et al., 2004).

We may draw several main conclusions, as follows (Putz, 2001a; Putz et al., 2011c):

- mutagenicity may be regarded as an electrophilic ligand-receptor interaction mechanism of covalent binding between the ligand molecule or SA and receptor;
- electronegativity and chemical hardness are crucial parameters in modeling the ligand-receptor interaction due to the EE and maximum chemical hardness principles, respectively;
- residual-QSAR is again shown to be reliable (Putz, 2011a) in its treatment of genotoxic carcinogenesis, as it better incorporates electronegativity and chemical hardness principles across the optimally selected pathways of organism cells' apoptosis;
- structural alert or molecular fragment analysis improves the residual-QSAR analysis with an enriched class of QSAR models that may be associated with molecular mechanisms of interaction in complex media.

3.3.7 SCREENING VS. STRUCTURAL SMILES QSAR (ON ANTI-HIV PYRIMIDINES)

3.3.7.1 Spectral-SAR Approach

In general, one should distinguish between two main directions in which a QSAR study may be conducted, namely; the present discussion follows (Putz & Dudaş, 2013b):

- *Drug design oriented,* which is generated through extensive database screening (Patlewicz et al., 2008a; Todeschini et al., 2007), similarity and domain considerations (Patlewicz et al., 2008b; Gallegos et al., 2008), producing QSAR models which should be then validated by internal (Patlewicz et al., 2007; Jaworska et al., 2005), external and read-across techniques (Jeliazkova et al., 2010) so that finally the molecules or molecular fragments predicted as most active or inhibitive depending on the endpoint target can be selected;
- *Mechanism oriented,* which consists mainly in the identification of the fundamental types of interaction that happen at the chemical-to-biological scale so that the structural properties of a compound constitute the causes that can be related to the manifest and recorded effects at a biological site (Pavan & Todeschini, 2008; Pavan et al., 2009; Roberts et al., 2008; Spycher et al., 2008; Benigni et al., 2007; Vracko et al., 2006);

In phenomenological terms, while the first direction is more related to technology and to the prescriptions for new synthesis, the second QSAR route is more on the scientific side due to the fundamental approach it involves; nevertheless, they both are related since after all, drug design is based on the desired or assumed mechanism of action specific to a given class of compounds, so knowing or revealing the mechanism of action for a given chemical-biological interaction only based on QSAR models remains as the first and probably the most important stage in drug design process itself.

Then, one faces with the true challenge, namely how to extract from a single or from a collection of QSAR models the “first causes” of a chemical-biological interaction. Fortunately, one may rely on the (multi) linear form of QSAR models since, when considered in terms of physico-chemical parameters with mechanistic interpretation at the nano-chemical scale, they provide just a manifestation of the quantum superposition principle (Dirac, 1947): while each structural parameter is associated with a given state or "chemical movement" specific to that state, their linear superposition combines into the macroscopic effect recorded as bio-/eco-/pharmaco-activity. Within this paradigm one has then the conceptual and computational freedom in establishing the "order" of the chemical states/movements toward the concerned endpoint. This direction has proven fruitful in assigning many useful QSAR tools thus enriching the related analysis and paving the way to *mechanistic drug design* through combination of various *in cerebro* (conceptual)—*in silico* (computational) approaches, such as (Putz & Dudaş, 2013b):

- Considering the elements of a QSAR model, i.e., both descriptors and activities as vectors in a multi-dimensional (chemical-biological) Banach-Hilbert (quantum) observable space (Putz, 2012a-b, 2013b; Putz et al., 2009b);
- Considering the descriptors of a QSAR model mainly with observable or physicochemical character, e.g., hydrophobicity for cellular wall transduction (the translation motion), the total energy for steric optimization (rotation motion), polarizability for molecular cloud deformation (vibrational motion) (Putz & Lacrămă, 2007; Putz & Putz, 2011), or more recently, through the chemical reactivity indices (electronegativity, chemical hardness and related quantities) for gaining more insight into the subtle bonding description (binding movement)—leading to the so-called chemical reactivity driven biological activity picture (which will be used also in the present work) (Putz & Putz, 2013b);
- Considering the systematical collection of QSAR models of descriptors in the previous entry along with their basic statistics, e.g., correlation factors, to be then employed either in an algebraic formulation of descriptor-activity correlations, proved to be always superior to the basic statistical one, or to entering in Euclidian paths among the computed endpoints (Putz et al., 2009a), thus involving the square form of the correlation factor, to produce and compare minimum distances toward the most comprehensive (superior in correlation)

QSAR model (in turn presumed to be the closest in the QSAR pool of models to the real/recorded activity). This approach, consecrated as Spectral-SAR (Putz & Lacrămă, 2007; Putz & Putz, 2011; Lacrămă et al., 2007; Putz et al., 2007, 2010), provides the mechanistic interpretation of biological action in terms of the hierarchy of structural causes (descriptors) along the least computed path across available QSAR models;

- Considering, more recently, the way of improving the previous entry by extensive use of the variational approach in all stages of Spectral-SAR, from screening (i.e., selecting the training set) from a set of toxicants, to assessing the minimum path by considering the molecular passage through cellular walls accompanied by the partial chemical bonds in molecules (Putz & Dudaş, 2003a), according with the Simplified Molecular-Input Line-Entry System (SMILES) (Weininger, 1988, 1990; Weininger et al., 1989; Helson, 2007);

This last point is from where the present work continues the idea of fully considering the SMILES structure in the computational development of QSARs, by calculating the associated descriptors and involving them in the mechanistic analysis. Actually, it was found that when using SMILES forms only for screening purposes, as in the present case for modeling the anti-HIV activity of selected uracil derivatives (Maruyama et al., 2003), the output mechanism provides an activated chemical-biological bonding not properly indicating the finalization of the ligand-receptor coupling to explain the anti-HIV activity. Therefore, the present report takes this concept one step further in order to complete the chemical-bonding picture by fully using the SMILES structures not only as a graphical tool but also considering them as an intermediate reality in the mechanistic picture of chemical ligand-biological receptor interaction yielding the recorded effect in the organism.

To this end, the above mechanistic-oriented framework will be unfolded, by applying the OECD-QSAR principles to the present purpose and conceptual-computational stages (Putz et al., 2011a), by combining Spectral-SAR methodology with variational principles of chemical reactivity driving biological activity and with the recursive minimization of paths across systematic QSAR with SMILES molecular (chemical reactivity) descriptors, to recognize the preferred hierarchy and the "first causes" that eventually result in the envisaged chemical binding and resulting anti-HIV activity.

This mechanism may be further used in a subsequent stage when extensive validation and drug design studies to recognize the molecular shape and structure (Putz et al., 2008a) which best accords with a particular mechanism of action can be envisaged.

For this point one considers the working molecules under the actually considered most likely form producing the considered end-point, namely the anti-HIV activity produced by uracil based pyrimidines (Maruyama et al., 2003; Chemical Identifier Resolver beta 4, 2013), along two forms of their SMILES structure, as presented in Table 3.46 (Putz & Dudaş, 2013b):

- the longest SMILES molecular chain (LoSMoC), when bonds are breaking on aromatic rings and moieties such that the resulting molecule displaying a sort of 2D form of original molecule along the "fractalic" chain, assumed to be the first stage in intermediary molecular defolding targeting the receptor; the maximum SMILES chain in LoSMoC are presumably responsible for best transport/transduction of ligand molecules through cellular (lipidic) walls, after which they may be released with a modified structure due to their further ionization applied by the interaction with cellular layers; accordingly, another form of SMIELS is generated and next considered, namely:
- the Branching SMILES (BraS), representing the second phase of molecular defolding and providing ligand bonds' breaking such that many "bays" to be formed, yet with consistent "arms" linking the short molecular "skeleton" aiming to favor the binding with receptor on its pockets; accordingly, the branching is not necessary in the same points of molecules through a series, but such as the maximum branching combined with equilibrium of branches to be obtained in the final BraS; for instance, a long branch adjacent to a short one will make not a strong enough "anchor" to bind the receptor pocket; therefore, the branching principle is to have the equilibrated anchor-clefs by themselves; as such, the branching up to fourth order are performed for molecules in Table 3.46.

However, one should note the fact that the most of drugs are ionized once immersed in the biological body is in accordance with the present two-steps of SMILES conformations, since in each of them more nucleophilic compounds are considered due to the successive bonding breaking

TABLE 3.46 Working Molecules (IUPAC Name and Molecular Weight MW Are Indicated) and Their Corresponding SMILES Topology, i.e., The Longest SMILES Molecular Chain (LoSMoC) As Upper Entry and the Branching SMILES (BraS) As Down Entry, For Each Pyrimidine Structure Considered, Along the Common Activity A=$\log_{10}(1/EC_{50})$ Employed From Half Maximal Effective Concentration (EC_{50}, μM) Antiviral Activity of 1,3-Disubstituted Uracils Against Human Immunodeficiency Virus (HIV-1) (Maruyama et al., 2003), With AIDS Code Indicated (Chemical Identifier Resolver beta4, 2013), Respectively*

No.	Structure 2D IUPAC name MW AIDS code	SMILES configurations LoSMoC BraS	 Code LoSMoC Code BraS	A	LogP	χ (eV)	η (eV)	π	ω (eV)
					… LoSMoC …				
					… BraS …				
1	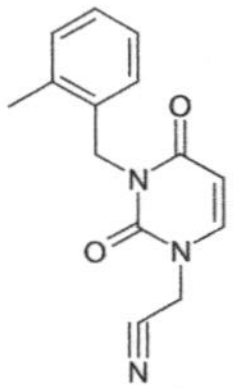	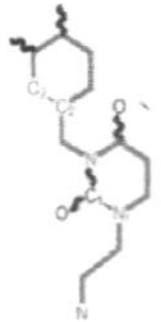	N#CCN1/C=C\C(=O) N(C1=O)Cc2ccc(C) c(C)c2	3.716698	0.91	23.107212	1.5817419	7.304356	168.78330
	[3-(2-Methyl-benzyl)-2,4-dioxo-3,4-dihydro-2H-pyrimidin-1-yl]-acetonitrile 255.28 AIDS352092	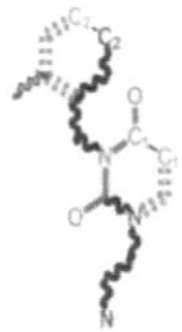	O=C1N(Cc(c(C)cc2) cc2) C(N(/C=C1\)CC#N)=O		0.44	13.240955	2.8324015	2.3374078	30.949511

TABLE 3.46 Continued

No.	Structure 2D / IUPAC name / MW / AIDS code	SMILES configurations: LoSMoC / BraS	Code LoSMoC / Code BraS	A	LogP	χ (eV)	η (eV)	π	ω (eV)
					… LoSMoC …				
					… BraS …				
2	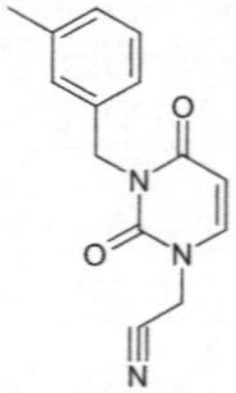 [3-(3-Methyl-benzyl)-2,4-dioxo-3,4-dihydro-2H-pyrimidin-1-yl]-acetonitrile 255.28 AIDS352093	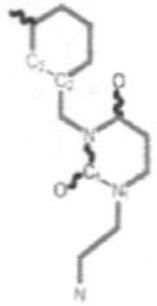	N#CCN1/C=C\C(=O) N(C1=O)Cc2cccc(C)c2	5.173925	0.47	22.812517	1.5937610	7.156819	163.26505
		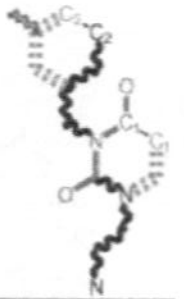	O=C1N(Cc(cc(C)c2) cc2) C(N(/C=C1\)CC#N)=O		0.44	13.043803	2.8273990	2.3066788	30.087865

TABLE 3.46 Continued

No.	Structure 2D / IUPAC name / MW / AIDS code	SMILES configurations: LoSMoC / BraS	Code LoSMoC / Code BraS	A	LogP	χ (eV)	η (eV)	π	ω (eV)
3	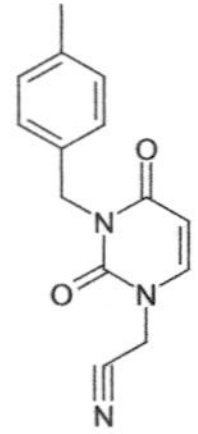 [3-(4-Methyl-benzyl)-2,4-dioxo-3,4-dihydro-2H-pyrimidin-1-yl]-acetonitrile 255.28 AIDS352094	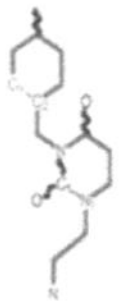	N#CCN1/C=C\C(=O)N(C1=O)Cc2ccc(C)cc2	4.023191	0.47	22.852718	1.5799314	7.232187	165.27512
		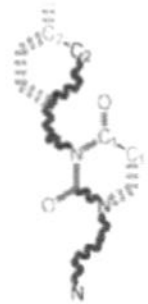	O=C1N(Cc(ccc2C)cc2)C(N(/C=C1\)CC#N)=O		0.88	13.149213	2.8323062	2.3212908	30.523148

(Columns LogP to ω (eV): upper row … LoSMoC …, lower row … BraS …)

TABLE 3.46 Continued

No.	Structure 2D IUPAC name MW AIDS code	SMILES configurations LoSMoC BraS	 Code LoSMoC Code BraS	A	LogP	χ (eV)	η (eV)	π	ω (eV)
					... LoSMoC ...				
					... BraS ...				
4	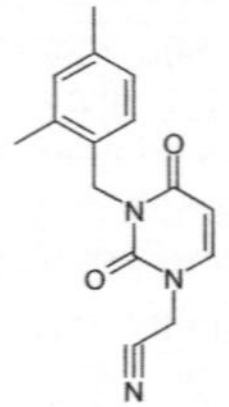	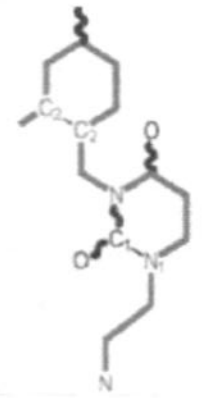	N#CCN1/C=C\C(=O) N(C1=O)Cc2ccc(C) cc2C	3.943095	1.06	22.695343	1.4889604	7.621204	172.96584
	[3-(2,4-Dimethyl-benzyl)-2,4-dioxo-3,4-dihydro-2H-pyrimidin-1-yl]-acetonitrile 269.30 AIDS352888	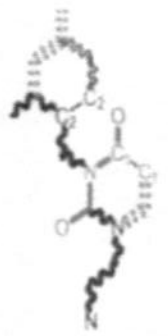	O=C1N(Cc2c(cc(cc2) C)C) C(N(/C=C1\)CC#N)=O		1.03	13.061603	2.7061581	2.4133112	31.521715

TABLE 3.46 Continued

No.	Structure 2D	SMILES configurations		A	LogP	χ (eV)	η (eV)	π	ω (eV)
	IUPAC name	LoSMoC	Code LoSMoC				... LoSMoC ...		
	MW	BraS	Code BraS				... BraS ...		
	AIDS code								
5	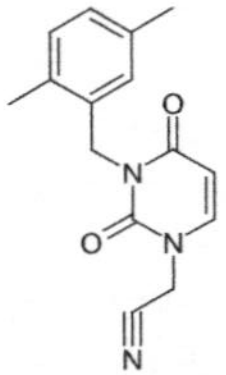	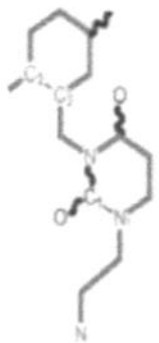	N#CCN1/C=C\C(=O) N(C1=O)Cc2cc(C) ccc2C	4.610833	1.06	22.961910	1.5967679	7.190121	165.09891
	[3-(2,5-Dimethyl-benzyl)-2,4-dioxo-3,4-dihydro-2H-pyrimidin-1-yl]-acetonitrile 269.30 AIDS352889	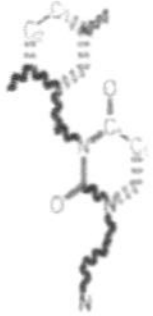	O=C1N(Cc(cc(C)c2) c(c2)C) C(N(/C=C1\)CC#N)=O		0.6	13.344068	2.8843065	2.3132194	30.867758

TABLE 3.46 Continued

No.	Structure 2D IUPAC name MW AIDS code	SMILES configurations LoSMoC BraS	 Code LoSMoC Code BraS	A	LogP	χ (eV)	η (eV)	π	ω (eV)
							… LoSMoC …		
							… BraS …		
6	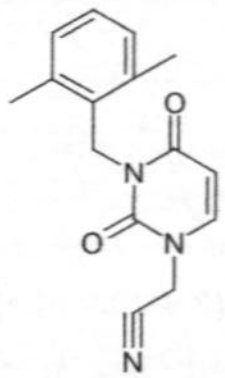	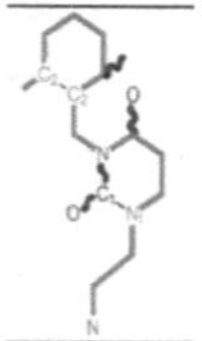	N#CCN1/C=C\C(=O) N(C1=O)Cc2c(C) cccc2C	3.707743	1.06	22.914792	1.5375402	7.45177	170.75577
	[3-(2,6-Dimethyl-benzyl)-2,4-dioxo-3,4-dihydro-2H-pyrimidin-1-yl]-acetonitrile 269.30 AIDS352890	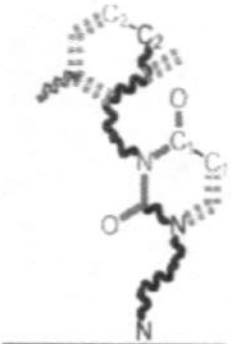	O=C1N(Cc(c(C)cc2) c(C)c2) C(N(/C=C1\)CC#N)=O		0.6	13.174123	2.7474378	2.3975289	31.585343

TABLE 3.46 Continued

No.	Structure 2D IUPAC name MW AIDS code	SMILES configurations LoSMoC / BraS	Code LoSMoC Code BraS	A	LogP	χ (eV)	η (eV)	π	ω (eV)
					... LoSMoC ...				
					... BraS ...				
7	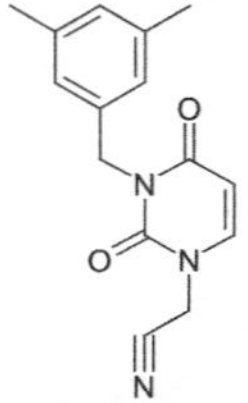 [3-(3,5-Dimethyl-benzyl)-2,4-dioxo-3,4-dihydro-2H-pyrimidin-1-yl]-acetonitrile 269.30 AIDS352095	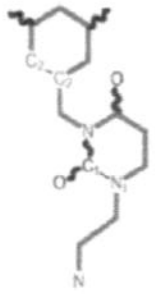	N#CCN1/C=C\C(=O) N(C1=O)Cc2cc(C) cc(C)c2	6.229147	0.63	22.322613	1.3441469	8.303636	185.35884
		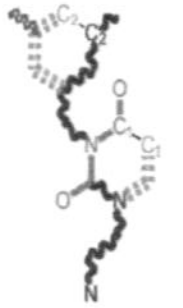	O=C1N(Cc(cc(C)c2) cc2C) C(N(/C=C1\)CC#N)=O		1.03	12.688503	2.5160717	2.5214906	31.993942

TABLE 3.46 Continued

No.	Structure 2D / IUPAC name / MW / AIDS code	LoSMoC / BraS	Code LoSMoC / Code BraS	A	LogP	χ (eV)	η (eV)	π	ω (eV)
		LoSMoC	Code LoSMoC				… LoSMoC …		
		BraS	Code BraS				… BraS …		
8	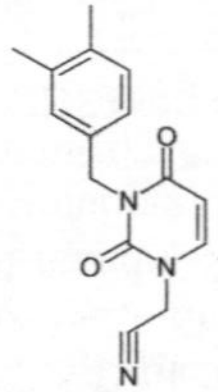	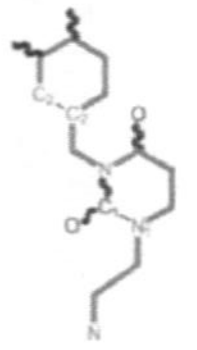	N#CCN1/C=C\C(=O) N(C1=O)Cc2ccc(C) c(C)c2	5.425968	0.63	22.513298	1.4966364	7.521298	169.32923
	[3-(3,4-Dimethyl-benzyl)-2,4-dioxo-3,4-dihydro-2H-pyrimidin-1-yl]-acetonitrile 269.30 AIDS352891	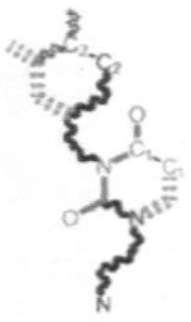	O=C1N(Cc(cc(c2C) C)cc2) C(N(/C=C1\)CC#N)=O		1.03	12.964034	2.7262701	2.3776137	30.823468

TABLE 3.46 Continued

No.	Structure 2D / IUPAC name / MW / AIDS code	SMILES configurations: LoSMoC / BraS	Code LoSMoC / Code BraS	A	LogP	χ (eV)	η (eV)	π	ω (eV)
9	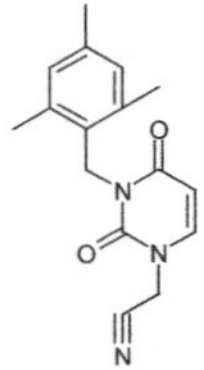	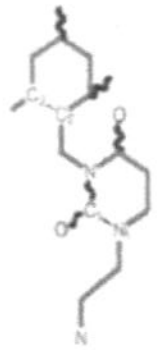	N#CCN1/C=C\C(=O) N(C1=O)Cc2c(C)cc(C) cc2C	3.716698	1.22	22.436637	1.3498377	8.310865	186.46785
	[3-(2,4,6-trimethyl-benzyl)- 2,4-dioxo-3,4-dihydro-2H-pyrimidin-1-yl]-acetonitrile 283.33 AIDS352892	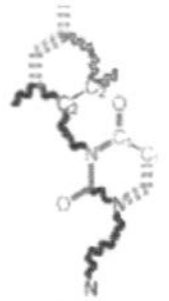	O=C1N(Cc2c(cc(cc2C) C)C) C(N(/C=C1\)CC#N)=O		1.62	12.848802	2.5836971	2.4865149	31.948740

TABLE 3.46 Continued

No.	Structure 2D / IUPAC name / MW / AIDS code	SMILES configurations		A	LogP	χ (eV)	η (eV)	π	ω (eV)
		LoSMoC	Code LoSMoC		... LoSMoC ...				
		BraS	Code BraS		... BraS ...				
10	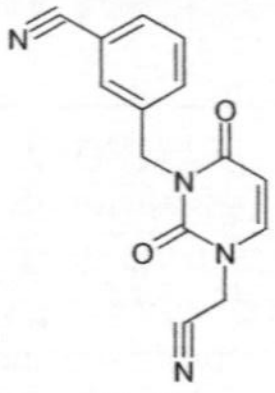	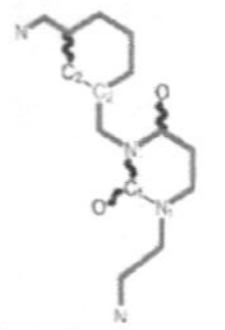	N#CCN1/C=C\C(=O) N(C1=O)Cc2cccc(c2) C#N	5.128427	0.04	22.981901	1.5807784	7.269172	167.05939
	[3-(3-cyanophenyl)methyl-2,4-dioxo-3,4-dihydro-2H-pyrimidin-1-yl]-acetonitrile 266.26 AIDS352893	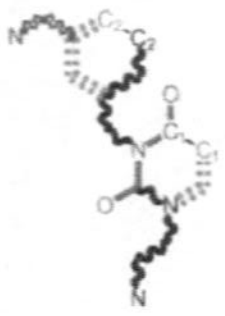	O=C1N(Cc(cc(C#N) c2)cc2) C(N(/C=C1\)CC#N)=O		0.01	12.984607	2.7188679	2.3878703	31.00556

TABLE 3.46 Continued

No.	Structure 2D / IUPAC name / MW / AIDS code	SMILES configurations: LoSMoC / BraS	Code LoSMoC / Code BraS	A	LogP	χ (eV)	η (eV)	π	ω (eV)
					… LoSMoC …				
11	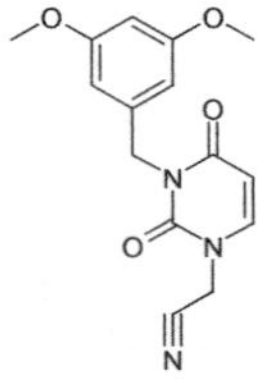	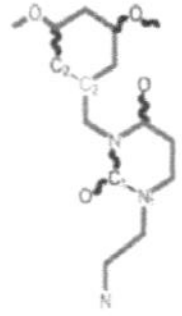	N#CCN1/C=C\C(=O)N(C1=O) Cc2cc(OC)cc(c2)OC	5.248720	–1.67	21.820275	1.0563595	10.32805	225.36097
					… BraS …				
	[3-(3,5-Dimethoxy-benzyl)-2,4-dioxo-3,4-dihydro-2H-pyrimidin-1-yl]-acetonitrile 301.30 AIDS352897	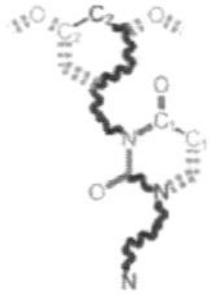	O=C1N(Cc(cc2OC)cc(OC)c2) C(N(/C=C1\)CC#N)=O		–0.72	12.366078	2.2360288	2.7651875	34.194524

TABLE 3.46 Continued

No.	Structure 2D / IUPAC name / MW / AIDS code	SMILES configurations		A	LogP	χ (eV)	η (eV)	π	ω (eV)
		LoSMoC	Code LoSMoC		… LoSMoC …				
		BraS	Code BraS		… BraS …				
12	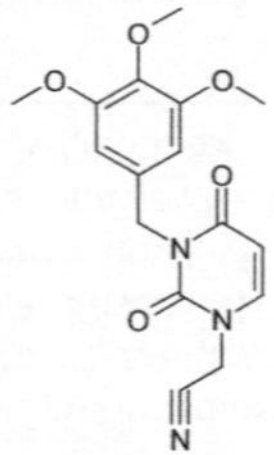 [3-(3,4,5-trimethoxy-benzyl)-2,4-dioxo-3,4-dihydro-2H-pyrimidin-1-yl]-acetonitrile 331.33 AIDS352898	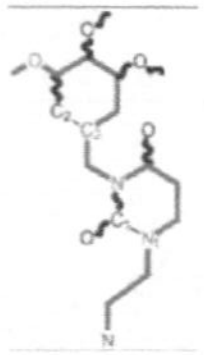	N#CCN1/C=C\C(=O)N(C1=O) Cc2cc(OC)c(OC)c(c2)OC	3.423658	–2.66	21.365171	1.0625102	10.0541	214.80760
		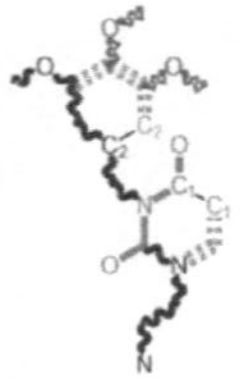	O=C1N(Cc2cc(c(OC)c(OC)c2)OC) C(N(/C=C1\)CC#N)=O		–2.26	12.143075	2.4593788	2.4687280	29.977950

TABLE 3.46 Continued

No.	Structure 2D	SMILES configurations		A	LogP	χ (eV)	η (eV)	π	ω (eV)
	IUPAC name	LoSMoC	Code LoSMoC				... LoSMoC ...		
	MW	BraS	Code BraS				... BraS ...		
	AIDS code								
13	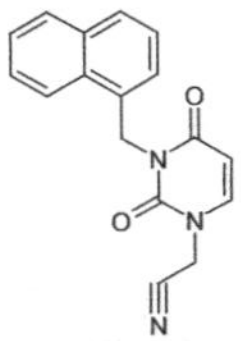	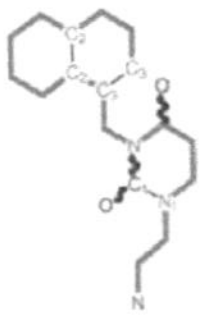	N#CCN1/C=C\C(=O)N (C1=O) Cc3c2ccccc2ccc3	5.268411	1.16	25.868615	1.4726275	8.78315	227.20792
	(3-Naphthalen-1-ylmethyl-2,4-dioxo-3,4-dihydro-2H-pyrimidin-1-yl)-acetonitrile 291.31 AIDS352899	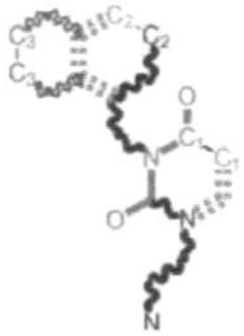	O=C1N(Cc(c(cc3) c(cc3)c2)cc2) C(N(/C=C1\)CC#N)=O		0.25	14.682316	2.7628433	2.6571026	39.012422

TABLE 3.46 Continued

No.	Structure 2D IUPAC name MW AIDS code	SMILES configurations LoSMoC BraS	 Code LoSMoC Code BraS	A	LogP	χ (eV)	η (eV)	π	ω (eV)
						… LoSMoC …			
						… BraS …			
14	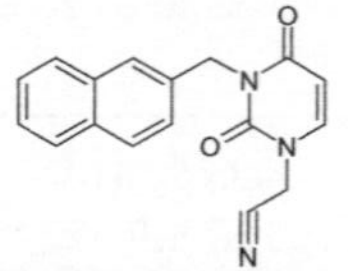 (3-Naphthalen-2-ylmethyl-2,4-dioxo-3,4-dihydro-2H-pyrimidin-1-yl)-acetonitrile 291.31 AIDS352900	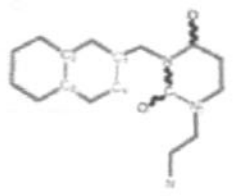	N#CCN1/C=C\C(=O)N (C1=O) Cc3cc2ccccc2cc3	4.435333	1.16	25.888824	1.3140309	9.850919	255.02871
		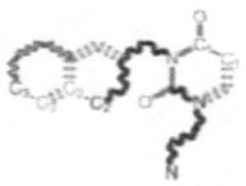	O=C1N(Cc(cc(ccc3) c2c3)cc2) C(N(/C=C1\)CC#N)=O		0.69	14.829177	2.6159392	2.8343888	42.031656

TABLE 3.46 Continued

No.	Structure 2D / IUPAC name / MW / AIDS code	SMILES configurations: LoSMoC / BraS	Code LoSMoC / Code BraS	A	LogP	χ (eV)	η (eV)	π	ω (eV)
15	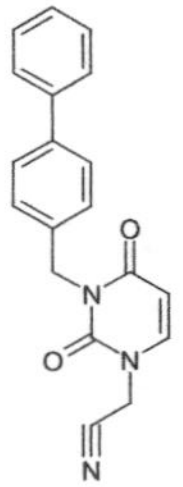 (3-Biphenyl-4-ylmethyl-2,4-dioxo-3,4-dihydro-2H-pyrimidin-1-yl)-acetonitrile 317.35 AIDS352901	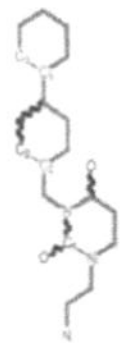	N#CCN1/C=C\C(=O)N (C1=O)Cc2ccc(cc2) c3ccccc3	4.236572	1.25	27.000458	1.2990428	10.39244	280.60074
		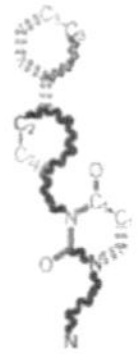	O=C1N(Cc(c2) ccc(c(cc3)ccc3)c2) C(N(/C=C1\)CC#N)=O		0.79	15.020930	2.3806514	3.1547941	47.387942

TABLE 3.46 Continued

No.	Structure 2D / IUPAC name / MW / AIDS code	SMILES configurations: LoSMoC / BraS	Code LoSMoC / Code BraS	A	LogP	χ (eV)	η (eV)	π	ω (eV)
						... LoSMoC ...			
						... BraS ...			
16	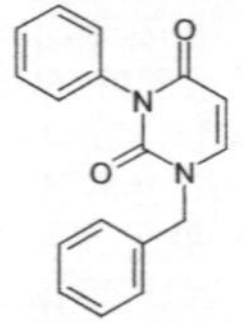	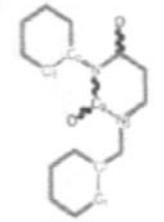	c1ccccc1CN2/C=C\ C(=O) N(C2=O)c3ccccc3	3.665546	1.55	28.617336	1.4763650	9.691822	277.35413
	1-Benzyl-3-phenyl-1H-pyrimidine-2,4-dione 278.31 AIDS352902	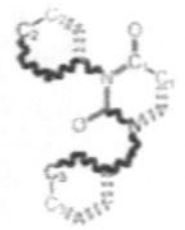	O=C1N(c(cc2)ccc2) C(N(/C=C1\) Cc(ccc3)cc3)=O		0.54	16.311764	2.7002385	3.0204302	49.268547

TABLE 3.46 Continued

No.	Structure 2D IUPAC name MW AIDS code	SMILES configurations LoSMoC BraS	 Code LoSMoC Code BraS	A	LogP	χ (eV)	η (eV) … LoSMoC … … BraS …	π	ω (eV)
17	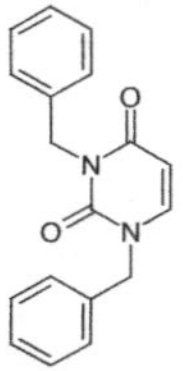 1,3-Dibenzyl-1H-pyrimidine-2,4-dione 292.34 AIDS352903	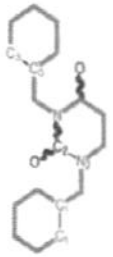	c1ccccc1CN2/C=C\C(=O)N(C2=O)Cc3ccccc3	4.954677	1.53	27.627131	1.4262804	9.685028	267.56953
		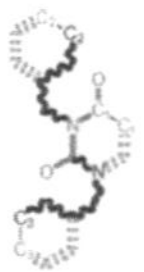	O=C1N(Cc(ccc2)cc2)C(N(/C=C1\)Cc(ccc3)cc3)=O		1.06	15.538736	2.6492805	2.9326332	45.569415

TABLE 3.46 Continued

No.	Structure 2D / IUPAC name / MW / AIDS code	LoSMoC / BraS	SMILES configurations: Code LoSMoC / Code BraS	A	LogP	χ (eV)	η (eV)	π	ω (eV)
							... LoSMoC ...		
							... BraS ...		
18	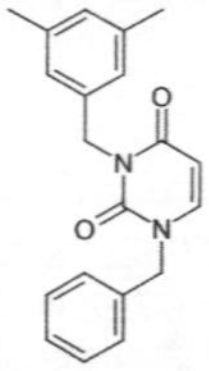 1-Benzyl-3-(3,5-dimethyl-benzyl)-1H-pyrimidine-2,4-dione 320.39 AIDS352096	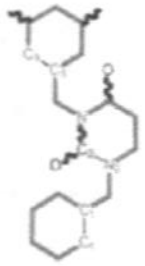	c1ccccc1CN2/C=C\C(=O) N(C2=O)Cc3cc(C)cc(C)c3	6.630784	1.84	25.860489	0.7302591	17.70638	457.89563
		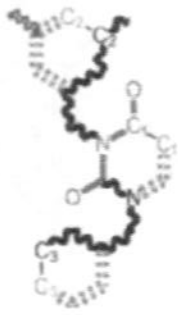	O=C1N(Cc(cc(C)c2)cc2C)C (N(/C=C1\)Cc(ccc3)cc3)=O		1.81	14.540931	1.5875011	4.5798175	66.594813

TABLE 3.46 Continued

No.	Structure 2D IUPAC name MW AIDS code	SMILES configurations LoSMoC BraS	 Code LoSMoC Code BraS	A	LogP	χ (eV)	η (eV)	π	ω (eV)
					… LoSMoC …				
					… BraS …				
19	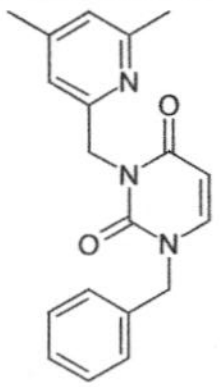 1-Benzyl-3-(4,6-dimethyl-pyridin-2-ylmethyl)-1H-pyrimidine-2,4-dione 321.38 AIDS352904	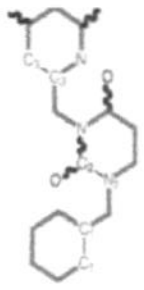	c1ccccc1CN2/C=C\C(=O) N(C2=O)Cc3nc(C)cc(C)c3	5.136082	0.41	26.114347	0.8253111	15.82091	413.15277
		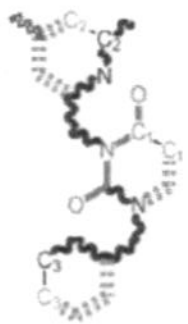	O=C1N(Cc(cc(C)c2)nc2C)C (N(/C=C1\)Cc(ccc3)cc3)=O		0.15	14.748792	1.7122755	4.3067812	63.519822

TABLE 3.46 Continued

No.	Structure 2D / IUPAC name / MW / AIDS code	SMILES configurations (LoSMoC / BraS)	Code LoSMoC / Code BraS	A	LogP	χ (eV)	η (eV)	π	ω (eV)
	IUPAC name	LoSMoC	Code LoSMoC			… LoSMoC …			
	MW	BraS	Code BraS			… BraS …			
	AIDS code								
20	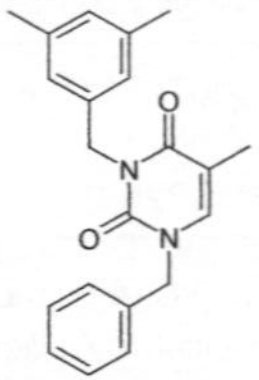	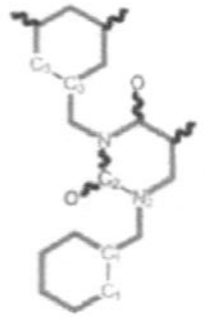	c1ccccc1CN2/C=C\(C)C(=O) N(C2=O)Cc3cc(C)cc(C)c3	5.841637	2.12	25.007275	1.0403700	12.01845	300.54873
	1-Benzyl-3-(3,5-dimethyl-benzyl)-5-methyl-1H-pyrimidine-2,4-dione 334.42 AIDS352905	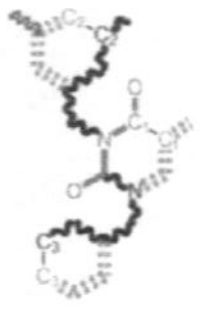	O=C1N(Cc(cc(C)c2)cc2C) C(N(/C=C1\C)Cc(ccc3)cc3)=O		2.39	14.063834	2.1272754	3.3055978	46.489379

TABLE 3.46 Continued

No.	Structure 2D / IUPAC name / MW / AIDS code	SMILES configurations: LoSMoC / BraS	Code LoSMoC / Code BraS	A	LogP	χ (eV)	η (eV)	π	ω (eV)
					... LoSMoC ...				
					... BraS ...				
21	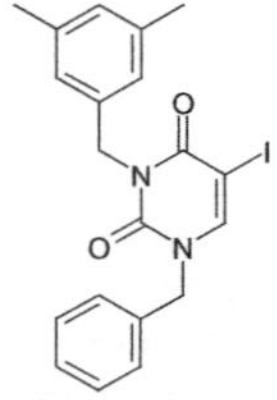 1-Benzyl-3-(3,5-dimethyl-benzyl)-5-iodo-1H-pyrimidine-2,4-dione 446.29 AIDS352906	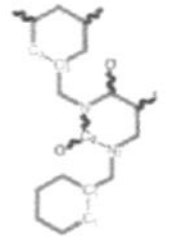	c1ccccc1CN2/C=C\(I)C(=O) N(C2=O)Cc3cc(C)cc(C)c3	4.379863	2.48	25.393186	0.8931783	14.21507	360.96592
		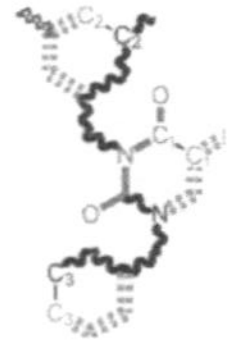	O=C1N(Cc(cc(C)c2)cc2C) C(N(/C=C1\I)Cc(ccc3)cc3)=O		2.53	13.656576	1.4894424	4.5844594	62.608023

TABLE 3.46 Continued

No.	Structure 2D / IUPAC name / MW / AIDS code	LoSMoC / BraS	SMILES configurations: Code LoSMoC / Code BraS	A	LogP	χ (eV)	η (eV)	π	ω (eV)
					... LoSMoC ...				
					... BraS ...				
22	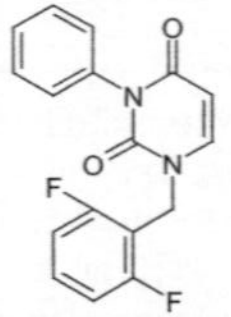	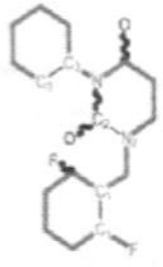	Fc1cccc(F)c1CN2/ C=C\C (=O)N(C2=O)c3ccccc3	3.690369	1.08	28.610234	1.4786792	9.674253	276.78264
	1-(2,6-Difluoro-benzyl)-3-phenyl-1H-pyrimidine-2,4-dione 314.29 AIDS352907	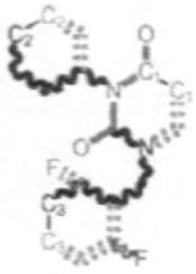	O=C1N(c(cc2)ccc2) C(N(/C=C1\) Cc(c(F)cc3)c(F)c3)=O		–0.66	16.175016	2.7665356	2.9233342	47.284980

TABLE 3.46 Continued

No.	Structure 2D	SMILES configurations		A	LogP	χ (eV)	η (eV)	π	ω (eV)
	IUPAC name	LoSMoC	Code LoSMoC		... LoSMoC ...				
	MW	BraS	Code BraS		... BraS ...				
	AIDS code								
23	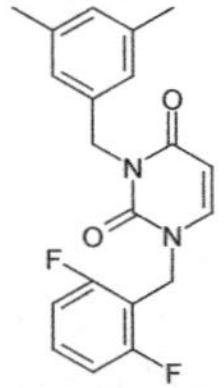 1-(2,6-Difluoro-benzyl)-3-(3,5-dimethyl-benzyl)-1H-pyrimidine-2,4-dione 356.37 AIDS352908	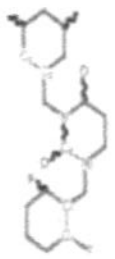	Fc1cccc(F)c1CN2/C=C\C(=O) N(C2=O)Cc3cc(C)cc(C)c3	6.939302	1.37	25.844444	0.7517152	17.19032	444.27415
		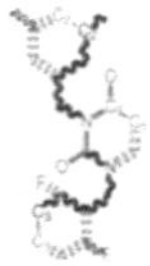	O=C1N(Cc(cc(C)c2)cc2C)C (N(/C=C1\)Cc(c(F)cc3)c(F)c3)=O		0.6	14.486247	1.5713578	4.6094680	66.773895

TABLE 3.46 Continued

No.	Structure 2D IUPAC name MW AIDS code	SMILES configurations LoSMoC BraS	 Code LoSMoC Code BraS	A	LogP	χ (eV)	η (eV)	π	ω (eV)
							… LoSMoC …		
							… BraS …		
24	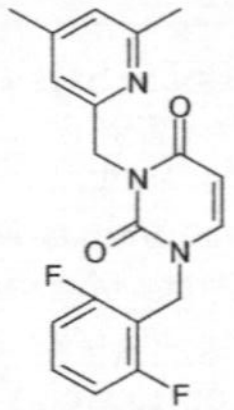 1-(2,6-Difluoro-benzyl)-3-(4,6-dimethyl-pyridin-2-ylmethyl)-1H-pyrimidine-2,4-dione 357.36 AIDS352909	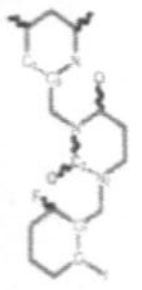	Fc1cccc(F)c1CN2/C=C\C(=O)N(C2=O)Cc3nc(C)cc(C)c3	5.193820	−0.06	26.085800	0.8406863	15.51458	404.71036
		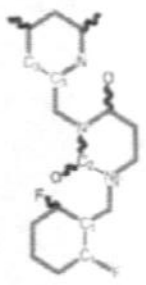	O=C1N(Cc(cc(C)c2)nc2C)C(N(/C=C1\)Cc(c(F)cc3)c(F)c3)=O		−1.05	14.690744	1.6779412	4.3776098	64.310348

TABLE 3.46 Continued

No.	Structure 2D / IUPAC name / MW / AIDS code	SMILES configurations: LoSMoC / BraS	Code LoSMoC / Code BraS	A	LogP	χ (eV)	η (eV)	π	ω (eV)
					… LoSMoC …				
					… BraS …				
25	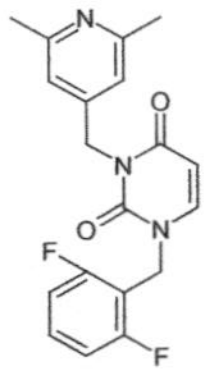 1-(2,6-Difluoro-benzyl)-3-(2,6-dimethyl-pyridin-4-ylmethyl)-1H-pyrimidine-2,4-dione 357.36 AIDS352910	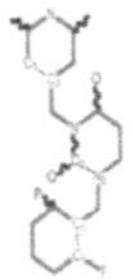	Fc1cccc(F)c1CN2/C=C\C(=O)N(C2=O)Cc3cc(C)nc(C)c3	3.886056	0.57	26.493803	0.9063530	14.61561	387.22308
		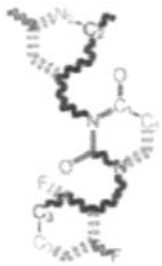	O=C1N(Cc(cc(C)n2)cc2C)C(N(/C=C1\)Cc(c(F)cc3)c(F)c3)=O		0.77	14.950333	1.7825743	4.1934669	62.693730

TABLE 3.46 Continued

No.	Structure 2D / IUPAC name / MW / AIDS code	SMILES configurations: LoSMoC / BraS	Code LoSMoC / Code BraS	A	LogP	χ (eV)	η (eV)	π	ω (eV)
		LoSMoC	Code LoSMoC		... LoSMoC ...				
		BraS	Code BraS		... BraS ...				
26	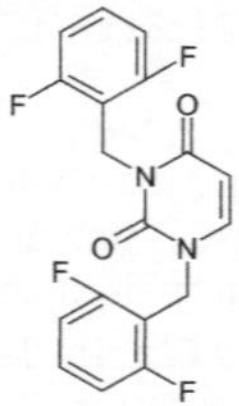 1,3-Bis-(2,6-difluoro-benzyl)-1H-pyrimidine-2,4-dione 364.30 AIDS352911	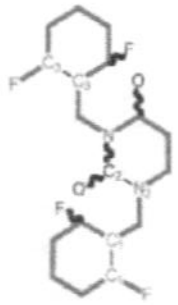	Fc1cccc(F)c1CN2/C=C\C(=O) N(C2=O)Cc3c(F)cccc3F	4.379863	0.59	27.958833	1.5546911	8.991764	251.39924
		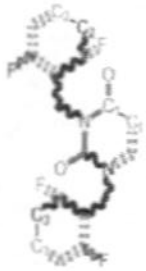	O=C1N(Cc(c(F)cc2)c(F)c2)C (N(/C=C1\)Cc(c(F)cc3)c(F)c3)=O		−1.34	15.611849	2.8690618	2.7207236	42.475527

TABLE 3.46 Continued

No.	Structure 2D / IUPAC name / MW / AIDS code	SMILES configurations: LoSMoC / BraS	Code LoSMoC / Code BraS	A	LogP	χ (eV)	η (eV)	π	ω (eV)
					… LoSMoC …				
					… BraS …				
27	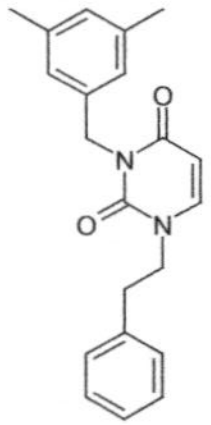 3-(3,5-Dimethyl-benzyl)-1-phenethyl-1H-pyrimidine-2,4-dione 334.42 AIDS352912	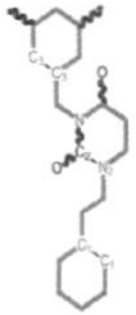	c1ccccc1CCN2/C=C\C(=O) N(C2=O)Cc3cc(C)cc(C)c3	5.206209	2.09	25.447501	0.8335692	15.26418	388.43520
		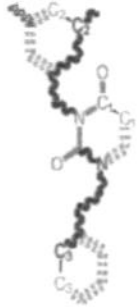	O=C1N(Cc(cc(C)c2)cc2C) C(N(/C=C1\)CCc(cccc3)c3)=O		2.06	14.410323	1.8477646	3.8993936	56.191522

TABLE 3.46 Continued

No.	Structure 2D IUPAC name MW AIDS code	SMILES configurations LoSMoC BraS	 Code LoSMoC Code BraS	A	LogP	χ (eV)	η (eV) … LoSMoC … … BraS …	π	ω (eV)
28	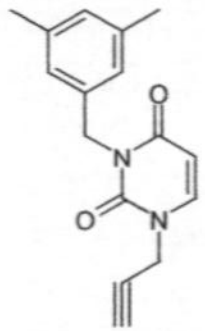	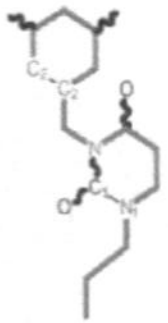	C#CCN1/C=C\C(=O) N(C1=O)Cc2cc(C) cc(C)c2	5.966576	0.77	21.628890	1.4603086	7.405589	160.17466
	3-(3,5-Dimethyl-benzyl)-1-prop-2-ynyl-1H-pyrimidine-2,4-dione 268.32 AIDS352913	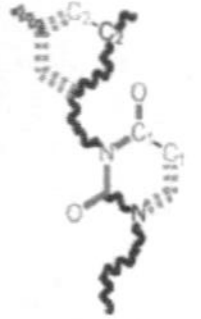	O=C1N(Cc(cc(C)c2) cc2C) C(N(/C=C1\)CC#C)=O		1.18	12.392809	2.5046350	2.4739751	30.659502

TABLE 3.46 Continued

No.	Structure 2D / IUPAC name / MW / AIDS code	SMILES configurations: LoSMoC / BraS	Code LoSMoC / Code BraS	A	LogP	χ (eV)	η (eV)	π	ω (eV)
					... LoSMoC ...				
					... BraS ...				
29	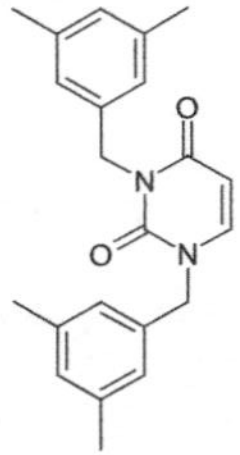 1,3-Bis-(3,5-dimethyl-benzyl)-1H-pyrimidine-2,4-dione 348.44 AIDS352914	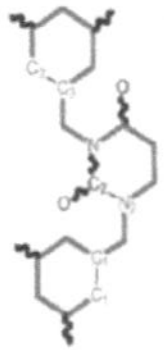	c1c(C)cc(C)cc1CN2/C=C\C(=O) N(C2=O)Cc3cc(C)cc(C)c3	6.283996	2.14	25.233546	0.8800182	14.33694	361.77196
		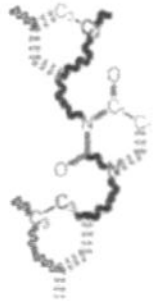	O=C1N(Cc(cc(C)c2)cc2C)C (N(/C=C1\)Cc(cc(cc3C)C)c3)=O		2.55	14.566107	1.9376961	3.7586149	54.748388

TABLE 3.46 Continued

No.	Structure 2D	SMILES configurations		A	LogP	χ (eV)	η (eV)	π	ω (eV)
	IUPAC name	**LoSMoC**	**Code LoSMoC**				**… LoSMoC …**		
	MW	**BraS**	**Code BraS**				**… BraS …**		
	AIDS code								
30	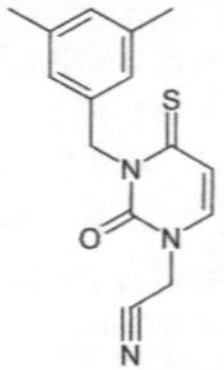	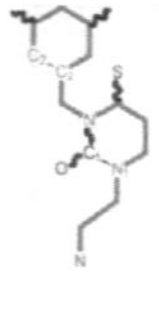	N#CCN1/C=C\C(=S)N (C1=O)Cc2cc(C)cc(C) c2	7.309803	1.28	21.897722	1.6182386	6.765913	148.15807
	[3-(3,5-Dimethyl-benzyl)-2-oxo-4-thioxo-3,4-dihydro-2H-pyrimidin-1-yl]-acetonitrile 285.36 AIDS352915	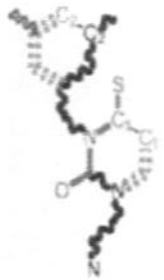	S=C1N(Cc(cc(C)c2) cc2C) C(N(/C=C1\)CC#N)=O		1.68	12.764862	3.0237637	2.1107572	26.943525

TABLE 3.46 Continued

No.	Structure 2D / IUPAC name / MW / AIDS code	LoSMoC / BraS	SMILES configurations / Code LoSMoC / Code BraS	A	LogP	χ (eV)	η (eV)	π	ω (eV)
					… LoSMoC …				
					… BraS …				
31	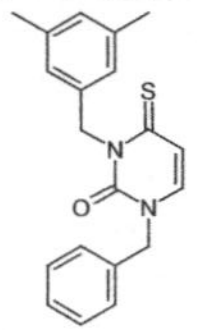 1-Benzyl-3-(3,5-dimethyl-benzyl)-4-thioxo-3,4-dihydro-1H-pyrimidin-2-one 336.45 AIDS352916	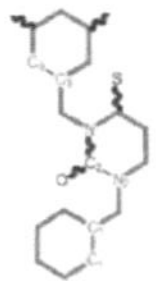	c1ccccc1CN2/C=C\C(=S) N(C2=O)Cc3cc(C)cc(C)c3	7.292429	2.49	25.217792	1.1471616	10.99139	277.17849
		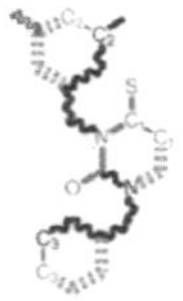	S=C1N(Cc(cc(C)c2)cc2C) C(N(/C=C1\)Cc(ccc3)cc3)=O		2.45	14.289267	2.4197012	2.9526925	42.191813

TABLE 3.46 Continued

No.	Structure 2D IUPAC name MW AIDS code	SMILES configurations LoSMoC BraS	Code LoSMoC Code BraS	A	LogP	χ (eV)	η (eV)	π	ω (eV)
					... LoSMoC ...				
					... BraS ...				
32	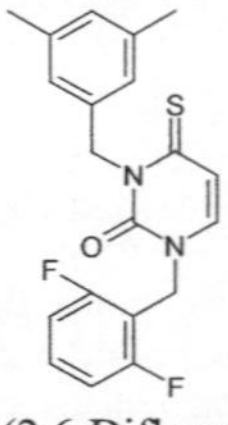 1-(2,6-Difluoro-benzyl)-3-(3,5-dimethyl-benzyl)-4-thioxo-3,4-dihydro-1H-pyrimidin-2-one 372.43 AIDS352917	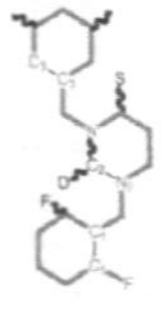	Fc1cccc(F)c1CN2/C=C\C(=S) N(C2=O)Cc3cc(C)cc(C)c3	7.229147	2.02	25.321304	1.0761564	11.76469	297.89740
		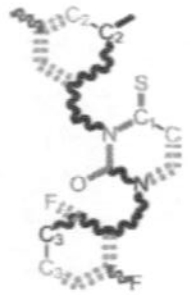	S=C1N(Cc(cc(C)c2)cc2C) C(N(/C=C1\)Cc(c(F)cc3)c(F)c3)=O		1.25	14.434969	2.3806265	3.0317586	43.763344

*The solubility parameter of lipophilicity (LogP), and the chemical reactivity parameters as electronegativity (χ) and chemical hardness (η), chemical power (π) and electrophilicity (ω) are considered within the semiempirical (AM1) framework (Polak-Ribiere conjugate gradient algorithm and geometry optimization till the root mean square RMS gradient was equal to or less than 0.01 kcal/Åmol) as provided by Hyperchem 7.01 computational environment (Hypercube, 2002), while the chemical reactivity values were computed in terms of HOMO and LUMO from the Eqs. (3.172) and (3.173) – see text and Tables 3.5 and 3.47, (3.165) and (3.168), respectively. SMILES legend is: ▬ principal SMILES chain; secondary SMILES branch; tertiary SMILES branch; quaternary SMILES branch; = double bond; # triple bond; /, \ directional bonds; () branch; C, N, F, S, I – atoms present in the molecule; c, n – atoms place in an aromatic ring; $C_{1/2/3}$, $N_{1/2}$, $c_{1/2/3}$, n_2 – connectivity points (Putz & Dudaş, 2013a, b).

and the lost of electronic pairs as the defolding goes from original to LoSMoC to BraS configuration.

These SMILES metabolic intermediates posing nucleophilic active sides are nevertheless confirmed at least for fused and non-fused diazines (Moldoveanu et al., 2009), among which are also those based on pyrimidines, already demonstrated with antiviral and anti-HIV acting in special (Gammon et al., 2008; Fan et al., 2009; De Clercq, 2005; Muhanji & Hunter, 2007) and antiinflamatory effects in general (Butnariu et al., 2007; Butnariu & Mangalagiu, 2009; Balan et al., 2009).

Electronegativity and chemical hardness values of Table 3.46 are based on HOMO and LUMO entries of Table 3.47 combined with Eqs. (3.172) and (3.173); they are further implemented in chemical power and electrophilicities Eqs. (3.165) and (3.168) to provide the respective LoSMoC and BraS results in Table 3.46 as well. Along the different hidrophobicities for LoSMoC and BraS molecules, these chemical-physical descriptors are further employed by QSAR modeling for explaining the chemical-biological binding of the actual series of pyrimidines to the reverse-transcriptase enzyme in HIV cell causing its inhibition for further action against the host organism's cells.

TABLE 3.47 The AM1 Computed Values (in electron-volts, eV) For the First Three Highest Occupied and Lowest Unoccupied Molecular Orbitals In Both Variants As the Longest SMILES Molecular Chain (LoSMoC) in Upper Entry and the Branching SMILES (BraS) in Down Entry, Employed For Computation of Electronegativity (χ), Chemical Hardness (η), Chemical Power (π) and Electrophilicity (ω), For Compounds of Table 3.46*

No.	HOMO1	LUMO1	HOMO2	LUMO2	HOMO3	LUMO3
			... LoSMoC ...			
			... BraS ...			
1	–24.49903	–19.06514	X	X	X	X
	–24.48801	–19.03451	–24.88237	–18.05411	–25.10602	–15.57611
2	–24.24188	–18.7667	X	X	X	X
	–24.23715	–18.75946	–24.69547	–17.93489	–24.84179	–15.32821
3	–24.25602	–18.82835	X	X	X	X
	–24.2567	–18.82977	–24.58191	–17.70927	–24.85183	–15.37954
4	–23.95141	–18.83626	X	X	X	X
	–23.95204	–18.83621	–24.28008	–17.60903	–24.88104	–15.38259

TABLE 3.47 Continued

No.	HOMO1	LUMO1	HOMO2	LUMO2	HOMO3	LUMO3
	... LoSMoC ...					
	... BraS ...					
5	–24.38787	–18.90236	X	X	X	X
	–24.38787	–18.90236	–24.42277	–17.3528	–24.90982	–15.43411
6	–24.24172	–18.95968	X	X	X	X
	–24.23569	–18.95218	–24.49526	–17.86779	–25.04196	–15.49452
7	–23.35131	–18.73365	X	X	X	X
	–23.35188	–18.73767	–24.22514	–17.79723	–24.54057	–15.31291
8	–23.79299	–18.65147	X	X	X	X
	–23.79239	–18.65293	–24.13254	–17.38683	–24.7122	–15.20959
9	–23.46857	–18.83136	X	X	X	X
	–23.46979	–18.83921	–24.28395	–17.49789	–24.46192	–15.37871
10	–24.37925	–18.94867	X	X	X	X
	–24.38	–18.9506	–24.99142	–18.7636	–25.14861	–15.67584
11	–22.38345	–18.75445	X	X	X	X
	–22.38345	–18.75445	–23.79029	–17.31787	–24.21747	–15.29094
12	–21.96501	–18.31488	X	X	X	X
	–21.96844	–18.31149	–23.85501.	–16.16856	–23.89945	–14.89846
13	–26.91465	–21.85561	X	X	X	X
	–26.91465	–21.85561	–27.7802	–20.69252	–28.33738	–18.9827
14	–26.66128	–22.14708	X	X	X	X
	–26.66128	–22.14708	–27.47999	–20.30796	–27.875	–19.37649
15	–27.68342	–23.22071	X	X	X	X
	–27.68553	–23.22033	–28.91865	–22.96564	–28.9519	–21.82943
16	–29.51216	–24.44028	X	X	X	X
	–29.52823	–24.42876	–29.75423	–23.06013	–30.9813	–22.86666
17	–28.49271	–23.59289	X	X	X	X
	–28.47523	–23.5581	–29.36592	–22.66404	–30.06262	–21.75203
18	–25.63183	–23.12311	X	X	X	X
	–25.62548	–23.11217	–26.88207	–22.20886	–27.55654	–20.01182
19	–26.0344	–23.19914	X	X	X	X
	–26.03953	–23.19181	–27.0533	–22.2293	–27.84065	–20.10159

TABLE 3.47 Continued

No.	HOMO1	LUMO1	HOMO2	LUMO2	HOMO3	LUMO3
	... LoSMoC ...					
	... BraS ...					
20	–25.36022	–21.78615	X	X	X	X
	–25.36493	–21.78792	–26.68329	–20.71338	–27.06831	–19.37157
21	–25.47117	–22.40276	X	X	X	X
	–24.47218	–22.40179	–26.77381	–21.92655	–27.2391	–19.67952
22	–29.50944	–24.42961	X	X	X	X
	–29.5088	–24.42942	–30.31708	–23.21535	–30.95683	–22.84796
23	–25.65356	–23.07113	X	X	X	X
	–25.6511	–23.06654	–26.89824	–22.43524	–27.60502	–19.97251
24	–26.0339	–23.14582	X	X	X	X
	–26.03578	–23.15325	–27.02319	–22.45168	–27.88159	–20.08076
25	–26.5313	–23.41763	X	X	X	X
	–26.55279	–23.43647	–27.38203	–22.60093	–28.51525	–20.85345
26	–29.02596	–23.685	X	X	X	X
	–28.90689	–23.43443	–29.78576	–22.74551	–29.8298	–21.98785
27	–25.41998	–22.55635	X	X	X	X
	–25.42157	–22.55341	–26.66657	–21.11333	–27.34483	–19.49103
28	–22.8969	–17.88018	X	X	X	X
	–22.90148	–17.87531	–22.96332	–17.29096	–24.06666	–14.20252
29	–25.29808	–22.27488	X	X	X	X
	–25.30234	–22.27582	–25.91112	–20.09649	–26.56577	–19.35705
30	–23.42159	–17.86232	X	X	X	X
	–23.42258	–17.86581	–23.58074	–15.8222	–23.9511	–15.31823
31	–25.7421	–21.80116	X	X	X	X
	–25.74458	–21.80102	–27.02937	–20.01162	–27.54453	–19.79035
32	–25.71771	–22.0207	X	X	X	X
	–25.71681	–22.02284	–27.02077	–19.81276	–27.47808	–19.75182

*Note that, in either LoSMoC or BraS forms, the overall compound was considered as carrying the [+2n] charge due to removed electronic pair out of each "broken bond" in SMILES configurations for compounds of Table 3.46. "X" indicates the truncation to the first order of HOMO and LUMO in LoSMoC calculations of electronegativity and chemical hardness of Eqs. (3.172) and (3.173), respectively (Putz & Dudaş, 2013b).

QSAR analysis requires a preliminary screening such that out of the available pool of molecules there are retained that ones that further fulfill certain similarity criteria to an increased degree of correlation.

This stage is presented in Table 3.48 separately for LoSMoC and BraS and for each such molecular defolding, and separately for SMILES central chain case (i) as well as for the N3-pyrimidine atom neighbors case; (ii) due to its central role in obtaining the spiroheterocyclic compounds and of their reaction pathways (Moldoveanu et al., 2009), which is presumably defolded also in the chemical-biological interaction; note that, consistent with the previous branching considerations the criteria for BraS are almost doubled respecting the LoSMoC.

The results of Table 3.48 leaves with two sets of molecules for each SMILES intermediate, while they are not necessary selected only upon the highest correlation factor recorded but through compromising the correlation factor with the number of chemical reactivity variables and with the number of compounds employed in correlation; as such, for each LoSMoC/BraS cases (i)/(ii) one should chose the molecular sets presenting the best combination among (Putz & Dudaş, 2013a, b):

- higher correlation factors;
- screening correlations having maximum of variables as descriptors;
- almost equal sets of compounds producing the precedent points;
- sets of compounds fulfilling the Topliss and Costello (1972) rule at least respecting the basic/independent descriptors of electronegativity and chemical hardness plus the hydrophobicity measure.

This way, the selected LoSMoC cases' variants are:

- the case (i)/V2 was chosen despite V1 since better fulfilling above criteria (e.g., being based on all variables and on 12 compounds and not on 4 variables and 11 compounds as in V1);
- the case (ii)/V6 was chose despite the versions V1 and V2 that have lesser compounds in the set, and to be closer to the previous case, for molecular sets' cardinals.

On the similarly grounds, the selected BraS cases' variants are:

- the case (i)/V5 despite variant V2 since having minimum 3 tertiary branching, while being in the similar correlation range, so that better fulfilling the "spirit" of molecular branching;

TABLE 3.48 Case (i): Screening Based On SMILES Central Chain and Case (ii): Screening Based On SMILES Central N-Atom Neighbors (N3 Atom of the Pyrimidine) For Chain Length and Atomic Neighboring in Longest SMILES Molecular Chain (LoSMoC) in Upper Entry and the Branching SMILES (BraS) in Down Entry For Various Versions (V's) of SMILES Based Screening Criteria Along the Molecules of Table 3.46, Respectively*

Index	Criteria	Case (i)		Case (ii)	
		Molecules	R_{QSAR}	Molecules	R_{QSAR}
V1 LoSMoC	Between 15–16 atoms LoSMoC	1–4, 6–11, 28	0.90371960 [(a)]	1–9, 28	0.92402295 [(c)]
V1 BraS	*Main chain and secondary branch with maximum 14 atoms*	*2–11, 13, 14, 16, 17, 22, 28*	*0.53158997*	*2, 3, 5–9, 13, 14, 16, 17, 22, 28, 29*	*0.70384894*
V2 LoSMoC	Between 18–21 atoms LoSMoC	13–17, 19, 21, 22, 24, 26, 31, 32	**0.75180080**	15–18, 21–23, 27, 29, 31, 32	0. 95150144 [(b)]
V2 BraS	*Main chain and secondary branch with minimum 14 atoms*	*7, 11, 12, 15–17, 19, 22, 24–26, 28, 30–32*	*0.95109419*	*7, 15–17, 20, 21, 22, 27, 28, 29, 30–32*	***0.87354213***
V3 LoSMoC	At least one triple bond in the main chain LoSMoC	1–7, 9–11, 13, 14, 28, 30	0.56411064	1–4, 6, 7, 9, 13, 15, 28, 30	0.49202776
V3 BraS	*Secondary and tertiary branches with maximum 14 atoms*	*2–10, 13, 14, 28*	*0.62469181*	*1–7, 9, 13, 14, 28*	*0.75756597*
V4 LoSMoC	More than three branches in the main chain LoSMoC	2–4, 6–11, 19, 21, 22, 24–26, 28, 30–32	0.43357261	2–4, 6, 7, 9, 15, 20–23, 27–32	0.61510478
V4 BraS	*Secondary and tertiary branches with minimum 14 atoms*	*11, 15–17, 19, 21–25, 31, 32*	*0.64694148*	*15–17, 20–23, 27, 29, 31, 32*	*0.94183439*
V5 LoSMoC	More than four branches in the main chain LoSMoC	7–9, 11, 19, 21, 24–26, 28, 30–32	0.47454364	7, 8, 20, 23, 27–32	0.71500251 [(d)]

TABLE 3.48 Continued

Index	Criteria	Case (i)		Case (ii)	
		Molecules	R_{QSAR}	Molecules	R_{QSAR}
V5 BraS	*Minimum 3 tertiary branches*	*6, 11, 15–17, 19, 22–26, 31, 32*	***0.94899619***	*6, 15–17, 20–23, 27, 29, 31, 32*	*0.64718879*
V6 LoSMoC	Ramifications of LoSMoC main chain containing groups formed only carbon and hydrogen atoms (except common = O, C = O)	2–4, 6–10, 19, 28, 30, 31	0.71050966 [(b)]	2–4, 6, 7, 9, 15, 20, 27–31	**0.64508095**
V6 BraS	*Minimum 1 quaternary branching*	*1, 2, 4, 6–8, 10,13–15, 19, 21–25, 28, 30–32*	*0.48549586*	*1, 2, 4, 6–8,13, 14, 20–23, 27–29, 30–32*	*0.63906586*
V7 LoSMoC	Ramifications of LoSMoC main chain containing groups consisting of a single atom or $-CH_3$ groups (except common = O, C = O)	2–7, 9, 10, 19, 22, 24–26, 28, 30–32	0.57636501	2–4, 6, 7, 9, 20–22, 27– 32	0.61600596 [(e)]
V7 BraS	*One of the secondary branches with minimum one triple bond*	*1–7, 9–11, 13–15, 28*	*0.63904635*	*1–7, 9, 13–15, 28*	*0.73556023* [(d)]
V8 LoSMoC	At least one branch for the last 6 points main chain LoSMoC	2–4, 6–11, 19, 23–25, 28, 30, 32	0. 51837657	2–4, 6, 7, 9, 20, 21, 27–32	0.69314160 [(d)]
V8 BraS	*The secondary branch linked with C2 of pyrimidinic nucleus with minimum 2 heteroatoms*	*1–6, 8–11, 13–15*	*0.58368204*	*1–6, 8, 9, 13–15*	*0.57765388* [(f)]
V9 LoSMoC	LoSMoC main chain contains after N3 atom of the pyrimidine nucleus (central main chain LoSMoC) a group $-CH_2-$	1–7, 9–11, 13–15, 19, 21, 24–26, 28, 30–32	0.37650771	1–8, 13–15, 20, 21, 27–32	0.63047473

TABLE 3.48 Continued

Index	Criteria	Case (i)		Case (ii)	
		Molecules	R_{QSAR}	**Molecules**	R_{QSAR}
V9 BraS	*The secondary branch linked with N3 of pyrimidinic nucleus contains only C and H atoms*	*1–8, 10, 11, 13–17, 25, 26, 28, 30–32*	*0.63881109*	*1–8, 13–17, 20, 21, 27–29, 30–32*	*0.72514327*
V10 BraS	The secondary branch linked with N3 of pyrimidinic nucleus contains 4 Carbon atoms	2–4, 6 8–10, 13, 14, 16, 19, 22, 24, 26	0.61480396	2–6, 8, 9, 13, 14, 16, 17, 22	0.53480139
V11 BraS	*The secondary branch linked with N3 of pyrimidinic nucleus contains 5–6 Carbon atoms*	*7, 12, 15, 18, 21, 23, 25, 28, 30–32*	*0.66627959*	*7, 15, 18, 20, 21, 23, 28, 29, 30–32*	*0.59914507*
V12 BraS	The tertiary branching are formed by maximum 3 atoms of C and H	2, 4–10, 13, 16, 19, 21–25, 28, 30–32	0.38470862	2, 4, 6–9, 13, 16–18, 20, 21, 28–32	0.61909773
V13 BraS	*The tertiary branches are formed only by C and H atoms*	*2–10, 13, 14, 16, 17, 19, 28, 30, 31*	*0.56415743*	*2–9, 13–16, 20, 27–31*	*0.64691170*
V14 BraS	Quaternary branching are contains only one C atom or CH3 group	1, 2, 5–7, 21–25, 28, 30–32	0.57731047	2, 5, 6, 20–23, 27, 28, 30–32	0.72850903
V15 BraS	*A single quaternary branching with maximum 2 atoms (C/O) and H*	*1, 2, 5–7, 19, 21, 22, 28, 30, 31*	*0.93051865*	*1, 2, 5, 6, 20–22, 27, 28, 30, 31*	*0.90565106*

(a) A=A(χ, η, ω, logP); (b) within statistical error tolerance 0.00002; (c) within statistical error tolerance $1E^{-25}$; (d) within statistical error tolerance 0.00004; (e) within statistical error tolerance 0.00008; (f) within statistical error tolerance 0.00003.

*The correlations' factors are given for full dependency of parameters of Table 3.46, i.e., A=A(χ, η, π, ω, logP), and for statistical error tolerance of 0.0001, unless otherwise indicated for the best correlation's combination such that the Topliss-Costello rule (Topliss & Costello, 1972) for ratio molecule-to-descriptors ≥4 to be generally respected (at least for χ and η as the main QSAR descriptors); the marked correlation corresponds with selected criteria and implicitly with the working molecular pool of Table 3.46 for each SMILES configuration (LoSMoC and BraS) and screening case (i and ii) further considered (see text) (Putz & Dudaş, 2013b).

- the case (ii)/V2 despite versions V4 and V15 (with lesser compounds in the set), being nevertheless in the same range of higher correlations and having the same cardinal of molecules in the set as its companion case (i)/V5.

They are further used for being integrated in appropriate measures towards establishing the (anti-HIV) mechanism of action.

Note that the Euclidean distance itself employs the square of the correlations factors, i.e., in a higher order statistics framework, which nevertheless may be further enriched with other statistical outputs and factors, although all directly or indirectly depend on the correlation factor (Putz et al., 2011b).

However, in the present study we are considering internal measures of the present QSAR models (unfolded in Table 3.49) by their minimal search – formally written as in Eq. (3.171) with the results presented in Table 3.50.

Worth also to remark that in the present uracils derivatives' anti-HIV analysis, the four-descriptors' dependency is not necessary in variational construct of Eq. (3.171) since not need in assessing the structural/reactivity parameters hierarchy in minimum variational path principle of Eq. (3.171) through being absorbed in the rest of correlations by means of the *transitivity chain rule* (Putz & Dudaş, 2013a, b):

- whenever two descriptors are common for adjacent activities' correlations – they will be considered as single common influence in chemical causes for the observed biological activity.

This way, the redundancies or double counting of models are avoided even with the price of "jumping" some intermediate models, as are the four-descriptors' endpoints. The results of Table 3.50 are interpreted in the sense of establishing the minimum of three path hierarchies, and then compared at the global level; note that more than three minimum paths will produce redundant information. Accordingly, the minimum paths, for LoSMoC and BraS cases (i)/(ii) separately, are (Putz & Dudaş, 2013b):

- For case LoSMoC/V2/(i):

$$\begin{aligned} &(\alpha)\text{: I2-II7-III8-V} \quad \delta[\alpha] = 0.029043119 \\ &(\beta)\text{: I2-II1-III8-V} \quad \delta[\beta] = 0.029516257 \\ &(\gamma)\text{: I2-II7-III7-V} \quad \delta[\gamma] = 0.029877668 \end{aligned} \tag{3.273}$$

TABLE 3.49 Statistical Correlation Results Obtained For Cases V2/(i) and V6/(ii) For Longest SMILES Molecular Chain (LoSMoC) and Respectively For the Cases V5/(i) and V2/(ii) For Branching SMILES (BraS) Selected Compounds' Sets Form Table 3.48 With Respective Molecules of Table 3.46 (Detailed Respective QSAR Models Dependencies On Chemical Reactivity Parameters Are Provided in Putz and Dudaş (2013b)

No.	A(x)	LoSMoC		BraS	
		$R_{Case\ V2/(i)}$	$R_{Case\ V6/(ii)}$	$R_{Case\ V5/(i)}$	$R_{Case\ V2/(ii)}$
I_1	A(logP)	0.36160241	0.43043863	*0.45645057*	*0.51687516*
I_2	A(χ)	0.70875308	0.04142206	*0.32832072*	*0.63329686*
I_3	A(η)	0.3850668	0.27082157	*0.3694801*	*0.10466918*
I_4	A(π)	0.20001171	0.23419593	*0.23910446*	*0.36217604*
I_5	A(ω)	0.0679732	0.21014	*0.12316764*	*0.52996859*
II_1	A(logP, χ)	0.72462236	0.54711991	*0.54563771*	*0.68322871*
II_2	A(logP, η)	0.53462981	0.45498598	*0.58822038*	*0.78078563*
II_3	A(logP, π)	0.4587341	0.47447182	*0.53086816*	*0.8624387*
II_4	A(logP, ω)	0.40635079	0.49281211	*0.48406183*	*0.85830581*
II_5	A(χ, η)	0.72042921	0.34882836	*0.44147923*	*0.65793015*
II_6	A(χ, π)	0.72662887	0.32861178	*0.42540934*	*0.67176394*
II_7	A(χ, ω)	0.72663277	0.33323936	*0.41607475*	*0.67165975*
II_8	A(η, π)	0.74023092	0.31232276	*0.46816571*	*0.69205634*
II_9	A(η, ω)	0.74918964	0.3278778	*0.47282745*	*0.6980058*
II_{10}	A(π, ω)	0.72422189	0.31072122	*0.4687647*	*0.66987725*
III_1	A(logP, χ, η)	0.72946153	0.54741756	*0.62478127*	*0.83591477*
III_2	A(logP, χ, π)	0.73229267	0.54735654	*0.62197159*	*0.86508134*
III_3	A(logP, χ, ω)	0.73214282	0.5471543	*0.61493693*	*0.86624574*
III_4	A(logP, η, π)	0.74609564	0.48854915	*0.62416978*	*0.87096819*
III_5	A(logP, η, ω)	0.751297	0.51239927	*0.63374038*	*0.86007179*
III_6	A(logP, π, ω)	0.72648755	0.52806785	*0.65025857*	*0.86552207*
III_7	A(χ, η, π)	0.75053661	0.35028746	*0.4752325*	*0.7019648*
III_8	A(χ, η, ω)	0.74939285	0.34885082	*0.52544907*	*0.70077495*
III_9	A(χ, π, ω)	0.72663285	0.35789332	*0.83429197*	*0.67179626*
III_{10}	A(η, π, ω)	0.74919138	0.33193549	*0.47362344*	*0.70165085*
V	A(logP, χ, η, π, ω)	0.7518008	0.64508095	*0.94899619*	*0.87354213*

TABLE 3.50 *Endpoint Paths* and Their Lengths (δ) Considered For the Best/Relevant QSAR's Correlations' Models of Table 3.49, in Cases V2/(i) and V6/(ii) For Longest SMILES Molecular Chain (LoSMoC) and In Cases V5/(i) and V2/(ii) For Branching SMILES (BraS) Selected Compounds' Sets From Table 3.48, Upon the Euclidian Metrics of Eq. (3.171) Applied on the First Four Shortest Intermediary QSAR Models of Table 3.49; the Overall First 3 Shortest Path-Lengths Are Identified In Each Configuration Case By Bolding and Labeling As Alpha (α), Beta (β) and Gamma (γ) Superscripts, Respectively (Putz & Dudaş, 2013b)

LoSMoC				BraS			
Path	$\delta_{V2/(i)}$	**Path**	$\delta_{V6/(ii)}$	**Path**	$\delta_{V5/(i)}$	**Path**	$\delta_{V2/(ii)}$
I1-II1-III5-V	0.363999003	I1-II1-III1-V	0.15216027	*I1-II1-III5-V*	*0.339267818*	*I1-II1-III3-V*	*0.247430746*
I1-II1-III7-V	0.363945917	I1-II1-III2-V	0.15219933	***I1-II1-III6-V***	**0.328852605**$^{\gamma}$	*I1-II1-III4-V*	*0.250851034*
I1-II1-III8-V	0.363872037	I1-II1-III3-V	0.15232909	***I1-II1-III9-V***	**0.323160465**$^{\beta}$	*I1-II1-III6-V*	*0.246918396*
I1-II7-III5-V	0.36586301	I1-II2-III1-V	0.13669055	*I1-II2-III5-V*	*0.344705061*	*I1-II2-III3-V*	*0.277498475*
I1-II7-III7-V	0.365814373	I1-II2-III2-V	0.13669292	*I1-II2-III6-V*	*0.332349493*	*I1-II2-III4-V*	*0.27890546*
I1-II7-III8-V	0.365747157	I1-II2-III3-V	0.13670114	***I1-II2-III9-V***	**0.301780663**$^{\alpha}$	*I1-II2-III6-V*	*0.277296451*
I1-II8-III5-V	0.378790523	I1-II3-III1-V	0.12960764	*I1-II3-III5-V*	*0.339863056*	*I1-II3-III3-V*	*0.345661527*
I1-II8-III7-V	0.378770846	I1-II3-III2-V	0.12961931	*I1-II3-III6-V*	*0.330206319*	*I1-II3-III4-V*	*0.345678373*
I1-II8-III8-V	0.378746997	I1-II3-III3-V	0.12965837	*I1-II3-III9-V*	*0.332807819*	*I1-II3-III6-V*	*0.345670347*

TABLE 3.50 Continued

LoSMoC				BraS			
Path	$\delta_{V2/(i)}$	**Path**	$\delta_{V6/(ii)}$	**Path**	$\delta_{V5/(i)}$	**Path**	$\delta_{V2/(ii)}$
I1-II9-III5-V	0.387593286	**I1-II4-III1-V**	**0.12810286**[α]	*I1-II4-III5-V*	*0.350074672*	*I1-II4-III3-V*	*0.341600891*
I1-II9-III7-V	0.387591632	**I1-II4-III2-V**	**0.1281234**[β]	*I1-II4-III6-V*	*0.342969246*	*I1-II4-III4-V*	*0.341675065*
I1-II9-III8-V	0.387594763	**I1-II4-III3-V**	**0.12819186**[γ]	*I1-II4-III9-V*	*0.369568114*	*I1-II4-III6-V*	*0.34160106*
I2-II1-III5-V	0.031042298	I2-II1-III1-V	0.51504227	*I2-II1-III5-V*	*0.392905816*	*I2-II1-III3-V*	*0.189846412*
I2-II1-III7-V	0.030413493	I2-II1-III2-V	0.51505381	*I2-II1-III6-V*	*0.383948387*	*I2-II1-III4-V*	*0.194283111*
I2-II1-III8-V	**0.029516257**[β]	I2-II1-III3-V	0.51509217	*I2-II1-III9-V*	*0.379084442*	*I2-II1-III6-V*	*0.18917817*
I2-II7-III5-V	0.030467382	I2-II2-III1-V	0.43487567	*I2-II2-III5-V*	*0.411103551*	***I2-II2-III3-V***	**0.170615371**[β]
I2-II7-III7-V	**0.029877668**[γ]	I2-II2-III2-V	0.43487642	*I2-II2-III6-V*	*0.40080012*	***I2-II2-III4-V***	**0.172894351**[γ]
I2-II7-III8-V	**0.029043119**[α]	I2-II2-III3-V	0.434879	*I2-II2-III9-V*	*0.37584055*	***I2-II2-III6-V***	**0.17028659**[α]
I2-II8-III5-V	0.033370142	I2-II3-III1-V	0.44987922	*I2-II3-III5-V*	*0.388579959*	*I2-II3-III3-V*	*0.229289585*
I2-II8-III7-V	0.033146038	I2-II3-III2-V	0.44988258	*I2-II3-III6-V*	*0.38016273*	*I2-II3-III4-V*	*0.22931498*

TABLE 3.50 Continued

LoSMoC				BraS			
Path	$\delta_{V2/(i)}$	**Path**	$\delta_{V6/(ii)}$	**Path**	$\delta_{V5/(i)}$	**Path**	$\delta_{V2/(ii)}$
I2-II8-III8-V	0.032872383	I2-II3-III3-V	0.44989384	*I2-II3-III9-V*	*0.382424544*	*I2-II3-III6-V*	*0.229302881*
I2-II9-III5-V	0.04049457	I2-II4-III1-V	0.46505147	*I2-II4-III5-V*	*0.382158589*	*I2-II4-III3-V*	*0.225267191*
I2-II9-III7-V	0.040478734	I2-II4-III2-V	0.46505713	*I2-II4-III6-V*	*0.375660505*	*I2-II4-III4-V*	*0.225379654*
I2-II9-III8-V	0.040508702	I2-II4-III3-V	0.46507599	*I2-II4-III9-V*	*0.400091867*	*I2-II4-III6-V*	*0.225267448*
I3-II1-III5-V	0.340602068	I3-II1-III1-V	0.29305119	*I3-II1-III5-V*	*0.371725449*	*I3-II1-III3-V*	*0.606860446*
I3-II1-III7-V	0.340545335	I3-II1-III2-V	0.29307147	*I3-II1-III6-V*	*0.36224466*	*I3-II1-III4-V*	*0.608262992*
I3-II1-III8-V	0.340466377	I3-II1-III3-V	0.29313888	*I3-II1-III9-V*	*0.357085205*	*I3-II1-III6-V*	*0.606651729*
I3-II7-III5-V	0.342455676	I3-II2-III1-V	0.22803128	*I3-II2-III5-V*	*0.386400836*	*I3-II2-III3-V*	*0.681535121*
I3-II7-III7-V	0.342403714	I3-II2-III2-V	0.2280327	*I3-II2-III6-V*	*0.375420048*	*I3-II2-III4-V*	*0.682109209*
I3-II7-III8-V	0.342331902	I3-II2-III3-V	0.22803762	*I3-II2-III9-V*	*0.34864824*	*I3-II2-III6-V*	*0.681452889*
I3-II8-III5-V	0.355336832	I3-II3-III1-V	0.23734499	*I3-II3-III5-V*	*0.368802149*	*I3-II3-III3-V*	*0.75781421*

TABLE 3.50 Continued

LoSMoC				BraS			
Path	$\delta_{V2/(i)}$	**Path**	$\delta_{V6/(ii)}$	**Path**	$\delta_{V5/(i)}$	**Path**	$\delta_{V2/(ii)}$
I3-II8-III7-V	0.355315856	I3-II3-III2-V	0.23735136	*I3-II3-III6-V*	*0.359922688*	*I3-II3-III4-V*	*0.757821894*
I3-II8-III8-V	0.355290432	I3-II3-III3-V	0.23737269	*I3-II3-III9-V*	*0.362310878*	*I3-II3-III6-V*	*0.757818233*
I3-II9-III5-V	0.364129287	I3-II4-III1-V	0.24859544	*I3-II4-III5-V*	*0.367313037*	*I3-II4-III3-V*	*0.753713772*
I3-II9-III7-V	0.364127526	I3-II4-III2-V	0.24860602	*I3-II4-III6-V*	*0.360547493*	*I3-II4-III4-V*	*0.753747392*
I3-II9-III8-V	0.364130859	I3-II4-III3-V	0.24864131	*I3-II4-III9-V*	*0.385936759*	*I3-II4-III6-V*	*0.753713849*
I4-II1-III5-V	0.525288611	I4-II1-III1-V	0.32781038	*I4-II1-III5-V*	*0.448453944*	*I4-II1-III3-V*	*0.369625875*
I4-II1-III7-V	0.525251826	I4-II1-III2-V	0.32782851	*I4-II1-III6-V*	*0.440627193*	*I4-II1-III4-V*	*0.371924125*
I4-II1-III8-V	0.525200637	I4-II1-III3-V	0.32788877	*I4-II1-III9-V*	*0.436395432*	*I4-II1-III6-V*	*0.369283099*
I4-II7-III5-V	0.527198557	I4-II2-III1-V	0.25851495	*I4-II2-III5-V*	*0.472588851*	*I4-II2-III3-V*	*0.42730628*
I4-II7-III7-V	0.527164806	I4-II2-III2-V	0.2585162	*I4-II2-III6-V*	*0.463653781*	*I4-II2-III4-V*	*0.428221331*
I4-II7-III8-V	0.527118165	I4-II2-III3-V	0.25852055	*I4-II2-III9-V*	*0.442255821*	*I4-II2-III6-V*	*0.42717511*

TABLE 3.50 Continued

LoSMoC				BraS			
Path	$\delta_{V2/(i)}$	**Path**	$\delta_{V6/(ii)}$	**Path**	$\delta_{V5/(i)}$	**Path**	$\delta_{V2/(ii)}$
I4-II8-III5-V	0.540332774	I4-II3-III1-V	0.26942851	*I4-II3-III5-V*	*0.441695569*	*I4-II3-III3-V*	*0.500330351*
I4-II8-III7-V	0.54031898	I4-II3-III2-V	0.26943412	*I4-II3-III6-V*	*0.434308982*	*I4-II3-III4-V*	*0.500341989*
I4-II8-III8-V	0.540302262	I4-II3-III3-V	0.26945292	*I4-II3-III9-V*	*0.436290182*	*I4-II3-III6-V*	*0.500336444*
I4-II9-III5-V	0.549182204	I4-II4-III1-V	0.281784	*I4-II4-III5-V*	*0.426373085*	*I4-II4-III3-V*	*0.496246943*
I4-II9-III7-V	0.549181037	I4-II4-III2-V	0.28179334	*I4-II4-III6-V*	*0.420558718*	*I4-II4-III4-V*	*0.496298005*
I4-II9-III8-V	0.549183247	I4-II4-III3-V	0.28182447	*I4-II4-III9-V*	*0.44251816*	*I4-II4-III6-V*	*0.49624706*
I5-II1-III5-V	0.657190923	I5-II1-III1-V	0.3508471	*I5-II1-III5-V*	*0.534442949*	*I5-II1-III3-V*	*0.238824486*
I5-II1-III7-V	0.657161522	I5-II1-III2-V	0.35086404	*I5-II1-III6-V*	*0.52789265*	*I5-II1-III4-V*	*0.242366256*
I5-II1-III8-V	0.657120609	I5-II1-III3-V	0.35092035	*I5-II1-III9-V*	*0.524365617*	*I5-II1-III6-V*	*0.238293632*
I5-II7-III5-V	0.65912139	I5-II2-III1-V	0.27934081	*I5-II2-III5-V*	*0.563677521*	*I5-II2-III3-V*	*0.265077074*
I5-II7-III7-V	0.659094395	I5-II2-III2-V	0.27934197	*I5-II2-III6-V*	*0.556207653*	*I5-II2-III4-V*	*0.266549633*

TABLE 3.50 Continued

LoSMoC				BraS			
Path	$\delta_{V2/(i)}$	**Path**	$\delta_{V6/(ii)}$	**Path**	$\delta_{V5/(i)}$	**Path**	$\delta_{V2/(ii)}$
I5-II7-III8-V	0.659057091	I5-II2-III3-V	0.27934599	*I5-II2-III9-V*	*0.53850008*	*I5-II2-III6-V*	*0.264865576*
I5-II8-III5-V	0.672348982	I5-II3-III1-V	0.29108509	*I5-II3-III5-V*	*0.52553652*	*I5-II3-III3-V*	*0.332571954*
I5-II8-III7-V	0.672337897	I5-II3-III2-V	0.29109028	*I5-II3-III6-V*	*0.519343768*	*I5-II3-III4-V*	*0.332589464*
I5-II8-III8-V	0.672324461	I5-II3-III3-V	0.29110768	*I5-II3-III9-V*	*0.521001709*	*I5-II3-III6-V*	*0.332581122*
I5-II9-III5-V	0.681219886	I5-II4-III1-V	0.30401219	*I5-II4-III5-V*	*0.502030388*	*I5-II4-III3-V*	*0.328514246*
I5-II9-III7-V	0.681218945	I5-II4-III2-V	0.30402085	*I5-II4-III6-V*	*0.497101738*	*I5-II4-III4-V*	*0.328591374*
I5-II9-III8-V	0.681220726	I5-II4-III3-V	0.30404971	*I5-II4-III9-V*	*0.515812781*	*I5-II4-III6-V*	*0.328514423*

- For case LoSMoC/V6/(ii):

$$\begin{aligned} (\alpha)&: \text{I1-II4-III1-V} \quad \delta[\alpha] = 0.12810286 \\ (\beta)&: \text{I1-II4-III2-V} \quad \delta[\beta] = 0.1281234 \\ (\gamma)&: \text{I1-II4-III3-V} \quad \delta[\gamma] = 0.12819186 \end{aligned} \tag{3.274}$$

- For case BraS/V5/(i):

$$\begin{aligned} (\alpha)&: \text{I1-II2-III9-V} \quad \delta[\alpha] = 0.301780663 \\ (\beta)&: \text{I1-II1-III9-V} \quad \delta[\beta] = 0.323160465 \\ (\gamma)&: \text{I1-II1-III6-V} \quad \delta[\gamma] = 0.328852605 \end{aligned} \tag{3.275}$$

- For case BraS/V2/(ii):

$$\begin{aligned} (\alpha)&: \text{I2-II2-III6-V} \quad \delta[\alpha] = 0.17028659 \\ (\beta)&: \text{I2-II2-III3-V} \quad \delta[\beta] = 0.170615371 \\ (\gamma)&: \text{I2-II2-III4-V} \quad \delta[\gamma] = 0.172894351 \end{aligned} \tag{3.276}$$

The variational results of Eqs. (3.274)–(3.276) are most intrigued in providing the reliability of the present approach because:

- All the LoSMoC least path lengths are shorter than those of BraS, this way confirming that the chain based SMILES intermediates are *prior* to those displaying branching SMILES conformations, i.e., in accordance with the steps [A]→[B] of Figure 3.30 in pyrimidine related uracil attack of reserve transcriptase;
- While passing from LoSMoC to BraS configurations in chemical-biological interaction of uracil derivatives – reverse transcriptase binding one notes the maintenance of the same criteria variants, namely V2 of Table 3.48,

$$\text{LoSMoC}/\boxed{\text{V2}}/\text{(i)} \rightarrow \text{BraS}/\boxed{\text{V2}}/\text{(ii)} \tag{3.277}$$

  meaning that the chain-to-branching passage seems to require the same features of the principal chain and of the secondary branch, equally;
- Looking now to the cases interchanged in transformation of Eq. (3.277) one also notes the passage from (i) case based on longest chain in SMILES configuration to the case (ii) based on the pyrimidinic N3 atom's neighbors, happens consistently: the mechanism of interaction is described as involving the trans-membrane transduction by means of longest chain of SMILES configuration; it is followed

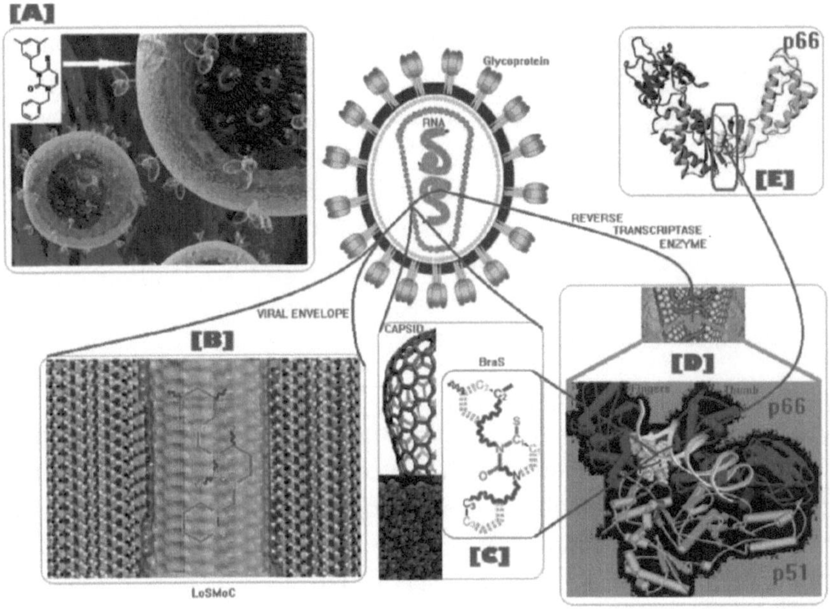

FIGURE 3.30 The mechanism of molecular interaction of the 1,3-disubstituted uracils, with the prototype no. (31) of Table 3.46 (since belonging to all selected QSAR-SMILES criteria and cases of Table 3.48) against human immunodeficiency virus (HIV-1), after ViiV (2012), through five stages: (A) the free molecular attack on the HIV viral envelope, after (Genetic Engineering & Biotechnology News, 2010); (B) the passage of the lipidic viral envelope of HIV under the form of longest SMILES molecular chain (LoSMoC) of Table 3.46, after (Putz & Dudaş, 2013a; Aksimentiev et al., 2004; Heng et al., 2004); (C) the transport though the protein layer of HIV capsid, after (Perilla et al., 2013; Ganser-Pornillos et al., 2008), yielding the Branching SMILES (BraS) configuration of Table 3.46 that further binds in (D) with the palm active region of p66 monomer of reverse transcriptase (RT), after (Picado, 2012; C.H.A.N.G.E., 2013), towards (E) the competitively inhibiting the RT by the formed ligand-receptor complex, after (Madrid et al., 1999), by means of chemical reactivity frontier electronic transfer as detailed in the Figures 3.31 and 3.32 (Putz & Dudaş, 2013b).

by the bonding stage centered on the N3 atom of pyrimidine ring nuclei as already proved to be specific for spirodiazine derivatives in their transformations towards anti-inflammatory recorded activities, anti-HIV included (Gammon et al., 2008).

When the chemical reactivity indices are the working QSAR descriptors one may expect the results obeying the established hierarchy for chemical binding scenario, see the Volume III of the present five-volume series (Putz, 2016c), and Putz (2011c, 2012a), according which a chemical reaction/interaction is triggered by electronegativity (its difference) and its minimizing principle, followed by chemical hardness maximization of the interacting frontier orbitals implying minimum charge exchange (parallel spins) followed by minimum activation by adducts' potential barrier tunneling (spin pairing), leaving with the conducted hierarchy (3.170).

However, worth noting that, when the SMIELS information is involved exclusively on screening procedures of Table 3.48, at the level of LoSMoC only, the application of the SPECTRAL-SAR multi-variationally algorithm (see Section 3.1.1) provides the chemical hierarchy of causes under succession

$$\eta \rightarrow \log P \rightarrow \pi \rightarrow \omega \rightarrow \chi \tag{3.278}$$

Accordingly, the series of 1,3-disubstituted uracil based molecules in case V2/(i) of Table 3.48 against the HIV infection is analyzed from the perspective of the mechanism of chemical-biological interaction towards inducing the ligand-receptor binding: it practically reverses the chemical reactivity "binding paradigm" (3.170), now having the electronegativity and chemical hardness at two ends in the mechanism of chemical-biological interaction. All in all, it associates with the following scenario (Figure 3.31) (Putz & Dudaş, 2013a):

- The adducts are considered the genuine molecular structure on one side, and the SMILES molecular structure projected ("virtually bound", *mutatis mutandis* as is in the electrostatic *image interaction* picture (Halliday et al., 1992; Griffiths, 1999)) on the targeted biological receptor (specifically on the V1/V2 loops of the gp120 site of the HIV reverse transcriptase) on the other side;
- The chemical-biological interaction will be driven by the electronic frontier dynamic reactivity between these two partners: the "pure chemical" molecule further called as *the ligand* (L) and the in situ embedded/projected SMILE molecule, further called as *the receptor* (R), towards inhibiting/blocking the biological site activity, here considered as anti-HIV action;
- The CB interaction initiate with chemical hardness adjustments between the HOMO-LUMO gaps of ligand and receptor in their maximum size such that fulfilling the variational (maximum hardness and HSAB) principles;
- Such largest energetic gap impedes further reactivity with HIV cellular walls through their penetration, in the second step of lipophilicity

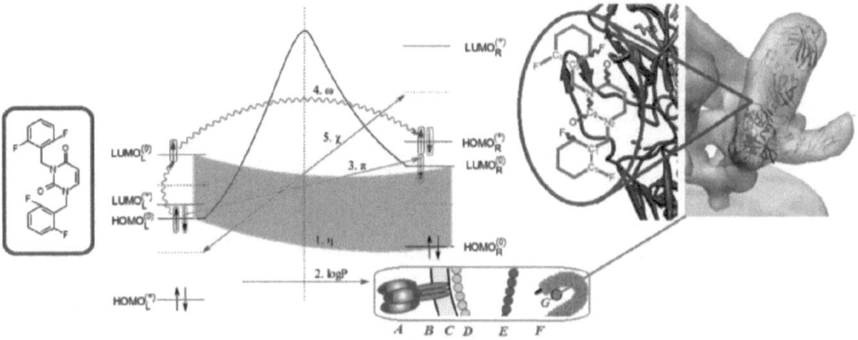

FIGURE 3.31 Representation of mechanistic QSAR results, see Eq. (3.278), for the uracil derivative compounds with HIV interaction through linking the genuine molecule (the ligand, L) to the SMILES molecule embedded/projected into receptor (R) site (on V1/V2 loops of gp120 site, in blue, from HIV, adapted from Refs. (Walker et al., 2009; RBVI, 2013) through variational principles of chemical reactivity of electronegativity, chemical hardness, chemical power, electrophilicity and lipophilicity, here resumed for molecule no. 26 of Table 3.46 since it belongs to the best selected QSAR V2/ (i) case by the SMILES screening of Table 3.2. Lipophilic (SMILES) molecular traveling is through successive HIV cell's membranes: (A) docking glycoprotein gp120; (B) transmembrane glycoprotein gp41; (C) lipid membrane; (D) matrix; (E) capsid; (F) RNA; till hitting (G) the reverse transcriptase (Putz & Dudaş, 2013a).

action (logP) hitting the reverse transcriptase's gp120 V1/V2 loop site [known as best host for anti-HIV drugs' actions (Walker et al., 2009; RBVI, 2013)];

- In the third instance, the charge transfer begins between chemical-L and biological-R, as above described, through the mixed reactivity indices; note that due to the presence of both the electronegativity and chemical hardness influence in the mixed reactivity indices (π and ω) their actions are accounted by two successive steps, depending on which fundamental reactivity index is activated; for electronegativity, the level-to-level (e.g., HOMO-to-HOMO) transitions are considered, while for chemical hardness the inter-levels, i.e., HOMO-to-LUMO, transitions are allowed; in the present case, since chemical hardness causes the triggering of the L-R interaction mechanism, the chemical power action (π) firstly drives the L-R charge transfer though $HOMO_L$-$LUMO_R$ transition, followed by the second $LUMO_R$-$HOMO_R$ excitation; in short, it is abbreviated as LRR electronic frontier charge transfer, without changing the spin, in accordance with the allied minimization principle, see the previous discussion of Eq. (3.167);

- The fourth stage of hierarchy (3.278) involves the electrophilicity (ω) transfer through minimization of the activation energy and L-R barrier tunneling according with its principle discussed around Eq. (3.169); combining this feature with above inter-level activated transfer by the chemical hardness, there results the electrophilicity action as the $HOMO_L$-$LUMO_L$ activation firstly, followed by the $LUMO_L$-$HOMO_R$ quantum tunneling charge transfer towards recomposing the electronic pair on the receptor side, yet on an activated $HOMO_R$ level; the present transfer will be thus labeled as of LLR type;

At the end of actual LR interaction mechanism the electronegativity aligns the ligand middy level of relaxed $HOMO_L$-$LUMO_L$ gap with that of the activated receptor $HOMO_R$-$LUMO_R$ forcing therefore the overall LR system into an activated state. Consequently, the anti-HIV effect remains unfinished, whereas further intermolecular and *in situ* interactions may decide on the total or partial inhibition of the HIV activity and cellular apoptosis.

Actually, when the SMILES involvement is completely considered in LoSMoC as well isBraS forms of Table 3.48, for the actual uracils' anti-HIV action, the transformation (3.278) is further transformed since being projected on the structural or chemical reactivity descriptors it encompasses for the shortest path lengths so that closing the variational QSAR modeling (Putz & Dudaş, 2013b):

$$\alpha_{LoSMoC/V2/(i)} \rightarrow \alpha_{BraS/V2/(ii)} \tag{3.279}$$

which is equivalently rewritten with the help of Eqs. (3.273) and (3.276)

$$[I2\text{-}II7\text{-}III8\text{-}V] \rightarrow [I2\text{-}II2\text{-}III6\text{-}V] \tag{3.280}$$

and even more with the help of endpoint identifications of Table 3.49, respectively as:

$$\begin{gathered}[(\chi)\rightarrow(\chi,\omega)\rightarrow(\chi,\omega,\eta)\rightarrow(\chi,\omega,\eta,\pi,\log P)]\\ \rightarrow[(\chi)\rightarrow(\eta,\log P)\rightarrow(\log P,\pi,\omega)\rightarrow(\chi,\eta,\log P,\pi,\omega)]\end{gathered} \tag{3.281}$$

Now, the solution of the structural/reactivity causes driving the ligand receptor binding mechanism in the present 1,3-disubstituted uracils against human immunodeficiency virus (HIV-1) action is given by combined the two variational principles noted before:

- transitivity chain rule, and
- minimization of redundancies

in structural/chemical reactivity dependencies; as such, the first alpha spectral-SAR hierarchy in (3.281) solves the first three causes

$$\chi \to \omega \to \eta \to (\pi, \log P) \tag{3.282}$$

while the second alpha path hierarchy of Eq. (3.281) solves the last degeneracy of (3.282) as next: one considers the already solved structural/reactivity causes of Eq. (3.282) from where there results that η follows χ; with this ordering back in Eq. (3.281) one yields that logP follows η; this should finally applied also in Eq. (3.282); all in all, the ordered causes of structural/reactivity influences in actual anti-HIV mechanism looks like (Putz & Dudaş, 2013b):

$$\chi \to \omega \to \eta \to \log P \to \pi \tag{3.283}$$

Equation (3.283) may be represented by the orbital based scheme of chemical reactivity driving biological (anti-HIV) activity as provided in Figure 3.32; it is explained in the light of chemical reactivity principles, see the Section 3.2.9 and the Volume III of the present five-volume work (Putz, 2016c), within the "time-space" framework fixed by the chemical reactivity- biological activity interaction:

- The developing time is not the physical one by an internal one related with the reaction coordinate, so that the reactivity-driven-activity steps are phenomenological ordered through being interrelated and inter-conditioned during the entire physical time of the binding (on a nano-seconds scale);
- The described interaction is spatially placed between the ligand (L) represented by the SMILES branched molecule resulted upon the HIV cell's transduction (at least of the viral envelope) and the receptor – the palm region of the p66 region of the reverse transcriptase.

In these conditions the found mechanism for uracil derivatives' anti-HIV activity goes as follows (Putz & Dudaş, 2013b):

- The first step is triggered by electronegativity (χ) and of its principle of minimization difference between ligand (L) and receptor

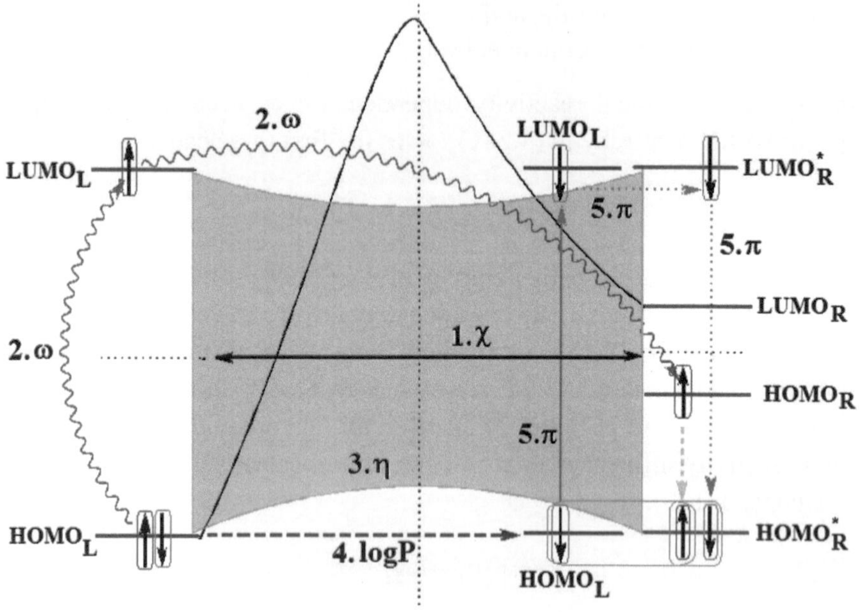

FIGURE 3.32 Representation of mechanistic molecular orbital interaction and bonding between the the uracil derivative compounds and HIV through binding the SMILES (essentially the BraS) molecule (the ligand, L) with the molecular pocket of the receptor (R) site, see the stages (D) & (E) of Figure 3.30, through variational principles of chemical reactivity of (1) electronegativity (χ), (2) electrophilicity (ω), (3) chemical hardness (η), (4) lipophilicity (logP) and (5) chemical power (π), according with the Spectral-QSAR analysis of Eqs. (3.281)–(3.283) (Putz & Dudaş, 2013b).

(R) HOMO-LUMO middy-levels, as provided by Eqs. (3.160) & (3.161); upon this stage the ligand and receptor are energetically aligned around a common electronegativity; it also associates with "preparation" of HOMO and LUMO states for ceding and accepting electrons by the accompanying interchanging charge;

- The second step accompanies the first one through the electrophilicity (ω) by putting in act the charge transfer by tunneling of L-R barrier for one electron of the $HOMO_L$ level passing to the $LUMO_L$ and then down to the $HOMO_R$ state by means of the LLR mechanism, see Figure 3.9(b); the minimization principle for electrophilicity, Eq. (3.169), further allows the relaxation of the transferred electron from the $HOMO_R$ to $HOMO_R^*$ level;
- The third step appears naturally "called" by the second one: the R to R* actually corresponds with the expansion of the $HOMO_R$-$LUMO_R$

gap to $HOMO_R^*$-$LUMO_R^*$ to be equal with $HOMO_L$-$LUMO_L$ one, in accordance with the maximum hardness principle, Eq. (3.164), being this step driven by chemical hardness;

- The forth step converts spatially the energetic HOMO-LUMO coupling of ligand-receptor by hydrophobicity/lipophilicity (logP) action eventually assuring also the capsid penetration; note that the previous charge transfer was realized through (quantum) tunneling, in accordance with electrophilicity driving action, so being consistent with earlier (second step) long-range action of the pyrimidines in the plasmidic region of HIV cell against its reverse transcriptase enzyme inside of the capsid, see Figure 3.30;
- The fifth and the last step is accomplished by chemical power (π) which assures the effective ligand-receptor binding (now also spatially in nature) by transferring the remaining electron of $HOMO_L$ to $LUMO_L$ and then to $LUMO_R^*$ by means of the LRR mechanisms of Figure 3.8(b); it nevertheless fulfils the minimization principle, Eq. (3.167), by undergoing the final $LUMO_R^*$ to $HOMO_R^*$ relaxation, when it pairs with the electron arrived from the electrophilicity step above.

Overall, the presented molecular mechanism fully explains the ligand-receptor binding in all respects (Putz & Dudaş, 2013b):

- Spatially (the molecule is placed in the pocket of HIV's reverse transcriptase);
- Energetically (all transitions compensate each other);
- By electronic pairing (assured by electrophicity and chemical power actions);
- By bonding on the relaxed $HOMO_R^*$ level.

This way the presented variational QSAR anti-HIV mechanism assures the stabilization of pyrimidine complex with the enzyme transcriptase receptor towards the concerned apoptosis of the HIV cells through inhibiting his enzymic activities for further actions (and replications) in organism.

This study complements by the orbital-diagramatic results of results Figure 3.32 those of Figure 3.31 (Putz & Dudaş, 2013a), by effectively employing the various forms of SMILES configuration for the ligand molecules, beyond the SMILES in screening only implication, with the satisfactory result the molecular anti-HIV mechanism appears to be reliable and self-consistently aided to the envisaged ligand-receptor bining.

However, while being aware of the importance the branching SMILES procedure has played in the actual endeavor, further study may be directed towards employing the topological branching information of the involved molecules, being this field equally rich and promising in QSAR chemical systems with high complexity (Balaban, 1976, 1997; Balaban et al., 2011; Basak et al., 2011).

3.3.7.2 SPECTRAL-Diagonal-SAR Approach

Here we like to test the performance of the SPECTRAL-Diagonal structure-activity relationship (SD-SAR) methodology of Section 3.2.2 (Putz, 2013a), against the ordinary QSAR analysis, see Section 2.4.3, for the anti-HIV class of chemical compounds of Section 3.3.7.1, either for genuine (as in gas-phase), or Simplified Molecular Input Line Entry System (SMILES) (as in cellular environment, see Section 3.2.9) molecular forms.

The working molecular cases were selected as the present discussion follows (Dudaş & Putz, 2014): *Genuine* (gas-phase)_*All* molecules of Table 3.46; *Genuine case 1 – version 2* with between 18–21 atoms for LoSMoC, see further (*Genuine_C1V2*: molecules **13, 14, 15, 16, 17, 19, 21, 22, 24, 26, 31, 32** of Table 3.46); *Genuine case 2 – version 7* when ramifications of LoSMoC main chain containing groups consisting of a single atom or –CH_3 groups except common =O, C=O (*Genuine_C2V7*: molecules **2, 3, 4, 6, 7, 9, 20, 21, 22, 27, 28, 29, 30, 31, 32** of Table 3.46); *LoSMoC* (cell-phase)_*All* molecules of Table 3.46; *LoSMoC case 1 – version 2* for molecules between 18–21 atoms LoSMoC (*LoSMoC_C1V2*: molecules **13, 14, 15, 16, 17, 19, 21, 22, 24, 26, 31, 32** of Table 3.46/LoSMoC version as in Figure 3.6); *LoSMoC* case *2 – version 6* when ramifications of LoSMoC main chain containing groups formed only carbon and hydrogen atoms except common =O, C=O (*LoSMoC_C2V6*: molecules **2, 3, 4, 6, 7, 9, 15, 20, 27, 28, 29, 30, 31** of Table 3.46/LoSMoC version as in Figure 3.6); *BraS* (cell-phase)_*All* molecules of Table 3.46; *BraS case 1 – version 5* for molecules with minimum 3 tertiary branches (*BraS_C1V5*: molecules **6, 11, 15, 16, 17, 19, 22, 23, 24, 25, 26, 31, 32** of Table 3.46/BraS version as in Figure 3.6); *BraS* case *2 – version 2* for molecules having main chain and secondary branch with minimum 14 atoms (*BraS_C2V2*: molecules **7, 15, 16, 17, 20, 21, 22, 27, 28, 29, 30, 31, 32** of Table 3.46/BraS version as in Figure 3.6);

All in all, the numerical results for all forms of molecules (gas-phase, LoSMoC, BraS) with all considered variants are employed into the actual Spectral-Diagonal-SAR algorithm of Eqs. (3.43a), (3.43b) as well as by ordinary QSAR equations (3.44a) and (3.44b) for all chemical reactivity indices considered correlating the anti-HIV activity of molecules of Table 3.46, with the results of Tables 3.51–3.54.

The results may be unfolded as following:

- The main striking finding is that in all case, despite running a different correlation equation, the SD-SAR provide exactly the same, so in general essentially the same, degree of correlation of the structural chemical information with the recorded biological activity, no mater which data set is employed of in which molecular form (gas-phase or cellular environment as LoSMoC and BraS are);
- This finding provides the fruitful information of present analytical SD-SAR modeling by Eqs. (3.43a) and (3.43b) for further multivariate analyses, when the algebraic formulation may offer better control in establishing the chemical-biological interacting mechanism than the "black-box" statistical running correlation;
- This generalization feature may offer the alternative in implementing the non-ambiguous algorithms in a custom QSAR assessment fro a class of compounds or toxicants, so fulfilling, along the establishing the chemical-biological mechanism of interaction, the main desiderates in QSAR-protocols, according with The Organisation for Economic Co-operation and Development (OECD) criteria (OECD, 2004, 2005, 2006, 2007).

Further interesting studies may approach the differences or the convergence between the actual SD-SAR and the recently developed Spectral-SAR algorithms in providing reliable OECD-QSAR models and chemical-biological interaction mechanism, since both equally recover the essential multivariate-statistics, although based on different algebraic arguments of orthogonalization: of the matrix of variables and of the variable (quantum) vectors, respectively.

However, by the present proof for equivalency of Spectral-DiagonalSAR with ordinary QSAR analysis may further use the actual SD-SAR method whenever analyticity in assessing structure-activity or reactivity-activity modeling interaction is required, i.e., for the final aim of a QSAR analysis through establishing the chemical–biological interaction mechanism.

TABLE 3.51 The QSAR and Spectral-Diagonal-SAR (SD-SAR) Mono-Variate Regressions, As Given By Eqs. (3.43a)–(3.43b) and (3.44a)–(3.44b), Along the Respective Correlation Factor (R) For the Molecules of Table 3.46, In All Genuine (Gas-Phase) and Cellular Assumed SMILES (Simplified Molecular Input Line Entry System) LoSMoC (Longest SMILES Molecular Chain) and BraS (Branching SMILES) Molecular Forms, See Figure 3.6 For a Given Example, Within the Relevant Variants As Derived From Best Screening Criteria-Variants Selections Data Sets Molecules, So Correlating the *Electronegativity* (χ) Chemical Reactivity Index With the Recorded anti-HIV Activity of Table 3.46, Respectively (Maruyama et al., 2003; Putz & Dudaş, 2013a, b; Dudaş & Putz, 2014)*

Working Case	QSAR_A(χ)	$R_{QSAR-A(\chi)}$	SD-SAR_A(χ)	$R_{SD-SAR-A(\chi)}$	QSAR_A(A_{SD-SAR})	$R_{A(A\_SD-SAR)}$
Genuine_All	$11.6023–1.3654 \times \chi$	0.18821014	$0.2241+1.0704 \times \chi$	0.18821014	$11.8881–1.2755 \times A_{SD-SAR}$	1
Genuine_C1V2	$13.6855–1.8221 \times \chi$	0.24093724	$0.2214+1.0549 \times \chi$	0.24093724	$14.068–1.7272 \times A_{SD-SAR}$	1
Genuine_C2V7	$18.8847–2.8483 \times \chi$	0.26291646	$0.2402+1.1412 \times \chi$	0.26291646	$19.4844–2.4957 \times A_{SD-SAR}$	1
LoSMoC_All	$6.1055–0.042 \times \chi$	0.07687813	$0.0088+0.2167 \times \chi$	0.07687813	$6.1072–0.1938 \times A_{SD-SAR}$	1
LoSMoC_C1V2	$22.6727–0.6637 \times \chi$	0.70875308	$0.0074+0.1981 \times \chi$	0.70875308	$22.6978–3.3498 \times A_{SD-SAR}$	1
LoSMoC_C2V6	$4.5512+0.0317 \times \chi$	0.04142206	$0.0099+0.236 \times \chi$	0.04142206	$4.5499+0.1345 \times A_{SD-SAR}$	1
BraS_All	$6.0922–0.073 \times \chi$	0.07048329	$0.0272+0.3804 \times \chi$	0.07048329	$6.0975–0.192 \times A_{SD-SAR}$	1
BraS_C1V5	$10.9355–0.3993 \times \chi$	0.32832072	$0.0248+0.3644 \times \chi$	0.32832072	$10.9627–1.0957 \times A_{SD-SAR}$	1
BraS_C2V2	$15.1127–0.6665 \times \chi$	0.63329686	$0.029+0.4109 \times \chi$	0.63329686	$15.1598–1.6217 \times A_{SD-SAR}$	1

*The last two columns serve for cross-check of perfect self-consistency (correlation factor goes as unity) in correlation when projecting with Eqs. (3.44a)–(3.44b) the experimentally recorded anti-HIV activity (A) onto the SD-SAR computed activity as predicted by Eqs. (3.43a)–(3.43b).

TABLE 3.52 The Same Kind of Analysis As in Table 3.51, Here For *Chemical Hardness* (η) As Chemical Reactivity Descriptor of Molecules of Table 3.46 (Putz & Duda, 2013a, b ; Dudaş & Putz, 2014)

Working Case	QSAR_A(η)	$R_{QSAR-A(\eta)}$	SD-SAR_A(η)	$R_{SD-SAR-A(\eta)}$	QSAR_A(A_{SD-SAR})	$R_{A(A\_SD-SAR)}$
Genuine_All	13.3934–1.914 × η	0.36438537	0.2702+1.1702 × η	0.36438537	13.8355–1.6355 × A_{SD-SAR}	1
Genuine_C1V2	17.6257–2.9498 × η	0.66661731	0.2759+1.1712 × η	0.66661731	18.3206–2.5185 × A_{SD-SAR}	1
Genuine_C2V7	17.4705–2.8105 × η	0.56228108	0.2928+1.2537 × η	0.56228108	18.127–2.2416 × A_{SD-SAR}	1
LoSMoC_All	6.6535–1.2567 × η	0.32158003	2.1087+2.6091 × η	0.32158003	7.6691–0.4816 × A_{SD-SAR}	1
LoSMoC_C1V2	7.0856–1.6999 × η	0.38506680	2.1274+2.5764 × η	0.38506680	8.4893–0.6597 × A_{SD-SAR}	1
LoSMoC_C2V6	7.031–1.3086 × η	0.27082157	2.0712+2.702 × η	0.27082157	8.0341–0.4843 × A_{SD-SAR}	1
BraS_All	6.6977–0.677 × η	0.26319217	0.8035+1.9086 × η	0.26319217	6.9827–0.3547 × A_{SD-SAR}	1
BraS_C1V5	7.4902–1.0641 × η	0.36948010	0.8808+1.9886 × η	0.36948010	7.9615–0.5351 × A_{SD-SAR}	1
BraS_C2V2	4.783+0.3287 × η	0.10466918	0.8809+2.0918 × η	0.10466918	4.6446+0.1571 × A_{SD-SAR}	1

TABLE 3.53 The Same Kind of Analysis As in Table 3.51, Here For *Chemical Power* (**π**) As Chemical Reactivity Descriptor of Molecules of Table 3.46 (Putz & Duda, 2013a, b; Dudaş & Putz, 2014)

Working Case	**QSAR_A(π)**	$\mathbf{R}_{QSAR-A(\pi)}$	**SD-SAR_A(π)**	$\mathbf{R}_{SD\text{-}SAR-A(\pi)}$	**QSAR_A($A_{SD\text{-}SAR}$)**	$\mathbf{R}_{A(A\_SD\text{-}SAR)}$
Genuine_All	$-0.0785+9.3408 \times \pi$	0.26137496	$4.0863+2.2607 \times \pi$	0.26137496	$-16.9621+4.1317 \times A_{SD-SAR}$	1
Genuine_C1V2	$-8.9003+24.872 \times \pi$	0.65944595	$3.9849+2.2432 \times \pi$	0.65944595	$-53.0822+11.0872 \times A_{SD-SAR}$	1
Genuine_C2V7	$-1.6523+12.6243 \times \pi$	0.41386360	$4.3178+2.4083 \times \pi$	0.41386360	$-24.2859+5.2418 \times A_{SD-SAR}$	1
LoSMoC_All	$3.9625+0.1054 \times \pi$	0.29741848	$0.0459+0.4928 \times \pi$	0.29741848	$3.9526+0.2139 \times A_{SD-SAR}$	1
LoSMoC_C1V2	$3.9217+0.0945 \times \pi$	0.20001171	$0.0401+0.4571 \times \pi$	0.20001171	$3.9134+0.2068 \times A_{SD-SAR}$	1
LoSMoC_C2V6	$4.2984+0.1058 \times \pi$	0.23419593	$0.0596+0.5747 \times \pi$	0.23419593	$4.2874+0.1842 \times A_{SD-SAR}$	1
BraS_All	$4.097+0.3196 \times \pi$	0.21488959	$0.5071+1.565 \times \pi$	0.21488959	$3.9934+0.2042 \times A_{SD-SAR}$	1
BraS_C1V5	$3.6123+0.4286 \times \pi$	0.23910446	$0.4319+1.4605 \times \pi$	0.23910446	$3.4855+0.2935 \times A_{SD-SAR}$	1
BraS_C2V2	$7.846-0.7305 \times \pi$	0.36217604	$0.5587+1.719 \times \pi$	0.36217604	$8.0834-0.4249 \times A_{SD-SAR}$	1

TABLE 3.54 The Same Kind of Analysis As In Table 3.51, Here For *Electrophilicity* (ω) As Chemical Reactivity Descriptor of Molecules of Table 3.46 (Putz & Duda, 2013a, b; Dudaş & Putz, 2014)

Working Case	QSAR_A(ω)	$R_{QSAR-A(\omega)}$	SD-SAR_A(ω)	$R_{SD-SAR-A(\omega)}$	QSAR_A(A_{SD-SAR})	$R_{A(A\_SD-SAR)}$
Genuine_All	3.3086+0.6679 × ω	0.12045179	0.6676+1.7656 × ω	0.12045179	3.056+0.3782 × A_{SD-SAR}	1
Genuine_C1V2	-2.8447+2.9398 × ω	0.43419210	0.6408+1.7182 × ω	0.43419210	-3.9412+1.7109 × A_{SD-SAR}	1
Genuine_C2V7	1.5818+1.4232 × ω	0.27391966	0.7043+1.8685 × ω	0.27391966	1.0453+0.7617 × A_{calc}	1
LoSMoC_All	4.2592+0.003 × ω	0.24972604	0.0000745+0.0199 × ω	0.24972604	4.2592+0.1552 × A_{SD-SAR}	1
LoSMoC_C1V2	4.5887+0.0013 × ω	0.06797320	0.000058+0.0174 × ω	0.06797320	4.5887+0.0763 × A_{SD-SAR}	1
LoSMoC_C2V6	4.55497+0.0032 × ω	0.21014000	0.0001+0.0241 × ω	0.21014000	4.55495+0.1358 × A_{SD-SAR}	1
BraS_All	4.4126+0.0153 × ω	0.16911195	0.0028+0.1226 × ω	0.16911195	4.4123+0.1248 × A_{SD-SAR}	1
BraS_C1V5	4.3406+0.0142 × ω	0.12316764	0.0021+0.1081 × ω	0.12316764	4.3403+0.1317 × A_{SD-SAR}	1
BraS_C2V2	8.6092–0.0677 × ω	0.52996859	0.003+0.1333 × ω	0.52996859	8.6108–0.5077 × A_{SD-SAR}	1

3.3.8 TOPO-REACTIVE QSAR (FOR PAHS' CARCINOGENESIS)

The colored algorithm is here illustrated for the list of working PAH structures of Table 3.55. Note that, for the shake of conceptual recitation, while for pentacene the colored recipe was illustrated in Figure 3.12, the same rule is available for the other compounds of Table 3.55, and for perylene is in Figure 3.33 drawn (Putz et al., 2013c).

Worth noting here that the PAH choice is based on the recent academic and industry interest regarding their environmental and health impact; the present discussion follows (Putz et al., 2013b-c):

- mixtures of PAHs are also known to cause skin irritation and inflammation. Anthracene, benzo(a)pyrene and naphthalene are direct skin irritants while anthracene and benzo(a)pyrene are reported to be skin sensitizers, i.e., cause an allergic skin response in animals and humans (INCHEM, 1998);

TABLE 3.55 Experimental Octanol-Water Partition Coefficients (logKOW) and the Retention Indices (RI) (van Den Dool & Kratz, 1963) For the Studied PAHs**

No.	Molecule	CAS	Formula	MW (D)[(a)]	$\log K_{OW}$[(b)]	RI
1.	Benzene	71–43–2	C_6H_6	78.11	2.13	663[(c)]
2.	Naphthalene	91–20–3	$C_{10}H_8$	128.17	3.33	1208[(c)]
3.	Anthracene	120–12–7	$C_{14}H_{10}$	178.23	4.54	1767[(d)]
4.	Phenanthrene	85–01–8	$C_{14}H_{10}$	178.23	4.55	1778.6[(e)]
5.	Tetracene	92–24–0	$C_{18}H_{12}$	228.29	5.62*	2426[(f)]
6.	Pyrene	129–00–0	$C_{16}H_{10}$	202.25	5.14	2113.4[(d)]
7.	Pentacene	135–48–8	$C_{22}H_{14}$	278.35	7.19	3125.02*
8.	Picene	213–46–7	$C_{22}H_{14}$	278.35	6.75*	3140[(f)]
9.	Perylene	198–55–0	$C_{20}H_{12}$	252.31	6.16*	2795[(f)]
10.	Benzo-a-pyrene	50–32–8	$C_{20}H_{12}$	252.31	6.16*	2812[(f)]
11.	Quaterrylene	188–73–8	$C_{40}H_2O$	500.59	11.74*	5970.35*
12.	Coronene	191–07–1	$C_{24}H_{12}$	300.35	6.75	3497[(g)]
13.	Hexabenzocoronene	190–24–9	$C_{42}H_{18}$	522.59	12.24*	6252.02*
14.	Dicoronylene	98570–53–7	$C_{48}H_2O$	596.67	13.91*	7200.47*

[(a)] From Chemical Book (2011); [(b)] from Duchowicz et al. (2007); [(c)] from Engel & Ratel (2007); [(d)] from Song et al. (2003); [(e)] from Zeng et al. (2007); [(f)] from Beernaert (1979); [(g)] from Oda et al. (1998).

**The data not available in the literature (marked with "*") were interpolated through correlations with the molecular weight (MW, Daltons), see Refs. (Putz et al., 2013b-c).

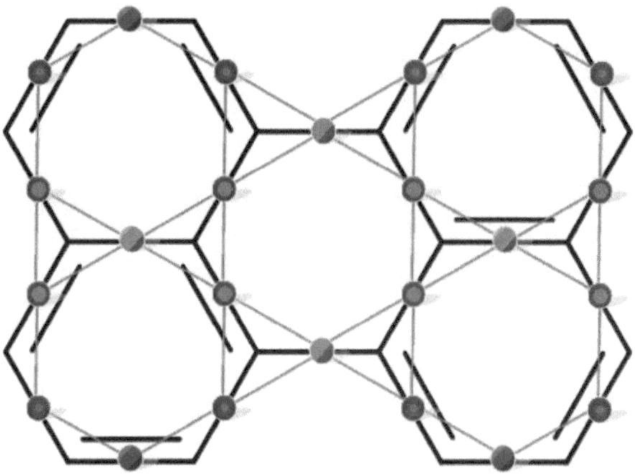

FIGURE 3.33 Illustration of the coloring of the perylene structure, respecting its most "dense" symmetrical (horizontal – middle) axis, following the Timisoara-Parma rule, as described in Figure 3.12 and Section 3.2.10.4 (Putz et al., 2013c).

- Benzo(a)pyrene is the most common PAH to cause cancer in animals and this compound is notable for being the first chemical carcinogen to be discovered. Based on the available evidence, both the International Agency for Research on Cancer (IARC, 2005) and US EPA (1994) classified a number of PAHs as carcinogenic to animals and some PAH-rich mixtures as carcinogenic to humans. The EPA has classified seven PAH compounds as probable human carcinogens: benz(a)anthracene, benzo(a)pyrene, benzo(b)fluoranthene, benzo(k)fluoranthene, chrysene, dibenz(ah)anthracene, and indeno(1,2,3-cd)pyrene;
- Most of the PAHs are not genotoxic by themselves and they need to be metabolised to the diol epoxides which react with DNA, thus inducing genotoxic damage. However, genotoxicity plays important role in the carcinogenicity process and maybe in some forms of developmental toxicity as well.

The electro-topological coloring algorithm presented in Section 3.2.10.4 is unfolded for the working PAHs following the successive steps (Putz et al., 2013b-c):

- Computing the first 3 HOMOs and LUMOs, reported in Table 3.56;
- Evaluating the compact finite differences of electronegativity and chemical hardness, employing the values of Table 3.56 in the

TABLE 3.56 Electronic Frontier Energetic (in electron-Volts, eV) Properties For Benzene and Paradigmatic PAH Molecules Computed Using the Semi-Empirical PM3 Method (Putz et al., 2013b)

No. crt.	Molecule	N_π	HOMO1	LUMO1	HOMO2	LUMO2	HOMO3	LUMO3
1.	Benzene	6	–9.751329	0.396204	–9.75139	0.396271	–12.3761	2.86574
2.	Naphthalene	10	–8.836945	–0.40645	–9.43547	0.064889	–10.678	1.081978
3.	Anthracene	14	–8.249425	–0.96838	–9.27205	–0.12648	–9.91016	0.458272
4.	Phenanthrene	14	–8.741241	–0.53429	–8.97665	–0.34949	–10.0386	0.523081
5.	Tetracene	18	–7.87106	–1.3526	–9.17744	–0.24918	–9.2819	–0.08773
6.	Pyrene	16	–8.249635	–1.00995	–9.04072	–0.33618	–9.94816	0.483801
7.	Pentacene	22	–7.611723	–1.62693	–8.8098	–0.5153	–9.11749	–0.33304
8.	Picene	22	–8.476872	–0.82467	–8.6867	–0.65607	–9.27715	–0.13945
9.	Perylene	20	–7.98719	–1.27454	–9.49631	0.007553	–9.54391	0.055699
10.	Benzo-a-pyrene	20	–8.041768	–1.22103	–8.90165	–0.48144	–9.63673	0.147193
11.	Quaterrylene	40	–7.425265	–1.90614	–8.394	–1.00441	–9.35793	–0.15158
12.	Coronene	24	–8.28931	–1.06324	–8.28931	–1.06324	–9.68944	0.170249
13.	Hexabenzocoronene	42	–8.05448	–1.36681	–8.05452	–1.36677	–8.71504	–0.7617
14.	Dicoronylene	48	–7.747883	–1.66594	–8.34093	–1.11796	–8.37042	–1.08884

equations (3.172) and (3.173) with the coefficients of Table 3.5, with the results in Tables 3.57 and 3.58, respectively;

- Considering the working PAHs upon the scheme of CFD hierarchy, as deduced by best fitting of the resulted parabolic energies with pi-energies, Eqs. (3.194) and (3.195), for the set of PAH molecules of Table 3.55 (Putz et al., 2013a-b), with the result as on the Figures 3.12, 3.13 and 3.33 and of Table 3.7;
- Coloring the topological descriptors with reactivity (frontier orbitals) contents: the conventional or newly proposed topological index is "translated" into a topo-reactive one by considering the topological matrix of that index and coloring it with molecular electronegativity values for nodes according with Timisoara-Parma rule; then take n-roots for any *n*-couplings reticules of whatever nodes within the molecule forming bonds and paths for given molecule or extended system;
- For given scheme of coloring, Eqs. (3.196), (3.197), (3.199), and (3.200) are applied with the computed electronegativity and chemical hardness values of Tables 3.57 and 3.58 to produce the parabolic counterpart – Eq. (3.198) – of the Wiener classical index, along other topological indices of interest, *topological efficiency index* ρ and the *extreme topological efficiency index* ρ^E:

$$\rho = W/Nw \quad \rho \geq 1 \tag{3.284}$$

$$\rho^E = w / w \quad \rho^E \geq 1 \tag{3.285}$$

Descriptor ρ has been firstly introduced by Ori and D'Mello (1993) on graphenic lattices, whereas ρ^E has been recently proposed by Cataldo et al. (2011b), and successfully applied by Sharma et al. (1997) to schwarzitic structures. By definition these invariants shall privilege, with some numerical differences, chemical structures growing in the most compact way around their minimal sites. In both cases of C_{60}-I_h buckminsterfullerene and benzene molecule, these indices reach the lower limit $\rho = \rho^E = 1$ evidencing all atoms are symmetry-equivalent. This result inspires the topological modeling ™ guiding criterion for stable systems: *the smallest is the topological efficiency index, the highest is the stability of the chemical structure under examination.* For a given vertex v_i of G its *eccentricity* ε_i is the largest distance between v_i and any other vertex of G, being the graph diameter M the maximum possible eccentricity, $M = \max\{\varepsilon_i\}$.

TABLE 3.57 Electronegativity Values (in electron-Volts, eV) For the Molecules in Table 3.56 Computed Using Eq. (3.172) and the Numerical Schemes in Table 3.5 (Putz et al., 2013b)

No.	χC2	χC4	χC6	χSP	χ6T	χ8T	χ8P	χ10P	χSLR
1.	4.677563	4.677563	4.68015	1.559189	2.945133	2.63113	2.54733	2.528442	2.783443
2.	4.621695	4.611096	4.610614	1.498169	2.880514	2.567244	2.483035	2.463587	2.716915
3.	4.608903	4.593842	4.588707	1.476058	2.860067	2.546385	2.465968	2.448904	2.705861
4.	4.637765	4.633547	4.635015	1.52905	2.908358	2.595825	2.512038	2.49291	2.746162
5.	4.611832	4.594919	4.587204	1.469626	2.856766	2.54236	2.464176	2.448584	2.708223
6.	4.629793	4.620017	4.617564	1.504159	2.887898	2.574318	2.492804	2.475133	2.731258
7.	4.619324	4.61212	4.611329	1.510958	2.888452	2.576307	2.494174	2.476	2.730185
8.	4.650773	4.647337	4.647194	1.536517	2.918722	2.605539	2.523899	2.506233	2.762146
9.	4.630863	4.611944	4.604367	1.467943	2.863175	2.54692	2.467261	2.450888	2.711043
10.	4.631397	4.621372	4.619137	1.503701	2.888219	2.574461	2.492529	2.474606	2.73053
11.	4.665703	4.660119	4.659737	1.532899	2.922154	2.607357	2.524708	2.506531	2.763124
12.	4.676277	4.676277	4.679054	1.558759	2.944323	2.630406	2.546393	2.527359	2.782077
13.	4.710646	4.710646	4.71157	1.570217	2.965963	2.649739	2.567422	2.54973	2.808405
14.	4.706913	4.703157	4.70166	1.553947	2.953178	2.636136	2.555042	2.538147	2.7986

TABLE 3.58 Chemical Hardness Values (in electron-Volts, eV) For the Molecules in Table 3.56 Computed Using Eq. (3.173) and the Numerical Schemes in Table 3.5 (Putz et al., 2013b)

No.	ηC2	ηC4	ηC6	ηSP	η6T	η8T	η8P	η10P	ηSLR
1.	5.073767	5.919389	4.214153	4.870808	6.483366	3.884713	3.781969	3.895765	4.808069
2.	4.21525	4.873214	3.431467	3.982449	5.307959	3.176384	3.128272	3.287641	4.180482
3.	3.640522	4.169587	2.902391	3.38303	4.52651	2.696762	2.685654	2.879618	3.777809
4.	4.103477	4.769881	3.380912	3.914125	5.197054	3.124154	3.058916	3.177355	3.956282
5.	3.259228	3.702025	2.550209	2.984271	4.004126	2.3781	2.392711	2.610879	3.51291
6.	3.619842	4.162113	2.911226	3.387157	4.527015	2.700341	2.676222	2.845643	3.687931
7.	2.992399	3.394895	2.335101	2.734121	3.673915	2.178059	2.193885	2.399796	3.243817
8.	3.8261	4.448015	3.153247	3.65035	4.838721	2.914825	2.855183	2.96529	3.683801
9.	3.356327	3.799414	2.606034	3.054601	4.105097	2.433524	2.458452	2.701619	3.67288
10.	3.410372	3.912122	2.728484	3.177989	4.25054	2.533383	2.517975	2.690792	3.513191
11.	2.759563	3.141553	2.170298	2.536952	3.416913	2.019469	2.022999	2.196567	2.949659
12.	3.613033	4.215205	3.000908	3.468512	4.605771	2.767983	2.697092	2.778787	3.420051
13.	3.343835	3.901137	2.777312	3.210076	4.249582	2.563713	2.50079	2.577189	3.16081
14.	3.040971	3.500256	2.451504	2.85087	3.792326	2.275403	2.255589	2.393534	3.076702

TABLE 3.59 Pi-Related Frontier Energetic Values (eV) Derived From Eqs. (3.195) With the Aid of PM3 Semiempirical Quantum Computation For E_{PI}React and the Data in Tables 3.57 and 3.58 For the Parabolic E_{PI}CFD of Eq. (3.194) (Putz et al., 2013b)

No.	E_{PI}REACT	E_pC2	E_pC4	E_pC6	E_pSP	E_p6T	E_p8T	E_p8P	E_p10P	E_pSLR
1.	746.569	154.5902	185.0326	123.6286	165.994	215.7304	124.0629	120.8669	125.0769	125.0769
2.	1218.214	375.3081	441.2105	297.0406	383.2632	501.9908	291.966	287.9968	304.1282	304.1282
3.	1690.023	649.0177	752.9253	504.6268	642.409	847.155	492.916	491.8647	530.1205	530.1205
4.	1689.735	739.3527	870.027	597.7684	745.7619	977.9056	575.9927	564.3791	587.8609	587.8609
5.	2161.905	972.9769	1116.747	743.698	940.4506	1245.915	724.7419	730.8832	801.8503	801.8503
6.	1912.512	852.6029	991.5807	671.3929	843.0456	1112.709	650.0983	645.2279	688.8826	688.8826
7.	2633.825	1346.696	1541.663	1028.74	1290.074	1714.629	997.5018	1006.968	1107.029	1107.029
8.	2632.882	1749.515	2050.598	1423.933	1732.966	2277.729	1353.453	1326.383	1380.063	1380.063
9.	2384.186	1249.914	1427.527	950.3261	1192.482	1584.775	922.4712	934.0354	1031.63	1031.63
10.	2384.171	1271.521	1472.422	999.0108	1241.121	1642.452	961.8639	957.3393	1026.825	1026.825
11.	4716.109	4228.672	4840.08	3286.087	3997.807	5350.174	3126.857	3135.809	3414.246	3414.246
12.	2829.185	1968.876	2315.728	1616.225	1960.453	2582.26	1531.228	1492.412	1539.925	1539.925
13.	4911.699	5700.677	6683.758	4701.292	5596.625	7371.691	4411.101	4303.562	4439.073	4439.073
14.	5606.158	6780.464	7838.839	5422.585	6493.815	8595.767	5115.994	5074.234	5392.87	5392.87

The *eccentric connectivity index* $\xi(N)$ of a graph G is a molecular descriptor defined as (Dureja & Madan, 2007):

$$\xi(N) = \tfrac{1}{2}\, \Sigma_i\, b_{il}\, \varepsilon_i \tag{3.286}$$

where b_{il} gives the number of bonds of atom v_i. Clearly, $b_{ik} = 0$ for $\varepsilon_i < k < M$. Recent papers (Kumar et al., 2004; Ashrafi et al., 2011; Došlić et al., 2010; Došlić et al., 2011) presents various applications of the topological invariant (A4) to the study of chemico-physical properties of crystallographic materials.

Topological indices $W(N)$, $\xi(N)$, ρ and ρ^E have been initially computed on monochromatic chemical graph.

- For further analysis purpose the structural-experimental values of octanol-water partition coefficients ($\log K_{OW}$), retention indices (RI) and molecular weight (MW) are also considered in Table 3.55.

The final test of reliability for our topological-reactivity coloring procedure is to perform the specific correlation between the structure-reactivity features in Table 3.60 and the experimental molecular properties including both physical [molecular weight MW, refractive index RI (Van Den Dool & Kratz, 1963)] and chemical (octanol-water coefficient $\log K_{OW}$) characteristics, which are provided in Table 3.55 from the available literature data (Chemical Book, 2011; Duchowicz et al., 2007; Engel & Ratel, 2007; Song et al., 2003; Zeng et al., 2007; Beernaert, 1979; Oda et al., 1998) along with the extrapolation-interpolation analysis that was performed (Putz et al., 2013b-c) to fill the data gaps. This procedure is in agreement with the current environmental and toxicological interest related to PAHs and to those studied herein. For example, anthracene, benzo-a-pyrene and naphthalene are direct skin irritants while anthracene and benzo(a)pyrene are reported to be skin sensitizers, i.e., cause an allergic skin response in animals and humans (PAH, 1998). In addition, benzo(a)pyrene is reported by the International Agency for Research on Cancer (PAH, 2005) as the most common PAH that causes cancer in animals and was the first chemical carcinogen discovered. For humans, several PAHs were identified as probable carcinogens including benz(a)anthracene, benzo(a)pyrene, benzo(b)fluoranthene, benzo(k)fluoranthene, chrysene, dibenz(ah)anthracene, and indeno(1,2,3-cd)pyrene. However, PAHs must be metabolized

TABLE 3.60 Topological Descriptors Are Provided For All of the Studied PAHs*

No.	N_π	W(N)	W-EL	W-CH	W-EP	ξ(N)	ρ	ρ_E
1.	6	27	3.413	5.458	176.022	36	1	1
2.	10	109	3.504	4.745	439.457	90	1.282	1.471
3.	14	279	3.501	4.0122	737.382	180	1.208	1.485
4.	14	271	3.396	4.309	797.006	170	1.335	1.690
5.	18	569	3.112	3.618	1116.126	298	1.290	1.653
6.	16	362	3.158	3.791	919.923	200	1.331	1.647
7.	22	1011	3.328	3.207	1478.855	448	1.259	1.658
8.	22	963	3.038	3.810	1777.159	424	1.347	1.862
9.	20	654	3.050	3.601	1379.215	286	1.258	1.462
10.	20	680	2.973	3.531	1353.045	318	1.360	1.840
11.	40	4544	2.937	2.744	4272.616	1136	1.321	1.779
12.	24	1002	3.369	3.493	1931.286	366	1.265	1.424
13.	42	4185	3.214	3.236	5574.144	966	1.303	1.588
14.	48	6712	3.172	2.853	6421.297	1412	1.319	1.745

*The number of π-electrons and the Wiener index *W(N)*, as well as its colored reactive forms, such as electronegativity (W-EL), chemical hardness (W-HD), and parabolic reactivity energy (W-E_P), are provided in the table. For comparison, the eccentric connectivity *ξ(N)* and both topological efficiency indices ρ and ρ^E are also included. All of these invariants are computed based on Eqs. (3.180) and (3.181) combined with Eqs. (3.196)–(3.200) and (3.284)–(3.286) (Putz et al., 2013b).

to the reactive diol epoxides to be genotoxic (Putz et al., 2013c; Putz & Putz, 2013b), see also the Section 3.3.9.

The correlation results are summarized in Table 3.61 and illustrated in Figures 3.34–3.36 exhibit the following characteristics (Putz et al., 2013b-c):

- The reactive-parabolic colored Wiener index (WEp) correlates very well with the classical Wiener index (W), as illustrated by Figure 3.34;
- In general, the reactive-parabolic colored Wiener index (WEp) better correlates the structural properties for both topological efficiency indices ρ and ρ^E (erho) and below of the classical Wiener (W) index with respect to the eccentric connectivity *ξ(N)* in Table 3.60 (see Figure 3.35);
- For the physical-chemical parameters in Table 3.60, the reactive-parabolic colored Wiener index (WEp) correlates the retention indices

TABLE 3.61 Synopsis of the Correlations of the Colored Parabolic Wiener (W-Ep) and Classical Wiener (W) Indices With Respect to the Topological-Structural and Experimental-Structural Indices in Tables 3.55 and 3.60, Respectively (Putz et al., 2013b)

Index		**Colored Wiener (W-Ep)**		**Wiener (W)**	
Properties		**Correlation Equation**	**R^2**	**Correlation Equation**	**R^2**
Topo	ξ	$-32.0058+4.55318\,\xi$	0.945958	$-683.175 + 4.88522\,\xi$	0.975885
	ρ	$N_\pi^2\,(8.70213-4.03572\,\rho)$	0.365523	$N_\pi^2\,(2.81505-3.60811\,\rho)$	0.279916
	erho	$N_\pi^2\,(6.42254-1.804\text{ erho})$	0.441993	$N_\pi^2\,(-1.0235 + 1.76758\text{ erho})$	0.406535
Exp	MW	$-1525.11 + 12.5101\text{ MW}$	0.96001	$-2134.64 + 12.8945\text{ MW}$	0.91427
	RI	$-1097.33 + 0.977391\text{ RI}$	0.961193	$-1693.13 + 1.00724\text{ RI}$	0.915063
	$\log K_{OW}$	$N_\pi^2\,(4.5632-0.147653\log K_{OW})$	0.716131	$N_\pi^2\,(0.616014 + 0.171197\log K_{OW})$	0.922355

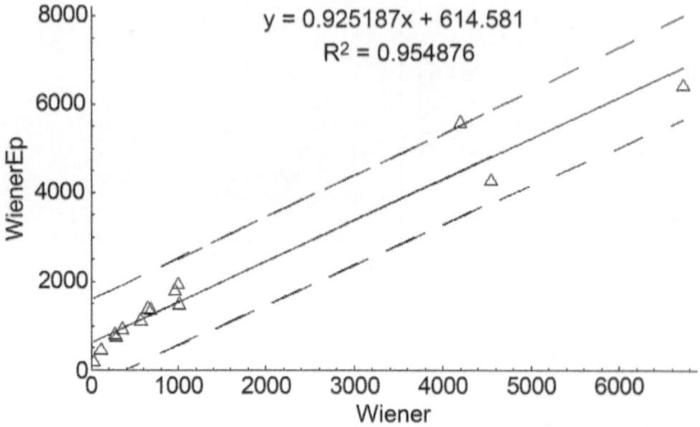

FIGURE 3.34 The linear correlation, with the confidence interval emphasized, between colored parabolic Wiener index (WienerEp) and the classical one (Wiener) for the PAH values in Table 3.60 (Putz et al., 2013b).

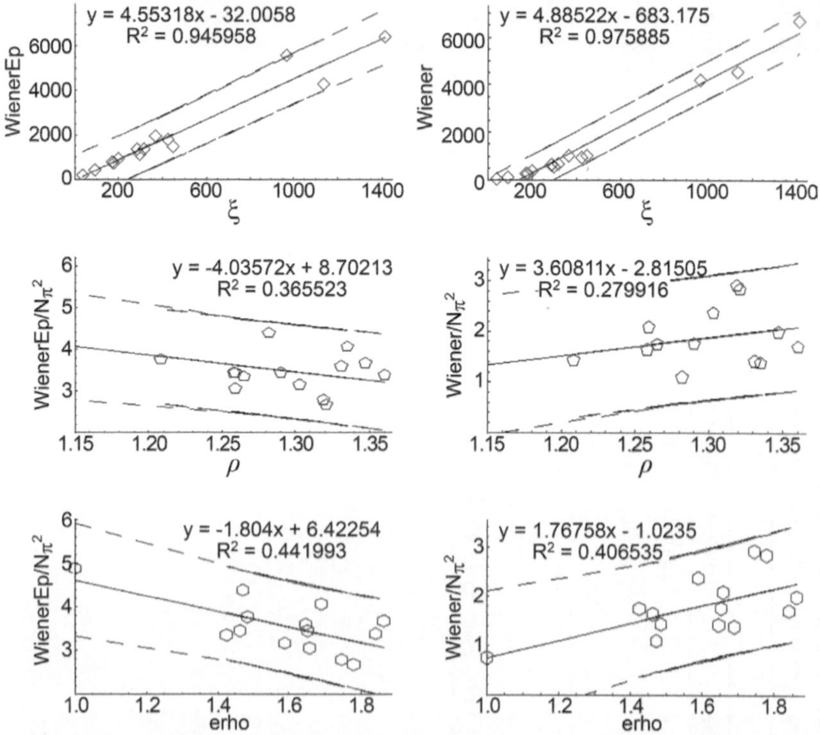

FIGURE 3.35 The linear correlations, with the confidence intervals emphasized, between the colored parabolic Wiener index (WienerEp) and the topo-structural PAH properties in Table 3.60 compared to the same type of correlations for the classical one (Wiener) (Putz et al., 2013b).

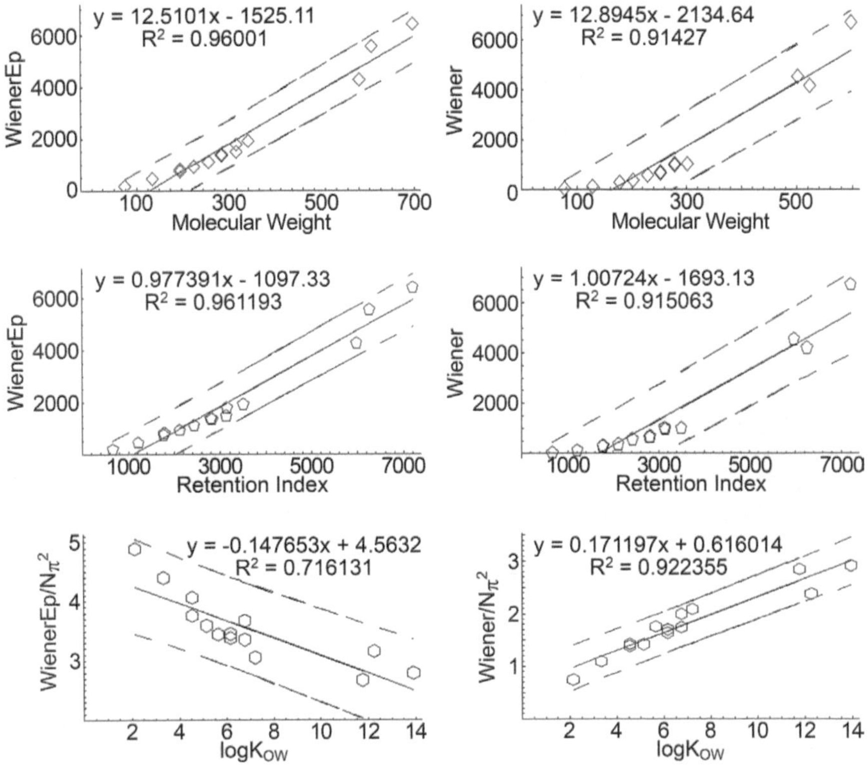

FIGURE 3.36 The linear correlations, with the confidence intervals emphasized, between the colored parabolic Wiener index (WienerEp) and the experimental-structural PAH properties in Table 3.55 compared to the same type of correlations for the classical one (Wiener) in Table 3.60 (Putz et al., 2013b).

(RI) and molecular weight (MW) better than the classical Wiener (W) index but does not correlate the octanol-water partition coefficients (logKOW) as well as the classical Wiener (W) index (see Figure 3.36).

Overall, the present topologic-reactive coloring scheme provides rich structure-stability information for the physical-chemical working indices of molecular systems and of PAHs in particular.

3.3.9 LOGISTIC QSAR (ON INTER-SPECIES TOXICITY)

Identifying chemical hazards without further laboratory testing stays as the most challenging aim of the current in silico methods in Quantitative

Structure Activity Relationship (QSAR); Accordingly, QSAR (Schultz et al., 2002) can minimize the in vivo and animal tests while predicting the safe use of drugs and other chemicals. Such goal mainly relays on predicting activity effects (ecotoxicity, toxicology, including carcinogenicity, etc.) by categorization on the available chemicals and then employing the experimental data through various regression (interpolation, extrapolation, read- across) techniques, see Figure 3.37. In particular, the read-across reveals to be most useful for most applications, since the presence or absence of a property is inferred from the established properties of one or more analogues. There are four general ways of estimating the missing data point when applying quantitative read-across; the present discussion follows (Putz et al., 2012):

(a) by using the endpoint value of a source chemical, for example, the closest analogue in a (sub)category;

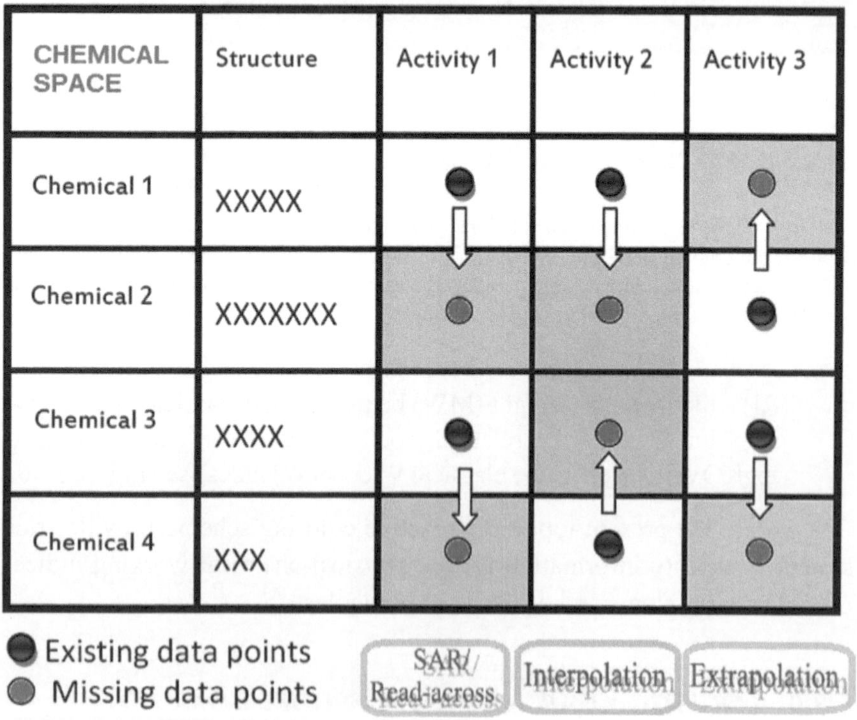

FIGURE 3.37 Graphical representation of a chemical category and some approaches for filling data gap (EU-QSAR, 2011; Putz et al., 2012).

(b) by using an internal QSAR to scale the available experimental results from two or more source chemicals to the target chemical;
(c) by processing the endpoint values from two or more source chemicals (e.g., by averaging, or by taking the most representative value).

However, both categorization process and selection of the analogues are depending on the mechanism used in hazard identification, and goes to assigning of the target chemical(s) to the same hazard category as the source chemical(s). Generically, the critical biological effect of the toxicant at the molecular or cellular level is been called as mechanism of toxic action.

There are some classes of mechanism of toxic action, such as: non-polar narcosis, polar narcosis, weak acid respiratory uncoupling, formation of free radicals, electrophilic reactions, toxic action by specific (receptor-mediated) mechanisms. However, the identification of the mechanism of toxic action is often a difficult task due to the complex nature of toxic activity (Schluz et al., 2003), see Table 3.62. For instance (Putz et al., 2012):

- the narcotic mechanism of toxic action is a result from non-specific non covalent reversible interaction with cell membranes (Schluz et al., 2003); non-polar narcotics, although can be included in the category of narcotic mechanism, are neutral non-reactive compounds (aliphatic alcohols, ketones and ethers), while polar narcotics are aromatic less inert and often posed an hydrogen donor moiety (phenols, anilines) (Veith & Broderius, 2003);
- some chemicals produce their toxic effects by forming free radicals;
- compound that can undergo direct electrophilic interaction may cause covalent charges in biological macromolecule (Hermes, 1990).
- there are also some compounds that may undergo metabolic reaction which result in more toxic form, etc.

Schultz and Cronin (2003) had identified the characteristics for ecotoxicological QSAR: reliable ecotoxicological data, high quality, interpretable and reproducible descriptors, of a number and type consistent with the endpoint being modelled; and the used of a statistical procedure allowing the development of a rigorous and transparent mathematical model.

For establish validity and accuracy criteria for accepting (Q)SARs, guidance and guidelines were needed. Only if regulatory authorities and industry can agree on the acceptable number of false negative and false

TABLE 3.62 Summary of Structural Criteria Used For Classifying Compounds According To Mechanism of Toxic Action (Hermes, 1990; Putz et al., 2012)

Mechanism of Action (MOA)	Structural Features
Non-polar narcosis	Saturated alkanes with, for example, halogen and/or alkoxy substituent (aliphatic alcohols, ketones, ether, amines); halogens and alkyl substituted benzenes
Polar narcosis	Phenols with pK_a greater than or equal to 6.0; phenols and anilines with three or fewer halogen atoms and/or alkyl substituent
Weak acid respiratory uncouples	Phenols and anilines with four or more halogen substituent or more than one nitro group, or single nitro group and more than one halogen group
Formation of free radicals	Phenol or aniline substituted with an electron releasing group (alkoxy, hydroxyl, more than one alkyl group)
Electrophile/proelectrophile	Activated unsaturated compounds; benzene rings without aniline or phenol substructures that have two nitro groups on one ring; phenols with single nitro group but not more than one halogen group; aromatic compounds with two or more hydroxy groups in the *orto* or *para* position and at least one unsubstituted aromatic carbon atom; quinines; aldehydes; compounds with halogens at α-position of an aromatic bond; ketenes; epoxides
Specific mechanism	Chemical interacting with specific biological macromolecules. For example, acetilcholinesterase inhibitors with organophosfate group.

positive results, animal testing could be reduced. But the regulatory authorities are primarily concerned with false negative results because if a (Q) SARs predict too many false negatives, animal testing will be asked for. On the other hand, industry will ask for more animal testing if too many false positives will be. So the validity and the available tools in QSARs have to be explored further (Hulzebos & Posthumus, 2003). Following the same line, Bradbury and co-workers developed a review of conceptual approaches for derivation of QSAR for ecotoxicological effects of organic chemicals. They spoke about the tentative assumptions about the modes of action for narcotics that have evolved from general perturbation of cellular membranes due to a nonspecific partitioning of xenobiotics, to mechanisms that invoke partitioning into specific membrane microsites or hydrophobic pockets of membrane bound proteins (Bradbury et al., 2003).

Hermens and Verhaar review the four classes of chemicals; the present discussion follows (Lacrămă et al., 2008; Putz, 2012b):

Class I, represented by relatively unreactive chemicals with a nonspecific mode of action and they represent the base-line toxicity or nonpolar narcosisi or simple narcosis (Bradbury et al., 2003; Hermens & Verhaar, 1995). *Narcosis* is a reversible state of kept activity of protoplasmic structures, which is a result of exposure to the xenobiotic. Also the terms narcosis and general anaesthesia are used as equivalently in the circumstance of intact organisms. It presumed that, when the xenobiotic blood concentration is at equilibrium with the aqueous exposure concentration the aqueous concentrations of narcotic chemicals are proportional to the concentrations at the site of action (Bradbury et al., 2003). In the class I of chemicals we found aliphatic and aromatic (halogenated hydrocarbons) (Hermens & Verhaar, 1995).

Class II of chemicals, *polar narcotics,* include: aromatics amine, nitroaromatics, anilines, phenols, and pyridines that do not have substituents associated with reactive or uncoupling modes of action. Polar narcotics act with a different molecular mechanism than baseline narcotics (Bradbury et al., 2003; Hermens & Verhaar, 1995).

Class III, the reactive chemicals, include epoxides, aldehydes, aziridines, quinones (generally, all alkylating agents) (Hermens & Verhaar, 1995). These chemicals (or their activated metabolites) react covalently with nucleophilic sites in cellular biomacromolecules (e.g., through nucleophilic substitution, Michael-type addition, or Schiff-base reactions) or gain an oxidative stress through redox cycling to derive toxic effects (Bradbury et al., 2003).

Class IV, the chemicals with a specific mode of action, include organophosphates, chlorinated, pesticides (Hermens & Verhaar, 1995).

Also Bradbury and co-workers review the oxidative phosphorylation uncoupling QSARs, which embody these chemicals that are typically weak acids and are represent by phenols, anilines, and pyridines but, different from polar narcotics, include multiple electronegative groups (i.e., more than one nitro substituents, more than three halogen substituents, or both bonded to the aromatic ring), see for instance the compounds in the series Figure 3.45 below.

Bradbury said that a mode of action domain must be clearly defined in terms of the biological model, endpoint, and exposure contour (Bradbury, 1995). The use of mode of action-based QSARs requires recognition of both: toxic mechanisms and the critical structural characteristics and properties of a chemical that rule its activity by a specific mechanism. The problem is that many chemical QSAR classes that historically were associated with a baseline narcosis include chemicals that act via a baseline narcosis mode of action, as well as chemicals that act through an electrophilic-based mode of action.

On the contrary, chemical classes not usually identified as acting by a *baseline narcosis* mode of action, such as the phenols, include chemicals that act through *baseline narcosis*, *polar narcosis*, *oxidative phosphorylation uncoupling*, or *electrophilic-based* modes of action (Bradbury et al., 2003).

One approach does develop QSAR equation is that of using congener series of chemicals. The initial research in the field of QSARs for ecological risk assessments predicated on the supposition chemicals from the same chemical class should behave in a toxicologically similar manner. Accordingly, homologous series of chemicals were used to develop structure–toxicity relationships and the premise was made that toxic effects were transmitting by common structural components used in chemical class task. Congener-derived QSARs are restricted because of the short structural domain of which they are grounded (Schultz & Cronin, 2003; Bradbury et al., 2003).

Further, the attention was to development of QSARs on the basis of mechanisms of action, from narcosis to reactive mode of action that, nevertheless is often a difficult task (Netzeva et al., 2004a).

The assumption was made that potency varied with chemical uptake, which correlated with the hydrophobicity of substituent moieties within the chemical class. Their opinion is that that the power of chemicals that act through mechanisms other than narcosis is, in part, related to their partitioning into hydrophobic biological compartments, and also is a fundamental principle for QSARs. Thus, octanol–water partition coefficients are typically working in QSARs as the chemical descriptor that comprise that aspect of variability in toxicity across chemicals that is exclusively characteristic to varying degrees of hydrophobicity, and uptake, of xenobiotics (Bradbury et al., 2003).

Biological activity for a given molecule (available or newly synthesized) interacting with a certain species is not always known or easily produced; determining these interactions can require extensive laboratory efforts and adherence to many environmental safety conditions.

The *fill-in-the-data-gaps* (FITDAGA) technique mainly features (OECD-QSAR, 2013):

- Giving a target molecule (existing or in silico designed) it provides the existing analogues whose experimental data are available respecting a studied "end-point" (activity to be modeled);
- In other words, it fills data required for the target molecule either by read-across (analogue approach) or by trend analysis (molecular similarity approach);
- All these implies an inner QSAR modeling specific with the profiled activity (ligand-receptor) hypothesized or known binding mechanism involving all available data (analogues and their endpoint measured effects).

The analogues are understood as chemicals with common behavior, toxicological effects, or whatever; Examples of *endpoints* may be (Putz et al., 2012):

- Sensitization potential for an untested compound (in cosmetics, for instance);
- The acute toxicity of the target molecule(s) to some species (daphnids, for instance, in ecotoxicology);
- The Ames mutagenicity potential, etc.

The analogues that are associated with the envisaged endpoint needs a categorization that includes the target molecule(s) too; *the category* can be, accordingly defined as (Putz et al., 2012):

- By the mechanistic grouping, for example, by protein binding, specific to all chemicals in that category in which case the mechanistic grouping is implemented, or by OASIS acute aquatic mechanism of action – MOA; note that, for Toolbox application, protein binding way of profiling stays as one of most realizable grouping method, very

suitable for modeling skin and respiratory sensitization, acute aquatic and inhalation toxicity, but also for chromosomal aberration; protein binding mechanism is built on conventional organic chemical mechanisms and relays on the electrophilic interaction mechanism of various structural functionalities (haloalkanes, isocyanates, isothiocyanates, diketones, aldehydes, acyl halides, alkyl sulfates, sulfonates, etc.);
- By structural grouping (e.g., by EcoSAR platform), reactivity grouping (e.g., Verhaar categorization), or by molecular similarity – in particular by organic functional groups;
- Refining by sub-category may be also considered.

While *trend analysis* simple scaling from the measured values to fill in the data gaps for a target molecule over a category relating with molecular mass, carbon chain length or other physical-chemical properties with a common mechanism, the *read-across* approach involves the category of a pool of selected analogues which may be regarded as more specific for the target molecule. It follows that the so-called *profiling* the target molecule vis-à-vis by its envisaged or researched molecular mechanism is of crucial importance in FITDATA; however, worth emphasizing that profiling is associated with the mechanism of action, and not to the endpoint, i.e., it does not refers to the environmental fate, ecotoxicity and toxicity data. For instance, while referring to mutagenesis it means (Putz et al., 2012):

- Changes in nucleotide sequence by base-substitutions or by small deletions/insertions (the so-called frame-shifts) – generically known as “point mutations”;
- It is triggered by mutagens that create DNA damage;
- It is a process at the cellular level, requiring enzyme and/or DNA replication (therefore, cells create mutations and not the mutagens!);
- It corresponds with the Ames mutagenesis scheme and endpoint based on strains of *Salmonella typhimurium*.

Unlike the profiling, the endpoint are fully corresponding to the toxicity data available (and stored in databases, Toolbox included, for example, OASIS Skin Sensitization – for cosmetic toxicity; ECETOX, ECOTOX, OASIS Aquatic – for ecotoxicology; ISSCAN Genotox or OASIS Genotox – for mutagenicity); however gathering the analogues data in endpoint databases will be driven by the envisaged/hypothesis profiling previously assumed.

Still, when *fill-in-the-data-gaps* are necessary the feed-back with profiling category follows in order to effectively select the analogs to be used in trend/read-across analysis. So the further selection of the category for analogs' data gathering should overlap with the endpoint output, i.e., having the same mechanism as common information, for profiling/categorization and for the way the observed effect is produced at chemical-biological interaction level.

Nevertheless, it is this point, i.e., after having the pool of analogs-of-the-target-molecule detected, when further chose the type of analysis (trend or read-across) as well as the specific endpoint (the envisaged target molecule's activity or on certain species) may be set. Possible refinements may apply as well, for instance (Putz et al., 2012):

- the reported outliers (e.g., lone negative results in the Ames tests indicates non-activity/non-mutagenic potential and should be removed) that are subject of deleting (pruning) from the category;
- sub-categorization process, by adding additional profilers (structural or mechanistically properties) and eventually eliminating the identified chemicals with different sub-profiling out from the working category; this possibility is based on dissimilarities that can appear respecting the target chemical, i.e., considering the substance type (e.g., discrete chemicals, mixtures – hydrolyzing chemicals, polymers, inorganics, organometallics) or the mode of action (e.g., in OASIS categorization the analogues of target molecules are in the group of phenols and anilines), and allows selection or removal from initial list of analogues.

All in all, the working flow for FITDATA is schematically illustrated on the Figure 3.38 (Putz et al., 2012).

Turning on specific case studies, worth recalling that the emitted polycyclic aromatic hydrocarbons (PAHs) can bind to particles in the air. Particle size depends in part on the source of the PAHs, while ambient air PAH concentrations show seasonal variation (Bostrom et al., 2002; IPCS, 1998). PAHs are found in meat and in other foods as a result of smoking, grilling, broiling, or other high-temperature processing. Uncooked foods and vegetables also contain low levels of PAHs but can be contaminated by airborne particle deposition or growth in contaminated soil. Humans are usually exposed to PAH mixtures rather than to individual chemicals, and PAH mixture composition varies with the combustion source and temperature (ATSDR, 1995).

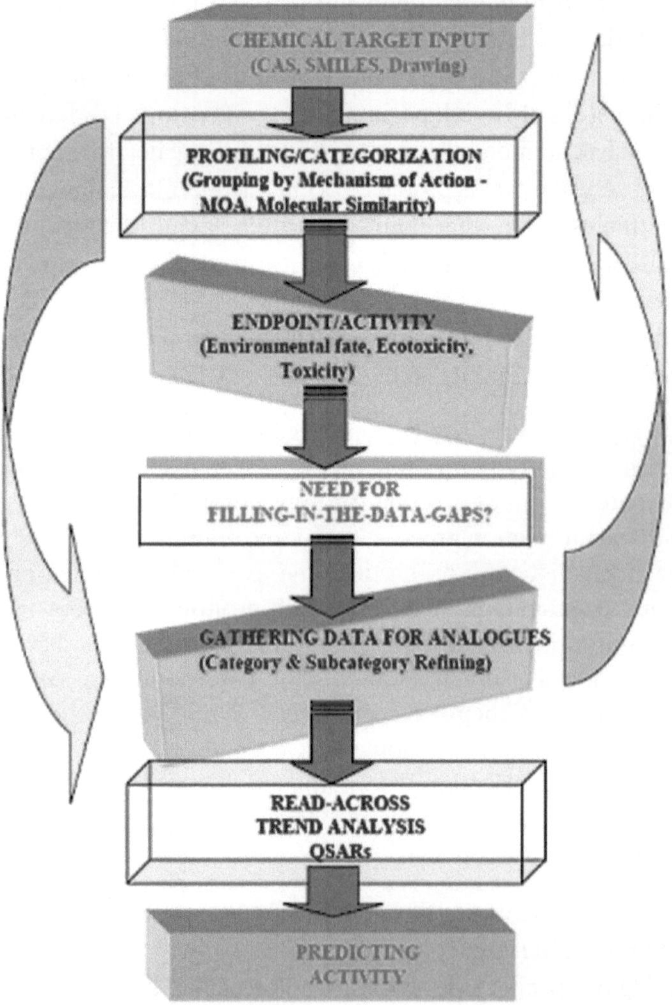

FIGURE 3.38 The *in-cerebro* scheme for *in-silico* evaluation of the environmental or toxicological activities for a given target chemical (under concern or newly designed or synthesized), following and explicating the Toolbox QSAR computational facility, after (OECD-QSAR, 2013; Putz et al., 2012).

ClPAHs are hybrids of dioxins and PAHs suspected of having similar toxicities (Ohura, 2007) and are generally known to be carcinogenic, mutagenic, and teratogenic, with greater mutagenicity, aryl-hydrocarbon receptor activity, and dioxin-like toxicity than the corresponding parent

PAHs (Kitazawa et al., 2006). Especially at the DNA interaction level, ClPAHs have the ability to bind to and activate the aryl hydrocarbon receptor (AhR), a cytosolic, ligand-activated transcription receptor.

The biological pathway involves translocation of the activated AhR to the nucleus. In the nucleus, the AhR binds to the AhR nuclear translator protein to form a heterodimer, leading to transcriptional modulation of genes and causing adverse changes in cellular processes and function (Blackenship et al., 2000). Several ClPAHs have been determined to be AhR-active. AhR-mediated toxicity (Figure 3.39) is activated by all embryotoxic HAH and PAH congeners (Scott, 2009; Denison & Heath-Pagliuso, 1998; Hahn, 1998) towards nuclear translocation, where the AhR heterodimerizes with the AhR nuclear translocator (ARNT).

The resulting ligand-AhR-ARNT complex further combines with various co-activators and promotes their expression through interacting in the promoter region of AhR-regulated genes with xenobiotic responsive elements (XREs) (Bock & Köhle, 2006). This activity plays a role in cell proliferation and differentiation (Nebert et al., 2000) and contributes to the biotransformation of xenobiotics, while also having a functional role in normal development and homeostasis (Fernandez-Salguero et al., 1997; Lund et al., 2003; Walisser et al., 2004). Furthermore, the role of the AhR in TCDD (2,3,7,8-Tetrachlorodibenzo-p-Dioxin) and unalkylated PAH toxicity has been assessed (Incardona et al., 2004, 2006), but its role in alkylated PAH toxicity has not (Scott, 2009).

Within this context, the present work unfold how the read-across prediction for target molecules may be unfolded by using the recent launched ToolBox Software delivered by OECD, with application to the challenging chlorinated polycyclic aromatic hydrocarbons (Cl-PAHs) paradigmatic structures at inter-species level of *Pymephales promelas* and *Rattus Norvegicus*.

The needed ecotoxicological data (bioaccumulation, toxicity, carcinogenicity) for the two additional species considered here (*Pymephales promelas* and *Rattus Norvegicus*) was provided through the input of the selected Cl-PAHs as chemical targets into the Toolbox in silico environment, with the results of Table 3.63, while their inter-activity correlations are reported in Figure 3.40. The results of Figure 3.40 lead with the worthy observations:

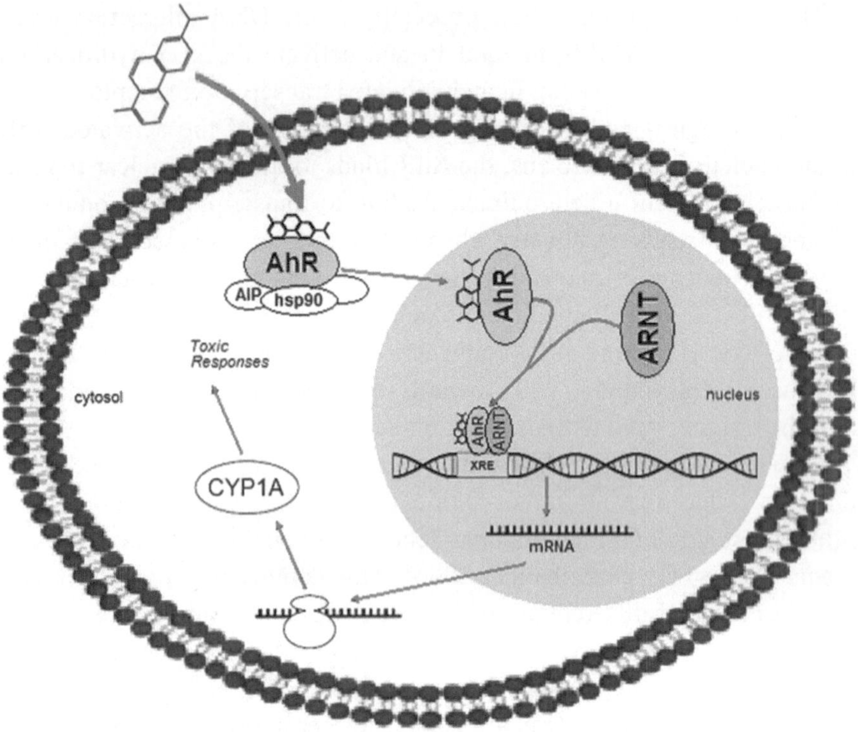

FIGURE 3.39 Generic mechanism of AhR-mediated toxicity (Putz & Putz, 2013b): AhR mediates signal transduction by dioxin-like ligands, which form a transcription factor complex with an aryl hydrocarbon nuclear translocator protein (ARNT). This heterodimer recognizes specific DNA sequences, namely dioxin responsive elements (DREs), and leads to induction of several genes forming the so-called Ah gene battery. In this process, the elevated levels of the protein products are assumed to be involved in the toxic action of AhR ligands—AIP: AhR inhibitory protein; hsp90: 90 kDa heat shock protein; ARNT: AhR nuclear translocator; XRE: xenobiotic response element; CYP1A: cytochrome P450 1A gene/protein; adapted from Refs. (Safe, 1990; Scott, 2009).

- the ecological fate (bioaccumulation) is more related with ecotoxicology of the same species against Cl-PAHs;
- human and fish inter-species activities are more likely than human-rat inter-activity correlation, respecting the Cl-PAHs environmental hazard;
- although with general very low values, environmental fate rather than ecotoxicity of *Pimephales* is more closely related with rat's carcinogenicity.

TABLE 3.63 The Chlorinated Polycyclic Aromatic Hydrocarbons (ClPAH) Activities As: Bioaccumulation Aquatic, Aquatic Toxicity, and Carcinogenity On Rats (Putz et al., 2012); While the EROD (ethoxyresorufin-*O*-deethylase) Activity Is Taken From Literature As the Relative Intensity of ClPAH-Induced Cytochrome P450 (CYP) Activity In Human Breast Cancer MCF-7 Cells (Ohura et al., 2010)

Cl-PAH	Structure CAS (SMILES)	Activity			
		EROD	**Environmental Fate: [Bioaccumulation aquatic] on Pimephales Promelas 96h [10^3 L/kg wet]**	**Ecotoxicological Information: [Aquatic Toxicity] LC50 on Pimephales Promelas 96h [10^{-1} mg/L]**	**Carcinogenity on Rats TD50 [10^{-3} mol/kg]**
9-chloro phenanthrene (9-ClPhe)	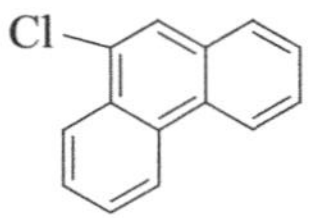 947–72–8 Clc2cc3c(c1c2c ccc1)cccc3	1.2	1.90	3.41	4.72
9,10-dichloro phenanthrene (9,10-Cl$_2$Phe)	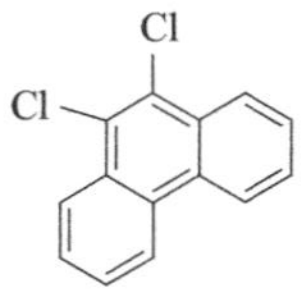 17219–94–2 Clc2c3c(c1c(ccc c1)c2Cl)cccc	1.4	1.74*	1.3	4.05

TABLE 3.63 Continued

Cl-PAH	Structure CAS (SMILES)	Activity			
		EROD	Environmental Fate: [Bioaccumulation aquatic] on Pimephales Promelas 96h [10^3 L/kg wet]	Ecotoxicological Information: [Aquatic Toxicity] LC50 on Pimephales Promelas 96h [10^{-1} mg/L]	Carcinogenity on Rats TD50 [10^{-3} mol/kg]
3,9,10-trichloro phenanthrene (3,9,10-Cl$_3$Phe)	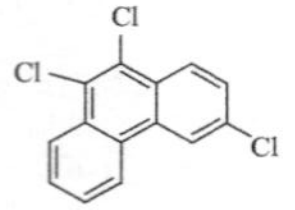 800409–57–8 Clc3ccc2c(Cl)c (Cl)c1ccccc1c2c3	4.4	4.22*	0.828**	3.84**
1-chloropyrene (1-ClPy)	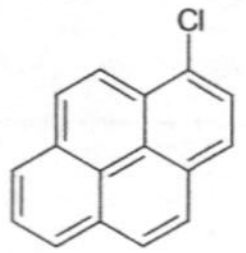 34244–14–9 Clc4ccc2ccc1cc cc3c1c2c4cc3	1.3	4.08*	1.3	0.00511

TABLE 3.63 Continued

Cl-PAH	Structure CAS (SMILES)	Activity			
		EROD	**Environmental Fate: [Bioaccumulation aquatic] on Pimephales Promelas 96h [10^3 L/kg wet]**	**Ecotoxicological Information: [Aquatic Toxicity] LC50 on Pimephales Promelas 96h [10^{-1} mg/L]**	**Carcinogenity on Rats TD50 [10^{-3} mol/kg]**
6-chlorobenzo[a] pyrene (6-ClBaP)	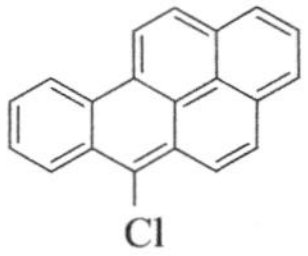 21248–01–1 Clc4c5c(c1ccc3 c2c1c4ccc2ccc3) cccc5	9	4.22*	0.33**	3.46**

*Predicted by Filling-in-the Data-Gap by Toolbox OECD facility [v.1.1.01/2009] with OASIS basesurface narcotis;

**By DNA binding of PAHs.

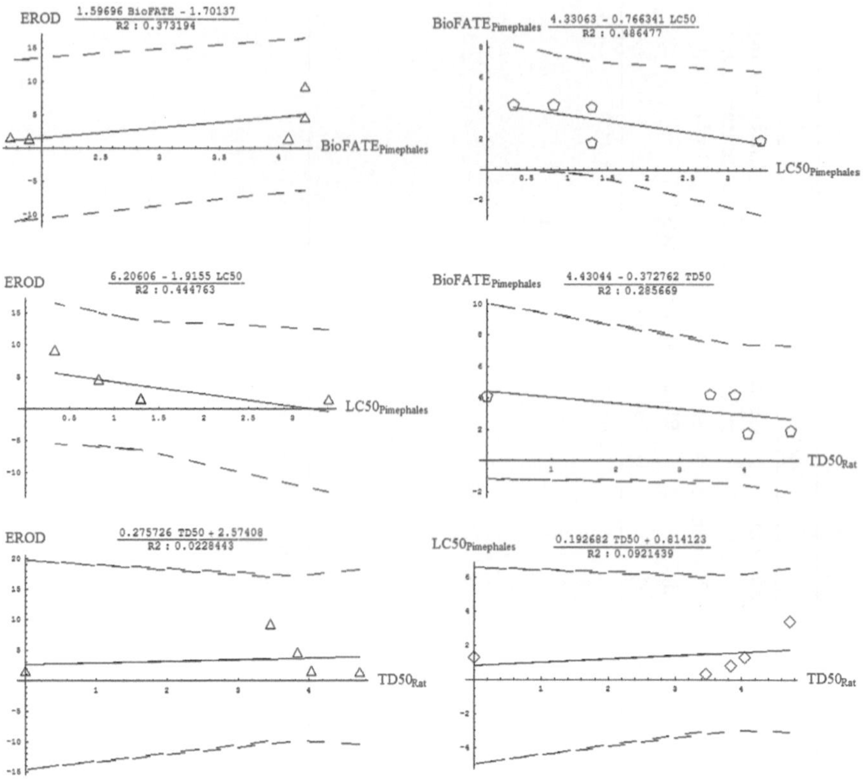

FIGURE 3.40 The inter-activity correlations (with the confidence intervals highlighted) for the data species of Table 3.63 (Putz et al., 2012).

The connection between the activity data of Table 3.63 and the molecular structural-frontier information from Tables 3.64–3.67 is made based on the *biological activity driven-chemical reactivity* algorithm, which is qualitatively presented in the scheme of Figure 3.14 and quantitatively represented by the logistic ligand progress curves of Eq. (3.221). It is clear that the present chemical-biological interaction, (ClPAH molecule-AhR-mediated toxicity, see Figure 3.39) is a specific realization of the generic ligand-receptor kinetics, which is modeled quantum mechanically and mostly using DFT methods.

The interaction involves the predicted norm of the respective chemical structure-biological activity correlation through the presence of the predicted initial (in time evolution of ligand-receptor kinetics) bound ligand to the receptor site, see Eq. (3.204), as well as the algebraic norm

TABLE 3.64 The Values of Electronegativity (in eV) For the ClPAHs of Table 3.63 Computed From the Frontier-Like Formula (3.160) With the HOMO and LUMO Frontier Orbital Energies That Were Evaluated in Various Quantum Mechanical Frameworks (Putz & Putz, 2013b): No-Exchange-No-Correlation (X0C0) to Hartree-Fock (HF), and Specific DFT Becke's Exchange-Correlation Forms (Becke97, Becke88-VWN, B3-PW91, B3-LYP)*

χ			***Quantum Chemical Framework***			
Cl-PAH	***X0C0***	***HF***	***Becke97***	***Becke88-VWN***	***B3-PW91***	***B3-LYP***
(I)	251.1668	161.6397	190.1459	191.0285	190.8256	190.7167
(II)	161.12	122.6308	155.8952	129.6772	155.9582	176.9095
(III)	16.58217	156.6598	78.37052	30.02489	233.2496	78.76642
(IV)	115.2805	62.72652	199.2066	220.7729	192.2411	194.9907
(V)	163.9368	256.0867	88.00236	98.01646	215.249	215.0238

*See the Volume I/Chapter 4 of the present five-volume book set (Putz, 2016a) and Putz & Mingos (2013), with large Gaussian basis functions (6–31G**) within HyperChem software (Hypercube, 2002).

TABLE 3.65 The Values of Chemical Hardness (in eV) For the ClPAHs of Table 3.63 Computed From the Frontier-Like Formula (3.163) With HOMO and LUMO Frontier Orbitals' Energies Evaluated As Explained in Table 3.64 (Putz & Putz, 2013b)

η			***Quantum Chemical Framework***			
Cl-PAH	***X0C0***	***HF***	***Becke97***	***Becke88-VWN***	***B3-PW91***	***B3-LYP***
(I)	1.183525	0.28125	2.195176	2.39061	2.329193	2.317177
(II)	2.44944	0.231994	0.448074	0.690163	0.385002	0.286087
(III)	0.065345	1.273979	1.074822	1.177084	0.214798	0.864036
(IV)	0.065396	0.389016	0.510063	0.298729	0.235435	0.381012
(V)	0.183769	0.43338	0.284748	0.567326	2.196144	1.941712

TABLE 3.66 The Values of Chemical Power For the ClPAHs of Table 3.63 Computed Using Definition (3.165) Applied to the Electronegativity and Chemical Hardness Values From Tables 3.60 and 3.61 (Putz & Putz, 2013b)

π			***Quantum Chemical Framework***			
Cl-PAH	***X0C0***	***HF***	***Becke97***	***Becke88-VWN***	***B3-PW91***	***B3-LYP***
(I)	106.1096	287.3595	43.30996	39.95393	40.96389	41.15281
(II)	32.88916	264.2969	173.9615	93.9468	202.542	309.1883
(III)	126.8827	61.48444	36.45746	12.75393	542.9511	45.58049
(IV)	881.4097	80.62214	195.2765	369.5204	408.2671	255.8852
(V)	446.0405	295.4529	154.5267	86.38468	49.00613	55.36966

TABLE 3.67 The Values of Electrophilicity for ClPAHs of Table 3.63 Computed Using Definition (3.168) Applied To the Electronegativity and Chemical Hardness Values From Tables 3.60 and 3.61 (Putz & Putz, 2013b)

ω	*Quantum Chemical Framework*					
Cl-PAH	***X0C0***	***HF***	***Becke97***	***Becke88-VWN***	***B3-PW91***	***B3-LYP***
(I)	26651.22	46448.71	8235.211	7632.341	7816.958	7848.526
(II)	5299.102	32410.95	27119.76	12182.76	31588.08	54698.35
(III)	2103.989	9632.138	2857.19	382.9353	126643.2	3590.212
(IV)	101609.3	5057.147	38900.38	81580.09	78485.72	49895.23
(V)	73122.46	75661.56	13598.72	8467.12	10548.52	11905.79

evaluation of the specific EC50, see Eq. (3.205), for each observed or recorded (experimentally or computationally by filling in the data gaps – see Figure 3.38) set of activities for the molecules of interest. At this point, one should note that the employed activities for bioaccumulation in *Pimephales promelas* (P.p.), ecotoxicology of P.p., and carcinogenity in rats shown in Table 3.63 were in fact the correspondent 50% read-across concentrations for aimed effects obtained using ToolBox in an *in silico* environment.

However, there turns out that when considering the EC50 and then extracting the activity relationships from the logarithmic forms, in each case, no significant distinction between the influences of the reactivity indices on the bio-and eco-toxicology activity correlation were recorded. Instead, the relationship cancels out all chemical information or mechanisms in producing biological effects.

This should be avoided (see, QSAR-OECD-5 principle of Section 3.2.5), so we consider the 50% read-across concentrations as the aimed effects and note this as peculiar *in silico* behavior for ToolBox that should be improved in the future. The biological-driving-chemical interaction results for ClPAHs-AhR-mediated toxicity for human breast cancer MCF-7 cells, aquatic bioaccumulation for *P.p.*, aquatic toxicity for *P.p.* and carcinogenity for rats are summarized in Tables 3.68–3.71 and in Figures 3.41–3.44; the present discussion also follows (Putz & Putz, 2013b).

This way, one can systematically identify which reactivity index (and correspondent principle thereof) is dominant in which quantum/DFT computational environment for the biological or ecological system

TABLE 3.68 The Synopsis of the Quantitative (χ, η, π, ω) Reactivity Indices-EROD Activity Relationships (EROD-QRAR), Including the Correlation Factor (R) and the Initial Predicted Concentration of Eq. (3.204), $L_0 = \exp\left(-\left\|\left|A^*\right\rangle\right\|\right)$ [μM], For Each Index/Quantum Chemical Method Considered in Tables 3.64–3.67 Against the EROD Activity of Table 3.63 (Putz & Putz, 2013b)

Quantum-method	Index	$\|A\rangle^*$ *FATE – QSAR – EQUATION*	*R*	L_0
X0C0	χ	$4.55\|1> - 7.7\times10^{-3}\|\chi>$	0.195	390.622
	η	$4.7\|1> - 1.57\|\eta>$	0.483	225.796
	π	$3.22\|1> + 7.63\times10^{-4}\|\pi>$	0.08	428.397
	ω	$2.83\|1> + 1.51\times10^{-5}\|\omega>$	0.197	389.811
HF	χ	$-2.78\|1> + 0.041\|\chi>$	0.856	64.0455
	η	$2.289\|1> + 2.243\|\eta>$	0.284	345.096
	π	$2.27\|1> + 6.02\times10^{-3}\|\pi>$	0.208	384.824
	ω	$0.96\|1> + 7.39\times10^{-5}\|\omega>$	0.631	146.062
Becke97	χ	$10.38\|1> - 0.049\|\chi>$	0.813	75.8095
	η	$5.08\|1> - 1.80\|\eta>$	0.417	265.837
	π	$3.39\|1> + 5.44\times10^{-4}\|\pi>$	0.012	436.279
	ω	$5.15\|1> - 9.32\times10^{-5}\|\omega>$	0.406	272.73
Becke88-VWN	χ	$6.99\|1> - 0.026\|\chi>$	0.592	165.435
	η	$4.67\|1> - 1.18\|\eta>$	0.288	342.959
	π	$4.33\|1> - 7.21\times10^{-3}\|\pi>$	0.306	332.855
	ω	$4.299\|1> - 3.8\times10^{-5}\|\omega>$	0.378	289.6
B3-PW91	χ	$-10.75\|1> + 0.072\|\chi>$	0.62	151.004
	η	$2.15\|1> + 1.22\|\eta>$	0.394	279.627
	π	$4.23\|1> - 3.1\times10^{-3}\|\pi>$	0.203	386.758
	ω	$3.83\|1> - 7.22\times10^{-6}\|\omega>$	0.109	421.53
B3-LYP	χ	$2.73\|1> + 4.25\times10^{-3}\|\chi>$	0.067	430.681
	η	$1.79\|1> + 1.44\|\eta>$	0.395	279.377
	π	$5.40\|1> - 0.014\|\pi>$	0.53	198.381
	ω	$5.24\|1> - 6.97\times10^{-5}\|\omega>$	0.508	211.162

(Putz & Mingos, 2013). As such, for the considered systems one finds the following (Putz & Putz, 2013b):

- For modeling the interaction of ClPAH ligands that bind human breast cancer MCF-7 cells (Table 3.68, Figure 3.41), it appears that

TABLE 3.69 The Same Synopsis As in Table 3.68 But For the FATE/Pp Activity of Table 3.63 (Putz & Putz, 2013b)

Quantum-method	Index	$\|A\rangle^*$ *FATE – QSAR – EQUATION*	*R*	L_0
X0C0	χ	$4.68\| 1> - 0.01\| \chi>$	0.677	589.933
	η	$4.132\| 1> - 1.139\| \eta>$	0.918	497.636
	π	$2.511\| 1> + 2.26\times10^{-3}\| \pi>$	0.617	610.868
	ω	$2.597\| 1> + 1.52\times10^{-5}\| \omega>$	0.517	643.122
HF	χ	$2.75\| 1> + 3.17\times10^{-3}\| \chi>$	0.173	716.799
	η	$2.31\| 1> + 1.77\| \eta>$	0.586	621.244
	π	$4.53\| 1> - 6.56\times10^{-3}\| \pi>$	0.592	619.391
	ω	$3.45\| 1> - 6.52\times10^{-6}\| \omega>$	0.145	719.657
Becke97	χ	$4.94\| 1> - 0.012\| \chi>$	0.523	641.171
	η	$3.9\| 1> - 0.744\| \eta>$	0.450	662.262
	π	$3.03\| 1> + 1.64\times10^{-3}\| \pi>$	0.095	723.676
	ω	$3.28\| 1> - 2.53\times10^{-6}\| \omega>$	0.029	726.44
Becke88-VWN	χ	$4.02\| 1> - 5.86\times10^{-3}\| \chi>$	0.343	688.379
	η	$4.07\| 1> - 0.817\| \eta>$	0.524	641.053
	π	$2.91\| 1> + 2.66\times10^{-3}\| \pi>$	0.295	698.247
	ω	$2.99\| 1> + 1.1\times10^{-5}\| \omega>$	0.286	699.851
B3-PW91	χ	$-3.71\| 1> + 0.035\| \chi>$	0.793	546.71
	η	$3.48\| 1> - 0.231\| \eta>$	0.195	714.148
	π	$2.51\| 1> + 2.91\times10^{-3}\| \pi>$	0.501	647.901
	ω	$2.52\| 1> + 1.4\times10^{-5}\| \omega>$	0.550	632.856
B3-LYP	χ	$4.165\| 1> - 5.45\times10^{-3}\| \chi>$	0.226	709.856
	η	$3.37\| 1> - 0.123\| \eta>$	0.088	724.144
	π	$3.65\| 1> - 2.94\times10^{-3}\| \pi>$	0.296	697.942
	ω	$3.596\| 1> - 1.42\times10^{-5}\| \omega>$	0.271	702.58

TABLE 3.70 The Same Synopsis As in Table 3.68 But For the Aqua-Toxicity/Pp Activity of Table 3.63 (Putz & Putz, 2013b)

Quantum-method	Index	$\|A\rangle^*$ *FATE – QSAR – EQUATION*	R	L_0
X0C0	χ	$0.132\| 1> + 9.19\times10^{-3}\| \chi>$	0.668	28156.8
	η	$1.11\| 1> + 0.407\| \eta>$	0.360	36315.3
	π	$1.74\| 1> - 9.7\times10^{-4}\| \pi>$	0.291	37716.6
	ω	$1.67\| 1> - 5.66\times10^{-6}\| \omega>$	0.211	39011.7

TABLE 3.70 Continued

Quantum-method	Index	$\vert A\rangle$* *FATE – QSAR – EQUATION*	R	L_0
HF	χ	$2.07\vert 1> - 4.19\times10^{-3}\vert \chi>$	0.250	38418.2
	η	$1.99\vert 1> - 1.06\vert \eta>$	0.386	35736.7
	π	$0.856\vert 1> + 2.92\times10^{-3}\vert \pi>$	0.289	37739.1
	ω	$1.46\vert 1> - 7.63\times10^{-7}\vert \omega>$	0.019	40521.2
Becke97	χ	$-0.62\vert 1> + 0.014\vert \chi>$	0.692	27454.6
	η	$0.234\vert 1> + 1.329\vert \eta>$	0.883	21941.5
	π	$2.268\vert 1> - 6.91\times10^{-3}\vert \pi>$	0.442	34392.8
	ω	$1.63\vert 1> - 1.07\times10^{-5}\vert \omega>$	0.134	39909.4
Becke88-VWN	χ	$0.239\vert 1> + 8.92\times10^{-3}\vert \chi>$	0.574	30855.6
	η	$0.215\vert 1> + 1.189\vert \eta>$	0.837	23264.0
	π	$1.56\vert 1> - 1.05\times10^{-3}\vert \pi>$	0.128	39968.1
	ω	$1.47\vert 1> - 1.56\times10^{-6}\vert \omega>$	0.044	40464.4
B3-PW91	χ	$4.24\vert 1> - 0.014\vert \chi>$	0.352	36491.9
	η	$0.99\vert 1> + 0.414\vert \eta>$	0.384	35785.8
	π	$1.92\vert 1> - 1.94\times10^{-3}\vert \pi>$	0.367	36155.2
	ω	$1.87\vert 1> - 8.46\times10^{-6}\vert \omega>$	0.366	36182.4
B3-LYP	χ	$0.801\vert 1> + 3.69\times10^{-3}\vert \chi>$	0.168	39563.0
	η	$0.814\vert 1> + 0.535\vert \eta>$	0.420	34927.0
	π	$1.606\vert 1> - 1.22\times10^{-3}\vert \pi>$	0.135	39900.5
	ω	$1.579\vert 1> - 5.683\times10^{-6}\vert \omega>$	0.119	40042.5

TABLE 3.71 The Same Synopsis As In Table 3.68 But For the Carcinogenity/Rats Activity of Table 3.63 (Putz & Putz, 2013b)

Quantum-method	Index	$\vert A\rangle$* *FATE – QSAR – EQUATION*	*R*	L_0
X0C0	χ	$2.279\vert 1> + 0.61\times10^{-3}\vert \chi>$	0.305	691.13
	η	$2.539\vert 1> + 0.856\vert \eta>$	0.481	607.539
	π	$4.79\vert 1> - 4.94\times10^{-3}\vert \pi>$	0.940	339.57
	ω	$4.639\vert 1> - 3.41\times10^{-5}\vert \omega>$	0.808	415.336
HF	χ	$0.872\vert 1> + 0.015\vert \chi>$	0.585	548.196
	η	$3.022\vert 1> + 0.369\vert \eta>$	0.085	749.612
	π	$1.37\vert 1> + 9.3\times10^{-3}\vert \pi>$	0.585	548.231
	ω	$2.143\vert 1> + 3.17\times10^{-5}\vert \omega>$	0.493	600.768

TABLE 3.71 Continued

Quantum-method	Index	$\|A\rangle^*$ *FATE – QSAR – EQUATION*	*R*	L_0
Becke97	χ	4.918| 1> – 0.012| χ>	0.365	665.551
	η	2.187| 1> + 1.129| η>	0.481	607.539
	π	5.113| 1> – 0.016| π>	0.639	516.73
	ω	4.994| 1> – 9.81×10⁻⁵| ω>	0.778	432.813
Becke88-VWN	χ	4.79| 1> – 0.012| χ>	0.482	606.849
	η	1.68| 1> + 1.49| η>	0.667	499.366
	π	4.712| 1> – 0.012| π>	0.960	328.542
	ω	4.38| 1> – 5.28×10⁻⁵| ω>	0.957	330.275
B3-PW91	χ	3.28| 1> – 3.42×10⁻⁴| χ>	0.005	754.817
	η	2.37| 1> + 0.79| η>	0.464	616.591
	π	4.134| 1> – 3.69×10⁻³| π>	0.443	627.51
	ω	3.866| 1> – 1.28×10⁻⁵| ω>	0.350	672.034
B3-LYP	χ	4.55| 1> – 7.8×10⁻³| χ>	0.225	719.275
	η	2.019| 1> + 1.03| η>	0.515	588.817
	π	4.172| 1> – 6.76×10⁻³| π>	0.476	610.099
	ω	4.244| 1> – 4.02×10⁻⁵| ω>	0.535	577.469

the B3-PW91 DFT exchange-correlation scheme recovers the consecrate chemical reactivity scheme (electronegativity > chemical hardness > chemical power > electrophilicity), which may called the I-BioAct-ChemReact scheme of biological activity-driven chemical reactivity realization. Almost the same situation appears for Becke97 computations, except for an inversion of chemical power with electrophilicity influence; in both cases, the prescribed reactivity hierarchy affirms the EE first, followed by chemical hardness principles (see Sections 3.2.9.1 and 3.2.9.2), while their combined equalization is naturally released with electrophilicity and then with chemical power equalizations at the level of ligand-receptor site molecular interactions

$$\textit{\textbf{I-BioAct-ChemReact}}\text{: } (\chi\wedge\eta)>(\pi\wedge\omega) \quad (3.287)$$

- The second scheme of biological-chemical interaction appears when either chemical power or electrophilicity indices and principles

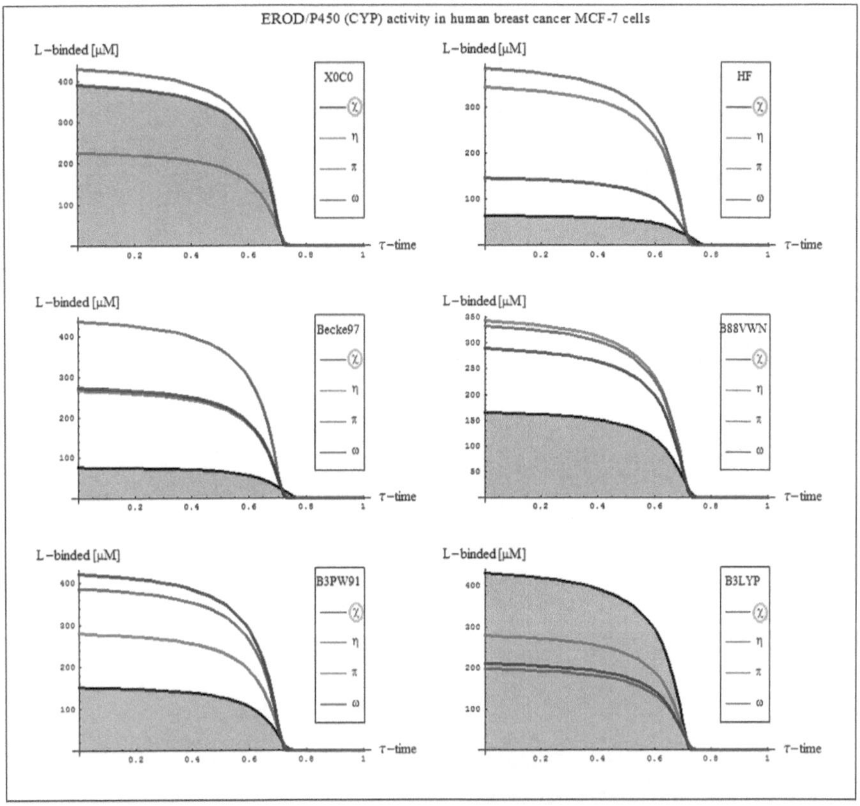

FIGURE 3.41 Comparative representation of the progress curves for the ClPAH ligands binding to human breast cancer MCF-7 cells, employing the recorded EROD/human-QRAR reactivity-activity information from Table 3.68 into logistic chemical-biological interactions modeled by Eq. (3.215), on the mapped unitary time scale of Eq. (3.208), for each index/quantum chemical method considered and for an EROD EC_{50}=34.696 μM norm parameter as computed with algebraic definition (3.205) and the EROD/human data of Table 3.63 (electronegativity curve is marked by the hatched area below it) (Putz & Putz, 2013b).

dominate those of electronegativity and chemical hardness, as is the case for B3-LYP computation

$$\textbf{\textit{II-BioAct-ChemReact}}\textbf{:}\ (\pi\wedge\omega)>(\chi\wedge\eta) \tag{3.288}$$

- while the third scheme is recorded when one of the basic chemical reactivity (electronegativity or chemical hardness) indices and

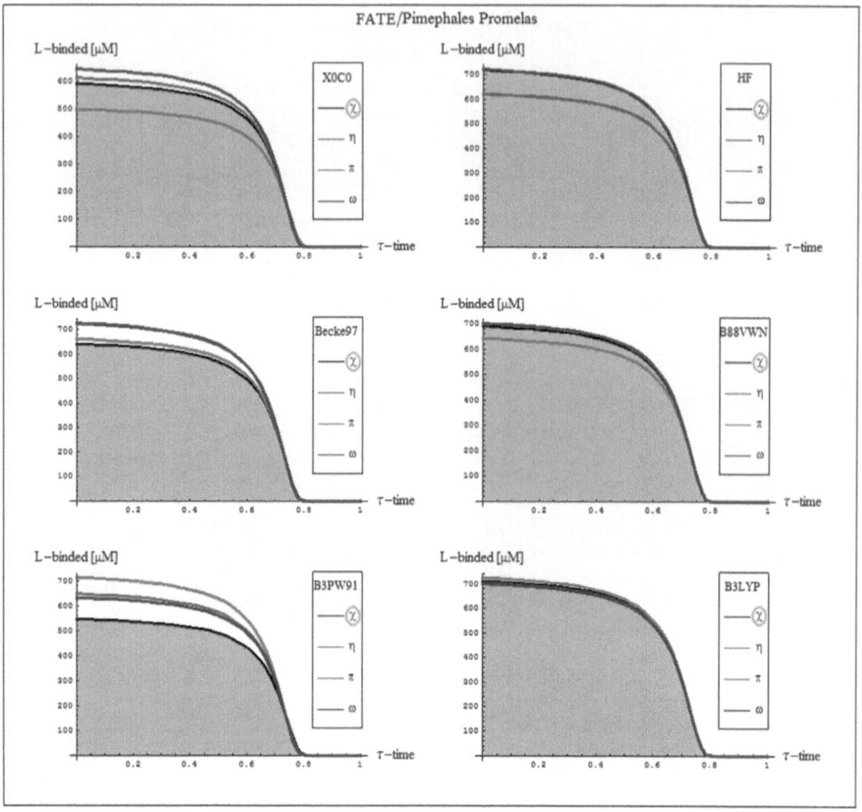

FIGURE 3.42 The same type of representations as in Figure 3.41 but for the ClPAH ligand – FATE/Pp QRAR reactivity-activity information from Table 3.69, with an FATE/Pp EC_{50}=464.437 μM norm parameter as computed using the algebraic definition (3.205) with the FATE/Pp data of Table 3.63 (electronegativity curve is marked by the hatched area below it) (Putz & Putz, 2013b)

principles acts in the first instance and is followed by one of the combined reactivity indices (chemical power or electrophilicity)

$$\textbf{\textit{III-BioAct-ChemReact}}\textbf{:}\ (\chi\wedge\eta)>(\pi\wedge\omega) \qquad (3.289)$$

In this last case, although one of the chemical reactivity principles acts first, it is then integrated into the combined reactivity as chemical power or electrophilicity, where implicitly, the other equalization principle for the remaining reactivity index naturally proceeds, as is the case for X0C0

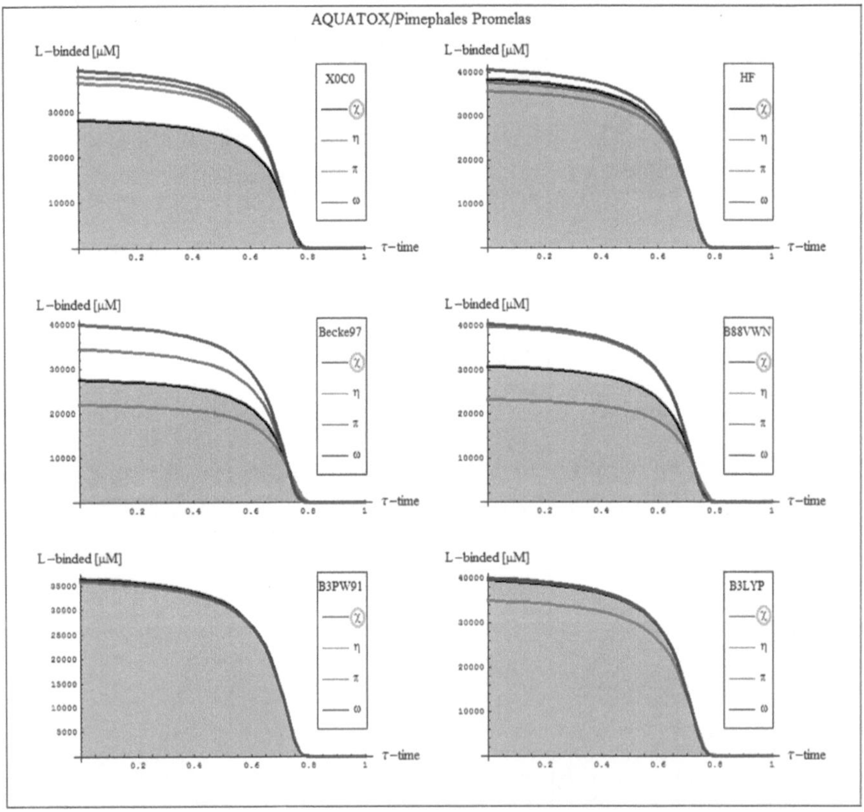

FIGURE 3.43 The same type of representation as in Figure 3.41 but for the ClPAH ligand –LC50/Pp QRAR reactivity-activity information from Table 3.70, with an EC_{50}= 18774.7 μM norm parameter as computed using the algebraic definition (3.205) with the LC50/Pp data of Table 3.63 (electronegativity curve is marked by the hatched area below it) (Putz & Putz, 2013b).

(no-exchange & no-correlation), HF and Becke-VWN computations. Note that the last two cases, Schemes II and III of the BioAct-ChemReact correlations, indicate a synergism between electronegativity and chemical hardness in the combined reactivity of chemical power and electrophilicity indices that, in this way, promote both indices as global equalization principles, so that electronegativity and chemical hardness are automatically fulfilled as well. For the EROD activity for human breast cancer MCF-7 cells, B3-PW91 and Becke 97 are in accord with an in-chain reactivity hierarchy as prescribed by chemical reactivity Eq. (3.170), while the

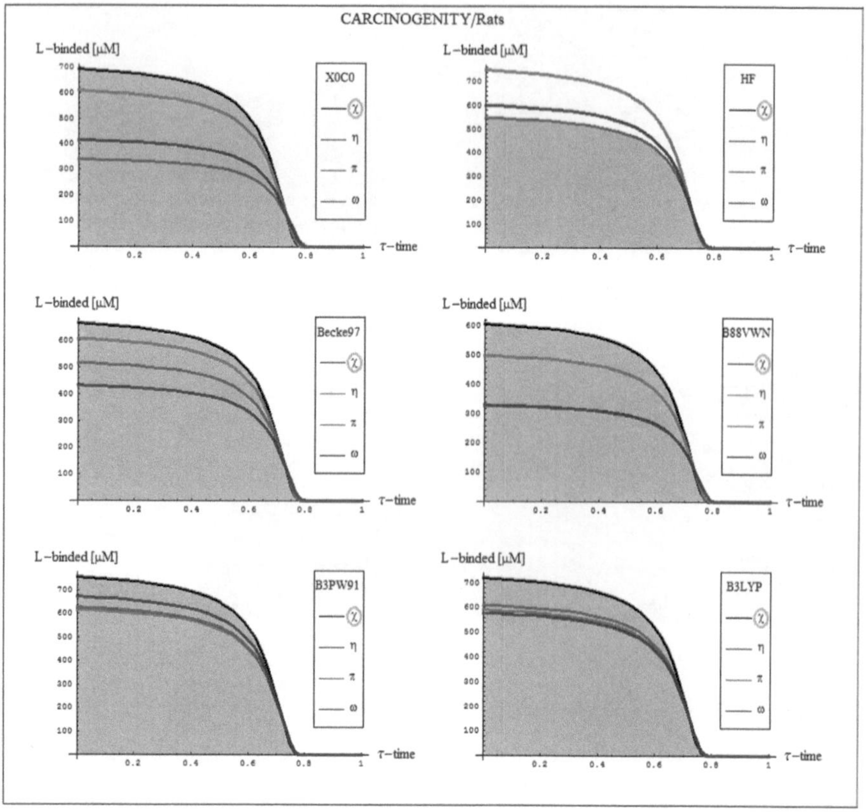

FIGURE 3.44 The same type of representation as in Figure 3.41 but for the ClPAH ligand –TD50/rats QRAR reactivity-activity information from Table 3.71, with an EC_{50}= 307.538 μM norm parameter as computed using the algebraic definition (3.205) with the TD50/Pp data of Table 3.63 (electronegativity curve is marked by the hatched area below it) (Putz & Putz, 2013b).

other two DFT computational frameworks (Becke88-VWN & B3-LYP) prescribe synergetics for these principles in combined indices and equalization principles of chemical power and electrophilicity (Putz & Mingos, 2013; Putz & Putz, 2013b):

- For the effect of the bioaccumulation of ClPAHs on the environmental fate of *Pimphales promelas* (Table 3.69, Figure 3.42), a case is made for the ***I-BioAct-ChemReact*** for X0C0 along Becke97 and Becke88-VWN computational schemes, or for ***II-BioAct-ChemReact*** for HF and B3-LYP computations, while the third case ***III-BioAct-ChemReact*** is found only for the B3-PW91 *in silico* approach;

- When the ecotoxicology of ClPAHs on the environmental fate of *Pimphales promelas* (Table 3.70, Figure 3.43) is considered to be the driving scheme, ***I-BioAct-ChemReact*** dominates the computational quantum frameworks because it is specific to Becke97, Becke88-VWN, B3-LYP, and even X0C0, while the schemes HF and B3-PW91 appear specific to the ***III-BioAct-ChemReact*** mechanism of the ligand-receptor interaction. Therefore, this chemical-biological system clearly favors chemical reactivity principles as triggering the chemical biological specific interactions;
- Finally, when ClPAH ligands mediated AhR toxicity in rats, carcinogenic effects are produced (Table 3.71, Figure 3.44). No specific ***I-BioAct-ChemReact*** mechanism of action was recorded because the ***II-BioAct-ChemReact*** hierarchy dominates this system, covering the X0C0, Becke97, Becke88-VWN, and B3-LYP computations, followed by the ***III-BioAct-ChemReact*** hierarchies for the HF and B3-PW91 *in silico* methods. Consequently, the synergistic mechanism of joint electronegativity and chemical hardness equalization in chemical power and electrophilicity is the driving mechanism specific for carcinogenesis.

One can nevertheless consider also a slightly different logistical-spectral implementation and analysis, namely based on the following algorithm; the present discussion also follows (Lacrămă et al., 2008; Putz, 2012b):

- for the activities of Table 3.72, their vectorial form is achieved through applying the specialized Spectral-SAR algorithm restrained to the Hansch structural descriptors of (3.63)

$$|Y_i\rangle^{ENDPOINT} = b_0|X_0\rangle + b_1|X_1\rangle + b_2|X_2\rangle + b_3|X_3\rangle,\ i=\overline{1,3};; \qquad (3.290)$$

  the *predicted* spectral norm can be draw, $\left\| A_i^* = Y_i \rangle \right\|,\ i=\overline{1,3}$, for each envisaged species;

- the initial chemical concentration of logistical chemical-biological progress curve equation (3.215) is identified with the predicted S-SAR activity norms

$$[L_{0(i)}] \rightarrow \left\| |Y_i\rangle^{ENDPOINT} \right\|, i=\overline{1,3} \qquad (3.291)$$

  based on idea that the evolution of the chemical concentration producing a biological effect starts evolving from the predicted

TABLE 3.72 Hydrophobic LogP, Electronic POL, and Optimized (Steric) Total Energy E_{TOT} Parameters, Computed Within HyperChem 7.0 Environment (Hypercube, 2002), For the Series of Compounds of Figure 3.45 Tested Upon *Chlorella vulgaris*, *Vibrio fischeri*, and *Pimephales promelas* Species (Lacrămă et al., 2008; Putz, 2012b), Providing Experimental Log(1/EC_{50}) Activities, A_1, A_2, and A_3, Respectively (Cronin et al., 2004)

No.	Parameters compound	A_1	A_2	A_3	1	*LogP*	*POL*	E_{TOT}
		$\|Y_1\rangle$	$\|Y_2\rangle$	$\|Y_3\rangle$	$\|X_0\rangle$	$\|X_1\rangle$	$\|X_2\rangle$	$\|X_3\rangle$
1	CH_3OH	–4.06	–3.21	–2.96	1	–0.27	3.25	–11622.9
2	C_2H_5OH	–3.32	–2.7	–2.49	1	0.08	5.08	–15215.4
3	C_4H_9OH	–2.73	–1.64	–1.37	1	0.94	8.75	–22402.8
4	C_4H_8O	–2.51	–1.76	–1.65	1	1.01	8.2	–21751.8
5	$C_5H_{10}O$	–2.23	–0.99	–1.25	1	1.64	10.04	–25344.6
6	C_6H_5OH	–1.46	0.5	0.51	1	1.76	11.07	–27003.1
7	$C_6H_5NH_2$	–1.34	–0.3	–0.16	1	1.26	11.79	–24705.9
8	CH_3-C_6H_4-OH	–1.01	1.16	0.29	1	2.23	12.91	–30597.6
9	OH-C_6H_4-O-CH_3	–0.97	1.46	0.05	1	1.51	13.54	–37976.3
10	OH-C_6H_4-NH_2	–0.91	–0.28	1.65	1	0.98	12.42	–32095.4
11	C_6H_5-CHO	–0.81	1.32	1.14	1	1.72	12.36	–29946.9
12	CH_3-C_6H_4-OH	–0.81	0.74	0.89	1	2.23	12.91	–30597.2
13	$C_6H_3(CH_3)_2OH$	–0.65	2.4	0.94	1	2.7	14.74	–34190.8
14	CH_3-C_6H_4-NO_2	–0.5	1.54	0.73	1	0.94	13.98	–42365.1
15	C_6H_5-O-Cl	–0.42	1.15	1.32	1	2.28	13	–35307.6
16	$C_6H_3(NO_2)NH_2$	–0.36	0.56	1.07	1	–1.75	15.22	–63030.2
17	$C_{11}H_8O_2$	0.16	2.81	3.19	1	2.39	20.99	–49768.3
18	$C_6H_4Cl_2$	0.37	1.56	1.19	1	3.08	14.29	–36217.2
19	$C_6H_3(NO_2)OH$	0.4	1.28	1.14	1	1.67	14.5	–65318
20	$C_6H_4N_2O_4$	0.41	3.07	2.37	1	1.95	13.86	–57926.7
21	$C_7H_6(NO_2)_2$	0.7	0.55	0.87	1	2.42	15.7	–61520.7
22	$C_{15}H_{23}OH$	1.45	1.41	2.78	1	5.48	27.59	–59316.5
23	$C_6H_3NCl_4$	1.48	2.15	2.93	1	3.34	19.5	–57920.2
24	C_6Cl_5OH	1.69	2.45	3.08	1	–0.54	20.71	–68512.4
25	$C_{12}H_{10}N_2O$	2.16	2.41	2.23	1	4.06	22.79	–55488.9
26	C_6Br_5OH	3.1	2.74	3.72	1	5.72	24.2	–66151.5

(computed) activity and diminished in time under the environmental and bio-degrability effects;

- in the same heuristically line the real maximum biological effect in chemical-biological equation (3.215) would be seen as the positive reminiscence of the predicted S-SAR activity against the measured activity:

$$\beta_{\max(i)} \rightarrow \sqrt{\left(\left\| \left|Y_i\right\rangle^{ENDPOINT}\right\| - \left\| \left|A_i\right\rangle^{EXP}\right\|\right)^2}, i = \overline{1,3} \quad (3.292)$$

- in these conditions, the computational EC_{50} parameter of chemical-biological equation (3.215) is as well considered as the positive reminiscence of the of the predicted S-SAR activity against its average activity:

$$EC_{50(i)} \rightarrow \sqrt{\left(\left\| \left|Y_i\right\rangle\right\| - \left|\overline{Y_i}\right\rangle\right)^2}, i = \overline{1,3} \quad (3.293)$$

- the progress curves of the so constructed biological-chemical activities (3.215) are computed and from them the cut-off time extracted from Eq. (1.56):

$$\tau_{\infty(i)} = \lim_{\beta_{(i)} \to 0} \left[1 - \frac{1}{\ln(t + e)}\right], i = \overline{1,3} \quad (3.294)$$

for each envisaged species in ecotoxicological battery;
- the obtained times are grouped, for each concerned species, and the inter-species mechanistic hierarchies are constructed according with the *least-action activity principle* of predicted norm-cut-off time paths

$$\left[A_{(i)}, B_{(i)}\right] = \sqrt{\left(\left\| \left|Y_{(i)}^B\right\rangle\right\| - \left\| \left|Y_{(i)}^A\right\rangle\right\|\right)^2 + \left(\tau_{\infty(i)}^B - \tau_{\infty(i)}^A\right)^2}, i = \overline{1,3}; A, B : ENDPOINTS \quad (3.295)$$

providing that *the shortest path the faster biological-chemical interaction is activated, thus assuring the more effective (toxico) chemical effect on biological system.*

The employed toxicity data for the chemical set of compounds of Figure 3.45 on *Chlorella vulgaris*, *Vibrio fischeri* and *Pimephales promelas* species are presented in the Table 3.72 and meet most of the criteria

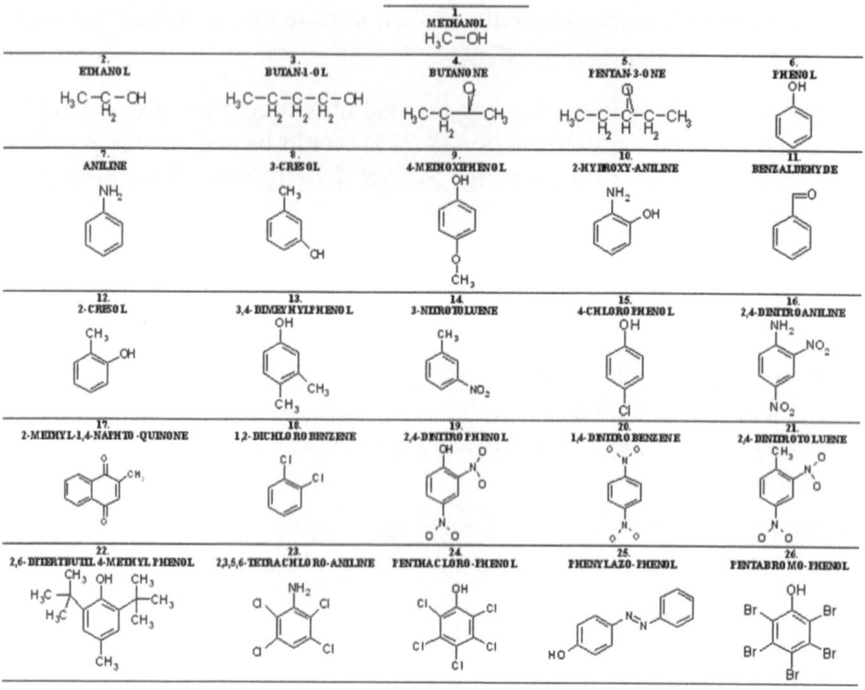

FIGURE 3.45 The series of chemical compound used for assessing toxicity at the level of biological species within the SPECTRAL-eco-battery modelling of the present case study (Lacrămă et al., 2008; Putz, 2012b).

for high quality data, i.e., it has been produced to a standard protocol, in a single laboratory, by a single worker. These toxicological data have previously been evaluated (and undergone process of pre-validation) by the development of QSARs for nonpolar and polar narcosis (Worgan et al., 2003) and investigated by QSARs with other species such as *T. pyriformis* (Netzeva et al., 2004a, b). However, the acute toxicity, assessed in short and low-cost unicellular tests, is also considered to be a surrogate for the prediction of toxicity to higher aquatic organisms (Schultz, 1997).

In this discussion, the biologically activities were took from the recent work of Cronin and collaborators (Cronin et al., 2004), which select diverse chemical structures incorporating narcosis, as well as other more specific mechanism of toxic action. Also the chemicals were required to span a enough range of hydrophobicity. Actually, the compounds of Figure 3.45 fully satisfy both dissimilarity- and similarity- based selection criteria. However, the workable activities stand as the toxicity data $Log(1/EC_{50})$. In the case of the unicellular alga (Cronin et al., 2004) (*Chlorella vulgaris*) they were collected in a biochemical assay (fifteen minutes) by the method which has been found to be an excellent predictor of the toxicity of chemicals to other species (Worgan et al., 2003), so with good application in ecotoxicology.

On the other side, the chemical structures of the molecules of Figure 3.45 were computed by the Hyper Chem Software (7 Release version) (Hypercube, 2002) providing that their 3D structures were optimized by AM1 semiempirical calculation while molecular dynamics involved the Polak-Rebier algorithm to reach to 0.01 root mean square gradient. The results are as well displayed in Table 3.72 as *LogP*, *POL*, and E_{TOT} for the hydrophobic, electronic and steric parameters of Eq. (11), respectively.

The previously presented S-SAR-logistical algorithm is now applied upon the data of Table 3.72 providing, in the first stage the results reported in Tables 3.73–3.75 for *Chlorella vulgaris*, *Vibrio fischeri* and *Pimephales promelas* species, respectively. At first glance, from Tables 3.73–3.75

TABLE 3.73 Spectral Structure Activity Relationships (S-SAR) Predicted With All Possible Correlation Models (Endpoints) Considered From Data of Table 3.72, Together With the Experimental or Measured Activity of *Chlorella vulgaris* Species, Paralleling the Corresponding Correlation Factor, Spectral Norm (3.291), and the Asymptotic Cut-Off Time (3.294) of the Associated Logistical-Spectral Model Through Eqs. (3.215) and (3.291)–(3.294) (Lacrămă et al., 2008; Putz, 2012b)

Activity (S-SAR Equation) Model (Endpoint)	r	$\lVert \lvert Y_1> \rVert$	τ_∞
$\lvert Y_1>^{Ia} = -1.65895 \lvert X_0> + 0.634092 \lvert X_1>$	0.618968	5.76725	0.436918
$\lvert Y_1>^{Ib} = -4.48035 \lvert X_0> + 0.279384 \lvert X_2>$	0.923589	8.18957	0.440029
$\lvert Y_1>^{Ic} = -4.00411 \lvert X_0> - 0.0000865456 \lvert X_3>$	0.887337	7.89593	0.439784
$\lvert Y_1>^{IIa} = -4.46015 \lvert X_0> + 0.0222605 \lvert X_1> + 0.275066 \lvert X_2>$	0.923735	8.19075	0.44003
$\lvert Y_1>^{IIb} = -4.24209 \lvert X_0> + 0.380779 \lvert X_1> -0.0000748672 \lvert X_3>$	0.954562	8.44123	0.440222
$\lvert Y_1>^{IIc} = -4.65267 \lvert X_0> + 0.180771 \lvert X_2> -0.0000388795 \lvert X_3>$	0.951655	8.41758	0.440204
$\lvert Y_1>^{III} = -4.50442 \lvert X_0> + 0.234126 \lvert X_1> + 0.0986645 \lvert X_2> - 0.00000533489 \lvert X_3>$	0.963357	8.51281	0.440274
$\lvert A_1>^{EXP}$	1.000000	8.81159	0.440478

TABLE 3.74 Spectral Structure Activity Relationships (S-SAR) Predicted With All Possible Correlation Models (Endpoints) Considered From Data of Table 3.72, Together With the Experimental or Measured Activity of *Vibrio fischeri* Species, Paralleling the Corresponding Correlation Factor, Spectral Norm (3.291), and the Asymptotic Cut-Off Time (3.294) of the Associated Logistical-Spectral Model Through Eqs. (3.215) and (3.291)–(3.294) (Lacrămă et al., 2008; Putz, 2012b)

Activity (S-SAR Equation) $^{\text{Model (Endpoint)}}$	r	$\lVert \lvert Y_2\rangle \rVert$	τ_∞
$\lvert Y_2\rangle^{Ia} = -0.201262\,\lvert X_0\rangle + 0.52453\,\lvert X_1\rangle$	0.51196	5.90227	0.436419
$\lvert Y_2\rangle^{Ib} = -2.64591\,\lvert X_0\rangle + 0.238822\,\lvert X_2\rangle$	0.789409	7.79875	0.440075
$\lvert Y_2\rangle^{Ic} = -2.15959\,\lvert X_0\rangle - 0.0000720417\,\lvert X_3\rangle$	0.738546	7.43153	0.439601
$\lvert Y_2\rangle^{IIa} = -2.65657\,\lvert X_0\rangle - 0.0117534\,\lvert X_1\rangle + 0.241101\,\lvert X_2\rangle$	0.789456	7.79909	0.440076
$\lvert Y_2\rangle^{IIb} = -2.35539\,\lvert X_0\rangle + 0.313287\,\lvert X_1\rangle - 0.0000624332\,\lvert X_3\rangle$	0.793251	7.82676	0.440109
$\lvert Y_2\rangle^{IIc} = -2.76727\,\lvert X_0\rangle + 0.169375\,\lvert X_2\rangle - 0.0000273805\,\lvert X_3\rangle$	0.805768	7.91823	0.440214
$\lvert Y_2\rangle^{III} = -2.68588\,\lvert X_0\rangle + 0.128529\,\lvert X_1\rangle + 0.124301\,\lvert X_2\rangle - 0.00000353238\,\lvert X_3\rangle$	0.809947	7.94886	0.440248
$\lvert A_2\rangle^{EXP}$	1.000000	9.37758	0.441463

TABLE 3.75 Spectral Structure Activity Relationships (S-SAR) Predicted With All Possible Correlation Models (Endpoints) Considered From Data of Table 3.72, Together With the Experimental or Measured Activity of *Pimephales promelas* Species, Paralleling the Corresponding Correlation Factor, Spectral Norm (3.291), and the Asymptotic Cut-Off Time (3.294) of the Associated Logistical-Spectral Model Through Eqs. (3.215) and (3.291)–(3.294) (Lacrămă et al., 2008; Putz, 2012b)

Activity (S-SAR Equation) $^{\text{Model (Endpoint)}}$	r	$\lVert \lvert Y_3\rangle \rVert$	τ_∞
$\lvert Y_3\rangle^{Ia} = -0.209645\,\lvert X_0\rangle + 0.566471\,\lvert X_1\rangle$	0.53973	6.4009	0.437342
$\lvert Y_3\rangle^{Ib} = -3.12624\,\lvert X_0\rangle + 0.277169\,\lvert X_2\rangle$	0.894346	8.90932	0.440991
$\lvert Y_3\rangle^{Ic} = -2.47683\,\lvert X_0\rangle - 0.0000815288\,\lvert X_3\rangle$	0.815901	8.32132	0.440454
$\lvert Y_3\rangle^{IIa} = -3.20607\,\lvert X_0\rangle - 0.0880001\,\lvert X_1\rangle + 0.294236\,\lvert X_2\rangle$	0.896578	8.92624	0.441005
$\lvert Y_3\rangle^{IIb} = -2.67949\,\lvert X_0\rangle + 0.324268\,\lvert X_1\rangle - 0.0000715835\,\lvert X_3\rangle$	0.866744	8.70086	0.440815
$\lvert Y_3\rangle^{IIc} = -3.23924\,\lvert X_0\rangle + 0.212504\,\lvert X_2\rangle - 0.0000254953\,\lvert X_3\rangle$	0.90632	9.00023	0.441064
$\lvert Y_3\rangle^{III} = -3.22812\,\lvert X_0\rangle + 0.0175597\,\lvert X_1\rangle + 0.206346\,\lvert X_2\rangle - 0.0000265806\,\lvert X_3\rangle$	0.906386	9.00074	0.441065
$\lvert A_3\rangle^{EXP}$	1.000000	9.72062	0.441561

(Lacrămă et al., 2008), there is clear that the correlation factor parallels the spectral norm and temporal cut-off limits, in each reported case, including the experimental or measured situation. That is, as correlation factor increase as both spectral norm and cut-off time increase as well.

Nevertheless, when thinking to a mechanistic interpretation of the computed models and of the associate endpoints the least, or minimum, action principle within the *spectral norm-cut-off time* abstract space has to be consider. That would require that shortest distance between two endpoints' spectral norm and cut-off times implies the first chemical-biological interaction. Therefore caution has to be paid to not confound the maximum-correlation factor-maximum-spectral norm-and-cut-off time behaviour with the minimum distance between such spectral norm-cut-off time "points" in the abstract endpoint space.

To further emphasize on the minimum path principle, in Figure 3.46 there are displayed the decay on chemical-biological interactions, according with Eq. (3.215) and the replacement rules (3.291)–(3.293), for all considered endpoints (S-SAR predicted models along the experimental activity) for all species of Table 3.72, involved in the present ecotoxicological battery. It follows that since the shapes of all considered species-end-points display the same features their difference should be evidenced only through their extremes, i.e., the initial chemical concentration (dose) and by their effectiveness in biological uptake (the time of action).

This way, through correspondence (3.291) and by Eq. (3.294), in fact, the main roles in establishing the endpoints parameters, when compared with others, refer to the spectral norm and cut-off time, respectively.

From these considerations follows the necessity of considering the spectral-norm-cut-off-time path (3.295) combination when one likes to assess through a single number (i.e., by a quantification procedure) the abstract, yet mechanistically, distance between two considered endpoints. However, even from representations of Figure 3.46 there comes out that *Pimephales promelas* species is characterized by the highest and most intensive as well as the most extensive chemical-biological interaction respecting the other two species in battery. There is therefore already clear that in an environmental context in which all three species exists and are affected by the same set of chemical compounds of Table 3.72 the *Pimephales promelas'* biological uptake comes firstly into act, i.e., will be the first one affected.

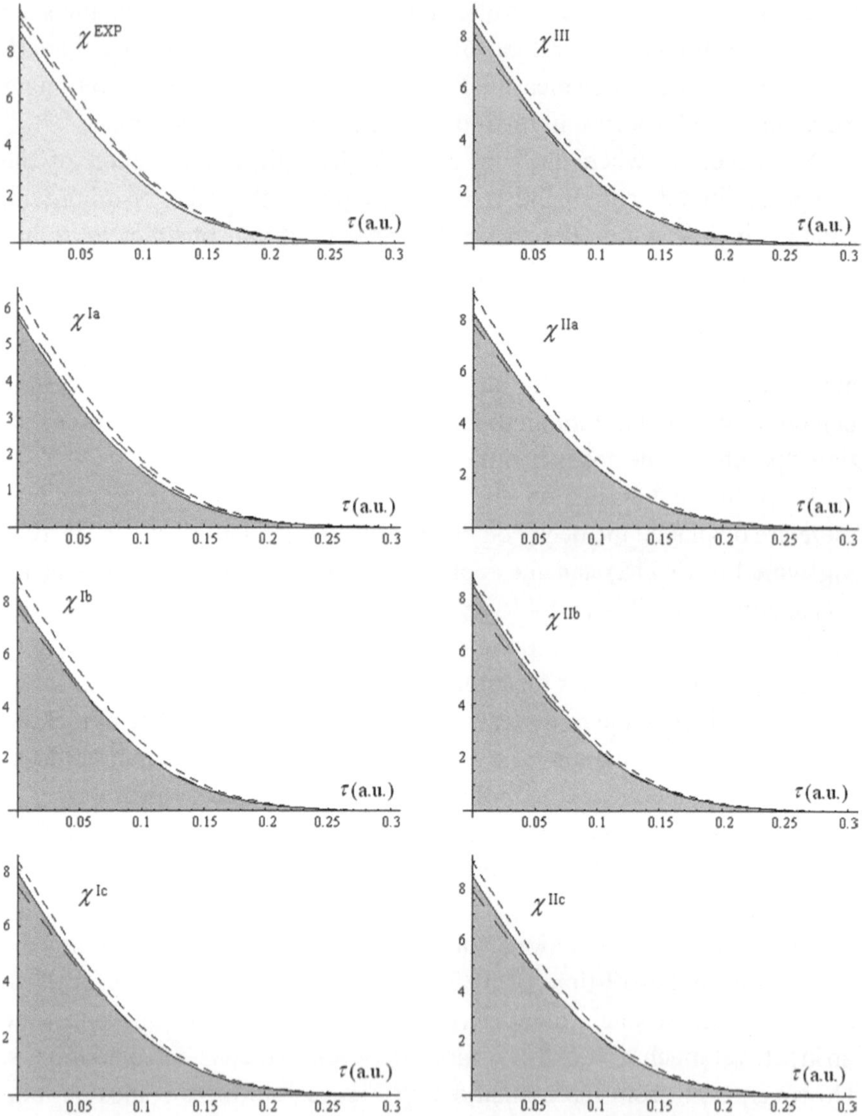

FIGURE 3.46 Temporal logistic-spectral representations of the Chlorella vulgaris (full line), Vibrio fischeri (interrupted line), and Pimephales promelas (dashed line) species for all endpoints of Tables 3.73–3.75 based on the equation (3.215) with the algorithm (3.291)–(3.293); after (Lacrămă et al., 2008; Putz, 2012b).

Nevertheless, to see the detailed mechanistically graph of dynamical actions and interactions intra- and inter-species of the ecotoxicological battery the least principle of chemical-biological interaction will be employed within the spectral norm-cut-off time with the help of the present S-SAR-logistical data of Table 3.76 (Lacrămă et al., 2008; Putz, 2012b).

The construction of such diagrams follows some preliminary or formal steps. Firstly, there is chosen that the species' endpoint states are represented by the height of the associated spectral norm (the equally choice would be performed in terms of cut-off time since parallels the spectral norm in all cases).

Then, a grid of the endpoints is considered according with the general tendency from minimum to maximum paths of Table 3.76 such that that once a model or endpoint is nominated across a path it will be not repeated across other.

For instance, identifying that the shortest path across all endpoints, when from each class of SAR models a single model is picked up or touched by chemical-biological evolution, one gets that the paths *Ib-IIa-III* and *Ib-IIc-III* are the two possible candidates. Further chose will be made in the light of the so-called *local* minimum path principle: the *Ib-IIa-III* path will be preferred since the state *Ib* is closer to *IIa* rather to the *IIc* one for all species of Tables 3.73–3.75. Thus, the chart will start with *Ib*

TABLE 3.76 Synopsis of The Values of Spectral Norm-Cut-Off Time Paths (3.295) Connecting the Endpoints of S-SAR models of Tables 3.73–3.75 Within the Spectral Norm-Cut-Off Time Abstract Space ($\| |Y_i> \|$ vs. $\tau\infty_{(i)}$) For the Species of Table 3.72 (Lacrămă et al., 2008; Putz, 2012b)

Paths among endpoints	Values for Species		
	Chlorella v.	*Vibrio f.*	*Pimephales p.*
Ia-IIa-III	2.74556	2.04659	2.59984
Ia-IIb-III	2.74556	2.04659	2.59984
Ia-IIc-III	2.74556	2.04659	2.59984
Ib-IIa-III	**0.323242**	**0.15011**	**0.0914192**
Ib-IIb-III	**0.323242**	**0.15011**	**0.508336**
Ib-IIc-III	**0.323242**	**0.15011**	**0.0914192**
Ic-IIa-III	0.616885	0.51733	0.679419
Ic-IIb-III	0.616885	0.51733	0.679419
Ic-IIc-III	0.616885	0.51733	0.679419

followed by *IIa* endpoint regions across ecotoxicological battery. Next we go to the next group of path from where the routes *Ic-IIa-III*, *Ic-IIb-III*, and *Ic-IIc-III* are the candidates.

The route *Ic-IIa-III* is now out of discussion since we already accommodated *IIa* after *Ib* on the planned ecotoxicological chart. Thus there remains to chose between *Ic* followed by *IIb* or by *IIc*; analyzing the data of Tables 3.73–3.75 for all species we found out that from three concerned species for two of them, namely for *Vibrio fischeri* and *Pimephales promelas*, the spectral norms of *IIb* endpoints rather that those of *IIc* situates closer to *Ic*. As such we conclude that the next order will be *Ic* followed by *IIb*.

At the end from the third group of paths in Table 3.76 it follows that *Ia* can only be followed by the remaining *IIc* set of endpoints since all other were previously accommodated in the chart. Worth noted that this chart ordering was identically with that considered in previous study of *Tetrahymena pyriformis* in relation with the same set of chemical compounds, see Section 3.3.1 and Putz & Lacrămă (2007). However, will all these considerations the spectral chart of Figure 3.47 can be drawn.

In Figure 3.47 appears that each endpoint region contains all *Chlorella vulgaris*, *Vibrio fischeri*, and *Pimephales promelas* species with their relative spectral norms. Now, they will be all connected through paths according with minimum path principles applied intra-species this time for the paths of Table 3.76. In this respect, it follows that the *Pimephales promelas* species provides the first, the most preeminent path, as already anticipated from Figure 3.46, from the one-parameter model (*Ib* or *POL*arizability based model) touching *IIa* one (based on electronic-*POL* followed by hydrophobicity-*Log* model) until the measured endpoint (i.e., the experimental one), not before that the steric effects (comprised in the *III*) to enter ultimately in action. Similar mechanistic analyses can be performed for all other connected endpoints (Lacrămă et al., 2008; Putz, 2012b).

However, there is clear from Figure 3.47 that three major hierarchies can be formulated: the primary one that starts from *Ib* (*POL*), the secondary one that starts from *Ic* (E_{TOT}) and the tertiary one that starts from *Ia* (*Log*) SAR models for all three species, respectively. The difference between these three hierarchies stands in their efficacies in action so that the longer path the more subtle (or less intensive) chemical-biological

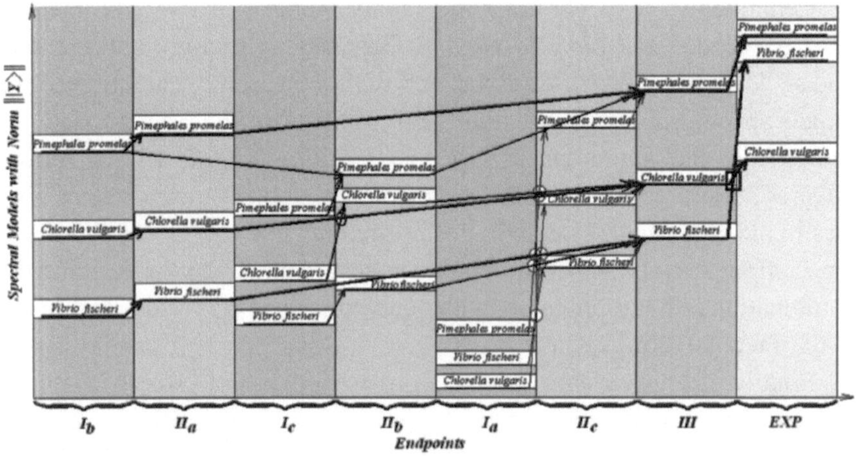

FIGURE 3.47 The spectral representation of the chemical-biological interaction paths across the S-SAR to experimental endpoints for the Chlorella vulgaris, Vibrio fischeri, and Pimephales promelas species, according with least path rule within the spectral norm-cut-off time space applied on Table 3.76 results, respectively. The primary, secondary and tertiary least path hierarchies of Table 3.76 are represented by decreasing the thickness of the connecting lines, while the interferences between species interactions is indicated by a rectangular mark for primary hierarchy and by circle marks for secondary and tertiary hierarchies; after (Lacrămă et al., 2008; Putz, 2012b).

interaction associates with. Still, these subtle interaction becomes important in endpoint-chart regions where their register interferences with other superior hierarchies, as is the case in Figure 3.47 also.

The inter-path crossings are marked in Figure 3.47 with a rectangular mark for primary hierarchy and by circles for the rest. The interference between the paths of primary hierarchy appears in region *III* between *Chlorella vulgaris* and *Vibrio fischeri* species only. That means that since *Pimephales promelas* species is firstly affected (attacked) by the set of compounds of Table 3.72 the other two species interchange their chemical-biological effects only after all hydrophobic, electronic, and steric effects are consumed. In other words, primarily, the effects on *Pimephales promelas* does not affect the other two species while the effects on one of the *Chlorella vulgaris* and *Vibrio fischeri* species induce (or transfer) similar toxic effects on the other. Moreover, in this stage, the *Pimephales promelas* species is once more isolated from the rest of species through additional path *Ib-IIb-III* that do not interfere with other species' paths.

Remarkably, when about the secondary hierarchies the only interference is recoded at *Chlorella vulgaris* between its primary and the secondary *toxicological waves*, associated with the second generation of the longer spectral-norm-cut-off time paths. Therefore this second stage of ecotoxicological action tell us that the *Chlorella vulgaris* will be further affected, however not transferring the effects upon it to the rest of species in the ecosystem. Moreover this reinforcing toxicological effect takes places from *Ic* to *IIb* endpoint regions in accordance with the fact that the combination of hydrophobicity with doubled steric effects (Lacrămă et al., 2008; Putz, 2012b).

Finally, the third stage, or wave, of toxicological effects (the endpoints) suggest that even being associated with the less effective chemical-biological action the tertiary hierarchy becomes important by means of the numerous interference that produce with both primary and secondary effects. For instance, the tertiary effects on the *Chlorella vulgaris* jointly affect the primary and secondary effects on *Vibrio fischeri* species thus sustaining in this more subtle way their ecotoxicological activity. Moreover, is in this third stage that, for the first time, the path of *Pimephales promelas* species interferes with those of *Chlorella vulgaris* and *Vibrio fischeri* species in both their secondary and primary stages of toxically action. This behaviour stands also as a computational confirmation of the empirical rule according with, indeed, in an ecosystem once one species is affected all other existing ones are soon or later, directly or by interference of effects, affected as well. Nevertheless, the present S-SAR-logistical model combined with the least principle of action gives a detailed map of these inter-actions both at the inter- and intra-species level. At the same time, the chart of Figure 3.47 has the important advantage that suggest by its mechanistically interpretation at which level and upon which structural property (*Log*, *POL* or E_{TOT}) one can act in order to promote or to inhibit a certain action through co-existing species. It furnishes, this way, in fact, a model battery of ecotoxicological action, here, in its SPECTRAL-ecotoxico-logistical version.

With all these, one may widely concluding that the presented SPECTRAL-SAR methodology provides the possibility of analytical characterizing the bio-and eco- activity of other species against given a set of trained or new synthesized chemicals and for the inter-species correlations; it has, beyond giving quantitatively similar results as the already traditional regression QSAR methods (Miller & Miller, 2000), many practical advantages, namely (Lacrămă et al., 2008; Putz, 2012b):

- it has the strength of no dependency on the way in which the input data are considered, thus being largely independent of the outliers detection (Putz & Lacrămă, 2007; Putz, 2012b);
- it uses the algebraically instead of statistically recipe to furnish a generalized view for the "intensity" of chemical-biological interaction, through the vectors, predicted norms $\left\| Y \right\rangle \|$, and of their properties in generalized multi-dimensional orthogonal spaces (Putz & Putz, 2011);
- it is also easily applicable to the case where the number of structural parameters exceeds those of the available biological activities, a situation more often observed in actual practice but being still an open problem in QSAR, due to the statistically forbidden condition that such situations imply (Putz et al., 2009a);
- it is also able to furnish the key in treating the so-called spectral analysis of the activity itself through *action norm* and its *least activity path principle* (i.e., $\delta \left\| Y \right\rangle \| = 0$ over many possible predicted end-points), thus providing the appropriate mechanistic picture of the envisaged ecotoxicity (Putz & Lacrămă, 2007; Putz, 2012b).

However, further studies on different species with diverse computational schemes and parameters are required, especially those that employ the quantum and algebraic features of the general SPECTRAL operatorial action througfully implemented these case studies. Furthermore, by involving the enzyme-substrate modeling for receptor-effector interactions for biological activity driving chemical reactivity, as was recently reported (Putz & Putz, 2013b), the present case study can conceptually assessed by computational implementation a definitive theory of inter- and intra- mode of boil/eco/toxico/pharma action such that it can be used for being integrated over a wide range of organisms toward designing specific eco-toxicological batteries, see Figure 3.48 and Refs. (Lacrămă et al., 2008; Putz, 2012b).

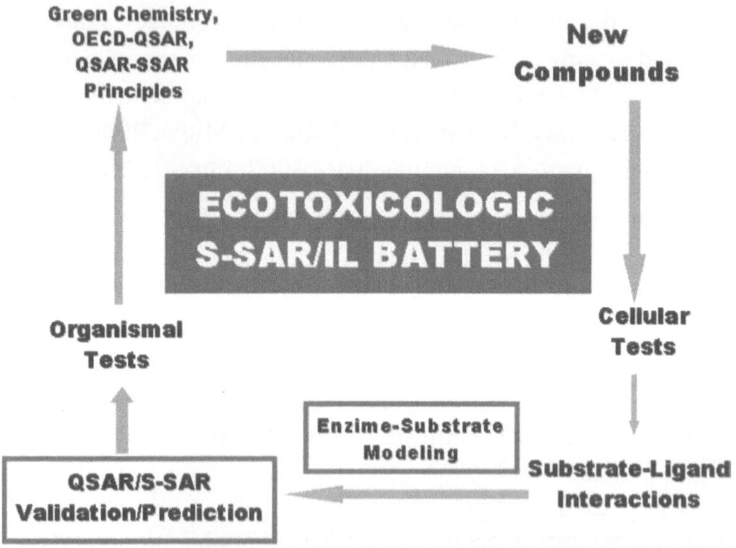

FIGURE 3.48 Conceptual description of an ecotoxicological battery constructed with the aid of QSAR/Spectral-SAR modeling of ligand interactions over cellular and later on organismal substrates to validate the toxicological safety of the newly synthesized compounds in accordance with the green chemistry, OECD-QSAR and, eventually, the QSAR/SSAR ecotoxicological principles (Putz & Putz, 2013a).

3.4 CONCLUSION

There is a noted fact the matter organization at the biological level cannot be described in the same way, i.e., with the same analytical quantum tools and at least not in the same form, as the physical or chemical systems. This because, when atoms combines in molecules and molecules fuses in biomolecules that folds in 3D flexible structures; also self-interaction is omnipresent along the structure demanding a special quantum with statistical treatment of the resulted system. Therefore, modeling the bio-chemical structure in interaction requires a special quantum with statistical combination sensible different by the quantum statistics of not-too-many-body particles' systems of common atoms and molecules. However, both quantum and statistics provide their essential tools in approaching the complex bio-chemical dynamics, namely (Putz et al., 2011a, 2012a):

- *The superposition principle* generally assumed in quantum mechanics, physics and chemistry; at biological side it is transposed by superimposition of different primer or molecular causes of which most preeminent are the electronic, hydrophobic and steric behaviors of molecules or ligands or substrates approaching the site, receptor or enzymes in organisms' cells, respectively. This principle allows the combination of these causes in the so-called registered effect (or activity or toxicity) even they are essentially different in nature (actually this a prerequisite demand in the *orthogonal* multidimensional space of interaction and reactivity)

- *The correlation principle*, as coming from mathematical statistics, between the computed activity effects as provided by the above superposition principle and the observed or recorded counterpart activity; usually such correlation is a natural consequence of the resulted multi-variate regression, although this is not compulsory to reflect the best superimposed quantum effects of descriptors, nor it may be always associated with a definite molecular mechanism of action in order to understand the intermediate steps of the analyzed bio-chemical interaction.

As a consequence the celebrated quantitative structure-activity relationships (QSAR) technique appears as an ingenious tool in producing analytical equations that are the conceptual-computational substitute for the natural laws that acts on the biological level of matter organization and interaction. The problem is that such model is not unique, nor various models for the same interaction cannot be said to be disjoint in describing the same phenomena.

All in all, the balance in modeling the chemical compounds with biological activity resides in choosing the best molecular descriptors (that can be of pure quantum, or topological or of physico-chemical nature in general) that produce the best statistical correlation of computed activity against the observed one; and even supposing one finds such a model out of the plethora of structure descriptors and of their combinations (currently counting of the thousand orders) there still remains the problem all this travail is mostly done on a given (finite) set of molecules, usually about five time larger than the cardinal or the descriptors family (the so-called Topliss and Costello (1972) rule). As a consequence, in fact, no universal QSAR model can be assessed for a given class of molecules. Therefore arises the need of a guidance or regulatory protocol according which a relevant QSAR study may be performed targeting the modeling for a specific behavior of a given chemical class of compounds within a biological or pharmacological or (eco)toxicological environment.

In responding to this increasing demand for benchmark principles to be followed by a reliable QSAR research, the Organization for Economic Co-operation and Development (OECD) advanced a set of standard principles for the validation, for regulatory purposes, of (quantitative) structure-activity relationship models (OECD, 2006, 2007, 2012; OECD-QSAR, 2013; Putz et al., 2011a). They are summarized in short as:

- QSAR-1: a defined endpoint
- QSAR-2: an unambiguous algorithm
- QSAR-3: a defined domain of applicability
- QSAR-4: appropriate measures of goodness-of–fit, robustness and predictivity
- QSAR-5: a mechanistic interpretation, if possible

and realized through the present extended case studies as:

- *Principle 1: a defined endpoint:* biological activity, then generalized to the final product rate in the ligand-receptor interaction, as based on the Michaelis-Menten enzyme-substrate kinetics;
- *Principle 2: an unambiguous algorithm:* the Spectral-, Diagonal-, Projective-, Quantum Amplitude-, Residual-, Alert-, Catastrophe-, SMILES- and topo-reactive QSAR variants;
- *Principle 3: a defined domain of applicability: for example,* the most preeminent ecotoxicities of the species, along mutagenicity and carcinogenicity by non-congeneric molecular series towards HIV inhibition by congeneric molecules;
- *Principle 4: appropriate measures of goodness-of–fit, robustness and predictivity*: the principle of minimum action, spectral paths, reactivity principles (minimum electronegativity, maximum chemical hardness, minimum of polarizability, etc.), along the Gaussian-quasi Gaussian and of ergodic/transitivity screening for the distribution of molecular activities for trial and testing series;
- *Principle 5: a mechanistic interpretation, if possible*: hierarchy of minimum paths across a *spectrum* of computed/predicted end-points towards identifying the specific influence of molecular descriptors within the orthogonal space of chemical-biological complex interaction by molecular structural characters/indices and of their algebraic behavior and predicted statistics.

Moreover, the OECD-QSAR can be still corresponding with the chemical orthogonal space (post-modern) accommodation of the Kantian interrogations of the being (Putz et al., 2011a):

- Qu-SAR-1: What/For What/Why do we model?
- Qu-SAR-2: How do we model?
- Qu-SAR-3: Which means do we model?
- Qu-SAR-4: There is good output of the model?
- Qu-SAR-5: Does the model connect causes with effects, past with the future?

Within such context, the present chapter reviews and largely illustrate how these OECD-QSAR principles may be unfolded for a recent developed alternative algebraic-quantum SAR correlation, as the Spectral-SAR was developed and reported in various bio-, eco- and pharmacological applications. It aims to explore the possibility expanding or grounding the OECD-QSAR principles by fundamental physico-mathematical relationships on orthogonal (eventually quantum or Hilbert) spaces towards providing more universal values for the specific chemoinformatics analyses, at the quantum level of SAR, aiming the unified Qu-SAR theory of chemical-biological complex/specific interactions.

KEYWORDS

- **algebraic correlation factor**
- **Aliphatic Amines' toxicity**
- **anti-HIV pyrimidines**
- **breast anti-cancer activity**
- **catastrophe QSAR**
- **chemical hardness**
- **chemical orthogonal space**
- **chemical power**
- **Daphnia magna**
- **ecotoxicology**
- **electronegativity**

- **electrophilicity**
- **inter-species toxicity**
- **Ionic liquids**
- **logistic QSAR**
- **PAHs' carcinogenesis**
- **projective QSAR**
- **QSAR=quantitative structure activity relationships**
- **quantum amplitude**
- **rats' toxicology**
- **Residual QSAR**
- **SMILES QSAR**
- **Spectral-Diagonal-SAR**
- **Spectral-SAR**
- **structural alerts**
- ***T. Pyriformis***
- **topo-reactive QSAR**

REFERENCES

AUTHOR'S MAIN REFERENCES

Chicu, S. A., Putz, M. V. (2009). Köln-Timişoara molecular activity combined models toward interspecies toxicity assessment. *Int. J. Mol. Sci.* 10(10), 4474–4497 (DOI: doi:10.3390/ijms10104474).

Duda-Seiman, C., Duda-Seiman, D., Dragoş, D., Medeleanu, M., Careja, V., Putz, M. V., Lacrămă, A.-M., Chiriac, A., Nuţiu, R., Ciubotariu, D. (2006). Design of anti-HIV ligands by means of minimal topological difference (MTD) method. *Int. J. Mol. Sci.* 7(11), 537–555 (DOI: 10.3390/i7110537).

Dudaş, N. A., Putz, M. V. (2014). Pyrimidine Derivatives with Biological Activity in Anti-HIV Therapy. The Spectral-Diagonal-SAR Approach. *Int. J. Chem. Model.* 6 (1), 95–114.

Lacrămă, A. M., Putz, M. V., Ostafe, V. (2008). Designing a spectral structure-activity ecotoxico-logistical battery. Putz, M. V. (Ed.) *Advances in Quantum Chemical Bonding Structures*, Transworld Research Network, Kerala, Chapter 16, pp. 389–419.

Lacrămă, A.-M., Putz, M. V., Ostafe, V. (2007). A Spectral-SAR model for the anionic-cationic interaction in ionic liquids: application to *Vibrio fischeri* ecotoxicity. *Int. J. Mol. Sci.* 8(8), 842–863 (DOI: 10.3390/i8080842).

Putz, M. V. (2016a). *Quantum Nanochemistry. A Fully Integrated Approach: Vol. I. Quantum Theory and Observability.* Apple Academic Press & CRC Press, Toronto-New Jersey, Canada-USA.

Putz, M. V. (2016b). *Quantum Nanochemistry. A Fully Integrated Approach: Vol. II. Quantum Atoms and Periodicity.* Apple Academic Press & CRC Press, Toronto-New Jersey, Canada-USA.

Putz, M. V. (2016c). *Quantum Nanochemistry. A Fully Integrated Approach: Vol. III. Quantum Molecules and Reactivity.* Apple Academic Press & CRC Press, Toronto-New Jersey, Canada-USA.

Putz, A. M., Putz, M. V. (2013a). Spectral-Structure Activity relationship (Spectral-SAR) assessment of ionic liquids' in silico ecotoxicity. In: Kadokawa, J. (Ed.) *Ionic Liquids – New Aspects for The Future*, InTech, Inc., Rijeka-New York-Shanghai, Chapter 4, pp. 85–126 (DOI:10.5772/51657).

Putz, M. V. (2013a). Spectral-diagonal approach of structure-property (activity) relationships: SD-QSP(A)R. The general formalism, *Int. J. Chem. Model.* 5(2/3), 357–367.

Putz, M. V. (2013b). Chemical orthogonal spaces (COSs), from structure to reactivity to biological activity. *Int. J. Chem. Model.* 5(1), 1–33.

Putz, M. V. (2013c). Bonding in orthogonal space of chemical structure: from in cerebro to in silico. *Int. J. Chem. Model.* 5(4), 369–395.

Putz, M. V., Dudaş, N. A. (2013a). Variational principles for mechanistic quantitative structure–activity relationship (QSAR) studies: application on uracil derivatives' anti-HIV action. *Struct. Chem.* 24(6), 1873–1893 (DOI: 10.1007/s11224–013–0249–6).

Putz, M. V., Dudaş, N. A. (2013b). Determining chemical reactivity driving biological activity from SMILES transformations: The bonding mechanism of anti-HIV pyrimidines. *Molecules* 18(8), 9061–9116 (DOI: 10.3390/molecules18089061).

Putz, M. V., Tudoran, M. A., Putz, A. M. (2013a). Structure properties and chemical-bio/ecological of PAH interactions: from synthesis to cosmic spectral lines, nanochemistry, and lipophilicity-driven reactivity. *Curr. Org. Chem.* 17(23), 2845–2871 (DOI: 10.2174/13852728113179990130).

Putz, M. V., Ori, O., Cataldo, F., Putz, A. M. (2013b). Parabolic reactivity "coloring" molecular topology: Application to carcinogenic PAHs. *Curr. Org. Chem.* 17(23), 2816–2830 (DOI: 10.2174/13852728113179990128).

Putz, M. V., Putz, A. M. (2013b). DFT Chemical Reactivity Driven by Biological Activity: Applications for the Toxicological Fate of Chlorinated PAHs. *Structure and Bonding* 150 (2013). 181–232 (DOI: 10.1007/978–3-642–32750–6_6).

Putz, M. V. (2012a). *Chemical Orthogonal Spaces*, Mathematical Chemistry Monographs Vol. 14, University of Kragujevac.

Putz, M. V., Ed. (2012b). *QSAR & SPECTRAL-SAR in Computational Ecotoxicology*, Apple Academics, Toronto.

Putz, M. V. (2012c). *Quantum Theory: Density, Condensation, And Bonding*, Apple Academics: Ontario-New Jersey.

Putz, M. V. (2011a). Residual-QSAR. Implications for genotoxic carcinogenesis. *Chem. Cent. J.* 5, 29/11 pages (DOI: 10.1186/1752–153X-5–29).

Putz, M. V. (2011b). Electronegativity and chemical hardness: different patterns in quantum chemistry. *Curr. Phys. Chem.* 1(2), 111–139 (DOI:10.2174/1877946811101020111).

Putz, M. V. (2011c). Quantum parabolic effects of electronegativity and chemical hardness on carbon π-systems, In *Carbon Bonding And Structures: Advances In Physics And Chemistry*; Putz, M. V., Eds., Springer Verlag: London, UK, 2011; pp. 1–32.

Putz, M. V. (2011d). Chemical action concept and principle. *MATCH Commun. Math. Comput. Chem.* 66(1), 35–63.

Putz, M. V., Lazea, M., Putz, A. M., Seiman-Duda, C. (2011b). Introducing catastrophe-QSAR. Application on modeling molecular mechanisms of pyridinone derivative-type HIV non-nucleoside reverse transcriptase inhibitors. *Int. J. Mol. Sci.* 12(12), 9533–9569 (DOI: 10.3390/ijms12129533).

Putz, M. V., Putz, A. M., Barou, R. (2011a). Spectral-SAR realization of OECD-QSAR principles. *Int. J. Chem. Model.* 3(3), 173–190.

Putz, M. V. (2010a). Cosmos, order and obligations: the big CO_2. *Int. J. Environ Sci.* 1 (January–June), 1–8.

Putz, M. V. (2010b). Chemical hardness: quantum observable? *Chemia-Studia Univ. Babeş-Bolyai* 55(2), 47 – 50.

Putz, M. V. (2009a). Path integrals for electronic densities, reactivity indices, and localization functions in quantum systems. *Int. J. Mol. Sci.* 2009, 10(11), 4816–4940 (DOI: 10.3390/ijms10114816).

Putz, M. V. (2009b). Electronegativity: quantum observable. *Int. J. Quantum Chem.* 109(4), 733–738.

Putz, M. V., Putz, A. M., Lazea, M., Chiriac, A. (2009a). Spectral vs. statistic approach of structure-activity relationship. application on ecotoxicity of aliphatic amines. *J. Theor. Comput. Chem.* 8(6), 1235–1251 (DOI: 10.1142/$SO_2$19633609005453).

Putz, M. V., Putz, A. M., Lazea, M., Ienciu, L., Chiriac, A. (2009b). Quantum-SAR extension of the Spectral-SAR algorithm. Application to polyphenolic anticancer bioactivity. *Int. J. Mol. Sci.* 10(3), 1193–1214 (DOI: 10.3390/ijms10031193).

Putz, M. V. (2008a). *Absolute and Chemical Electronegativity and Hardness*, Nova Publishers Inc., New York.

Putz, M. V. (2008b). Maximum hardness index of quantum acid-base bonding. *MATCH Commun. Math. Comput. Chem.* 60(3), 845–868.

Putz, M. V. (2003). *Contributions within density functional theory with applications in chemical reactivity theory and electronegativity*, Dissertation.com: Parkland, Florida.

Putz, M. V., Russo, N., Sicilia, E. (2004). On the application of the HSAB principle through the use of improved computational schemes for chemical hardness evaluation. *J. Comput. Chem.* 25(7), 994–1003 (DOI: 10.1002/jcc.20027).

Putz, M. V., Lacrămă, A. M. (2007). Introducing Spectral Structure Activity Relationship (S-SAR) analysis. Application to ecotoxicology", *Int. J. Mol. Sci.* 8(5), 363–391 (DOI: 10.3390/i8050363).

Putz, M. V., Lacrămă, A. M., Ostafe, V. (2007). Spectral-SAR ecotoxicology of ionic liquids. The *Daphnia magna* case. *Int. J. Ecol. (former Res. Lett. Ecol.)* Article ID12813/5 pages (DOI: 10.1155/2007/12813).

Putz, M. V., Putz (Lacrămă) A.-M. (2008). Spectral-SAR: old wine in new bottle. *Chemia-Studia Universitatis Babeş-Bolyai* 53(2), 73–81.

Putz, M. V., Putz, A. M., Ostafe, V., Chiriac, A. (2010). Spectral-SAR ecotoxicology of ionic liquids-acetylcholine interaction on, *E. Electricus* Species. *Int. J. Chem. Model.* 2(1), 85–96.

Putz, M. V., Putz, A. M. (2011). Timişoara Spectral – Structure Activity Relationship (Spectral-SAR) algorithm: from statistical and algebraic fundamentals to quantum consequences. InŞ Putz, M. V. (Ed.) *Quantum Frontiers of Atoms and Molecules*, NOVA Science Publishers, Inc., New York, Chapter 21, pp. 539–580.

Putz, M. V., Tudoran, M. A., Putz, A. M. (2012). Modeling chlorinated polycyclic aromatic hydrocarbons (Cl-PAH) eco- and toxico-logy by QSAR-OECD ToolBox facility. *Int. J. Chem. Model.* 4(4), 533–543.

Putz, M. V., Mingos, D. M. P., Eds. (2013). *Applications of Density Functional Theory to Biological and Bioinorganic Chemistry*, Structure and Bonding Series Vol. 150, Springer Verlag, Heidelberg-Berlin.

Putz, M. V., Duda-Seiman, C., Duda-Seiman, D. M., Putz, A. M. (2008a). Turning SPECTRAL-SAR into 3D-QSAR analysis. Application on H^+K^+-ATPase inhibitory activity. *Int. J. Chem. Model.* 1(1), 45–62.

Putz, M. V., Duda-Seiman, D., Mancaş, S., Duda-Seiman, C., Lacrămă, A. M. (2008b). Quantum and topological impact on HMG-CoA reductase inhibitors. Putz, M. V. (Ed.) *Advances in Quantum Chemical Bonding Structures*, Transworld Research Network, Kerala, Chapter 15, pp. 355–387.

Putz, M. V., Ionaşcu, C., Putz, A. M., Ostafe, V. (2011c). Alert-QSAR. Implications for electrophilic theory of chemical carcinogenesis. *Int. J. Mol. Sci.* 12(8), 5098–5134 (DOI: 10.3390/ijms12085098).

Putz, M. V., Ori, O., De Corato, M., Putz, A. M., Benedek, G., Cataldo, F., Graovac, A. (2013c). Introducing "colored" molecular topology by reactivity indices of electronegativity and chemical hardness. In: Ashrafi, A. R., Cataldo, F., Iranmanesh, A., Ori, O. (Eds.) *Topological Modeling of Nanostructures and Extended Systems*, Springer Verlag, Dordrecht, Chapter 9, pp. 265–286 (DOI: 10.1007/978–94–007–6413–2_9).

Tarko, L., Putz, M. V. (2012). On Quantitative Structure-Toxicity Relationships (QSTR) using High Chemical Diversity Molecules Group, *J. Theor. Comput. Chem* 11(2), 265–272 (DOI: 10.1142/$SO_2$19633612500174).

SPECIFIC REFERENCES

Aerts, D., Czachor, M., Gabora, L., Kuna, M., Posiewnik, A., Pykacz, J., Syty, M. (2003). Quantum morphogenesis: a variation on Thom's catastrophe theory. *Phys. Rev. E* 67, 051926.

Afantitis, A., Melagraki, G., Sarimveis, H., Koutentis, P. A., Markopoulos, J., Igglessi-Markopoulou, O. (2006). A novel simple QSAR model for the prediction of anti-HIV activity using multiple linear regression analysis. *Mol. Divers.* 10, 405–414.

Aksimentiev, A., Heng, J. B., Timp, G., Schulten, K. (2004). Microscopic kinetics of DNA translocation through synthetic nanopores. *Biophys. J.* 87, 2086–2097.

Amić, D., Davidović-Amić, D., Trinajstić, N. (1995). Calculation of retention times of anthocyanins with orthogonalized topological indices. *J. Chem. Inf. Comput. Sci.* 35, 136–139.

Anastas, P. T., Levy, I. J., Parent, K. E. (Eds.) (2009). *Green Chemistry Education: Changing the Course of Chemistry*, ACS Symposium Series 1011, American Chemical Society, Washington DC.

Anastas, P. T., Warner, J. C. (1998). *Green Chemistry Theory and Practice*, Oxford University Press, New York.

Anastas, P. T., Zimmerman, J. B. (2003). Design through the twelve principles of green engineering. *Environ. Sci. Technol.* 37, 94A-101A.

Anderson, T. W. (1958). *An Introduction to Multivariate Statistical Methods*, New Wiley, York.

Arcos, J. C., Argus, M. F. (1995). *Multifactor interaction network of carcinogenesis—a "tour guide"*; In: Arcos, J. C., Argus, M. F., Woo, Y. T. (Eds.) *Chemical Induction of Cancer. Modulation and Combination Effects*, Birkhauser, Boston, pp.1–20.

Arnold, V. I. (1976). Local normal forms of functions. *Invent. Math.* 35, 87–109.

Ashrafi, A. R., Došlić, T., Saheli, M. (2011). The eccentric connectivity index of TUC4C8(R) nanotubes. *MATCH Commun. Math. Comput. Chem.* 65, 221–230.

Atanasiu, M. (1971). Order in Complication (in Romanian), Lyceum Series No. 96, Albatros Publishing House, Bucharest.

ATSDR (1995). Agency for Toxic Substances and Disease Registry: Toxicological profile for polycyclic aromatic hydrocarbons. Available at URL: http://www.atsdr.cdc.gov/toxprofiles/tp.asp?id=122&tid=25

Ayers, P. W., Parr, R. G. (2000). Variational principles for describing chemical reactions: the Fukui function and chemical hardness revisited. *J. Am. Chem. Soc.* 122, 2010–2018.

Bader, R. F. W. (1990). *Atoms In Molecules – A Quantum Theory*, Oxford University Press, Oxford.

Bader, R. F. W. (1994). Principle of stationary action and the definition of a proper open system. *Phys. Rev. B*, 49, 13348–13356.

Bader, R. F. W. (1998). A bond path: a universal indicator of bonded interactions. *J. Phys. Chem. A* 102, 7314–7323.

Bak, A., Polanski, J. (2006). A 4D-QSAR study on anti-HIV HEPT analogues. *Bioorg. Med. Chem.* 14, 273–279.

Balaban, A. T. (1976). *Chemical Applications of Graph Theory*, Academic Press, London.

Balaban, A. T. (1982). Distance Connectivity Index. *Chem. Phys. Lett.* 89, 399–404.

Balaban, A. T. (1982). Highly discriminating distance-based topological index. *Chem. Phys. Lett.* 89, 399–404.

Balaban, A. T. (1997). From chemical topology to 3D geometry. *J. Chem. Inf. Comput. Sci.* 37, 645–650.

Balaban, A. T., Harary, F. (1971). The characteristic polynomial does not uniquely determine the topology of a molecule. *J. Chem. Docum.* 11, 258–259.

Balan, A. M., Florea, O., Moldoveanu, C., Zbancioc, G., Iurea, D., Mangalagiu, I. I. (2009). Diazinium salts with dihydroxyacetophenone skeleton: syntheses and antimicrobial activity. *Eur. J. Med. Chem.* 44, 2275–2279.

Basak, S. C., Mills, D. R., Balaban, A. T., Gute, B. D. (2011). Prediction of mutagenicity of aromatic and heteroaromatic amines from structure: a hierarchical QSAR approach. *J. Chem. Inf. Comput. Sci.* 41, 671–678.

Beernaert, H. (1979). Gas Chromatographic analysis of polyclylic aromatic hydrocarbons. *J. Chromatogr.* 173, 109–118.

Benfenati E (Ed) (2010). CAESAR QSAR models for REACH. *Chem Central J*, 4 (Suppl 1), S1-S5.

Benigni, R., Bossa, C., Jeliazkova, N., Netzeva, T., Worth, A. (2008). *The Benigni/Bossa Rules for Mutagenicity and CarcInogenicity—A Module of Toxtree*. European Commission report EUR 23241 EN; IdeaConsult Ltd.: Sofia. http://toxtree.sourceforge.net/carc.html (accessed on 3 August 2011).

Benigni, R., Bossa, C., Netzeva, T. I., Rodomonte, A., Tsakovska, I. (2007). Mechanistic QSAR of aromatic amines: New models for discriminating between homocyclic mutagens and nonmutagens, and validation of models for carcinogens. *Environ. Mol. Mutagen.* 48, 754–771.

Benigni, R., Bossa, C., Tcheremenskaia, O., Worth, A. (2009). Development of structural alerts for the in vivo micronucleus assay in rodents. *EUR 23844 EN* pp. 1–43.

Benigni, R., Netzeva, T. I., Benfenati, E., Bossa, C., Franke, R., Helma, C., Hulzebos, E., Marchant, C., Richard, A., Woo, Y. T., Yang, C. (2007). The expanding role of predictive toxicology: an update on the (Q)SAR models for mutagens and carcinogens. *J Environ Sci Health C* 25, 53–97.

Benigni, R., Netzeva, T. I., Benfenati, E., Bossa, C., Franke, R., Helma, C., Hulzebos, E., Marchant, C., Richard, A., Woo, Y. T., Yanc, C. (2007). The expanding role of predictive toxicology: An update on the (Q)SAR models for mutagens and carcinogens. *J. Environ. Sci. Health C* 25, 53–97.

Bernot, R. J., Brueseke, M. A., Evans-White, M. A., Lamberti, G. A. (2005). acute and chronic toxicity of imidazolium-based ionic liquids on Daphnia magna. *Environ. Toxicol. Chem.* 24, 87–92.

Blackenship, A. L., Kannan, K., Villalobos, S.A; Villeneuve, D. L., Falandysz, J., Imagawa, T., Jakobsson, E., Giesy, J. P. (2000). Relative potenties of individual polychlorinated naphthalenes and halowad mixtures to induce Ah receptor-mediated reponse. *Environ. Sci. Technol.* 34, 3153–3158.

Blair, C. M., Henze, H. R. (1932). The number of stereoisomeric and non-stereoisomeric mono-substitution products of the paraffins. *J. Amer. Chem. Soc.* 54, 1098; idem. The number of stereoisomeric and non-stereoisomeric paraffin hydrocarbons. 1538–1545.

Bock, K. W., Köhle, C. (2006). Ah receptor: dioxin-mediated toxic responses as hints to deregulated physiologic functions. *Biochem Pharmacol* 72, 393–404.

Bostrom, C. E., Gerde, P., Hanberg, A., Jernstrom, B., Johansson, C., Kyrklund, T., Rannug, A., Tornqvist, M., Victorin, K., Westerholm, R. (2002). Cancer risk assessment, indicators, and guidelines for polycyclic aromatic hydrocarbons in the ambient air. *Environ. Health. Perspect.* 110, 451–488.

Box, G. E. P., Hunter, W. G., Hunter, J. S. (1978). *Statistics for Experimenters,* John-Wiley, New York.

Bradbury, S. P. (1995). Quantitative structure activity relationship and ecological risk assessment: an overview of predictive aquatic toxicology research. *Toxicol. Lett.* 79, 229–237.

Bradbury, S. P., Russom, C. L., Ankley, G. T., Schultz, T. W., Walker, J. D. (2003). Overview of data and conceptual approaches for derivation of quantitative structure-activity relationships for ecotoxicological effects of organic chemicals. *Environ. Toxicol. Chem.* 22(8), 1789.-1798.

Brown AJ (1892). XXXI.—Influence of oxygen and concentration on alcoholic fermentation. J. Chem. Soc. Trans. 61, 369–385.

Brown AJ (1902). XXXVI.—Enzyme action. *J. Chem. Soc. Trans.* 81, 373–388.

Butnariu, R., Caprosu, M., Bejan, V., Ungureanu, M., Poiata, A., Tuchilus, C., Florescu, M., Mangalagiu, I. I. (2007). Pyridazine and phthalazine derivatives with potential antimicrobial activity. *J. Heterocycl. Chem.* 44, 1149–1152.

Butnariu, R., Mangalagiu, I. I. (2009). New pyridazine derivatives: synthesis, chemistry and biological activity. *Bioorg. Med. Chem.* 17, 2823–2829.

C.H.A.N.G.E. (2013). Counteracting HIV/AIDS through new gene enhancement. Available online: http://dev.nsta.org/evwebs/577/Present_Technology_Page.html (accessed on 5 April 2013).

Caspase-2 Antibody (2011). IHC WORLD, LLC: Woodstock, MD, USA. Available online: http://www.ihcworld.com/products/antibody-datasheets/Caspase2.IW-PA1113.htm (accessed on 3 August 2011).

Cataldo, F., Ori, O., Graovac, A. (2011a). Graphene topological modifications. *Int. J. Chem. Model.* 3, 45–63.

Cataldo, F., Ori, O., Iglesias-Groth, S. (2010). Topological lattice descriptors of graphene sheets with fullerene-like nanostructures. *Mol. Simul.* 36, 341–353.

Cataldo, F., Ori, O., Vukicevic, D., Graovac, A. (2011b). In: Cataldo, F., Graovac, A., Ori, O. (eds.) *The Mathematics and Topology of Fullerenes*, Series of Carbon Materials: Chemistry and Physics, Springer Science+Business Media, B. V., Dordrecht, 205–216.

Cayley, A. (1857). On the theory of the analytical formes called threes. *Philos. Mag.* 13, 19–30.

Cayley, A. (1874). On the mathematical theory of isomers. *Philos. Mag.* 67, 444–446.

Cayley, A. (1875). Über die analytischen figuren, welche in der mathematik bäume genannt werden und ihre anwendung auf die theorie chemischer verbindungen. *Ber. Dtsch. Chem. Ges.* 8, 1056–1059.

Cayley, A. (1881). On the analytic forms called Trees. *Amer. J. Math.* 4, 266–268.

Cerf, R. (2006). Catastrophe theory enables moves to be detected towards and away from self – organization: the example of epileptic seizure onset. *Biol. Cybern.* 94, 459–468.

Chamorro, E., Chattaraj, P. K., Fuentealba, P. (2003). Variation of the electrophilicity index along the reaction path. *J. Phys. Chem. A.* 107, 7068–7072.

Chattaraj, P. K., Lee, H., Parr, R. G. (1991). Principle of maximum hardness. *J. Am. Chem. Soc.* 113, 1854–1855.

Chattaraj, P. K., Liu, G. H., Parr, R. G. (1995). The maximum hardness principle in the Gyftpoulos-Hatsopoulos three-level model for an atomic or molecular species and its positive and negative ions. *Chem. Phys. Lett.* 237, 171–176.

Chattaraj, P. K., Maiti, B. (2003). HSAB principle applied to the time evolution of chemical reactions. *J. Am. Chem. Soc.* 125, 2705–2710.

Chattaraj, P. K., Schleyer, P. V. R. (1994). An ab initio study resulting in a greater understanding of the HSAB principle. *J. Am. Chem. Soc.* 116, 1067–1071.

Chattaraj, P. K., Sengupta, S. (1996). Popular electronic structure principles in a dynamical context. *J. Phys. Chem.* 100, 16126–16130.

Chatterjee, S., Hadi, A. S., Price, B. (2000). Regression Analysis by Examples (3rd Ed.) John-Wiley, New-York.

Chemical Book (2011), http://www.chemicalbook.com (Accessed August 19, 2011).

Chemical Identifier Resolver *beta 4* (2013), http://cactus.nci.nih.gov/chemical/structure.

Chen, X., Zhan, P., Li, D., De Clercq, E., Liu, X. (2011). Recent advances in DAPYs and related analogues as HIV-1 NNRTIs. *Curr. Med. Chem.* 18, 359–376.

Coffman, D. D., Blair, C. M., Henze, H. R. (1933). The Number of Structurally Isomeric Hydrocarbons of the Acetylene Series. *J. Amer. Chem. Soc.* 55, 252–253.

Copeland, R. A. (2000). Enzymes, Wiley-VCH, New York

Cornish-Bowden A (1999). *Fundamentals of Enzyme Kinetics*, Butterworths, London.

Couling, D. J., Bernot, A. R., Docherty, K. M., Dixon, J. K., Maginn, E. J. (2006). assessing the factors responsible for ionic liquid toxicity to aquatic organisms via quantitative structure– property relationship modeling. *Green Chem.* 8, 82–90.

Crampin, E. J., Schnell, S., McSharry, P. E. (2004). Mathematical and computational techniques to deduce complex biochemical reaction mechanisms. *Prog. Biophys. Mol. Biol.* 86, 77–112.

Croce, C. M. (2008). Oncogenes and cancer. *N. Engl. J. Med.* 358, 502–511.

Cronin, D., Mark, T. (2006). The role of hydrophobicity in toxicity prediction. *Curr. Comput. Aided Drug Design* 2, 405–413.

Cronin, M. T. D., Netzeva, T. I., Dearden, J. C., Edwards, R., Worgan, A. D. P. (2004). Assessment and modeling of the toxicity of organic chemicals to *Chlorella vulgaris*: development of a novel database. *Chem. Res. Toxicol.* 17, 545–554.

Cronin, T. D., Aptula, A. O., Duffy, J. C., Netzeva, T. I., Rowe, P. H., Valkova, I. V., Wayne-Schultz, T. (2002). Comparative assessment of methods to develop QSARs for the prediction of the toxicity of phenols to Tetrahymena pyriformis. *Chemosphere* 49, 1201–1221.

Crum-Brown, A. (1864). On the theory of isomeric confounds. *Trans. Roy. Soc. Edinburgh* 23, 707.

Danaei, G., Vander Hoorn, S., Lopez, A. D., Murray, C. J., Ezzati, M. (2005). Causes of cancer in the world: comparative risk assessment of nine behavioural and environmental risk factors. *Lancet* 366, 1784–1793.

Daudel, R., Leroy, G., Peeters, D., Sana, M. (1983). *Quantum Chemistry*, John Wiley & Sons, New York.

De Clercq, E. (2005). New approaches toward Anti-HIV chemotherapy. *J. Med. Chem.* 48, 1297–1313.

De Clercq, E. (2009a). Anti-HIV drugs: 25 compounds approved within 25 years after the discovery of HIV. *Int. J. of Antimicrob. Agents* 33, 307–320.

De Clercq, E. (2009b). The history of antiretrovirals: Key discoveries over the past 25 years. *Rev. Med. Virol.* 19, 287–299.

Denison, M. S. Heath-Pagliuso, S. (1998). The Ah receptor: a regulator of the biochemical and toxicological actions of structurally diverse chemicals. *Bull. Environ. Contam. Toxicol.* 61, 557–568.

Dingli, D., Nowak, M. A. (2006). Cancer biology: infectious tumour cells. *Nature (London)* 443, 35–36.

Dirac, P. A. M. (1947). *The Principles of Quantum Mechanics*, 2nd ed., Clarendon Press: Oxford.

Dittrich, W., Reuter, M. (1992). *Classical and Quantum Dynamics. From Classical Paths to Path Integrals*, Springer-Verlag, Berlin.

Docherty, K. M., Hebbeler, S. Z., Kulpa, C. F.Jr. (2006). An assessment of ionic liquid mutagenicity using the ames test. *Green Chem.* 8, 560–567.

Docherty, K. M., Kulpa, C. F.Jr. (2005). Toxicity and antimicrobial activity of imidazolium and pyridinium ionic liquids. *Green Chem.* 7, 185–189.

Domingo, L. R., Asensio, A., Arroyo, P. (2002a). Density functional theory study of the Lewis acid-catalyzed Diels–Alder reaction of nitroalkenes with vinyl ethers using aluminum derivatives. *J. Phys. Org. Chem.* 15, 660–666.

Domingo, L. R., Aurell, M. J., Pérez, P., Contreras, R. (2002b). Quantitative characterization of the global electrophilicity power of common diene/dienophile pairs in Diels–Alder reactions. *Tetrahedron* 58, 4417–4423.

Došlić, T., Graovac, A., Ori, O. (2011). Eccentric connectivity index of hexagonal belts and chains. *MATCH Commun. Math. Comput. Chem.* 65, 745−752.

Došlić, T., Saheli, M., Vukičević, D. (2010). Eccentric connectivity index: extremal graphs and values. *Iranian, J. Math. Chem. 1,* 45−55.

Draper, N. R., Smith, H. (1998). *Applied Regression Analysis* (3rd ed.). John Wiley.

Drefahl, A. (2011). CurlySMILES: a chemical language to customize and annotate encodings of molecular and nanodevice structures. *J. Cheminformat.* 3, 1.

Duchowicz, P. R., Bucknum, M. J., Castro, E. A. (2007). New molecular descriptors based upon the Euler equations for chemical graphs. *J. Math. Chem.* 41, 193–208.

Dureja, H., Madan, A. K. (2007). Superaugmented eccentric connectivity indices: newgeneration highly discriminating topological descriptors for QSAR/QSPR modeling. *Med. Chem. Res.* 16, 331−341.

EBI (2011). The European Bioinformatics Institute. Available online: http://www.ebi.ac.uk/pdbsum/2BE2 (accessed on 11 September 2011).

Eckert, M., Fleischmann, G., Jira, R., Hermann, M. B., Golka, K. (2006). *Acetaldehyde*; In: *Ullmann's Encyclopedia of Industrial Chemistry*, Wiley-VCH, Weinheim.

El Safadi, Y., Vivet-Boudou, V., Marquet, R. (2007). HIV-1 reverse transcriptase inhibitors. *Appl. Microbiol. Biotechnol.* 75, 723–737.

Engel, E., Ratel, J. (2007). Correction of the data generated by mass spectrometry analyses of biological tissues: Application to food authentication. *J. Chromatogr. A* 1154, 331–341.

EPA (1994a). Chemical summary for acetaldehyde. US Environmental Protection Agency: Washington, DC; Available online: http://www.epa.gov/chemfact/s_acetal.txt (accessed on 3 August 2011).

EPA (1994b). Chemicals in the environment: Acetaldehyde (CAS no. 75–07–0). US Environmental Protection Agency: Washington, DC, USA, 1994. Available online: http://www.epa.gov/chemfact/f_acetal.txt (accessed on 3 August 2011).

Ertl P (2010). Molecular structure input on the web. J Cheminformat. 2, 1.

Esposito, F., Corona, A., Tramontano, E. (2012). HIV-1 reverse transcriptase still remains a new drug target: structure, function, classical inhibitors, and new inhibitors with innovative mechanisms of actions. *Mol. Biol. Int.* 2012, 586401.

Euler, L. (1736). Comment. Acad. Sci. I. Petropolitane 8, 128; translated (1853). in The Konisberg bridge *Sci. Amer.* 189, 66–70.

EU-QSAR (2011). European Chemicals Agency, 2011, http://www.qsartoolbox.org

Fadeeva, V. N. (1959). *Computational Methods of Linear Algebra*, Dover Publications, New York.

Fan, S. Y., Zheng, Z. B., Mi, C. L., Zhou, X. B., Yan, H., Gong, Z. H., Li, S. (2009). Synthesis and evaluation of novel chloropyridazine derivatives as potent human rhinovirus (HRV) capsid-binding inhibitors. *Bioorg. Med. Chem.* 17, 621–624.

Fernandez-Salguero, P. M., Ward, J. M. Sundberg, J. P. Gonzales, F. J. (1997). Lesions of aryl-hydrocarbon receptor-deficient mice. *Vet. Pathol.* 34, 605–614.

Fjodorova, N., Vračko, M., Novič, M., Roncaglioni, A., Benfenati, E. (2010). New public QSAR model for carcinogenicity. *Chem. Cent. J. 4 (Suppl 1)* S3, 1–S3, 15.

Franke, R., Gruska, A. (2003). General introduction to QSAR; In: Benigni, R. (Ed.) *Quantitative Structure-Activity Relationhsip (QSAR) Models of Mutagens and Carcinogens.* CRC Press, Boca Raton, pp.1–40.

Freemantle, M. (2007). New frontiers for ionic liquids. *Chem. Eng. News* 1, 23–26.

Gallegos, A., Poater, A., Jeliazkova, N., Patlewicz, G., Worth, A. P. (2008). Toxmatch—A chemical classification and activity prediction tool based on similarity measures. *Regul. Toxicol. Pharm.* 52, 77–84.

Gammon, D. B., Snoeck, R., Fiten, P., Krecmerova, M., Holy, A., De Clercq, E., Opdenakker, G., Evans, D. H., Andrei, G. (2008). Mechanism of antiviral drug resistance of Vaccinia virus: identification of residues in the viral DNA polymerase conferring differential resistance to Antipoxvirus drugs. *J. Virol.* 82, 12520–12534.

Ganser-Pornillos, B. K., Yeager, M., Sundquist, W. I. (2008). The structural biology of HIV assembly. *Curr. Opin. Struct. Biol.* 18, 203–217.

Garg, R., Gupta, S. P., Gao, H., Babu, M. S., Debnath, A. K., Hansch, C. (1999). QSAR studies on anti HIV-1 drugs. *Chem. Rev.* 99, 3525–3601.

Genetic Engineering and Biotechnology News Pfizer Inks Deal with K. U. Leuven for HIV drugs with new mechanism of action. (2010). Available online: http://www.genengnews.com/gen-news-highlights/pfizer-inks-deal-with-k-u-leuven-for-hiv-drugs-with-new-mechanism-of-action/81243931/ (accessed on 5 April 2013).

Gillespie, R. J., Hargittai, I. (1991). *The VSEPR Model Of Molecular Geometry*. Allyn and Bacon. Boston.

Gillespie, R. J., Nyholm, R. S. (1957). Inorganic stereochemistry. *Quart. Rev. Chem. Soc.* 11, 339–380.

Gold, L. S. (2011). The carcinogenic potency project. The Carcinogenic Potency Database (CPDB), Berkeley, CA, USA. Available online: http://potency.berkeley.edu/index.html (accessed on 3 August 2011).

Green, J. R., Margerison, D. (1978). Statistical Treatment of Experimental Data, Elsevier, New York.

Greim, H., Bury, D., Klimisch, H.-J., Oeben-Negele, M., Ziegler-Skylakakis, K. (1998). Toxicity of aliphatic amines: structure-activity relationship. *Chemosphere* 36, 271.

Griffiths DJ (1999). *Introduction to electrodynamics*, Prentice Hall, New Jersey.

Gupta, S. P. (2002). Advances in QSAR studies of HIV-1 reverse transcriptase inhibitors. *Prog. Drug Res.* 58, 223–264.

Hahn, M. E. (1998). The aryl hydrocarbon receptor: A comparative perspective. *Comp. Biochem. Physiol. C Pharmacol. Toxicol. Endocrinol.* 121, 23–53.

Haldane, J. B. S. (1930). *The Enzymes*, Longmans-Green, London.

Halliday D., Resnick, R., Krane, K. S. (1992). *Physics,* John Wiley & Sons, New York.

Hansch, C. A. (1969). A quantitative approach to biological-structure activity relationships. *Acta Chem. Res.* 2, 232–239.

Hansch, C., Kurup, A., Garg, R., Gao, H. (2001). Chem-bioinformatics and QSAR: A review of QSAR lacking positive hydrophobic terms. *Chem. Rev.* 101, 619–672.

Harary, F. (1969). *Graph Theory*, Addison-Wesley, Reading Mass.

Hassani, S. (2008). *Mathematical Methods: For Students of Physics and Related Fields 2nd edition*, Springer-Verlag, New York.

Havsteen, B. H. (2002). The biochemistry and medical significance of the flavonoids. *Pharmacol. Ther.* 96, 67–202.

Heitler, W., London, F. (1927). Wechselwirkung neutraler atome und homöopolare bindung nach der quantenmechanik. *Z. Phys.* 44, 455–472.

Helson, H. E. (2007). *Structure Diagram Generation*; In: Lipkowitz, K. B., Boyd, D. B. (Eds.) *Reviews in Computational Chemistry*, John Wiley & Sons Inc, Hoboken, NJ, Vol. 13, pp. 313–398.

Heng, J. B., Ho, C., Kim, T., Timp, R., Aksimentiev, A., Grinkova, Y. V., Sligar, S., Schulten, K., Timp, G. (2004). Sizing DNA using a nanometer-diameter pore. Biophys. J. 87, 2905–2911.

Henri, V. (1901). Uber das gesez der wirkung des invertins. Z. Phys. Chem. 39, 194–216

Henri, V. (1902). Theorie generale de l'action de quelques diastases. *C. R. Hebd. Acad. Sci.* 135, 916–919.

Henze, H. R., Blair, C. M. (1933). The number of structurally isomeric hydrocarbons of the ethylene series. *J. Amer. Chem. Soc.* 55, 680–686.

Henze, H. R., Blair, C. M. (1934). The number of structural isomers of the more important types of aliphatic compounds. *J. Amer. Chem. Soc.* 56, 157–157.

Hermens, J. L. M., Verhaar, H. J. M. (1995). *QSAR in Environmental Toxicology and Chemistry*; In: *Classical and Three-Dimensional QSAR in Agrochemistry*, American Chemical Society, Washington, DC.

Hermes, J. L. M. (1990). Electrophile and acute toxicity to fish. *Environ. Health Perspect.* 87, 219–255.

Himmel, D. M., Das, K., Clark, A. D., Hughes, S. H., Benjahad, A., Oumouch, S., Guillemont, J., Coupa, S., Poncelet, A., Csoka, I., et al. (2005). Crystal structures for HIV-1 reverse transcriptase in complexes with three pyridinone derivatives: A new class of non-nucleoside inhibitors effective against a broad range of drug-resistant strains. *J. Med. Chem.* 48, 7582–7591.

Hoffmann, R. (1963). An extended Hückel theory. I. Hydrocarbons. *J. Chem. Phys.* 39(6), 1397–1412.

Hosoya, H. (1971). A newly proposed quantity characterizing the topological nature of structural isomers of saturated hydrocarbons. *Bull. Chem. Soc. Japan* 44, 2332–2339.

Hosoya, H., Murakami, M. (1975). Topological index as applied to π-electronic systems. II. Topological bond order. *Bull. Chem. Soc. Japan* 48, 3512–3517.

Hückel, E. (1931a). Quantentheoretische Beiträge zum Benzolproblem. I. *Z. Physik* 71, 204–286.

Huijbregts, M. A. J., Rombouts, L. J. A., Ragas Ad. M. J., van de Meent, D. (2005). Human-toxicological effect and damage factors of carcinogenic and noncarcinogenic chemicals for life cycle impact assessment. *Integr. Environ. Assess. Manage.* 1, 181–244.

Hulzebos, E. M., Posthumus, R. (2003). (Q)SARs: Gatekeepers against risk on chemicals? SAR. SAR QSAR Environ. Res.14(4), 285–316.

Husserl, E. (1980). *Ideas Pertaining to a Pure Phenomenology and to a Phenomenological Philosophy – Third Book: Phenomenology and the Foundations of the Sciences*, Translated by Klein, T. E., Pohl, W. E. Dordrecht, Kluwer.

Hypercube (2002). *Program Package, HyperChem 7.01,* Hypercube Inc. Gainesville, FL.

Hückel, E.(1931b). Quantentheoretische beiträge zum benzolproblem. II, *Z. Physik* 72, 310–337.

IARC (2005). *International Agency for Research on Cancer Monographs on the Evaluation of Carcinogenic Risks to Humans;* In: Some Non-heterocyclic Polycyclic Aromatic Hydrocarbons and Some Related Exposures, Lyon, Vol. 92.

Incardona, J. P., Collier, T. K., Scholz, N. L. (2004). Defects in cardiac function precede morphological abnormalities in fish embryos exposed to polycyclic aromatic hydrocarbons. *Toxicol. Appl. Pharmacol.* 196, 191–205.

Incardona, J. P., Day, H. L., Collier, T. K., Scholz, N. L. (2006). Developmental toxicity of 4-ring polycyclic aromatic hydrocarbons in zebrafish is differentially dependent on AH receptor isoforms and hepatic cytochrome P4501A metabolism. *Toxicol. Appl. Pharmacol.* 217, 308–321

INCHEM (1998). International Programme On Chemical Safety. Polycyclic aromatic hydrocarbons, selected non-heterocyclic. Environmental Health Criteria 202, World Health Organization, Geneva [web: http://www.inchem.org/documents/ehc/ehc/ehc202.htm]

IPCS (1998). International Programme on Chemical Safety: Selected non-heterocyclic policyclic aromatic hydrocarbons. Environmental Health Criteria 202 Available URL: http://www.inchem.org/documents/ehc/ehc/ehc202.htm)

Iranmanesh, A., Ashrafi, A. R., Graovac, A., Cataldo, F., Ori, O. (2012). *Distance in Molecular Graphs*; In: Gutman, I., Furtula, B., (Eds.) *Mathematical Chemistry Monographs Serie,* Faculty of Science, Kragujevac, *Vol. 13*, pp. 135–155.

Ivetac A; McCammon, J. A. (2009). Elucidating the inhibition mechanism of HIV-1 non-nucleoside reverse transcriptase inhibitors through multi-copy molecular dynamics simulations. *J. Mol. Biol.* 388, 644–658.

Jastorff, B., Molter, K., Behrend, P., Bottin-Weber, U., Filser, J., Heimers, A., Ondurschka, B., Ranke, J., Scaefer, M., Schroder, H., Stark, A., Stepnowski, P., Stock, F., Stormann, R., Stolte, S., Welz-Biermann, U., Ziegert, S., Thoming, J. (2005). Progress in evaluation of risk potential of ionic liquids—basis for an eco-design of sustainable products. *Green Chem.* 7, 362–372.

Jastorff, B., Stormann, R., Ranke, J., Molter, K., Stock, F., Oberheitmann, B., Hoffmann, W., Hoffmann, J., Nuchter, M., Ondruschka, B., Filser, J. (2003). how hazardous are ionic liquids? Structure – activity relationship and biologic testing as important elements for sustainability evaluation. *Green Chem.* 5, 136–142.

Jaworska J; Nikolova-Jeliazkova N; Aldenberg, T. (2005). QSAR applicability domain estimation by projection of the training set descriptor space: a review. *Altern. Lab. Anim.* 33, 445–459.

Jeliazkova, N., Jaworska, J., Worth, A. P. (2010). Open source tools for read across and category formation; In: Cronin, M. T. D., Madden, J. (Eds.) *In Silico Toxicology. Principles and Applications*, Royal Society of Chemistry, Cambridge, pp. 408–445.

Kaufmann, G. R., Cooper, D. A. (2000). Antiretroviral therapy of HIV-1 infection: established treatment strategies and new therapeutic options. *Curr. Opin. Microbiol.* 3, 508–514.

Kier, L. B., Hall, L. H. (1986). *Molecular Connectivity in Structure-Activity Analysis*, Research Studies Press, Letchworth.

Kirchhoff, G. (1842). Über die auflösung der gleichungen, auf welche man bei der untersuchung der linearen verteilung galvanischer ströme geführt wird. *Ann. Phys. Chem.* 72, 497–508.
Kitazawa, A., Amagai, T., Ohura, T. (2006). Temproal trends and relationships of particulate chlorinated polycyclic aromatic hydrocarbons and their parent compounds in urban air. *Environ. Sci. Technol.* 40, 4592–4598.
Klein, D. J., Randić, M., Babić, D., Lučić, B., Nikolić, S., Trinajstić, N. (1997). Hierarchical orthogonalization of descriptors. *Int. J. Quant. Chem.* 63, 215–222.
Klopman, G., Zhang, Z., Woodgate, S. D., Rosenkranz, H. S. (1995). The structure-toxicity relationship challenge at hazardous waste sites. *Chemosphere* 31, 2511–2519.
Knudson, A. G. (2001). Two genetic hits (more or less) to cancer. *Nat. Rev. Cancer.* 1, 157–162.
Kohn, W., Becke, A. D., Parr, R. G. (1996). Density functional theory of electronic structure. *J. Phys. Chem.* 100, 12974–12980.
Koopmans, T. (1934). Uber die zuordnung von wellen funktionen und eigenwerter zu den einzelnen elektronen eines atom. *Physica* 1, 104–113.
Krausslich, H. G., Bartenschlager, R. (Eds.) (2010). *Antiviral strategies;* In: *Handbook of Experimental Pharmacology,* Springer-Verlag, Berlin-Heidelberg, Vol. 189.
Krokidis X; Noury S; Silvi, B. (1997). Characterization of elementary chemical processes by catastrophe theory. *J. Phys. Chem. A* 101, 7277–7282.
Kumar, V., Sardana, S., Madan, A. K. (2004). Predicting anti-HIV activity of 2,3-diary l-1,3-thiazolidin-4-ones: computational approaches using reformed eccentric connectivity index. *J. Mol. Model.* 10, 399–407.
Lacorre, P. (1997). Predation and generation processes through a new representation of the cusp catastrophe. *Acta Biotheor.* 45, 93–115.
Lele, S. K. (1992). Compact finite difference schemes with spectral-like resolution. *J. Comput. Phys.* 103, 16–42.
Leo, A., Hansch, C., Elkins, D. (1971). Partition coefficients and their uses. *Chem. Rev.* 71, 525–616.
Lessigiariska, I., Worth, A. P., Cronin, M. T. D. (2010). *Structure-Activity Relationships for Pharmacotoxicological Endpoints,* Lambert Academic Publishing, Saarbrucken, pp. 232.
Lewis, D. F. V., Bird, M. G., Jacobs, M. N. (2002). Human carcinogens: an evaluation study via the COMPACT and HazardExpert procedures. *Hum. Exp. Toxicol.* 21, 115–122.
Lewis, G. N. (1916). The atom and the molecule. *J. Am. Chem. Soc.* 38, 762–785.
Löwdin, P.-O. (1950). On the nonorthogonality problem connected with the use of atomic wave functions in the theory of molecules and crystals. *J. Chem. Phys.* 18, 367–370.
Lu, X.-F., Chen, Z.-W. (2010). The development of anti-HIV-1 drugs. *Acta Pharm. Sin.* 45, 165–176.
Lučić, B., Nikolić, S., Trinajstić, N., Juretić, D. (1995). The structure-property models can be improved using the orthogonalized descriptors. *J. Chem. Inf. Comput. Sci.* 35, 532–538.
Lund, A. K., Goens, M. B., Kanagy, N. L., Walker, M. K. (2003). Cardiac hypertrophy in aryl hydrocarbon (ahr) null mice is associated with elevated angiotensin ii, endothelin-1 and mean arterial blood pressure. *Toxicol. Appl. Pharmacol.* 193, 177–187.
Löwdin, P.-O. (1970). On the nonorthogonality problem. *Adv. Quantum Chem.* 5, 185–199.

Madrid, M., Jacobo-Molina, A., Ding, J., Arnold, E. (1999). Major subdomain rearrangement in HIV-1 reverse transcriptase simulated by molecular dynamics. *Proteins Struct. Funct. Bioinf.* 35, 332–337.

Mandal, A. S., Roy, K. (2009). Predictive QSAR modeling of HIV reverse transcriptase inhibitor TIBO derivatives. *Eur. J. Med. Chem.* 44, 1509–1524.

Marchant, C. A. (1996). Prediction of rodent carcinogenicity using the DEREK system for 30 chemicals currently being tested by the National Toxicology Program. The DEREK Collaborative Group. *Environ Health Perspect* 104(Suppl 5), 1065–1073.

Marino, D. J. G., Castro, E. A., Toropov, A. (2006). Improved QSAR modeling of anti-HIV-1 activities by means of the optimized correlation weights of local graphs invariants. *Central Eur. J. Chem.* 4, 135–148.

Maruyama, T., Kozai, S., Yamasaki, T., Witvrouw, M., Pannecouque, C., Balzarini, J., Snoeck, R., Andrei, G., De Clercq, E. (2003). Synthesis and antiviral activity of 1,3-disubstituted uracils against HIV-1 and HCMV *Antivir. Chem.Chemoth.* 14, 271–279.

Masterton, W. L., Slowinski, E. J., Stanitski, C. L. (1983). *Chemical Principles*, CBS College Publishing, Philadelphia.

Matthews, E. J., Contrera, J. F. (1998). A new hightly specific method for predicting the carcinogenic potential of pharmaceuticals in rodents using enhanced MCASEQSAR-ES software. *Regul Toxicol Pharmacol* 28, 242–264.

McQueen, D. J., Post, J. R., Mills, E. L., Fish, C. J. (1986). Trophic relationships in freshwater pelagic eco-systems. *Can. J. Fish. Aquat. Sci.* 43, 1571–1581.

Medina-Franco, J. L., Rodríguez-Morales, S., Juárez-Gordiano, C., Hernández-Campos, A., Castillo, R. (2004). Docking-based CoMFA and CoMSIA studies of non-nucleoside reverse transcriptase inhibitors of the pyridinone derivative type. *J. Comput. Aided Mol. Des.* 18, 345–360.

Mehellou, Y., De Clercq, E. (2010). Twenty-six years of anti-HIV drug discovery: where do we stand and where do we go? *J. Med. Chem.* 53(2), 521–538.

Meneses, L., Fuentealba, P., Contreras, R. (2006). On the variations of electronic chemical potential and chemical hardness induced by solvent effects. *Chem. Phys. Lett.* 433, 54–57.

Merlo, L. M., Pepper, J. W., Reid, B. J., Maley, C. C. (2006). Cancer as an evolutionary and ecological process. *Nat. Rev. Cancer.* 6, 924–935.

Michaelis, L., Menten, M. L. (1913). Die Kinetik der Invertinwirkung. *Biochem, Z.* 49, 333–369.

Middleton, E., Jr., Kandaswami, C., Theoharides, T. C. (2000). The effects of plant flavonoids on mammalian cells: implications for inflammation, heart disease, and cancer. *Pharmacol. Rev.* 52, 673–751.

Mihalić, Z., Veljan, D., Nikolić, S., Plavšić, D., Trinajstić, N. (1992). The distance matrix in chemistry. *J. Math. Chem.* 11, 223–258.

Miller, E. C. Miller, J. A. (1981). Searches for ultimate chemical carcinogens and their reactions with cellular macromolecules. *Cancer* 47, 2327–2345.

Miller, J. A., Miller, E. (1977). Ultimate chemical carcinogens as reactive mutagenic electrophiles; In: Hiatt, H. H., Watson, J. D., Winsten, J.A (Eds.) *Origins of Human Cancer*, Cold Spring Harbor Laboratory, Cold Spring Harbor, pp. 605–628.

Miller, J. N., Miller, J. C. (2000). *Statistics and Chemometrics for Analytical Chemistry* (fourth edition), Pretience Hall, Harlow.

Mineva, T., Sicilia, E., Russo, N. (1998). Density functional approach to hardness evaluation and its use in the study of the maximum hardness principle. *J. Am. Chem. Soc.* 120, 9053–9058.

Mohar, B., Pisanski, T. (1988). How to compute the Wiener index of a graph. *J. Math. Chem.* 2, 267–277.

Moldoveanu, C. C., Jones, P. G., Mangalagiu, I. I. (2009). Spiroheterocyclic compounds: old stories with new outcomes. *Tetrahedron Lett.* 50, 7205–7208.

Morse, M. (1931). The critical points of a functional on n variables. *Trans. Am. Math. Soc.* 33, 72–91.

Mortier, W. J., Genechten, K. V., Gasteiger, J. (1985). Electronegativity equalization: application and parametrization. *J. Am. Chem. Soc.* 107, 829–835.

Muhanji, C. I., Hunter, R. (2007). Current developments in the synthesis and biological activity of HIV-1 double-drug inhibitors. *Curr. Med. Chem.* 14, 1207–1222.

Nebert, D. W., Roe, A. L., Dieter, M. Z., Solis, W. A., Yang, Y., Dalton, T. P. (2000). Role of the aromatic hydrocarbon receptor and [Ah] gene battery in the oxidative stress response, cell cycle control, and apoptosis. *Biochem. Pharmacol.* 59, 65–85.

Nelson, S. D., in: *Bioactivation of Foreign Compounds*, M. W. Anders (Ed.), 349–374. Academic Press, Inc., Orlando, USA, 1985.

Netzeva, T. I., Dearden, J. C., Edwards, R., Worgan, A. D. P., Cronin, M. T. D. (2004a). QSAR analysis of the toxicity of aromatic compounds to Chlorella vulgaris in a novel short-term assay. *J. Chem. Inf. Comput. Sci.* 44(1), 258–265.

Netzeva, T. I., Worgan, A. D. P., Dearden, J. C., Edwards, R., Cronin, M. T. D. (2004b). Toxicological evaluation and QSAR modelling of aromatic amines to Chlorella vulgaris. *Bull. Environ. Contam. Toxicol.* 73, 385–391.

Noble, D. (2002). The rise of computational biology. Nat. Rev. Mol. Cell. Biol. 3, 459–463

NTP & NIEHS (2004), National Toxicology Program (NTP) and National Institute of Environmental Health Sciences (NIEHS). Review of Toxicological Literature for Ionic Liquids. Prepared By Integrated Laboratory Systems Inc., Research Triangle Park, NC.

Oda, J., Yasuhara, A., Matsunaga, K., Saito, Y. (1998). Identification of polycyclic aromatic hydrocarbons of the particulate accumulated in the tunnel duct of freeway and generation of their oxygenated derivatives. *Jpn. J. Toxicol. Environ. Health* 44, 334–351.

OECD (2004). Report from the Expert Group on (Quantitative) Structure-Activity Relationships [(Q)SARs] on the Principles for the Validation of (Q)SARs, Series on Testing and Assessment, No. 49, OECD, Paris; 2004. pp. 206 (http://www.oecd.org/document/30/0,2340, en_2649_34365_1916638_1_1_1_1, 00.html, accessed 3 March 2011).

OECD (2005). Guidance Document on the Validation and International Acceptance of New or Updated Test Methods for Hazard Assessment, Series on Testing and Assessment, No. 34, OECD, Paris; 2005. pp. 96 (http://www.oecd.org/document/30/0,2340). en_2649_34365_1916638_1_1_1_1,00.html, accessed 3 March 2011).

OECD (2006). Report on the Regulatory Uses and Applications in OECD Member Countries of (Quantitative) Structure-Activity Relationship [(Q)SAR] Models in the Assessment of New and Existing Chemicals, Series on Testing and Assessment, No. 58, OECD, Paris, pp. 79 (http://www.oecd.org/document/30/0,2340, en_2649_34365_191 6638_1_1_1_1,00.html, accessed 3 March 2011).

OECD (2007). Guidance Document on the Validation of (Quantitative) Structure-Activity Relationship [(Q)SAR] Models, Series on Testing and Assessment, No. 69,

OECD, Paris; pp. 154 (http://www.oecd.org/document/30/0,2340, en_2649_34365_1916638_1_1_1_1,00.html, accessed 3 March 2011).

OECD Toolbox (2012), Guidance Document for using the (Q)SAR Application Toolbox to develop chemical categories according to the OECD Guidance on Grouping of Chemicals [http://www.oecd.org/env/ehs/risk-assessment/theoecdqsartoolbox.htm].

OECD-QSAR (2013), www.oecd.org/env/existingchemicals/qsar

Ohura, T. (2007). Environmental behavior, sources, and effects of chlorinated polycyclic aromatic hydrocarbons. *The Scientific World* 7, 372–380/

Ohura, T., Morita, M., Kuruto-Niwa, R., Amagai, T., Sakakibara, H., Shimoi, K. (2010). Differential action of chlorinated polycyclic aromatic hydrocarbons on aryl hydrocarbon receptor-mediated signaling in breast cancer cells. *Environ. Toxicol.* 25, 180–187.

Ori, O., Cataldo, F., Vukicevic, D., Graovac, A. (2011). In: Cataldo, F., Graovac, A., Ori, O. (Eds.) *The Mathematics and Topology of Fullerenes*, Springer Science+Business Media, B. V., Dordrecht, *Carbon Materials: Chemistry and Physics Series,* Vol. 4, pp. 205–216.

Ori, O., D'Mello, M. (1992). A topological study of the structure of the C_{76} fullerene. *Chem. Phys. Lett.* 197, 49–54.

Ori, O., D'Mello, M. (1993). Analysis of the structure of the C_{78} fullerene: A topological approach. *Appl. Phys. A Solids Surfaces* 56, 35–39.

Padmanabhan, J., Parthasarathi, R., Subramanian, V., Chattaraj, P. K. (2005b). Molecular structure, reactivity, and toxicity of the complete series of chlorinated benzenes. *J. Phys. Chem. A.* 109, 11043–11049.

Padmanabhan, J., Parthasarathi, R., Subramanian, V., Chattaraj, P. K. (2006a). Chemical reactivity indices for the complete series of chlorinated benzenes: solvent effect. *J. Phys. Chem. A.* 110, 2739–2745.

Padmanabhan, J., Parthasarathi, R., Subramanian, V., Chattaraj, P. K. (2006b). Group philicity and electrophilicity as possible descriptors for modeling ecotoxicity applied to chlorophenols. *Chem. Res. Toxicol.* 19, 356–364.

Pagano, J. S., Blaser, M., Buendia, M. A., Damania, B., Khalili, K., Raab-Traub, N., Roizman, B. (2004). Infectious agents and cancer: criteria for a causal relation. *Semin. Cancer Biol.* 14, 453–471.

PAH (1998). International Programme on Chemical Safety. Polycyclic aromatic hydrocarbons, selected non-heterocyclic. *Environmental Health Criteria 202*, World Health Organization: Geneva; http://www.inchem.org/documents/ehc/ehc/ehc202.htm (Accessed August 22, 2012).

PAH (2005). *Some Non-heterocyclic Polycyclic Aromatic Hydrocarbons and Some Related Exposures*; In: *International Agency for Research on Cancer Monographs on the Evaluation of Carcinogenic Risks to Humans*, Lyon, France, Vol. 92.

Parr, R. G., Donnelly, R. A., Levy, M., Palke, W. E. (1978). Electronegativity: the density functional viewpoint. *J. Chem. Phys.* 68, 3801–3808.

Parr, R. G., Szentpaly, L. V., Liu, S. (1999). Electrophilicity index. *J. Am. Chem. Soc.* 121(9), 1922–1924.

Parr, R. G., Yang, W. (1984). Density functional approach to the frontier electron theory of chemical reactivity. *J. Am. Chem. Soc.* 106, 4049–4050.

Parr, R. G., Yang, W. (1989). *Density-Functional Theory Of Atoms And Molecules*, Oxford University Press, New York.

Parthasarathi, R., Elango, M., Subramanian, V., Chattaraj, P. K. (2005a). Variation of electrophilicity during molecular vibrations and internal rotations. *Theor. Chem. Acc.* 113, 257–266.

Parthasarathi, R., Padmanabhan, J., Subramanian, V., Maiti, B., Chattaraj, P. K. (2003b). Chemical reactivity profiles of two selected polychlorinated biphenyls. *J. Phys. Chem. A* 107, 10346–10352.

Parthasarathi, R., Subramanian, V., Chattaraj, P. K. (2003a). Effect of electric field on the global and local reactivity indices. *Chem. Phys. Lett.* 382, 48–56.

Parthasarathi, R., Subramanian, V., Roy, D. R., Chattaraj, P. K. (2004). Electrophilicity index as a possible descriptor of biological activity. *Bioorg. Med. Chem.* 12, 5533–5543.

Patlewicz, G., Aptula, A. O., Uriarte, E., Roberts, D. W., Kern, P. S., Gerberick, G. F., Kimber, I., Dearman, R. J., Ryan, C. A., Basketter, D. A. (2007). An evaluation of selected global (Q)SARs/expert systems for the prediction of skin sensitisation potential. *SAR QSAR Environ. Res.* 18, 515–541.

Patlewicz, G., Jeliazkova, N., Gallegos, A., Worth, A. P. (2008a). Toxmatch—A new software tool to aid in the development and evaluation of chemically similar groups. *SAR QSAR Environ. Res.* 19, 397–412.

Patlewicz, G., Jeliazkova, N., Safford, R. J., Worth, A. P., Aleksiev, B. (2008). An evaluation of the implementation of the Cramer classification scheme in the Toxtree software. *SAR QSAR Environ. Res.* 19, 495–524.

Pauling L (1946). Molecular architecture and biological reaction. *Chem. Eng. News.* 24, 1375–1377

Pauling, L., Sherman, J. (1933). The nature of the chemical bond. VI. The calculation from thermochemical data of the energy of resonance of molecules among several electronic structures. *J. Chem. Phys.* 1, 606–618.

Pauling, L., Wheland, G. W. (1933). The nature of the chemical bond. V. The quantum-mechanical calculation of the resonance energy of benzene and naphthalene and the hydrocarbon free radicals. *J. Chem. Phys.* 1, 362–375.

Pauling, L., Wilson, E. B. (1935). *Introduction To Quantum Mechanics With Applications To Chemistry*, McGraw-Hill, New York.

Pavan, M., Todeschini, R. (2009). Multi-criteria decision making methods; In: Brown, S., Walczak, B., Tauler, R. (Eds.) *Comprehensive Chemometrics*, Elsevier, Amsterdam, Vol. 1, pp. 591–629.

Pavan, M., Todeschini, R. (Eds.) (2008). Scientific data ranking methods: theory and applications. Data handling. In: *Science and Technology*, 1st ed, Elsevier, Oxford, Vol. 27.

Pearson, R. G. (1985). Absolute electronegativity and absolute hardness of Lewis acids and bases. *J. Am. Chem. Soc.* 107, 6801–6806.

Pearson, R. G. (1990). Hard and soft acids and bases—the evolution of a chemical concept. *Coord. Chem. Rev.* 100, 403–425.

Pearson, R. G. (1997). *Chemical Hardness*, Wiley-VCH, Weinheim.

Pérez, P., Toro-Labbé, A., Aizman, A., Contreras, R. (2002). Comparison between experimental and theoretical scales of electrophilicity in benzhydryl cations. *J. Org. Chem.* 67, 4747–2752.

Perilla, J. R., Zhao, G., Chandler, D., Gronenborn, A., Zhang, P., Schulten, K. (2013). Refinement of Atomic Models of HIV-1 Oligomers. Available online: http://www.ks.uiuc.edu/ (accessed on 5 April 2013).

Pernak, J. Chwala, P. (2003). Synthesis and anti-microbial activities of choline-like quaternary ammonium chlorides. *Eur. J. Med. Chem.* 38, 1035–1042.
Pernak, J., Sobaszkiewicz, K., Mirska, I. (2003). Antimicrobial Activities of Ionic Liquids. Green Chem. 5, 52–56.
Picado, M. J. (2012). Avances en la diseminación del VIH, La Ciencia y sus Demonios, 30 April 2012. Available online: http://lacienciaysusdemonios.com/2012/04/30/avances-en-la-diseminacion-del-vih/#more-24038 (accessed on 5 April 2013).
Platt, J. R. (1952). Prediction of isomeric differences in paraffin properties, *J. Phys. Chem.* 56(3), 328–336.
Pérez, P., Toro-Labbé, A., Contreras, R. (2001). Solvent effects on electrophilicity. *J. Am. Chem. Soc.* 123, 5527–5531.
Poston, T., Stewart, I. (1978). *Catastrophe Theory and Its Applications*, Pitman Publishing, Boston.
Prabhakar, Y. S., Solomon, V. R., Gupta, M. K., Katti, S. B. (2006). QSAR studies on thiazolidines: A biologically privileged scaffold. *Top. Heterocycl. Chem.* 4, 161–249.
Prajapati, D. G., Ramajayam, R., Yadav, M. R., Giridhar, R. (2009). The search for potent, small molecule NNRTIs: A review. *Bioorg Med. Chem.* 17, 5744–5762.
Price, N. (2008). Hail Caesar. *Chemistry & Industry* 15, 18–19.
Pólya, G. (1937). Kombinatorische anzahlbestimmungen für gruppen, graphen, und chemische verbindungen. *Acta Math.* 68, 145.-254
Quashie, P. K., Sloan, R. D., Wainberg, M. A. (2012). Novel therapeutic strategies targeting HIV integrase. *BMC Medicine* 10, 34.
Ramos, E. U., Vaes, W. H. J., Verhaar, H. J. M., Hermens, J. L. M. (1997). Polar narcosis: Designing a suitable training set for QSAR studies. *Environ. Sci. & Pollut. Res.* 4(2), 83–90.
Randić, M. (1975). Characterization of molecular branching. *J. Am. Chem. Soc.* 97, 6609–6615.
Randić, M. (1984a). On molecular identification numbers. *J. Chem. Inf. Comput. Sci.* 24, 164–175.
Randić, M. (1984b). Nonempirical aproach to Structure-Activity Studies. *Int. J. Quantum Chem.: Quantum Biol. Symp.* 11, 137–153.
Randić, M. (1986). Molecular ID Numbers by Design. *J. Chem. Inf. Comput. Sci.* 26, 134–136.
Randić, M. (1992). Similarity based on extended basis descriptors. *J. Chem. Inf. Comput. Sci.* 32, 686–692.
RBVI (2013). Resource for Biocomputing, Visualization, and Informatics: http://www.cgl.ucsf.edu/chimera/data/hiv09/hiv-demo.html. (Accessed 23 February 2013).
Rebehmed, J., Barbault, F., Teixeira, C., Maurel, F. (2008). 2D and 3D QSAR studies of diarylpyrimidine HIV-1 reverse transcriptase inhibitors. *J. Comput. Aided Mol. Des.* 22, 831–841.
Ritter, S. K. (2008). Calling all chemists. *Chem. Eng. News* 18, 59–68.
Roberts, D. W., Aptula, A. O., Patlewicz, G., Pease, C. (2008). Chemical reactivity indices and mechanism-based read-across for non-animal based assessment of skin sensitisation potential. *J. Appl. Toxicol.* 28, 443–454.
Rong, C., Lian, S., Yin, D., Zhong, A., Zhang, R., Liu, S. (2007). Effective simulation of biological systems: choice of density functional and basis set for heme-containing complexes. *Chem. Phys. Lett.* 434, 149–154.

Roukos, D. H., (2009). Genome-wide association studies: how predictable is a person's cancer risk? *Expert Rev. Anticancer Ther.* 9, 389–392.

Rouvray, D. H. (1976). In: *Chemical Applications of Graph Theory*, A. T. Balaban (Ed.), Academia Press, London.

Rouvray, D. H. (1986). *Mathematics and Computational Concepts in Chemistry*, N. Trinajstić (Ed.), Ellis Horwood, New York.

Roy, D. R., Pal, N., Mitra, A., Bultinck, P., Parthasarathi, R., Subramanian, V., Chattaraj, P. K. (2007). An atom counting strategy towards analyzing the biological activity of sex hormones. *Eur. J. Med. Chem.* 42, 1365–1369.

Safe, S. (1990). Polychlorinated Biphenyls (PCBs), Dibenzo-p-Dioxins (PCDDs), Dibenzofurans (PCDFs), and related compounds: environmental and mechanistic considerations which support the development of Toxic Equivalency Factors (TEFs). *Crit. Rev. Toxicol.* 21, 51–88.

Sanderson, R. T. (1988). Principles of electronegativity Part, I. General nature. *J. Chem. Edu.* 65, 112–119.

Sargent, J. M., Williamson, C. J., Maliepaard, M., Elgie, A. W., Scheper, R. J., Taylor, C. G. (2001). Breast cancer resistance protein expression and resistance to daunorubicin in blast cells from patients with acute myeloid leukaemia. *Br. J. Haematol.* 115, 257–262.

Savageau, M. A. (1969). Biochemical systems analysis: I. Some mathematical properties of the rate law for the component enzymatic reactions. *J. Theor. Biol.* 25, 365–369.

Savageau, M. A. (1999). *Biochemical System Analysis: A Study of Function and Design in Molecular Biology*, Addison-Wesley, Reading.

Schluz, T. W., Cronin, M. T. D., Netzeva, T. I., Aptula, A. O. (2003). Quantitative structure activity relationship (QSAR) in toxicology: a historical perspective. *J. Mol. Struct. THEOCHEM* 622, 1–22.

Schnell, S., Maini, P. K. (2003). A century of enzyme kinetics: reliability of the K_m and V_{max} estimates. *Comm. Theor. Biol.* 8, 169–187.

Schnell, S., Turner, T. E. (2004). Reaction kinetics in intracellular environments with macromolecular crowding: simulations and rate laws. *Prog. Biophys. Mol. Biol.* 85, 235–260

Schultz, T. W. (1997). TETRATOX: Tetrahymena pyriformis population growth impairment endpoint-A surrogate for fish lethality. *Toxicol. Methods* 7, 289–309.

Schultz, T. W. (1997). TETRATOX: the *Tetrahymena pyriformis* population growth impairment endpoints. A surrogate for fish lethality. *Toxicol. Methods* 7, 289–309.

Schultz, T. W. (1999). Structure-toxicity relationships for benzene evaluated with *Tetrahymena pyriformis*. *Chem. Res. Toxicol.* 12, 1262–1267.

Schultz, T. W., Cronin, M. T. D. (2003). Essential and desirable characteristics of ecotoxicity quantitative structure-activity relationships. *Environ. Toxicol. Chem.* 22, 599–607.

Schultz, T. W., Cronin, M. T. D., Netzeva, T. I., Aptula, A. O. (2002). Structure-toxicity relationships for aliphatic chemicals evaluated with Tetrahymena pyriformis. *Chem. Res. Toxicol.* 15, 1602–1609.

Schultz, T. W., Cronin, M. T. D., Netzeva, T. I., Aptula, A. O. (2002). Structure-toxicity relationships for aliphatic chemicals evaluated with *Tetrahymena pyriformis*. *Chem. Res. Toxicol.* 15, 1602–1609.

Schultz, T. W., Netzeva, T. I., Cronin, M. T. D. (2003). Selection of data sets for QSARs: analyses of *Tetrahymena* toxicity from aromatic compounds. *SAR QSAR Environ. Res.* 14, 59–81.

Scott JA (2009). *The mechanism of retene toxicity in the early life stages of fish.* PhD Thesis, Queen's University, Kingston, Ontario

Selassie, C. D. (2003). *History of Quantitative Structure-Activity Relationships*; In: Abraham, D. J. (Ed.) *Burger's Medicinal Chemistry and Drug Discovery*, 6th ed., Wiley, New York, pp. 1–48.

Senior, J. K. (1951). Unimerism. *J. Chem. Phys.* 19, 865.

Seyfel, J. K. (1985). *QSAR and Strategies in the Design of Bioactive Compounds*, VCH Weinheim, New York.

Sharma, V., Goswami, R., Madan, A. K. (1997). Eccentric connectivity index: a novel highly discriminating topological descriptor for structure-property and structure-activity studies. *J. Chem. Inf. Comput. Sci.* 37, 273−282.

Sheldon, R. A. (2005). Green solvents for sustainable organic synthesis: state of the art. *Green Chem.* 7, 267–278.

Siegmund-Schultze, R. (1986). Der beweis des Hilbert-Schmidt theorem. *Archive for History of Exact Sciences* 36, 251–270.

Silvi, B., Savin, A. (1994). Classification of chemical bonds based on topological analysis of electron localization functions. *Nature* 371, 683–686.

Song, C., Lai, W.-C., Madhusudan Reddy, K., Wei, B. (2003). In: *Analytical Advances for Hydrocarbon Research*, HsuC. S. (Ed.) Kluwer Academic/Plenum Publishers, New York, pp. 147–193.

Šoškić, M., Plavšić, D., Trinajstić, N. (1996). Link between orthogonal and standard multiple linear regression models. *J. Chem. Inf. Comp. Sci.* 36, 829–832.

Sowin, T. J., Melcher, L. M. (2004). *Acetaldehyde*; In: Paquette, L. (Ed.) *Encyclopedia of Reagents for Organic Synthesis*, Wiley & Sons, New York.

Spearman, C. (1927). *The Abilities of Man*, MacMillan, London.

Spycher, S., Netzeva, T. I., Worth, A. P., Escher, B. I. (2008). Mode of action-based classification and prediction of activity of uncouplers for the screening of chemical inventories. *SAR QSAR Environ. Res.* 19, 433–463.

Statistica (1995). StatSoft, Inc. *STATISTICA for Windows, Computer program and manual.*

Steen, L. A. (1973). Highlights in the history of spectral theory. *Am. Math. Monthly* 80, 359–381.

Steiger, J. H., Schonemann, P. H. (1978). A history of factor indeterminacy; In: Shye, S. (Ed.) *Theory Construction and Data Analysis in the Behavioural Science*, Jossey-Bass Publishers, San Francisco.

Tachibana, A. (1987). Density functional rationale of chemical reaction coordinate. *Int. J. Quantum Chem.* 21, 181–190.

Tachibana, A., Nakamura, K., Sakata, K., Morisaki, T. (1999). Application of the regional density functional theory: the chemical potential inequality in the HeH^+ system. *Int. J. Quantum Chem.* 74, 669–679.

Tachibana, A., Parr, R. G. (1992). On the redistribution of electrons for chemical reaction systems. *Int. J. Quantum. Chem.* 41, 527–555.

Tarko, L., Lupescu, I., Groposila – Constantinescu, D. (2005). Sweetness power QSARs by PRECLAV software. *ARKIVOC X*:254–271.

Thom, R. (1973). *Stabilitè Structurelle et Morphogènése*, Benjamin-Addison-Wesley, New York.

Todeschini, R., Ballabio, D., Consonni, V., Mauri, A., Pavan, M. (2007). CAIMAN (Classification And Influence Matrix Analysis), A new approach to the classification based on leverage-scaled functions. *Chemom. Intell. Lab.* 87, 3–17.

Todeschini, R., Consonni, V. (2000). *Handbook of Molecular Descriptors*, Wiley-VCH, Weinheim.

Topliss, J. (1983). *Quantitative Structure-Activity Relationships of Drugs*, Academic Press, New York.

Topliss, J. G., Costello, R. J. (1972). Chance correlation in structure-activity studies using multiple regression analysis. *J. Med. Chem.* 15, 1066–1068.

Topliss, J. G., Edwards, R. P. (1979). Chance factors in studies of quantitative structure-activity relationships. *J. Med. Chem.* 22, 1238–1244.

Torrent-Sucarrat, M., Solà, M. (2006). Gas-phase structures, rotational barriers, and conformational properties of hydroxyl and mercapto derivatives of cyclohexa-2,5-dienone and cyclohexa-2,5-dienthione. *J. Phys. Chem. A.* 110, 8901–8911.

Turner, T. E., Schnell, S., Burrage, K. (2004). Stochastic approaches for modelling in vivo reactions. *Comput. Biol. Chem.* 28, 165–178.

Van Den Dool, H., Kratz, P. D. (1963). A generalization of the retention index system including linear temperature programmed gas—liquid partition chromatography. *J. Chromatogr.* 11, 463–471.

van Wezel, A. P., Opperhuizen, A. (1995). Narcosis due to environmental poluants in aquatic organisms: residue-based toxicity, mechanisms, and membrane burdens. *Crit. Rev. Toxicol.* 25, 255.

Veith, G. D., Broderius, S. J. (2003). Rules for distinguish toxicants that cause type I and type II narcosis syndromes. Environ. Health Perspect. 87, 207–2011.

ViiV (2012). Healthcare Group of Companies. The biology of HIV. Available online: http://www.apositivelife.com/forasos/biology-of-hiv.html (accessed on 4 May 2013)

Viret, J. (1994). Reaction of the organism to stress: the survival attractor concept. *Acta Biotheor.* 42, 99–109.

Viret, J. (2010). Topological approach of Jungian psychology. *Acta Biotheor.* 58, 233–245.

Voet, D., Voet, J. G. (1995). *Biochemistry* (2nd ed.) Wiley, New York.

Vracko, M., Bandelj, V., Barbieri, P., Benfenati, E., Chaudhry, Q., Cronin, M. T. D., Devillers, J., Gallegos, A., Gini, G., Gramatica, P., *et al.* (2006). Validation of counter propagation neural network models for predictive toxicology according to the OECD principles: A case study. *SAR QSAR Environ. Res.* 17, 265–284.

Walisser, J. A., Bunder, M. K., Glover, E., Bradfield, C. A. (2004). Gestational exposure of Ahr and Arnt hypomorphs to dioxin rescues vascular development. *Proc. Natl. Acad. Sci.* 101, 16677–16682.

Walker, L. M., Phogat, S. K., Chan-Hui, P. Y., Wagner, D., Phung, P., Goss, J. L., Wrin, T., Simek, M. D., Fling, S., Mitcham, J. L., Lehrman, J. K., Priddy, F. H., Olsen, O. A., Frey, S. M., Hammond, P. W., Protocol G Principal Investigators, Kaminsky, S., Zamb, T., Moyle, M., Koff, W. C., Poignard, P., Burton, D. R. (2009). Broad and potent neutralizing antibodies from an african donor reveal a new HIV-1 vaccine target. *Science* 326, 285–289

Ward, E. M., Thun, M. J., Hannan, L. M., Jemal, A. (2006). Interpreting cancer trends. *Ann. NY. Acad. Sci.* 1076, 29–53.

Weininger, D. (1988). SMILES, a chemical language and information system. 1. Introduction to methodology and encoding rules. *J. Chem. Inf. Comput. Sci.* 28, 31–36

Weininger, D. (1990). SMILES. 3. DEPICT. Graphical depiction of chemical structures. *J. Chem. Inf. Model.* 30, 237–243.

Weininger, D., Weininger, A., Weininger, J. L. (1989). SMILES. 2. Algorithm for generation of unique SMILES notation. *J. Chem. Inf. Comput. Sci.* 29, 97–101.

Wheland, G. W. (1944). *The Theory Of Resonance And Its Application To Organic Chemistry*, Wiley, New York.

WHO (1998). *Alcohol Drinking*. World Health Organization, International Agency for Research on Cancer: Lyon, France, 1988. Available online: http://monographs.iarc.fr/ENG/Monographs/vol44/volume44.pdf (accessed on 3 August 2011).

Wiener, H. (1947). Structural determination of paraffin boiling points. *J. Am. Chem. Soc.* 69, 17–20. b) idem., Correlation of heats of isomerization, and differences in heats of vaporization of isomers, among the paraffin hydrocarbons. 69, 2636–2638.

Wilfred, K. (2003). *Advanced Calculus*, Pearson Education. Inc, Essex.

Wilson, E. B. (1928). Review of the abilities of man, their nature and measurement, by Spearman, C. *Science* 67, 244–248.

Wilson, E. B., Hilferty, M. M. (1931). The distribution of chi-square. *Proc. Nat. Acad. Sci.* 17, 684.

Wilson, E. B., Worcester, J. (1939). A note on factor analysis. *Psychometrika* 4, 133–148.

Woo, Y. T. (2003). Mechanisms of action of chemical carcinogens, and their role in Structure-Activity Relationships (SAR) analysis and risk assessment. In *Quantitative Structure-Activity Relationship (QSAR) Models of Mutagens and Carcinogens*, Benigni, R. (Ed.) Boca Raton, CRC Press, pp. 41–80.

Woo, Y. T., Lai, D. Y. (2005). *OncoLogic: a mechanism-based expert system for predicting the carcinogenic potential of chemicals*; In: Helma, C. (Ed.) *Predictive Toxicology*, Boca Raton, CRC Press, pp. 385–413.

Worgan, A. D. P., Dearden, J. C., Edwards, R., Netzeva, T. I., Cronin, M. T. D. (2003). Evaluation of a novel short-term algal toxicity assay by the development of QSARs and inter-species relationships for narcotic chemicals. *QSAR Comb. Sci.* 22, 204–209.

Worth, A. P., Bassan, A., de Brujin, J., Gallegos Saliner, A., Netzeva, T., Patlewicz, G., Pavan, M., Tsakovska, I., Eisenreich, S. (2007). The role of the European chemicals bureau in promoting the regulatory use of (Q)SAR methods. *SAR QSAR Environ Res* 18, 111–125.

Yang, W., Parr, R. G., Pucci, R. (1984). Electron density, Kohn-Sham frontier orbitals, and Fukui functions. *J. Chem. Phys.* 81, 2862–2863.

Zeeman, E. C. (1976). Catastrophe theory. *Scient. Am.* 234, 65–83.

Zeng, Y.-X., Zhao, C.-X., Liang, Y.-Z., Yang, H., Fang, H.-Z., Yi, L.-Z., Zeng, Z.-D. (2007). Comparative analysis of volatile components from Clematis species growing in China. *Anal. Chim. Acta.* 595, 328–339.

Zhan, P., Chen, X., Li, D., Fang, Z., de Clercq, E., Liu, X. (2011). HIV-1 NNRTIs: Structural diversity, pharmacophore similarity, and implications for drug design. *Med. Res. Rev.* 33(S1), E1-E71.

Zhang, S., Yang, X., Coburn, R. A., Morris, M. E. (2005). Structure activity relationships and quantitative structure activity relationships for the flavonoid-mediated inhibition of breast cancer resistance protein. *Biochem. Pharmacol. 70,* 627–639.

Zhang, S., Yang, X., Morris, M. E. (2004a). Combined effects of multiple flavonoids on breast cancer resistance protein (ABCG2)-mediated transport. *Pharm. Res.* 21, 1263–1273.

Zhang, S., Yang, X., Morris, M. E. (2004b). Flavonoids are inhibitors of breast cancer resistance protein (ABCG2)-mediated transport. *Mol. Pharmacol.* 65, 1208–1216.

EPILOGUE TO QUANTUM NANOCHEMISTRY

"For Those Preparing for Healthy Science, We Salute You!"

From this five-volume book, we learned more about the first principles of substance, i.e., quantum chemistry, and thus the XXI scholars should not forget the silent war the Science is engaged to win:

- to preserve and prolong life itself;
- to safely and sustainable transform matter's constituents.

Both represent the contemporary and necessary need for the future translation of the alchemic double dream/vision:

- the life elixir to *immortality*; and
- the transmutation of elements to *gold!*

Accordingly, since the 2013 author's scientific visit to the Erice mountain in Sicily (with its fabulous mythology as being the son of Neptune and Venus on earth), we like to make us further aware of the Erice Spirit (Erice's Geist), including the creation of The World Federation of Scientists (WFS)[1], fully engaged in promoting the 15 planetary emergencies:

- the water;
- the soil;
- the food;
- the energy;
- the pollution;
- the limits of development;
- the climatic changes;
- the global monitoring of planet;
- the new military threats in the multipolar world;

[1]http://www.federationofscientists.org/

- the science and technology for developing countries to avoid a north-south environmental holocaust;
- the problem of organ substitution;
- the infectious diseases;
- the cultural pollution;
- the common defense against cosmic objects; and
- the huge military investments.

All these bring the HH Pope John-Paul II[2] to visit the *Ettore Majorana Foundation and Centre for Scientific Culture*[3] (founded on 1961–1963) and its activities, making the decisive statement in 1979 by recognizing that "Science and Faith are both God's Gifts", and in original as "Scienza e Fede sono entrambe doni di Dio", as was repeated by his famous III phrase to WFS, at his 1993 visit at Erice.

In the same spirit he also placed his famous VIII phrase saying that "the man can perish by the technology that himself developed, and not by the truths discovered through the scientific research" (in original as "L'uomo può perire per effecto della tecnica che egli stesso sviluppa, non della verità che egli scopre mediante la ricerca scientifica").

So, the principles create and do not destroy, so are the first principles of *Quantum Nanochemistry*, in fact—*the first principles of life inside the Life*! The fact that Science is harmless and technology can offend in some (un)expected and even silent ways stays as the big next challenge of the twenty-first century generation of scholars: they are are called to direct the fundamental principles of Nature, and the quantum principles especially, to produce such technology that improves and solves the above two main desiderates of Life and the 15 commandments of Mundus Paxis so that the current and future centuries have sustainable Life on Earth!

Overall, don't forget that "Only the healthy ideas can fly all over the World(s)!"

We Salute You to the Endless Light of the Day!

---

[2]http://www.vatican.va/holy_father/john_paul_ii/speeches/1993/may/documents/hf_jp-ii_spe_19930508_scienziati-erice_it.html
[3]http://www.ccsem.infn.it/

INDEX

C

D

E

F

I

K

L

Q

R

S

T

U

V

W

X

Z